P9-DXN-434

THE HUMAN BODY

Editors   David Heidenstam, Ann Kramer, Ruth Midgley, Susan Sturrock

Contributors   Ruth Berenbaum, Kati Boland, Rosie Boycott, Jefferson Cann, Cornelius Cardew, Minakshi Compton, David Lambert, Robert Royston, Irene Staunton, Judy Todd, Yvonne Wicken, Elizabeth Wilhide

Editorial assistants   Rosemary Chamberlain, Kathryn Dunn, Catherine Groom, Susan Leith, Mary Ling, Barbara Schofield, Julie Vosburgh

Art directors   Robin Crane, Roger Kohn, Kathleen McDougall

Artists   Jeff Alger, Eileen Batterberry, Alison Blythe, Steven Clark, Robert Galvin, Peter Golding, Richard Hummerstone, Susan Kinsey, Pavel Kostal, Janos Marffy, Graham Rosewarne, Diane Taylor

Art assistants   Julie Briggs, Carlton Facey, Kim Hall, Brian Hewson, Jim Kane

THE HUMAN BODY

The Diagram Group

Facts On File
460 Park Avenue South
New York, N.Y. 10016

THE HUMAN BODY

Copyright © 1980 Diagram Visual Information Ltd.

All rights reserved. No part of this book may be reproduced or utilized in any form or by any means, electronic or mechanical, including photocopying, recording or by any information storage and retrieval system, without permission in writing from the Publisher.

First published in one volume in the United States in 1980 by Facts On File, Inc.

Picture research by Annie Horton

Library of Congress Cataloging in Publication Data

Diagram Group.
  The human body

  A combination and adaptation of the Group's Man's body, c1976, Woman's body, published in 1977, and Child's body, c1977.
  Includes index.
  1. Health. 2. Human physiology. 3. Children—Care and hygiene. 4. Child development. I. Title.
RA776.D46   613   79-21673
ISBN 0-87196-309-4

Printed in the United States of America
10 9 8 7 6 5 4 3 2

FOREWORD

THE HUMAN BODY is designed to provide clear, straightforward and unbiased explanations of every aspect of the body's functioning, care and development. Its aim is to provide everyone with enough basic knowledge about the human body to be able to make intelligent decisions about the physical well-being of each and every member of the family.

Because the language of medical experts is often complex and outside the experience of many men and women, we have tried to synthesize a vast amount of up-to-date medical research and statistical data and translate it into everyday terms, so that even the most complicated body system is easily understandable. To make the information even more accessible, we have included well over 2,000 illustrations, charts and diagrams. Finally, carefully planned chapters, clearly titled panels of information, and a comprehensive table of contents and index should enable the reader to find answers to particular questions quickly and easily.

THE HUMAN BODY is divided into two sections. Part I, "Man's Body, Woman's Body," charts the course of our lives from conception through old age, and includes chapters on life and death; parts of the body; illness and disease; food and fitness; stress and drugs; sex and sexuality; contraception, infertility and abortion; pregnancy; and growing older. Part II, "Child's Body," contains practical child-care advice as well as information on the physical, emotional, social and intellectual development of the child from birth through adolescence.

All the material in THE HUMAN BODY has been presented to a team of practicing physicians, gynecologists, pediatricians and child-care experts for their advice and review. Because medical opinions and theories vary and sometimes contradict each other, we have attempted to remain free of bias and present as many points of view as possible.

The words average and typical are often used in THE HUMAN BODY. These are reference points only, derived from statistical findings of scientific surveys and should not be made the basis of any judgment or personal assessment. The terms refer generally to what is, or what happens, in a large number of cases; not to what is necessarily best or appropriate for a specific individual.

Similarly, although we offer a great deal of practical advice in the "Child's Body" section, we would like to emphasize that considerable flexibility is called for on the part of every parent. There are no fixed rules for bringing up and caring for children, and each of us — parent or child — has individual strengths and weaknesses, talents and needs, that must be taken into account.

Written and illustrated with the entire family in mind, we have created THE HUMAN BODY with the belief that a clearer understanding of the development and functioning of the body will lead people to a fuller, more confident appreciation of their bodies and their lives.

The Diagram Group
1980

CONTENTS
PART 1

MAN'S BODY, WOMAN'S BODY

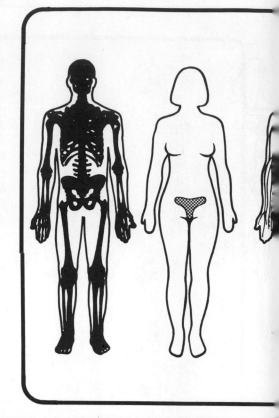

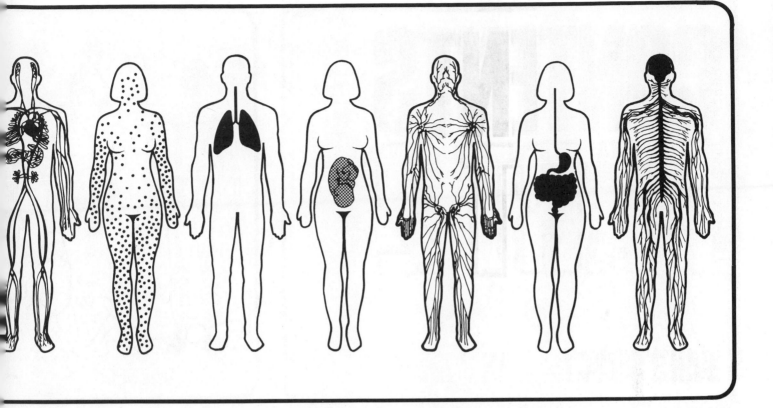

CONTENTS
PART 2

CHILD'S BODY

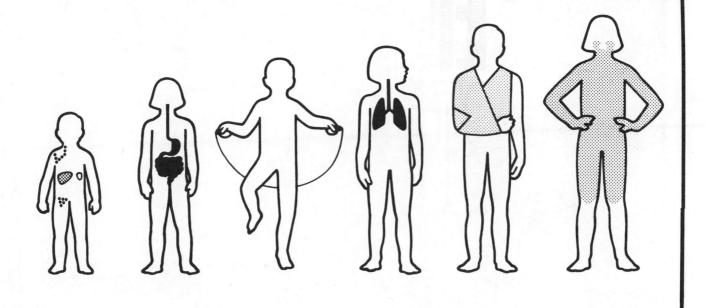

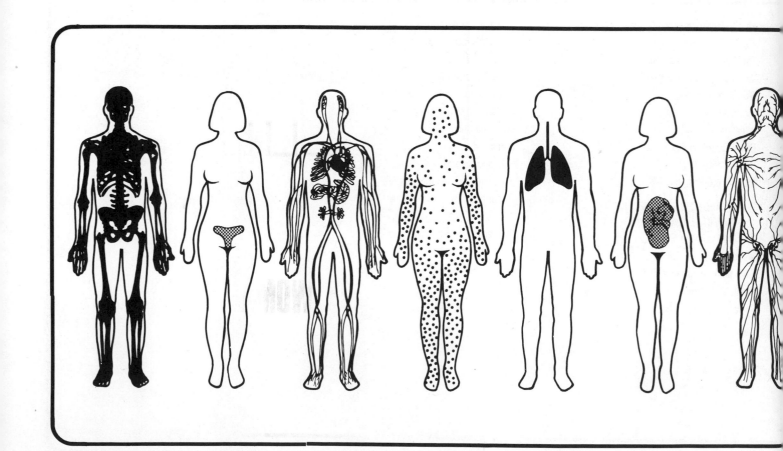

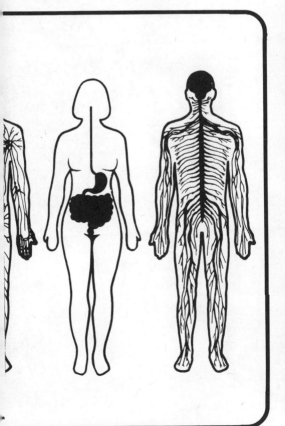

PART 1

MAN'S BODY, WOMAN'S BODY

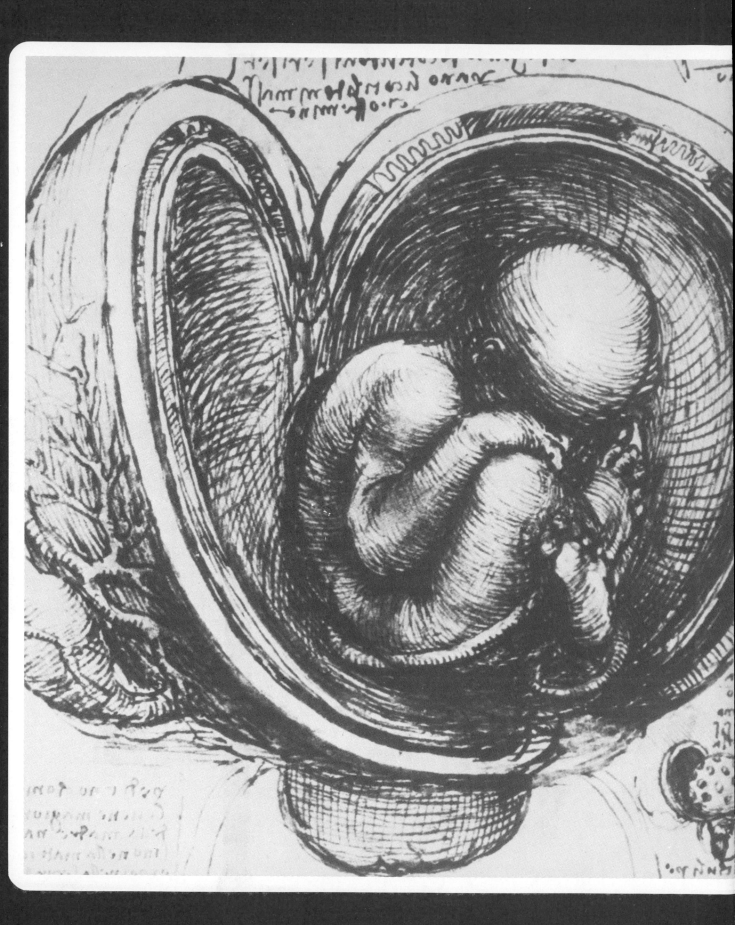

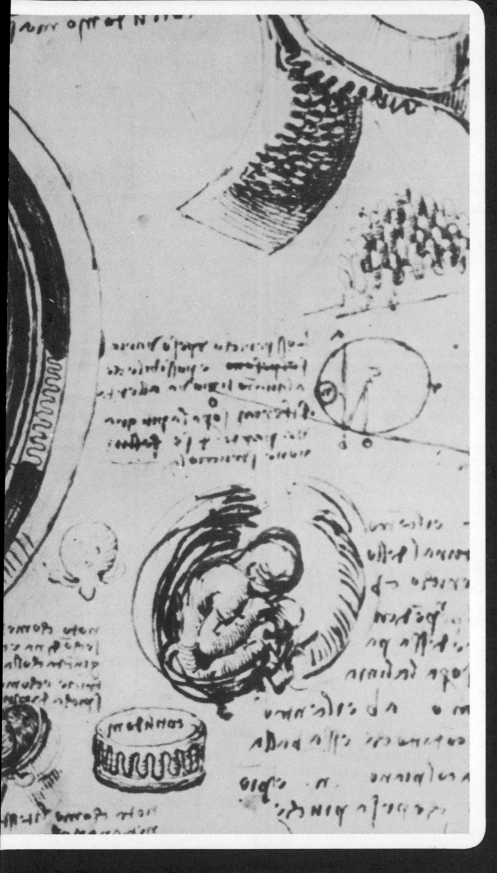

1 Life and death

Each of us is a unique human being, yet from the moment of our conception to the moment of our death, our bodies develop along certain similar and predictable paths.

Left: Leonardo da Vinci drawings of the developing fetus (Reproduced by gracious permission of Her Majesty Queen Elizabeth II)

CONCEPTION

SEX DETERMINATION

WHY SEXUAL REPRODUCTION?
Some creatures - such as the
amoeba - reproduce by just
splitting in two. That may be
convenient! But the resulting
offspring are very predictable. One
amoeba can father endless
generations: but the youngest will
still be identical with the first.
Sexual reproduction, in contrast,
gives almost infinite variety, for
the offsprings' characteristics are
a jumbled mixture from both
parents' family trees. Some have
some talents, and some have
others - and some manage to
combine many talents: all of which
helps a species as complex as man
to survive, in a complex and
demanding world.

SEXUAL INHERITANCE
Every cell in the human body
contains a "blueprint" of
information, on the basis of which
it was constructed. This
information is contained in 23 pairs
of chromosomes, which lie in the
nucleus of the cell. The chromosomes determine the output of
protein, the basic building unit:
which proteins the cell manufactures,
when, and how. This in turn determines the cell's characteristics and
activities.
When a cell divides to make the
body grow, the pairs of chromosomes double up before the division.
This means both new cells still get
23 pairs, and still get every piece
of information that the old cell had.
But when the body produces cells
for sexual reproduction, it divides
cells without doubling their
chromosomes first. The pairs
separate and each sexual cell gets
only 23 single chromosomes, one of
each pair. So each of these cells
has only half the information that
goes into a normal cell.
The sperm is the male example of
such a cell, and the ovum is the
female example. When the two

unite, the new fertilized cell, from
which the offspring grows, contains
23 pairs of chromosomes again -
each parent having contributed half.

BOY OR GIRL?
Of the 23 pairs of chromosomes in a
body cell, 22 are always matching
pairs - each chromosome in the pair
is similar. Women also have their
23rd pair identical: they are both
called X chromosomes. But men,
instead, have two chromosomes
that do not match. One is an X,
as in women; the other is
different, and is called a Y
chromosome. It is these final
chromosomes that contain the code
that determines sex. The XX
combination occurs in, and
produces, a female; the XY
combination, a male.
When the female cell splits, for
sexual reproduction, the sexual
cell that results contains one
chromosome from each of the 23
pairs - including one X chromosome.
When the male cell splits, for
sexual reproduction, the sexual cell
contains one chromosome from each
of the 22 identical pairs, plus
either the X or the Y - but not
both. Which it is determines the
offspring's sex. If it is an X it forms
a pair with the female's X - a
female child occurs. If it is a
Y, it gives the XY non-pair that
is typical of, and forms, a man.
More males are conceived than
females- but males die off at a
faster rate, even in the womb.
There are more male stillbirths
than female stillbirths, and the
sooner the stillbirth occurs in the
term of pregnancy, the more likely
it is to be male. At four months
there are about two male stillbirths
for every female; at full term it is
more like 1.5 to 1. As a result, the
the ratio of fetus males to females
declines, until at full term, in
advanced societies, it is about 105
or 106 males for every 100 females.

BOY OR GIRL?

Spermatogonia

Secondary spermatocyte

Male cell division (above) results
in two spermatozoa, one with an
X chromosome (a), the other with
a Y chromosome (b). (See shaded
boxes.)
Female cell division (below)
results in one ovum, with an X
chromosome (c) (the other cells
degenerate).
The ovum is fertilized (d) by the
sperm with the Y chromosome, so
the child is male.

Secondary oocyte

Oogonia

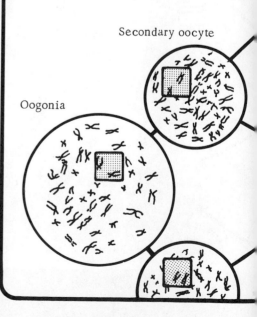

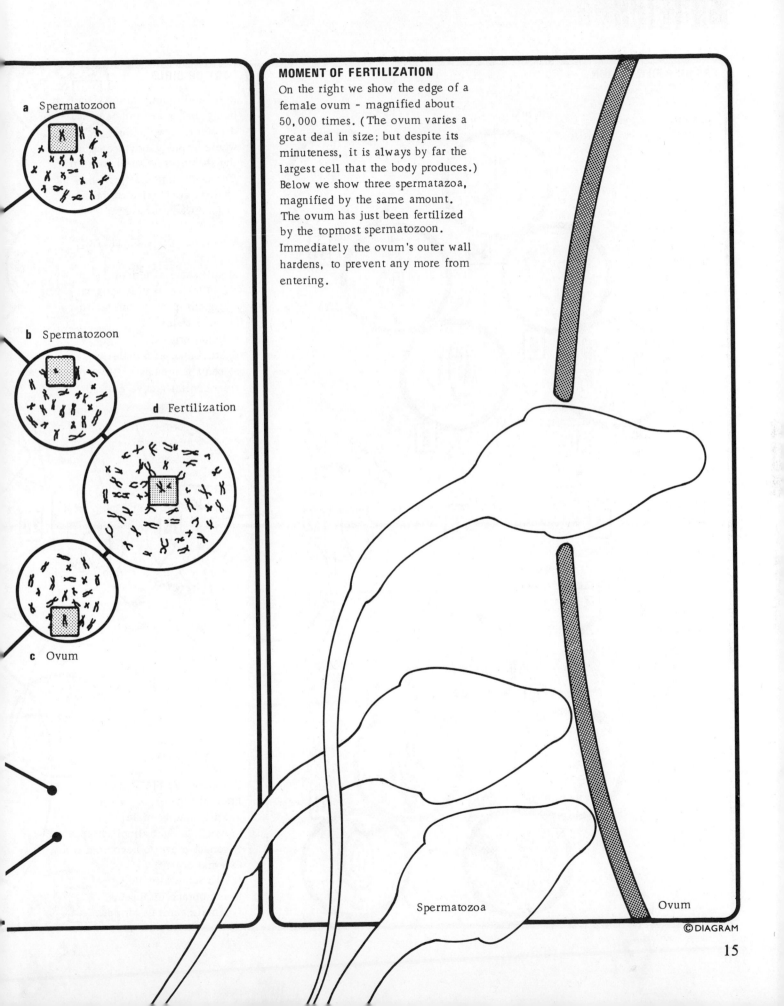

a Spermatozoon

b Spermatozoon

d Fertilization

c Ovum

MOMENT OF FERTILIZATION

On the right we show the edge of a
female ovum - magnified about
50,000 times. (The ovum varies a
great deal in size; but despite its
minuteness, it is always by far the
largest cell that the body produces.)
Below we show three spermatazoa,
magnified by the same amount.
The ovum has just been fertilized
by the topmost spermatozoon.
Immediately the ovum's outer wall
hardens, to prevent any more from
entering.

Spermatozoa   Ovum

©DIAGRAM

THE FETUS

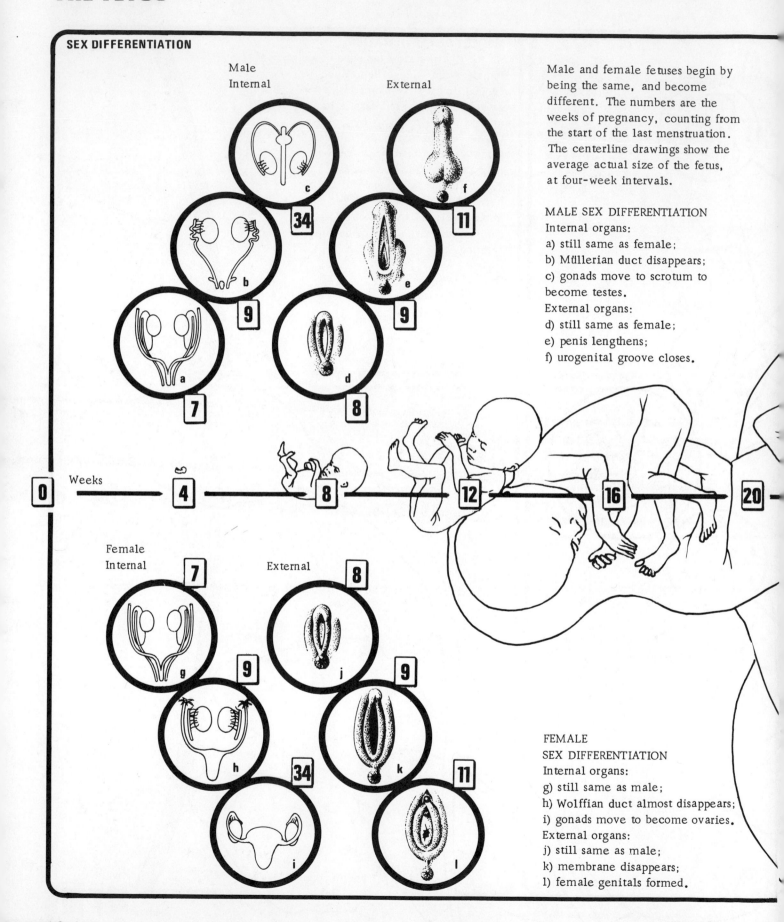

SEX DIFFERENTIATION

Male
Internal

External

c

34

b

9

e

9

f

11

a

7

d

8

Male and female fetuses begin by
being the same, and become
different. The numbers are the
weeks of pregnancy, counting from
the start of the last menstruation.
The centerline drawings show the
average actual size of the fetus,
at four-week intervals.

MALE SEX DIFFERENTIATION
Internal organs:
a) still same as female;
b) Müllerian duct disappears;
c) gonads move to scrotum to
become testes.
External organs:
d) still same as female;
e) penis lengthens;
f) urogenital groove closes.

Weeks

0 | **4** | **8** | **12** | **16** | **20**

Female
Internal

7

External

8

g

9

j

9

h

34

k

11

i

l

FEMALE
SEX DIFFERENTIATION
Internal organs:
g) still same as male;
h) Wolffian duct almost disappears;
i) gonads move to become ovaries.
External organs:
j) still same as male;
k) membrane disappears;
l) female genitals formed.

16

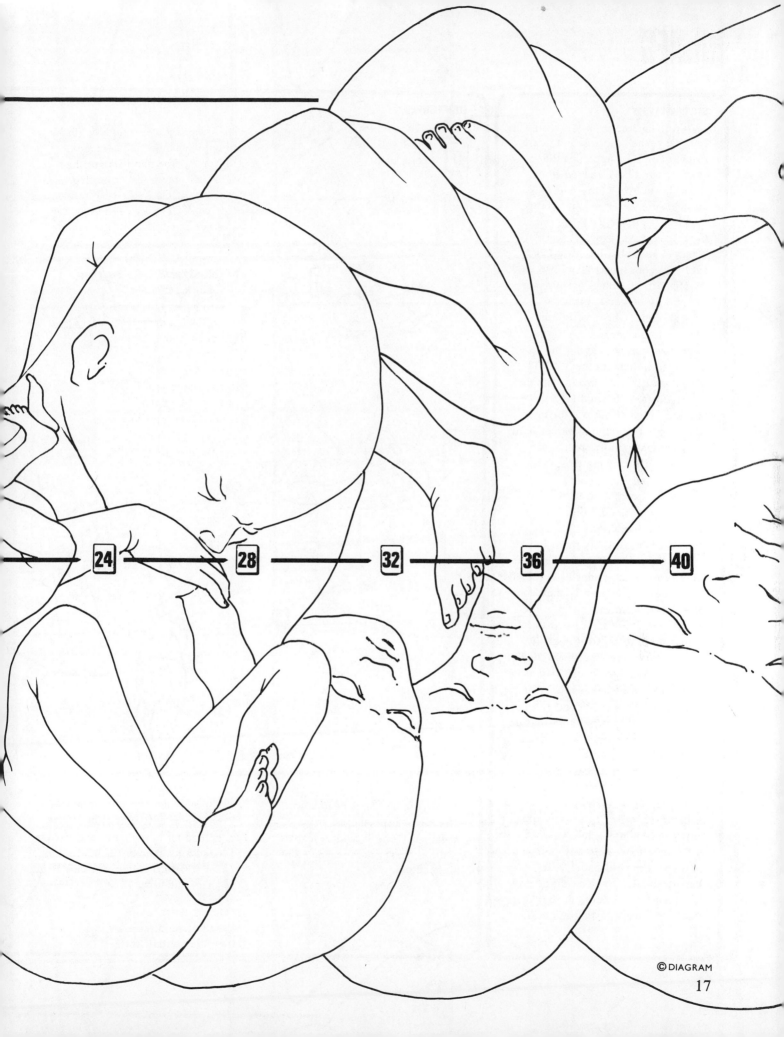

24 28 32 36 40

©DIAGRAM

BABIES

BIRTHWEIGHT

The 6ft kangaroo has a less than 1gm baby; the blue whale a nearly 10 ton one. Human babies that have survived have ranged from under 2lb to over 29lb - but it is far healthier just to be an average 7lb 4oz.

In fact the boys' average is slightly higher (7½lb), and the girls'(just over 7lb) correspondingly lower. Boys' hearts and lungs are already marginally bigger at birth too (though their livers are lighter).

All this is not because boys are later than girls in leaving the womb (in which case boys would have more time to develop before birth). In fact, if anything, there is a very slight tendency for there to be more boys among babies born after unusually short pregnancies, and more girls among those born after unusually long ones.

However, "premature babies" are quite often defined by birthweight, rather than length of pregnancy. On this criterion, slightly more female babies are termed "premature", as slightly more are under 5½lb (2.5 kg) in weight. But really they are often "full-term low birthweight". Whether underweight or overweight, babies that are far from the average have less likelihood of survival. Average-weight babies have under a 2% death rate; 6 or 9 lb babies a 3% one; 4½ or 10½ pounders a 10% rate. Those that do survive are also more likely to be handicapped.

Perfectly normal babies vary greatly in rate of development. Sitting up for a few moments without support can start any time between 5 months and a year - walking without help any time between 8 months and 4 years. Parents should not think that delay is always very serious, or that it is likely to have a lasting effect.

DEVELOPMENT

 Age in months

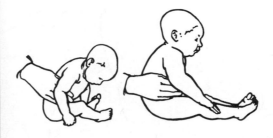

Crawling

Sitting

Walking

A newborn baby lies head down, hips high, knees tucked under abdomen. If he is held in a sitting position, his back is rounded and his head droops.

Between one and three months, he begins to lift his chin off the ground for a moment, and lift his head for a moment if held sitting. But if held standing, he sags at the knees and hips.

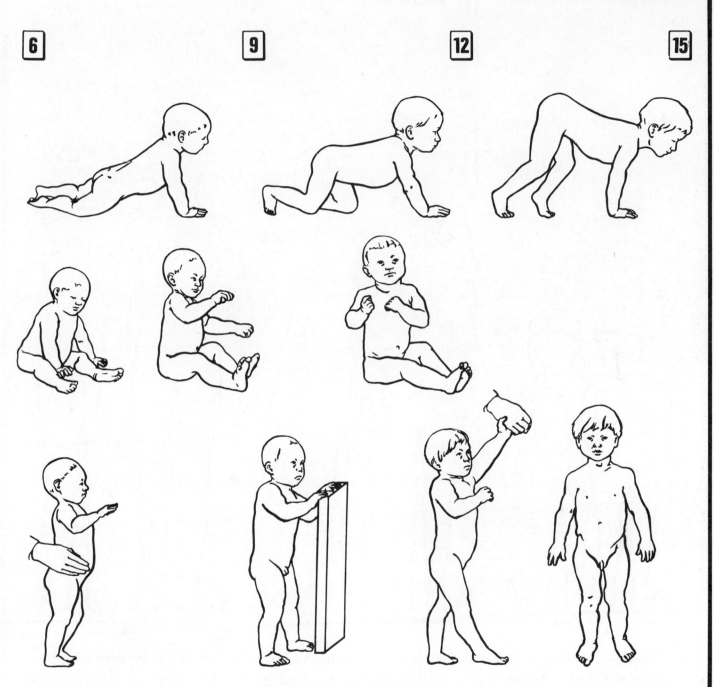

6 **9** **12** **15**

By about six months, he can support himself on his arms, lying or sitting, and can bear his own weight if held standing.

Between eight and ten months, he begins to be able to crawl on hands and knees, to sit and lean forward without support, and to hold himself upright.

At a year he can creep like a bear, on hands and feet, turn around as he sits, and walk with one hand held. At 13 months, he can walk alone.

©DIAGRAM

GROWTH

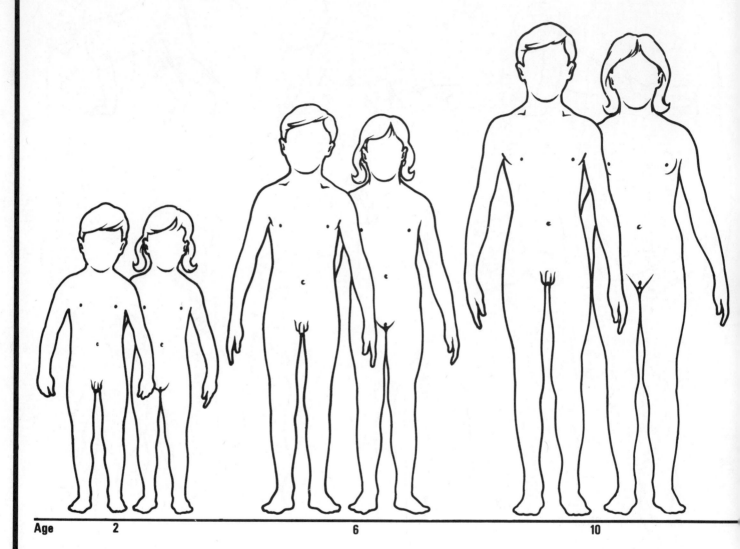

Age	2	6	10

a 2ft 10in, 25 lb (2ft 10in, 27 lb)
 .86m, 11.3kg (.86m, 12.2kg)
b 53% (49%)
c 20% (17%)

a 3ft 8in, 42 lb (3ft 9in, 46 lb)
 1.12m, 19.05kg (1.14m, 20.9kg)
b 69% (65%)
c 33% (30%)

a 4ft 5in, 68 lb (4ft 6in, 70 lb)
 1.35m, 30.8kg (1.37m, 31.7kg)
b 83% (78%)
c 53% (45%)

The first set of figures (a) gives typical heights and weights for each age (figures for boys in brackets). The second set of figures (b) indicates how much of a child's eventual height it is likely to have

achieved at each age, eg 83% at age 10. (Boys' figures in brackets.) The third set of figures (c) indicates how much of its eventual weight a child is likely to have achieved at each age, eg 53% at age 10. (Boys' figures in brackets.)

Of course, such predictions are only averages. There are two reasons why a child may be taller (shorter) than average for its age. It may be going to be a tall (short) adult. Or it may be advancing faster (slower) than

20

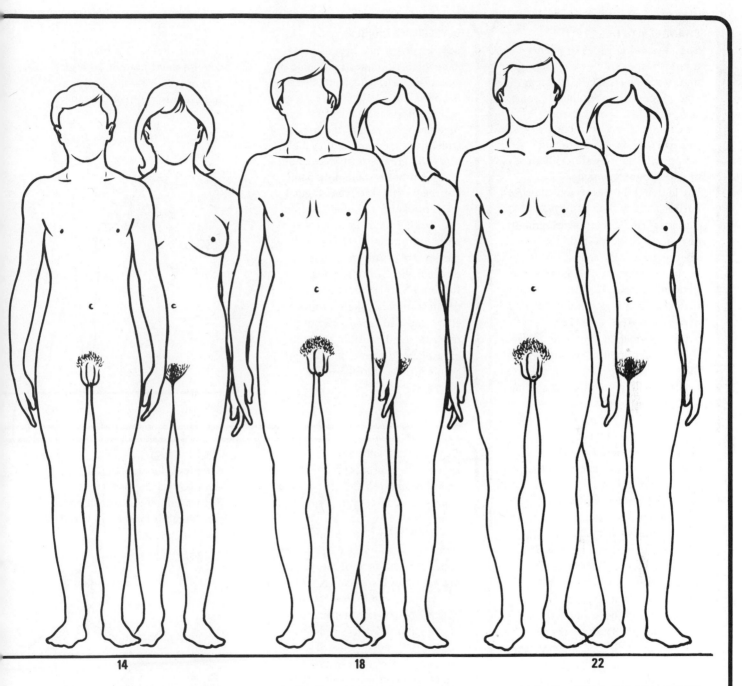

14   18   22

a 5ft 2in, 109 lb (5ft 3in, 108 lb)
   1.57m, 49.4kg (1.60m, 49kg)
b 97% (91%)
c 85% (70%)

usual to an eventual average
height: that is, advanced (or
behind) in its general development
for its age.
The development of a child's
permanent teeth gives a rough
guide to this general rate of

a 5ft 4in, 125 lb (5ft 8in, 143 lb)
   1.63m, 56.7kg (1.75m, 64.9kg)
b 100% (99%)
c 98% (92%)

development. Compare the ages
at which the teeth appear with the
ages on p. 52. If they appear at the
earlier age, the development is
advanced; if at the later age, slow;
if between, average.
It is interesting to note how much

a 5ft 4in, 128 lb (5ft 9in, 155 lb)
   1.63m, 58.1kg (1.75m, 70.3kg)
b 100% (100%)
c 100% (100%)

more slowly a child moves toward
its eventual weight than its eventual
height. At the age of 2, for instance
a child is already about one-half its
but only adult height,
one-fifth its adult weight.

©DIAGRAM

PUBERTY

CHANGES AT PUBERTY

The physical changes of puberty transform the body of a child into that of an adult. These changes take place over a number of years, and occur in response to the production of sex hormones (testosterone in the male and estrogen in the female). The age at which puberty occurs varies, but in general it starts and finishes earlier in girls than in boys.

The most significant development of puberty is the maturing of the sex organs . Girls begin to menstruate and, after the first few periods, begin to produce mature ova during the menstrual cycle . Usually, a single ovum is produced from one of the ovaries each month. Maturing of the male sex organs during puberty results in the production of sperm, which are sometimes emitted from the penis during sleep ("wet dreams"). It is important that adolescents should be fully informed of these developments before they occur.

Male and female hormones are also responsible for the development during puberty of what are called secondary sexual characteristics. Before puberty, except for very obvious genital differences, the physical appearance of boys and girls is fairly similar. During puberty, differences betweeen the two sexes are emphasized. Girls generally do not grow as tall as boys. Girls develop rounded contours and broad hips, compared with the angular, more muscular male physique. Both sexes develop body hair during puberty but it occurs in different areas and is heavier in males. Also at this time, a boy's larynx grows more than a girl's, making his Adam's apple more pronounced and his voice deeper .

DEVELOPMENT: GIRLS

Before puberty the breasts are undeveloped, there is no pubic or underarm hair, and the body shape is boyish. In early puberty - perhaps from age 11 to 13 - the face becomes fuller, the pelvis widens, fat is deposited on the hips, the breasts start to develop and the nipples stand out, pubic hair begins to grow, the vaginal walls thicken, and menstruation may begin. Later in puberty - perhaps from 14 to 17 - growth of the breasts continues, pubic hair thickens, menstruation begins if it has not already done so, the genitals mature, skeletal growth ends, and body shape becomes more rounded . Further rounding, breast development, and weight gain continue into the early twenties .

Illustrated here is a typical development pattern for a girl during puberty - obviously individuals vary considerably.

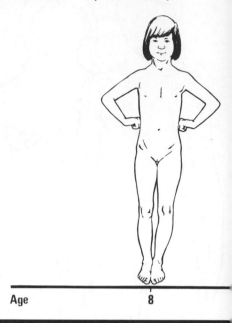

Age   8

DEVELOPMENT: BOYS

In a boy before puberty the penis and scrotum are small, and there is no pubic, underarm, or other coarse body hair. In early puberty - perhaps from 12 to 15 - the testes begin to enlarge, pubic hair appears at the base of the penis, the penis begins to grow, and there is a sudden, rapid increase in height. As puberty continues - perhaps from 15 to 18 - the shoulders broaden, the voice deepens , hair grows in the armpits and on the upper lip, penis growth continues, sperm are produced, pubic hair coarsens and spreads and other body hair grows, the prostate gland enlarges, height and weight increase and there is a sudden gain in strength. Height and weight typically continue to increase into the early twenties, and there is also further growth of coarse body hair.

A typical developmental pattern for a boy during puberty is illustrated here - as with girls, considerable individual variations obviously occur.

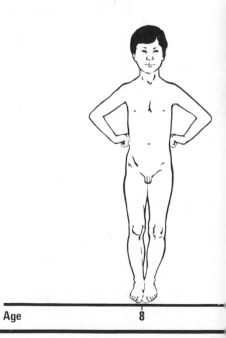

Age   8

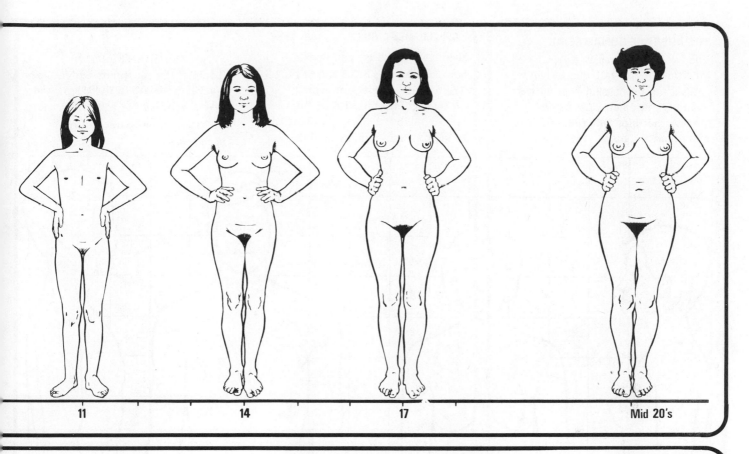

11   14   17   Mid 20's

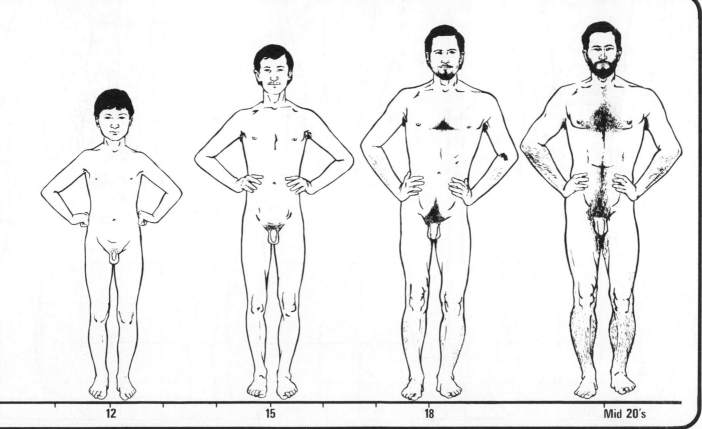

12   15   18   Mid 20's

© DIAGRAM

THE END PRODUCT

THE AVERAGE HUMAN BEING

The average woman is almost 5ft 3¾in (1.62m) tall: she weighs almost 135 lb (61.2kg), her bust is 35½in (89cm), her waist 29¼in (74cm), her hips 38in (96cm). The maximum weight she reaches is about 152 lb (69kg), and that is between the ages of 55 and 64. The average man is just over 5ft 9in (1.75m) tall. He weighs almost 162lb (73.5kg), his chest is 38¾in (98cm) round, his waist 31¾in (81cm), his hips 37¾in (96cm). The maximum weight he reaches is about 172 lb (78kg), and that is between the ages of 35 and 54. These figures are for people in the USA.

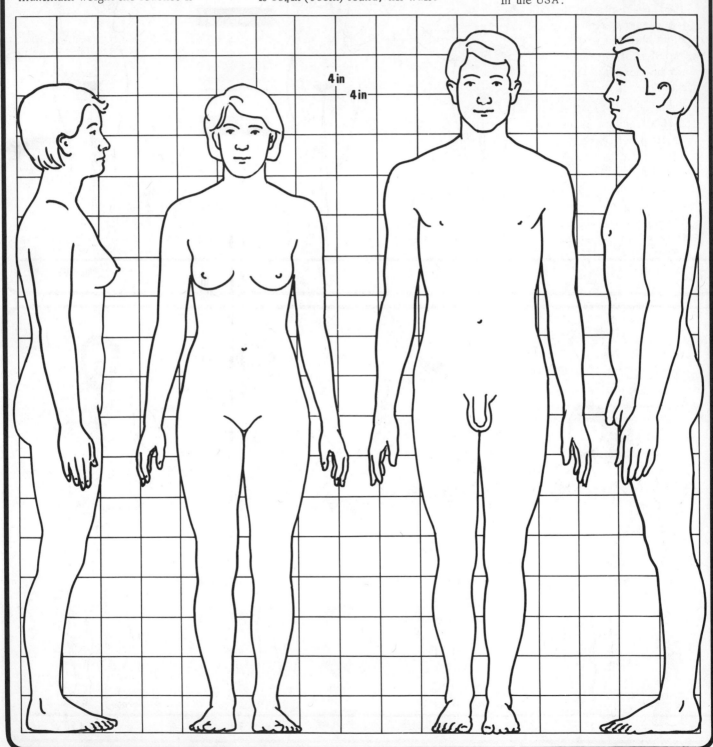

4 in
4 in

NORMAL AND ABNORMAL

The range of the normal is fairly small: the range of the possible is fairly wide. A convenient example is height. In every 100 women, 95 are between 4ft 10in (1.47m) and 5ft 8in (1.73m). But the tallest woman who has ever lived (whose height has been verified) was 7ft 11in (2.4m) at death (age 27), and the shortest $23\frac{1}{4}$in (59cm) (age 19).

In fact, the distribution of many physical characteristics in a population can be summed up in a "normal distribution" curve, as shown below. The range of the characteristic goes all the way from a to z, and there are people at every point between. But there are very few people at either of the extremes - and very many in the central area.

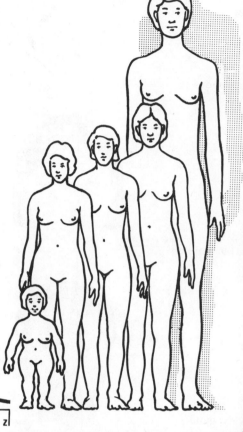

Population

a Characteristic (eg height) z

CELL LIFE

The full-grown body is still changing constantly: each day, millions of body cells die, and must be replaced. Below, we show some of their maximum life expectancies.

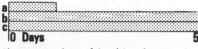

a
b
c
0 Days 5

a) scavenging white blood cells
b) intestinal cells
c) blood clotting cells

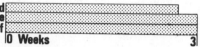

d
e
f
0 Weeks 3

d) skin cells e) kidney cells
f) female egg cells

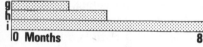

g
h
i
0 Months 8

g) male sperm cells h) eyelashes
i) liver cells

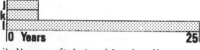

j
k
l
0 Years 25

j) disease-fighting blood cells
k) scalp hairs
l) bone cells

WOMAN AND MAN

This shows how some characteristics of the typical woman and man compare.

 Average brain weight

 Heart weight

 Quantity of blood

 Skin surface area

 Lung capacity (age 25)

©DIAGRAM

25

ETHNIC VARIATIONS

ETHNIC VARIATIONS

No one now is very happy with the word "race": it has been too much a part of man's inhumanity to man. But patterns of ethnic variation do, of course, exist - that is, fairly consistent differences in the physical characteristics of different peoples.

We are all aware of how people vary in stature, skin color, hair type, and facial features. But the ethnologist also notices such things as blood type, the ability to taste certain substances, and even the type of wax that forms in the ear. All these are part of the variety of human inheritance.

Three great ethnic groups - Caucasoid, Mongoloid, and Negroid - account between them for almost all of the world's population. We have tried to illustrate their typical characteristics. But sometimes the differences within each group are as large as those between them. Taking, for example, the old preoccupation, skin color, Negroids range from near black to sallow; Mongoloids from yellowish to flat white to deep bronze; and Caucasoids from fair pinkish in northern Europe to the dark brown people of southern India.

Another extreme variable is height. It is less totally inherited, and more immediately determined by environment, than other ethnic criteria. But not only does it range widely among individuals of a given people, but also between different peoples of the same ethnic group.

ETHNIC TYPES
Basic facial characteristics

Caucasoid

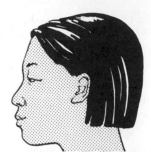

Mongoloid

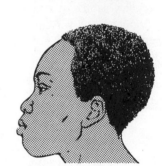

Negroid

BODY AND CLIMATE

Animals of a species differ in coat color, size, limb length, and location of fat deposits, according to climatic conditions where they live. Some human variations are also related to climate.

ETHNIC FEATURES
The environment did not make people acquire inheritable characteristics - but it did decide which characteristics flourished. Those people who flourished bred among themselves, and passed on these features (which are not, of course, lost if the inheritor moves to a new environment).

Skin color is a well-known example. The extra melanin in dark skins gives added protection against the sun. Where the sun is no problem, pale skin allows better vitamin D formation.

Yellow skin contains a dense keratin layer that reflects light well in deserts or snow. Dark eye color also protects against sunlight; and so do the thick, folded eyelids of Mongoloids. Negroid hair protects against heat on the scalp, but allows sweat loss from the neck. Straight hair, grown long, protects against the cold. Noses typically vary with air humidity. In dry conditions they are longer and narrower, so inhaled air is moistened. But the flat Mongoloid

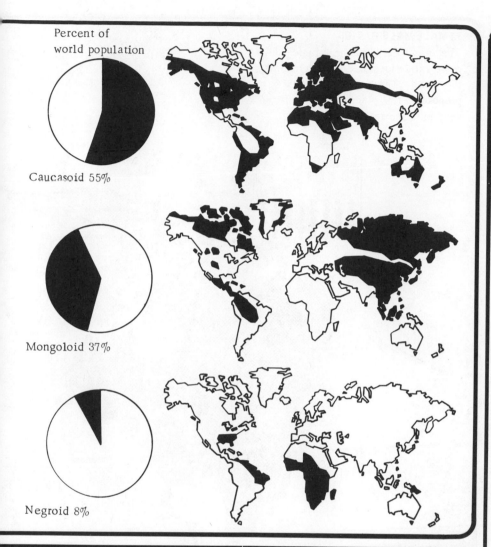

Percent of
world population

Caucasoid 55%

Mongoloid 37%

Negroid 8%

ETHNIC VARIETY

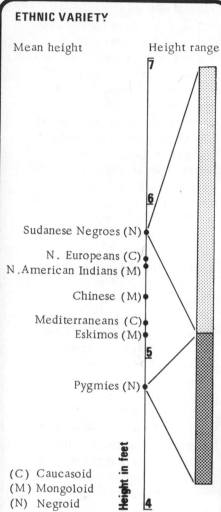

Mean height   Height range

Sudanese Negroes (N)
N. Europeans (C)
N.American Indians (M)

Chinese (M)

Mediterraneans (C)
Eskimos (M)

Pygmies (N)

Height in feet

(C) Caucasoid
(M) Mongoloid
(N) Negroid

face developed as protection against the cold, and here the nose is not prominent and exposed. The Eskimo have taken this further, by developing facial fat.

OTHER VARIABLES

Other features, not entirely inherited, also vary with climate. Average weight is greater the colder it is. For instance, the average Eskimo is considerably heavier than the average Spaniard. Body shape also varies: two bodies of the same weight can have very different surface areas. Body area is larger for weight, the hotter it is; a large area gives more skin from which to sweat and to radiate heat. Metabolic rate varies in the same way. A typical European has a "thermal equilibrium" of 77°F (25°C), i.e. with that temperature around him, naked, standing still, he shows no tendency to get hotter or colder. The Eskimo's metabolic rate is 15 to 30% higher than the European's, giving him a lower thermal equilibrium, while an Indian's, Brazilian's or Australian's metabolic rate is 10% lower than the European's.

Height relates partly to ethnic factors - but little to overall ethnic group. Above, left, average heights for a sample selection of peoples reveal a jumbled sequence of Negroids, Caucasoids, and Mongoloids (eg Negroid peoples are both shortest and tallest). Also, even within a people, other genetic and environmental variations prevent too great a consistency. Above, right, for example, the height range shows the tallest pygmy as tall as the shortest Sudanese Negro.

© DIAGRAM

WOMEN AND MEN

WOMEN AND MEN

World-wide, there are just over 100 women for every 100 men. But this ratio varies greatly from country to country. In most of Asia and the Middle East, and large parts of Africa, men predominate; in the United States and western Europe, women. In general, in primitive parts of the world, lack of contraception and medical facilities increases the maternal death rate in childbirth. Also the lower status given to women may still mean that more effort is made to save a male child. Yet some less developed countries show a female predominance.

A male predominance can also arise because of immigration, as in Alaska. Men are the first to go to new countries in search of their fortune. As standards rise, the numbers of women catch up. But Canada still has fewer women per 100 men than the United States; and in the USA itself, female predominance was not reached until the census of 1950.

THE AGE PATTERN

In most societies, fewer women than men are born. The ratio is generally about 100 female babies for every 105 male. But, in modern American and western European society, fewer females than males die at each age before old age is reached (except perhaps for a brief period during childhood). In other words, female life expectancy is longer. So the older one is, the more women of that age group there are for every 100 men. Parity between the sexes is usually reached between the ages of 30 and 40. After this, female predominance grows steadily, until at 95 a man is outnumbered 4 to 1.

FEMALE-MALE RATIOS

The map tints show the female-male ratios for most countries in the world. The lines of symbols pinpoint a few places that have very extreme ratios.

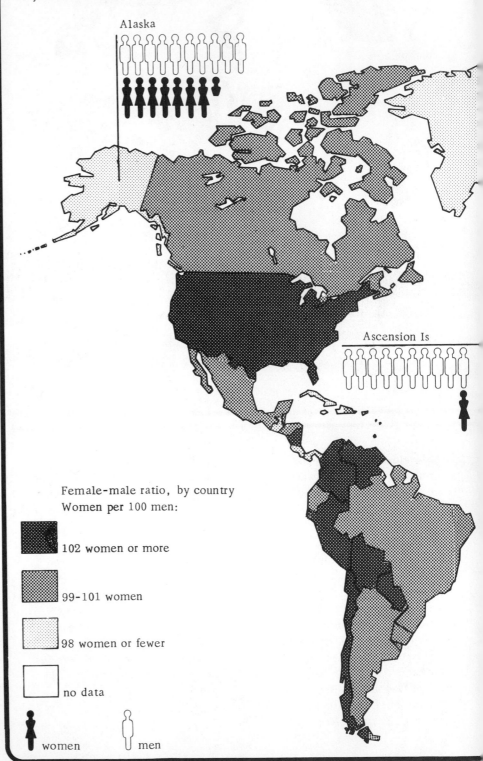

Female-male ratio, by country
Women per 100 men:

- 102 women or more
- 99-101 women
- 98 women or fewer
- no data

women   men

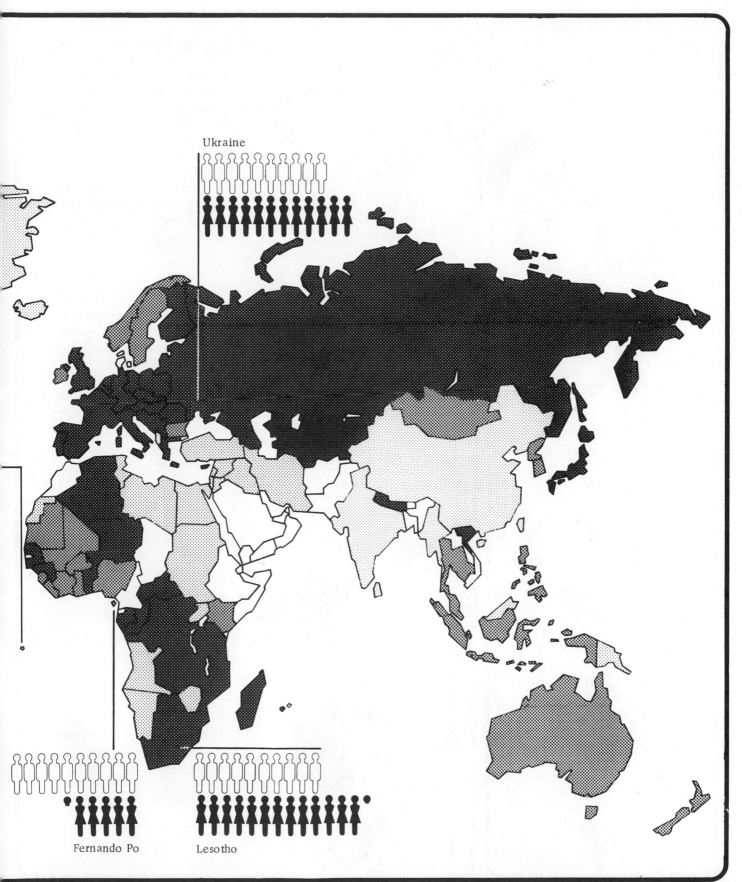

Ukraine

Fernando Po   Lesotho

LIFE EXPECTANCY 1

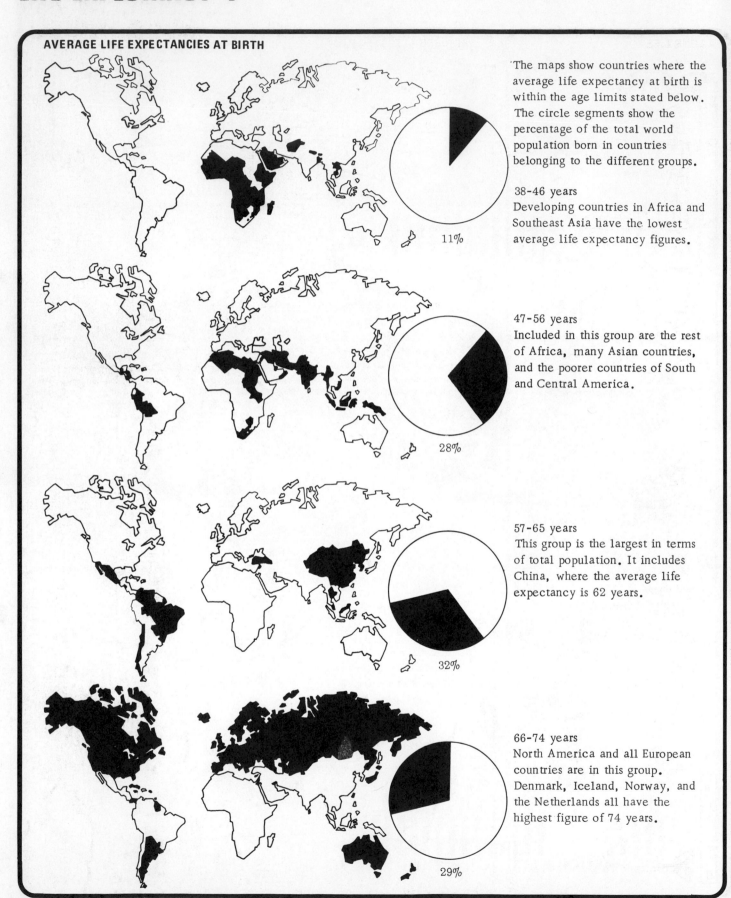

AVERAGE LIFE EXPECTANCIES AT BIRTH

The maps show countries where the average life expectancy at birth is within the age limits stated below. The circle segments show the percentage of the total world population born in countries belonging to the different groups.

38-46 years
Developing countries in Africa and Southeast Asia have the lowest average life expectancy figures.

11%

47-56 years
Included in this group are the rest of Africa, many Asian countries, and the poorer countries of South and Central America.

28%

57-65 years
This group is the largest in terms of total population. It includes China, where the average life expectancy is 62 years.

32%

66-74 years
North America and all European countries are in this group. Denmark, Iceland, Norway, and the Netherlands all have the highest figure of 74 years.

29%

A LONG LIFE

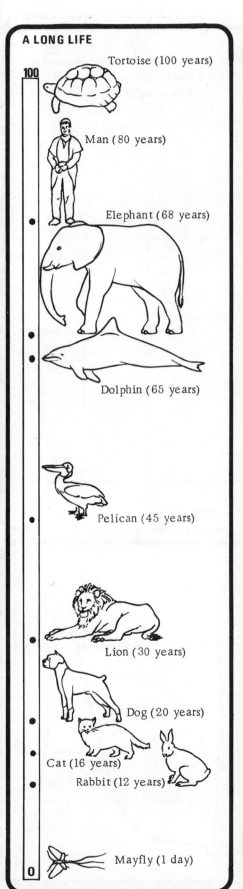

Tortoise (100 years)

Man (80 years)

Elephant (68 years)

Dolphin (65 years)

Pelican (45 years)

Lion (30 years)

Dog (20 years)

Cat (16 years)

Rabbit (12 years)

Mayfly (1 day)

PAST LIFE EXPECTANCIES

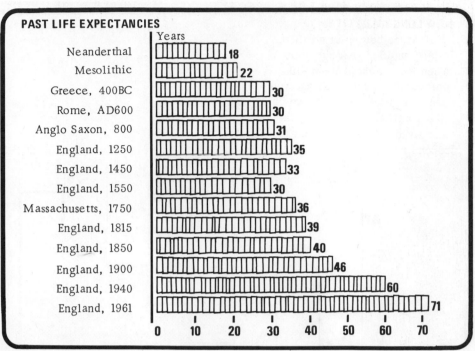

Years

	Years
Neanderthal	18
Mesolithic	22
Greece, 400BC	30
Rome, AD600	30
Anglo Saxon, 800	31
England, 1250	35
England, 1450	33
England, 1550	30
Massachusetts, 1750	36
England, 1815	39
England, 1850	40
England, 1900	46
England, 1940	60
England, 1961	71

MEN AND WOMEN

On average, women live longer than men in most countries of the world. (In the selection below, for example, only India is the exception.) But no one is quite sure why women live longer. It may be some factor in their physical constitution. Alternatively, it may be the consequence of the different types of work that men and women, at present, tend to do.

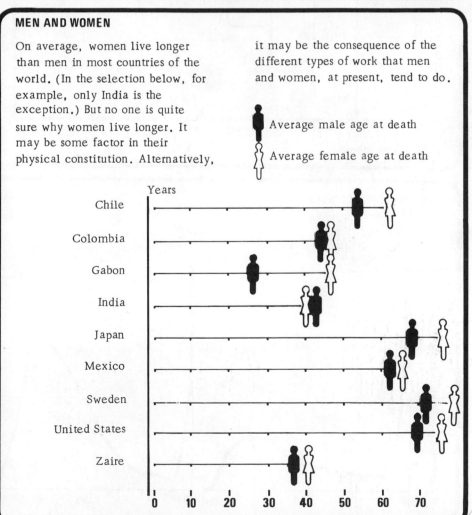

Average male age at death

Average female age at death

Years

Chile
Colombia
Gabon
India
Japan
Mexico
Sweden
United States
Zaire

©DIAGR

LIFE EXPECTANCY 2

HOW LONG WILL I LIVE?

Look at the bottom of this first table, and find your age now. Then look up the column - the number at the top is the age you can expect to live to. But you are likely to live longer if your father also lived to an old age - as the second table shows.

78 Women
78 Men

78 Father died under 50
78 Father died 50-79
78 Father died 80 or over

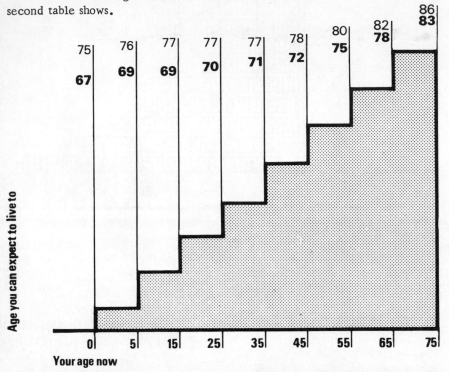

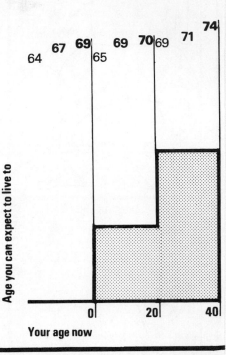

PAST PATTERNS OF DYING

in a developed country.

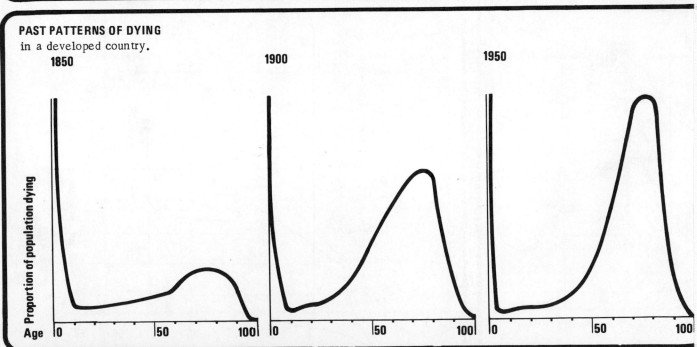

1850

1900

1950

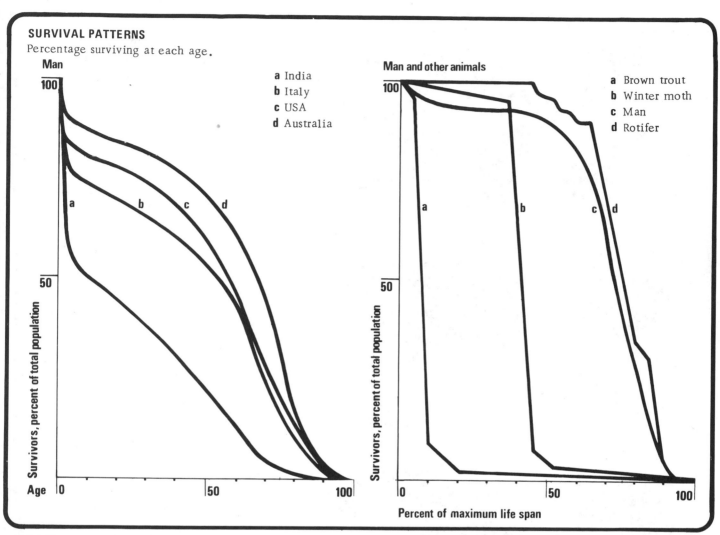

SURVIVAL PATTERNS

Percentage surviving at each age.

Man

a India
b Italy
c USA
d Australia

Survivors, percent of total population

Age

Man and other animals

a Brown trout
b Winter moth
c Man
d Rotifer

Survivors, percent of total population

Percent of maximum life span

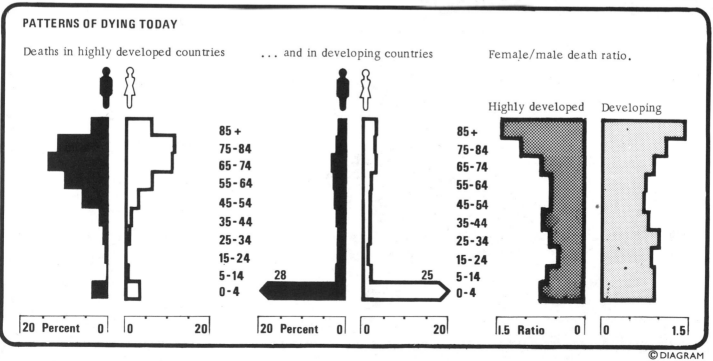

PATTERNS OF DYING TODAY

Deaths in highly developed countries

... and in developing countries

Female/male death ratio.

85 +
75 - 84
65 - 74
55 - 64
45 - 54
35 - 44
25 - 34
15 - 24
5 - 14
0 - 4

28

25

Highly developed   Developing

85 +
75 - 84
65 - 74
55 - 64
45 - 54
35 - 44
25 - 34
15 - 24
5 - 14
0 - 4

| 20 Percent 0 | 0 20 | 20 Percent 0 | 0 20 | 1.5 Ratio 0 | 0 1.5 |

©DIAGRAM

33

CAUSES OF DEATH: World

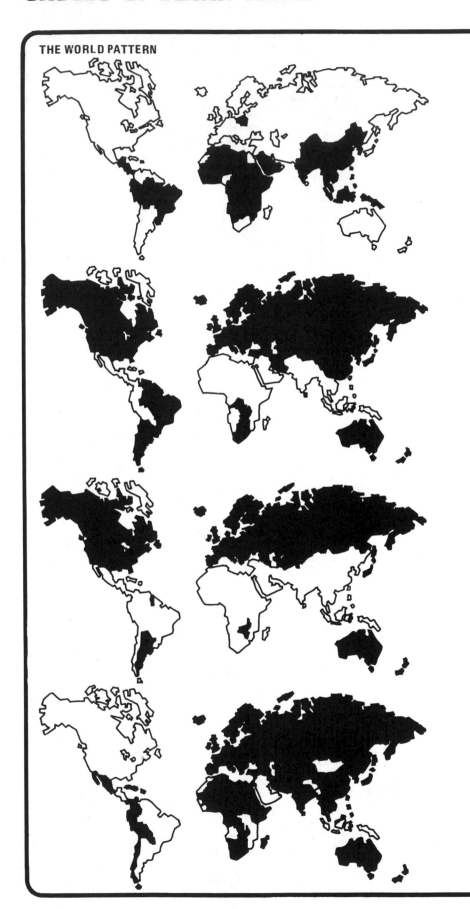

THE WORLD PATTERN

INFECTIOUS DISEASES
These are the killers in the poorer parts of the world: typhoid, paratyphoid, cholera, malaria, dysentery, tuberculosis, meningitis, smallpox, and yellow fever.

CARDIOVASCULAR DISORDERS
The affluent world can deal with infection, but its life-style creates its own problems: degeneration of the blood vessels, thromboses and embolisms, high blood pressure, and heart failure.

CANCER
Another of the main problems of western society: cancers of the lung, stomach, intestines, pharynx and larynx; cancers of the breast, cervix, and uterus in women; and leukemia.

RESPIRATORY DISORDERS
Scourges of the Old World rather than of the New: in poor countries, influenza; in rich but polluted ones, bronchitis and emphysema.

RICH DEATH AND POOR DEATH

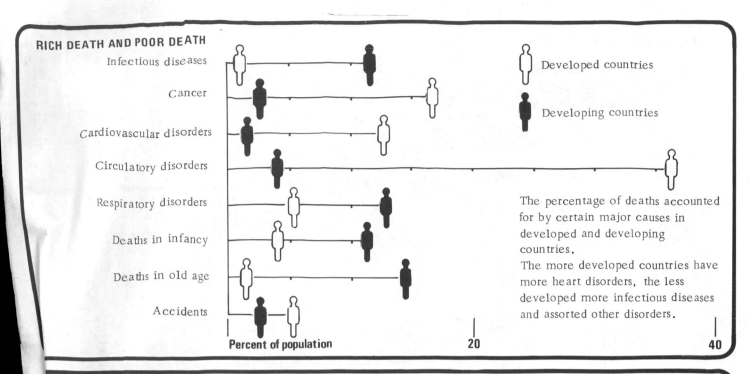

Infectious diseases

Cancer

Cardiovascular disorders

Circulatory disorders

Respiratory disorders

Deaths in infancy

Deaths in old age

Accidents

Percent of population　　　　　　　20　　　　　　40

Developed countries

Developing countries

The percentage of deaths accounted for by certain major causes in developed and developing countries.

The more developed countries have more heart disorders, the less developed more infectious diseases and assorted other disorders.

VERY TEN DEATHS

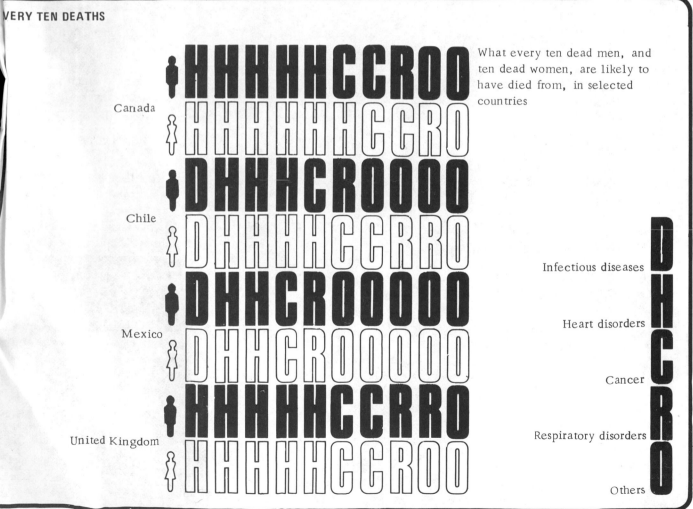

What every ten dead men, and ten dead women, are likely to have died from, in selected countries

Canada

Chile

Mexico

United Kingdom

Infectious diseases　D

Heart disorders　H

Cancer　C

Respiratory disorders　R

Others　O

©DIAGRAM

35

CAUSES OF DEATH: USA

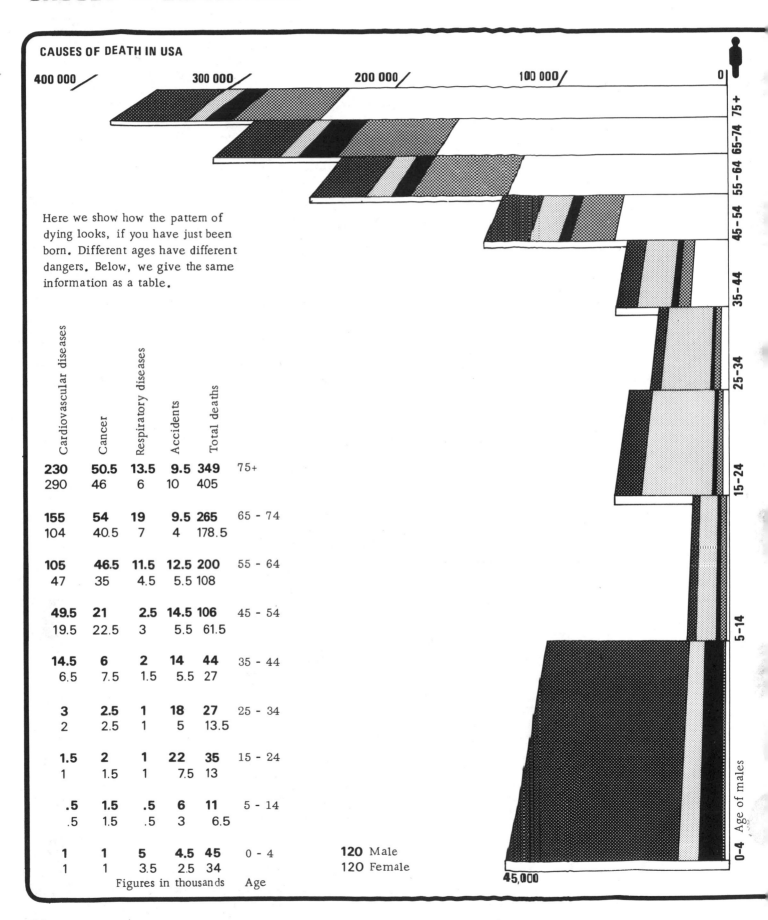

CAUSES OF DEATH IN USA

400 000 / 300 000 / 200 000 / 100 000 / 0

Here we show how the pattern of dying looks, if you have just been born. Different ages have different dangers. Below, we give the same information as a table.

Cardiovascular diseases	Cancer	Respiratory diseases	Accidents	Total deaths	Age
230	**50.5**	**13.5**	**9.5**	**349**	75+
290	46	6	10	405	
155	**54**	**19**	**9.5**	**265**	65 - 74
104	40.5	7	4	178.5	
105	**46.5**	**11.5**	**12.5**	**200**	55 - 64
47	35	4.5	5.5	108	
49.5	**21**	**2.5**	**14.5**	**106**	45 - 54
19.5	22.5	3	5.5	61.5	
14.5	**6**	**2**	**14**	**44**	35 - 44
6.5	7.5	1.5	5.5	27	
3	**2.5**	**1**	**18**	**27**	25 - 34
2	2.5	1	5	13.5	
1.5	**2**	**1**	**22**	**35**	15 - 24
1	1.5	1	7.5	13	
.5	**1.5**	**.5**	**6**	**11**	5 - 14
.5	1.5	.5	3	6.5	
1	**1**	**5**	**4.5**	**45**	0 - 4
1	1	3.5	2.5	34	

120 Male
120 Female

Figures in thousands   Age

Age of males

45,000

36

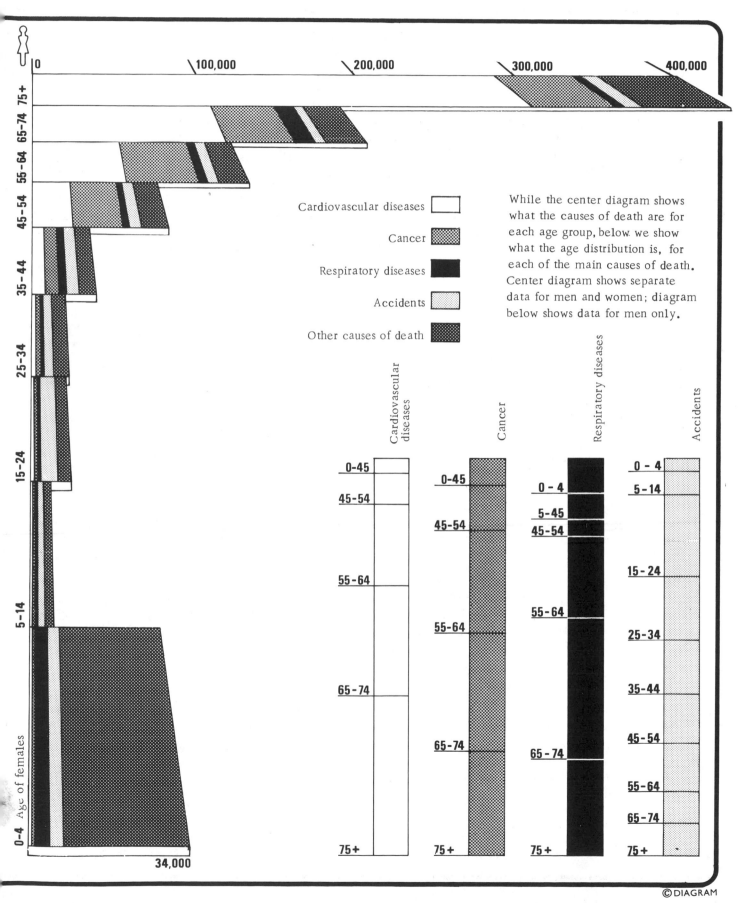

0   100,000   200,000   300,000   400,000

75+
65-74
55-64
45-54
35-44
25-34
15-24
5-14
0-4

Age of females

34,000

Cardiovascular diseases
Cancer
Respiratory diseases
Accidents
Other causes of death

While the center diagram shows
what the causes of death are for
each age group, below we show
what the age distribution is, for
each of the main causes of death.
Center diagram shows separate
data for men and women; diagram
below shows data for men only.

Cardiovascular diseases

0-45
45-54
55-64
65-74
75+

Cancer

0-45
45-54
55-64
65-74
75+

Respiratory diseases

0-4
5-45
45-54
55-64
65-74
75+

Accidents

0-4
5-14
15-24
25-34
35-44
45-54
55-64
65-74
75+

©DIAGRAM

37

ACCIDENTS

ACCIDENTS

Accidents are a leading cause of death in all modern societies. In the USA they claim over 100,000 fatalities every year, ranking only beneath heart disease, cancer, and strokes in the number of their victims. And in the age groups up to 24 years accidents are by far the most important cause of death, outnumbering all the other top ten causes put together.

Men are far more likely to suffer accidental death than women (69% of all accidental deaths are male) - and this is true for all causes and all ages. In the 15 to 24 age group, in fact, there are four such male deaths to every one female. Only in the over 75 age group are there rather more female accidental deaths in absolute numbers - and this is only because women so much outnumber men in this category. So, even at this age, an individual man remains far more susceptible to accidental death than an individual woman.

The overall distribution of fatal accidents by age is fairly even, rising only in the 75 and over and 15 to 24 age groups. (The great predominance of death from accident for all ages under 25 is, of course, because there are not many other things that young people often die from.)

CAUSES

By far the most important causes of accidental death are motor vehicle accidents. They account for almost half of all such deaths, and are the leading cause in all age categories except 75 and over. Motor vehicle accidents, in fact, account for over $\frac{2}{3}$ of accidental deaths in the 15 to 24 age group. The next main cause is accidental fall (16% of deaths) - and this by itself accounts for the significance of accidental death in those 75 and over. Fifty-five per cent of all fatal falls are in this age group, and they account for 63% of the group's accidental deaths.

The other leading causes of accidental death are: fires and burns (6% of deaths), drowning (6%), choking on food and objects, ($2\frac{1}{2}$%), firearm accidents (2%), poisoning by solids and liquids ($2\frac{1}{2}$%), and gas poisoning (1%). Fires and burns are notable for fatalities in the old and the under fives; choking in the under fives; and drowning in the 5 to 24 age group. Other causes are spread out fairly evenly over the age categories.

The male predominance reaches its highest in firearm deaths (86% male). It is lowest in falls (50% male), due to the large number of women over 75 dying this way. Comparing today with fifteen years ago, death rates from motor vehicle accidents have risen, and those from falls and fires have gone down.

MOTOR ACCIDENTS

The main types of fatal motor vehicle accident are, beginning with the most important: collisions with other motor vehicles (42%); overturning or going off the road (28%); hitting a pedestrian (17%); collisions with fixed objects (7%); collisions with trains ($2\frac{1}{2}$%); and collisions with bicycles ($1\frac{1}{2}$%). Most causes distribute over the age groups to coincide with the general age pattern of motor vehicle fatalities i.e. they are much more common in those of working age (15 to 65).

However, pedestrian deaths are spread much more evenly over all age groups - with the result that almost half of those killed under 14 are pedestrians. Also over a third of those 75 and over are pedestrian deaths - but even more, in this age group, are deaths in motor vehicle collisions.

Driver
a Exceeding speed limit
b Driving on wrong side of road
c Reckless driving
d No right of way
e Driving off highway
f Others

Pedestrian
a Crossing intersection with signal
b Walking on rural highway
c Crossing intersection against signal
d Crossing intersection, no signal
e Crossing from behind parked car
f Others
g Crossing between intersections
h Not on roadway
i Children playing in street

INJURY

Accidents also cause, every year, in every 100 people, about 31 injuries that limit activity for a time and/or need medical attention. The rate for men (37 per 100) is half again as high as that for women. Among these accidental injuries, those at work are least common (about 4 per 100), those at home next (about 12 per 100), and others, including motor vehicle accidents, highest (about 19 per 100). Comparing men and women, men are equally likely to be injured at home, more likely to be injured in a motor accident, and (because of job differences) over 6 times more likely to be injured at work.

HOME ACCIDENTS

Home accidents claim over 25,000 victims every year in the USA. About a third of these are 75 or over, and almost a quarter of the remainder are in the 0 to 4 age group.

The main types of accident in the home are, beginning with the most important: falls (36%); deaths associated with fire (21%); poisoning by solids and liquids ($9\frac{1}{2}$%); choking ($8\frac{1}{2}$%); firearms and poisoning by gas (each $4\frac{1}{2}$%); and "mechanical" suffocation (4%). Falls are most prevalent in the old; suffocation in the under fives; and poisoning with solids and liquids in the 25 to 44 age group, as well as in the under fives. Also, among other, unspecified causes of fatal home accidents, one third of the victims are children under five. Other causes distribute comparatively evenly over the age groups.

NATIONAL COMPARISONS

Full data for all countries on accidental death is not available, and what is available may not be strictly comparable. However, quoted death rates for about 45 countries (with almost no African, Middle Eastern, Asian, or Communist countries included) give the following observations. In the USA the death rate from all accidents is 51.7 deaths each year per 100,000 of the population. Notably higher are Chile (71.9), Austria (66.7), and France (62.5). Notably lower are the Dominican Republic (17.4), Singapore (21.2), and Hong Kong (23.6).

For motor vehicle accidents the US death rate (21.2) is exceeded in Austria, Germany, Australia, and especially South Africa (27.3). Lowest are Ceylon (2.0), the United Arab Republic (2.6), and Mexico (3.1).

The death rate from falls (the next main cause of accidental death in the USA, with a rate of 10.5) is considerably exceeded in Austria, Denmark, and Switzerland (all about 18 to 19), and especially in West Berlin (28.7). Lower rates occur in Poland (1.2) and Singapore (2.8).

Other odd observations from the statistics are:
the very high death rate from fire and explosion in the United Arab Republic (14.2, compared with 4.0 for the USA) - presumably a consequence of oil mining; the rather high death rates for drowning in Japan (7.6), Portugal (7.7), and Colombia (8.9), compared with 2.9 for the USA; the extremely high homicide death rate in El Salvador (30.3, compared with 7.3 for the USA); and the high rate for other unspecified accidents in the United Arab Republic (31.4), Columbia (17.2), and Belgium (18.4), as against 6.5 in the USA.

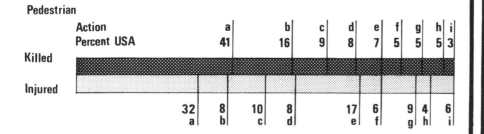

Driver

	Action	a	b	c	d	e	f
	Percent USA	41	16	15	13	10	4

Killed

Injured

		a	b	c		d	e	f
		42	7	19		19	7	6

Pedestrian

	Action	a	b	c	d	e	f	g	h	i
	Percent USA	41	16	9	8	7	5	5	5	3

Killed

Injured

		a	b	c	d		e	f	g	h	i
		32	8	10	8		17	6	9	4	6

©DIAGRAM

SUICIDE

THE TENTH CAUSE OF DEATH

Suicide is killing yourself. There are 365,000 suicides in the world each year - and about 3 to 4 million attempts. In the United States there is one suicide every twenty minutes, and about 25 to 30,000 a year i.e. about 13 to 14 per 100,000 people of all ages. It is the tenth leading cause of death.

SOCIETY AND SUICIDE

Suicide occurs in almost all societies - but in some far more than others, and in some eras more than others. Two things seem to explain the differences.

One is prosperity. Suicide is a phenomenon of prosperous countries, prosperous regions, even, within cities, prosperous neighborhoods - while in many poor lands suicide is so rare that the word, and the concept, are barely understood. Suicide does not occur when the outside world makes it a struggle for us to keep alive. It appears when the world leaves us alone with our consciousnesses.

The other factor is society's attitude to suicide - including the effect of religion and moral belief. Our decline in religious fear of suicide is hard to separate from the impact of prosperity: the most traditionally Christian countries are also among the most rural and most poor. In any case, when a suicide rate seems to rise, as attitudes change, it may only mean that suicide is more openly admitted. But the importance of society's attitude is more clearly seen when the story goes the other way. In imperial Rome, and modern Japan, suicide declined as prosperity rose, because an ancient code that valued suicide was broken.

THE INDIVIDUAL AND SUICIDE

Suicide is a product of social isolation - of loneliness and the sense of uselessness. Single people are more likely to kill themselves than married; widowed than single; and divorced than widowed. The products of broken homes are more likely to kill themselves than the products of happy homes; and those with no religious convictions than those who have.

Hence suicide accompanies modern prosperity. The familiar features of industrial society - the geographical and social mobility, the separation of young and aging adults from the family, the pressure to achieve, the lack of a role for those who can no longer work (where school is compulsory, grandparents are no longer needed to look after the children) - all these, and the other features that they create, such as the failure rate of modern marriages, leave individuals and even family units struggling to persuade themselves that they have a place that they belong to, and a value to others.

So suicide rates reach their highest in the managerial and professional classes, and also in cities.

And, despite the generally close link with prosperity, from one month and year to another the suicide rate keeps in step with the level of unemployment. Suicide is a hopeless admission of defeat: it happens when a isolated person cracks under the strain of his isolation, and it happens mostly to those who have already shown signs of their defeat, in psychic depression and perhaps in alcoholism. It is only rarely a dramatic gesture, and very rarely a rational preference to a painful death from incurable disease.

SEX AND SUICIDE

Successful suicide is a male achievment. Today, and historically, men are twice as likely to kill themselves as women are.

AGE AND SUICIDE

In some societies - such as imperial China - suicide among the old was rare, because the old were revered. But in most cultures the likelihood of suicide grows with age. It is rare in children. Then, in the 15 to 24 age group, it suddenly leaps to prominence as the fourth main cause of death. But this is because there is not much else - except road accidents - that they often die from. (The exception is among students, where the suicide rate is genuinely high.) Thereafter, the rate rises steadily, reaching a peak in men in the 75 to 84 years age group. (In women, the peak is earlier, in middle age.) Men over 65 have three times the suicide rate of male teenagers.

ATTEMPTED SUICIDE

There are ten times more attempted suicides than suicides; and most are not trying to kill themselves. Suicide is typical of men and of the old. Attempted suicide is typical of women and of the young - and especially of young women. Of every four men who try suicide, three kill themselves; of every four women, only one.

Of course, some genuinely wish to commit suicide, and fail - just as others wish only to attempt suicide, and unhappily succeed. But, in general, those who attempt suicide, and live, have different motives from those who kill themselves. Suicide happens among those who are socially isolated; attempted suicide among those who are socially - in fact, emotionally - involved. Suicide is a way of ending your pain for yourself. Many who genuinely think of suicide are deterred by the thought of the pain and grief they would leave behind. Attempted suicide is a way of trying to call on those emotional ties: it is an appeal for help, or a blackmail note for it.

But never ignore someone who talks of suicide. Talking of it does not mean he only wants to attempt it: talking of it need not be a plea for sympathy. Two-thirds of those who kill themselves have told someone beforehand what they intended to do; and those who have tried once are not safe from trying again and succeeding.

SUICIDE RATES

... for selected countries.

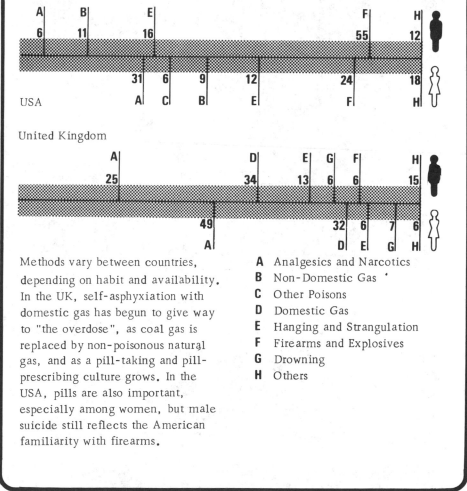

Rate per 100 000   25   50

METHODS OF SUICIDE

USA

United Kingdom

Methods vary between countries, depending on habit and availability. In the UK, self-asphyxiation with domestic gas has begun to give way to "the overdose", as coal gas is replaced by non-poisonous natural gas, and as a pill-taking and pill-prescribing culture grows. In the USA, pills are also important, especially among women, but male suicide still reflects the American familiarity with firearms.

A Analgesics and Narcotics
B Non-Domestic Gas
C Other Poisons
D Domestic Gas
E Hanging and Strangulation
F Firearms and Explosives
G Drowning
H Others

© DIAGRAM

41

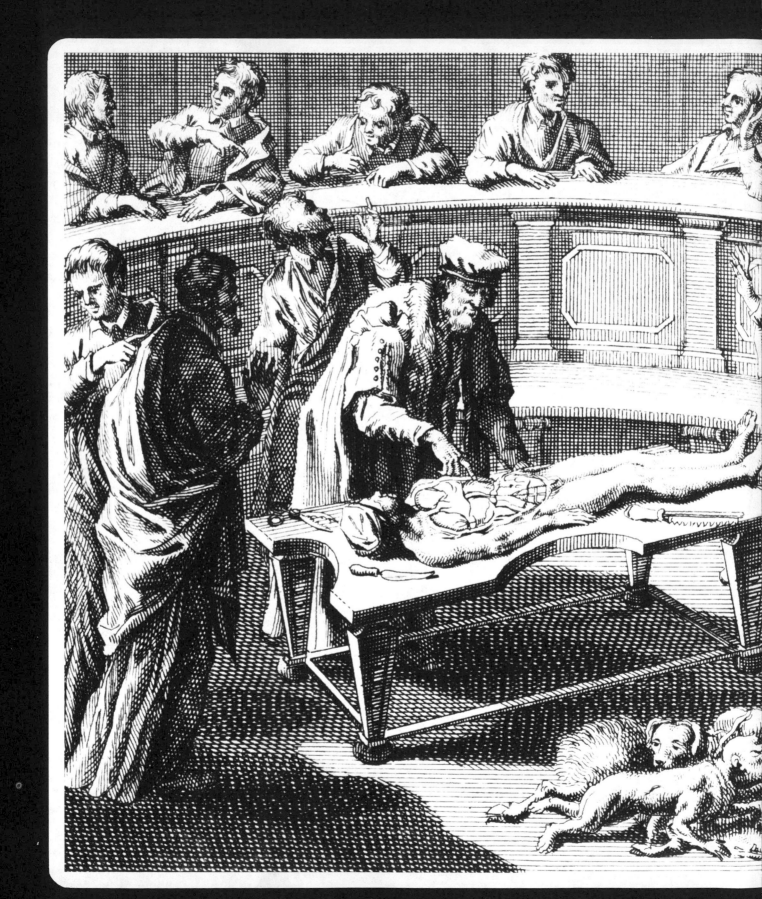

2 Parts of the body

A basic knowledge of how our bodies work can go a long way in equipping us to promote and safeguard the health of ourselves and our families.

Left: An 18th-century anatomical demonstration (from Tabulae Anatomicae by Bartolommeo Eustachi, courtesy of Ann Ronan Picture Library)

THE SKIN

The skin is the largest organ of the body. It covers an area of about 17sq ft (1.6sq m) in the average adult woman - compared with 20sq ft (1.9sq m) in the average man. It accounts for about 16% of the total body weight. Skin thickness over most of the body is about 1.2mm - compared with 0.5mm on the eyelids and 4-6mm on the palms and soles.

The skin is attached to the underlying tissues by elastic fibers, which give it relative flexibility to allow for free joint movement.

In old age the body bulk shrinks and the skin loses its elasticity, causing bagginess and wrinkles.

FUNCTIONS OF THE SKIN

The skin is a versatile organ with a variety of essential functions. Most important, the skin protects the more delicate internal organs, acting as a barrier against physical damage, harmful sun rays, and bacterial infection.

The skin also acts as a sensory organ, being more richly supplied with nerve endings than any other part of the body. Sensations of touch, pain, heat, and cold from the skin provide the brain with a continuous flow of information about the body's surroundings.

Also very important is the skin's role in the regulation of body temperature: 85% of body heat loss is through the skin. During exposure to heat, blood vessels near the skin surface dilate so that more blood flows near the surface to lose its heat. When it is cold, these blood vessels contract to reduce blood flow near the skin's surface.

Body heat is also reduced by the evaporation of perspiration (see G04) on the skin, while the chemical content of perspiration indicates the skin's function as an organ of excretion.

Finally, the skin plays a part in the manufacture of Vitamin D in sunlight (see H10), and even of some antibodies.

STRUCTURE OF THE SKIN

The skin consists of two distinct layers - the epidermis, or outer layer, and the dermis, or inner layer. The epidermis is covered by a thin layer of keratin - the horny protein material also found in hair and nails.

Deep in the dermis, just above the subcutaneous fatty layer, lie the sweat glands which secrete sweat through ducts, or pores, to the surface of the skin.

Also found in the dermis are nerves, and the blood capillaries which nourish the epidermal cells. Hairs are produced by specialized epidermal cells and grow from hair follicles which extend down into the dermal layer. Each hair has its own erector muscle, and a sebaceous gland which secretes grease, or sebum, to keep the skin supple.

SKIN COLOR

A person's skin color is due partly to color pigments found in the skin cells, and partly to the presence of tiny blood vessels near the surface of the skin.

Most important of the pigments that color the skin is melanin - a brown pigment present in skin cells known as melanoblasts.

The melanoblasts of dark-skinned people contain more melanin granules than those of people with fairer skins. The concentration of melanin in an individual's skin is largely determined by heredity - but can be considerably modified by exposure to sunlight (see G07)

There are a few individuals whose bodies contain no melanin pigment at all. Known as albinos, these people have white hair, light-colored eyes, and a pale skin tinted pink by blood vessels.

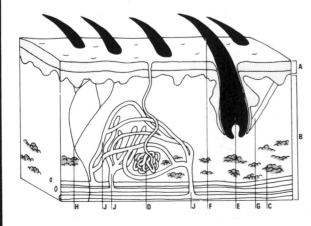

A Epidermis
B Dermis
C Subcutaneous fat
D Sweat gland
E Hair follicle
F Hair shaft
G Erector muscle
H Nerve
J Blood capillaries

PERSPIRATION

Perspiration is a term used to describe both the fluid produced by the sweat glands (sweat), and also the process during which this fluid is produced.

Sweat contains over 99% water, together with small amounts of salts, urea, and other waste products. An average person produces about $1\frac{1}{2}$ pints of sweat a day in temperate conditions.

The process of perspiration helps keep body temperature down because heat is lost when sweat evaporates. Some sweating, however occurs when the body is cool and the skin dry. In some areas of the body perspiration is increased by exercise or mental anxiety.

BODY ODOR

Fresh perspiration produces very little smell in a healthy person. Stale perspiration results in body odor because bacteria that live on the skin act on the sweat to produce substances that smell. Body odor problems are commonly associated with the underarm and genital areas, where perspiration contains fats attractive to bacteria. Here, too, body shape and clothing cause a build up of perspiration by slowing down the rate of evaporation. Foot odor is another common problem caused by perspiring into a constricted area.

Regular washing and changes of clothing help counteract body odor problems. Most people also use a deodorant and/or antiperspirant. Chemicals in deodorants and some soaps slow down the growth of bacteria. With soaps, any lasting effect comes only from soap that remains in the pores after washing. Deodorants are more effective because they dry on the skin and can be concentrated where needed. Antiperspirants reduce perspiration in areas where they are applied, although perspiration over the body as a whole is not reduced. They work by blocking the pores, or by swelling the surrounding area to shrink the pore size. Manufactured sprays, sticks, roll-ons, and creams usually contain both deodorant and antiperspirant.

SKIN CANCER

Skin cancers are abnormal growths in the epidermis, that invade other tissues. The cause of most of them is unknown, but one common cause is consistent exposure of an unprotected skin to strong sunlight. A high proportion is found in those engaged in outdoor work or sports, or in white people living in tropical climates. Melanin gives effective protection to a dark skin.

Skin cancers are usually soon noticed and slow to grow, and they can be destroyed in their early stages by careful surgery or radiotherapy. However, if they are neglected, they can spread beneath the skin, or in some cases metastasize to other parts of the body. Then they are much more dangerous.

Avoidance of excessive exposure to sunlight is the only known preventive technique.

SKIN AND SUN

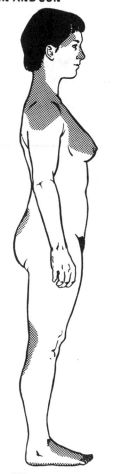

Sunburn danger areas

Exposure to the ultraviolet rays of the sun produces an increased concentration of melanin in the skin. In fair-skinned people this increase in melanin produces freckles and tanning.

Freckles are brown spots formed by patches of melanin. A suntan results from a more even increase in the skin's melanin content. Many people believe that they look more attractive when they have a suntan - and lying in the sun is a popular holiday pastime. Certainly the sun often produces an improvement in skin conditions such as acne, and sunbathing can produce feelings of relaxation and general well-being. If you are unused to the sun or have a very fair skin, it is essential to sunbathe with moderation. Painful sunburn - or even sunstroke - may be the price of over-zealous exposure to the sun's rays. Gradual building up of sunbathing time and the use of protective oils and creams are simple precautions that are well worth the trouble.

©DIAGRAM

BIRTHMARKS

Birthmarks are various types of skin blemish present at birth. They include strawberry marks, port wine stains, vitiligo, and liver spots.

Strawberry marks are red, slightly raised and spongy areas of skin containing enlarged blood vessels. They are usually fairly small and often disappear without treatment. If a strawberry mark persists, it may be shrunk by injections or removed surgically.

Port wine stains are dark red, flat areas of skin containing enlarged blood vessels. They tend to be extensive and often occur on the neck and face. Surgical removal may not be recommended because of the risk of unsightly scarring. Various treatments have been developed by dermatologists to make the mark less noticeable, and special cosmetics provide satisfactory concealment.

Vitiligo can be present at birth. It is a condition in which an area of skin always remains white whatever the color of the skin around it. It can be concealed, but there is no treatment.

Liver spots are dark patches of skin resembling large freckles. They are caused by concentrations of the brown pigment, melanin.

MOLES

Moles are raised brown skin blemishes comprising a mass of cells with a high concentration of melanin. They are sometimes present at birth or may develop later - pregnancy often causes an increase in their size or number. Some moles have a growth of hair which should not be plucked because of the risk of infection. If removal of a mole is considered, it is important to consult your doctor. Most moles are harmless, but occasionally a mole may become malignant. Medical advice should always be sought if a mole changes character and enlarges, ulcerates, or bleeds.

DERMATITIS (ECZEMA)

Dermatitis is a general term for inflammation of the skin. It is usually caused by exposure to a particular substance, but may also be of nervous origin.

Some substances usually have an irritative effect on the skin: others affect only those people who are hypersensitive, or "allergic" to them. Frequent culprits include cosmetics, paints, detergents, insecticides, metals, textiles, rubber, and some plants.

After contact with the offending substance, the blood vessels dilate and become porous. This allows fluid from the cells to collect in the skin and form blisters, which eventually burst. Later, the fluid dries out and the area becomes encrusted. The skin thickens around the sores and flakes off in scales. There is a serious risk of infection if the affected area is scratched or left untreated.

Recurrence can be prevented by identifying the condition's cause and then avoiding or protecting against the substance responsible.

HIVES

Hives is a common allergic reaction characterized by painful, irritating skin wheals. Also called nettle rash, its medical name is urticaria. It is most commonly caused by an allergy to a particular type of food - citrus fruits, shellfish, wheat products, and chocolate can all be troublemakers. Other causes include antibiotics, dust, pollen, and emotional stress.

An allergy produces hives because sensitive skin tissues react by releasing the chemical histamine, which dilates the blood vessels. This increases the flow of fluid to the skin, and produces wheals. The condition is not usually serious, and relief may be obtained by applying soothing lotions. Medical attention must be sought, however, if large swellings that may affect breathing occur in the area of the mouth and throat.

BOILS

Boils are painful, pus-filled lumps caused by bacterial infection of a hair follicle, a sebaceous or sweat gland, a cut, or some other break in the skin. They occur most commonly around sites of friction with clothing, such as the neck or wrists, and may be an indication that a person is run down. Only after the dead skin that forms the boil's core has been released will the boil disappear. Most boils require no more than a protective dressing. A doctor should be consulted if a boil is particularly painful, if several boils occur, or if the sufferer is very young or very old.

A   Epidermis
B   Dermis
C   Pus
D   Dead tissue

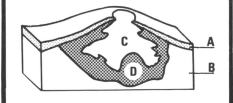

ACNE

Acne is an infection of the sebaceous glands resulting in pimples, blackheads, whiteheads, and sometimes boils and cysts. It characterstically develops in adolescence, when the sebaceous glands become more active. Face, neck, shoulders, chest, and back may all be affected. Most cases clear up if attention is paid to diet, hygiene, and choice of cosmetics. Treatments include lotions and creams to reduce the spread of infection, make the skin peel, and unblock the pores. Antibiotics may be needed in severe cases. Exposure to sunlight or ultraviolet rays can also help make the skin peel, and through tanning, hide the spots.

BLACKHEADS AND WHITEHEADS

A blackhead or whitehead appears when a skin pore becomes blocked by dust, dirt, or sebum. The waxy plug that blocks the pore is called a comedo. This forms a blackhead when it is exposed to the air: oxidation turns the head of the comedo black. If it is not open to the air, a whitehead is formed. A pore may be cleaned by gently pressing out its contents, but unless this is done soon after the plug is formed the spot is probably best left to take its own course. When cleaning out pores it is important to avoid damaging the skin or spreading infection. A preliminary wash with warm water will loosen the plugs.

PSORIASIS

Psoriasis is a chronic skin complaint characterized by red spots and patches covered with loose, silvery scales. The skin of the elbows, forearms, knees, legs, and scalp is most usually affected. The condition results from large-scale production of an abnormal type of keratin. It takes 28 days for normal skin to produce a mature keratin cell, but only 4 days for a person with psoriasis. The cause is unknown but there may be a genetic link. Psoriasis is not infectious and does not affect general health. The condition comes and goes intermittently but there is no cure. Various types of treatment bring some relief.

WARTS

Warts are small benign tumors of the skin. As well as the type common on the hands, plantar warts are common on the feet and moist warts occur on the genitals. Many vanish without treatment - otherwise they can be "frozen" or removed chemically.

CHILBLAINS

These occur on hands, feet, and ears. They are due to poor circulation in cold weather: skin tissue is injured by lack of blood, causing patches of swelling, redness, itchiness, and pain. The best prevention is avoidance of extreme cold; wearing of adequate clothing; and taking exercise to improve the circulation. If the skin does become white and numb through cold, it should be warmed slowly in water or by rubbing, and not by direct heat.

PLASTIC SURGERY

The main uses of plastic surgery are:
a) to correct congenital defects, such as cleft palate or hare lip;
b) to correct defects due to muscular atrophy, paralysis, and other illness;
c) to restore damaged skin and bones after burns and accidents; and
d) to alter the features of the face and body to make the owner appear younger or more presentable (cosmetic surgery).
Here we are mainly concerned with with the last of these.
FACELIFTS remove deep folds of skin around the jaws and mouth. An incision is made from the temple, down in front of the ear, to the nape of the neck. The flap of skin is pulled taut, the excess trimmed off, and the remainder stitched back in place.
EYELID OPERATIONS remove excess skin and fatty tissue from the upper and lower eyelids. The technique is similar to facelifts.
NOSE SURGERY can correct the bridge line, shorten the nose, build up a depressed nose, straighten a crooked one, or alter the shape of the tip. All involve shaving back or implanting extra bone or cartilage. The incisions are made inside the nose, and the final new shape takes about 6 . months to settle.
CHIN SURGERY can correct receding chins by implanting silicone, bone, or cartilage over the jaw to extend it.
EAR SURGERY can correct ears that stick out too far. An ellipse of skin is removed from the back of the ear, and the cartilage beneath is trimmed.
DERMABRASION is removal of skin that has been pock-marked by severe acne, smallpox, etc. The old skin is removed by a high speed rotating brush. Crusts form, and peel after 5 days to 2 weeks, revealing a new layer of pink skin beneath. Success depends on the depth and nature of the problem, and the skin's reaction to treatment.
All plastic surgery requires time for healing and for scars and bruises to disappear.

©DIAGRAM

HAIR 1

HAIR

Hair is found over the whole surface of the human body except the palms of the hands, soles of the feet, and parts of the genitals.

There are three types of hair: scalp hair, body hair, and sexual hair. Scalp hair resembles the body hair of other mammals. Human body hair is usually fine and light in color. Sexual hair develops around the genitals, the armpits, and (in men) the face. Its growth is dependent on the male sex hormone testosterone produced by both sexes at puberty.

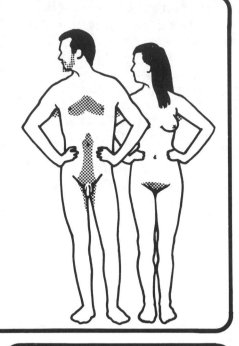

HAIR ROOT

Each hair, properly called hair shaft, grows from its own individual follicle, and each follicle has its own sebaceous (oil) gland, and tiny muscle. Capillaries supply nutrients from the blood stream.

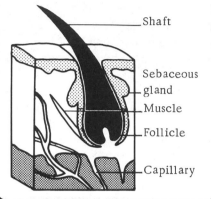

- Shaft
- Sebaceous gland
- Muscle
- Follicle
- Capillary

FUNCTION OF THE HAIR

Hair has 2 major functions: it acts as a protective barrier, and it acts to conserve heat.

The eyelashes protect the eyes, and the hairs in the nose and ears prevent the entry of foreign bodies. The eyebrows prevent sweat from dripping into the eyes.

Air trapped between hairs on the body insulates the skin and reduces heat loss. In the cold, or in danger, a tiny erector muscle attached to each hair follicle contracts to make the hair stand on end. The resulting "goose flesh" means that more air can be trapped, reducing even further the heat loss. Hair on the head is a particularly effective insulation.

Besides fulfilling these roles, hair is often considered an attractive bodily feature and it can play a part in sexual attraction.

GROWTH OF THE HAIR

Hair on the scalp grows at the rate of about $\frac{1}{2}$in (1.25cm) per month. (This means that the end of a hair measuring 18in (45cm) is about 3 years old!) The root is the only live part of the hair: it grows and pushes the dead shaft out above the skin. Hair growth is cyclical with a growth phase followed by a rest phase in which the hair is loosened. The loosened hair is then pushed out by a new hair growing in its place. In this way, up to 100 hairs are lost each day from a normal head of hair.

THICKNESS OF GROWTH

The thickness of the growth of the hair depends on the number of hair follicles. The follicles are established before birth and no new ones are formed later in life. The thickness of individual hairs is influenced by hereditary factors.

STRAIGHT OR CURLY?

The degree of curliness of the hair depends on the shape of the follicle from which it grows.

Straight hair grows from a more or less round follicle and is round in cross-section. This hair shape is characteristic of Mongoloids.

Curly hair is oval in cross-section. It grows from a very curved follicle which forces the growing hair into curls. This hair shape is characteristic of Negroes.

Wavy hair is kidney-shaped in cross-section. The extent of the curl depends in the curve of the follicle. This hair shape is characteristic of Caucasians.

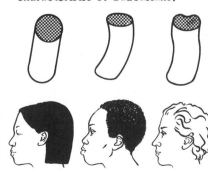

Mongoloid   Negroid   Caucasian

HAIR SHAFT

A cross-section through the hair shaft shows a hollow core (medulla) surrounded by an outer cortex, and covered by a thin coating of keratin - a sheet of horny cells which overlap one another. This coating is called the cuticle.

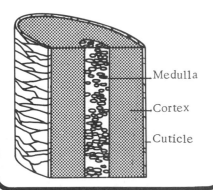

Medulla
Cortex
Cuticle

HAIR CARE

a) Wash your hair regularly. For normal hair, this means every 5-7 days, but if you have greasy hair (caused by overactive sebaceous glands) you will need to wash it more often. Remember, however, that washing can actually stimulate the glands through rubbing the scalp. Also the detergent in the shampoo can strip the hair of its natural oil and cause the glands to work overtime to replace it.

b) Always choose a mild shampoo. Lots of lather feels good but it is caused by detergents. Use an appropriate shampoo for your hair type: lemon-based for greasy hair and cream for dry hair.

c) Stick to a sensible diet (see H32).

d) Choose a good quality brush and comb. Sharp teeth or bristles can damage the structure of the hair.

e) Unclean brushes and combs spread infection and bacteria. Keep them clean and do not lend them to anyone else.

f) Remember that the condition of your hair reflects your state of health and general well-being. A balanced diet, plenty of sleep, and regular trimming, will do more to make your hair look good than anything else.

HAIR COLOR

The color of the hair is decided by heredity.

Special pigment cells at the base of the hair follicle give a hair its color. These cells inject colored granules of black, brown, or yellow into the hair.

If the cells receive no pigment, the cortex of each hair becomes transparent and the hair appears white.

"Gray" hair is the result of a mixture of dark and white hairs.

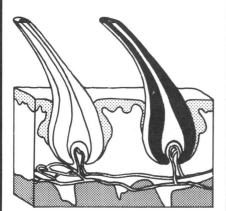

White hair
Cortex contains transparent cells

Normal hair
Cortex contains pigment cells

DIET

Protein, vitamins of the B complex, and certain minerals, are all essential for strong, healthy hair. The best sources of protein are meat, fish, milk, cheese, and eggs. Vitamin B is obtained from liver and from brewer's yeast - easily available in tablet form. Iron, copper, and iodine, are probably the most important minerals for healthy hair. Iron and copper are readily available in everyday foods like meat and green vegetables, and iodine is present in fish and shellfish. People with greasy hair should avoid fried and fatty foods, and concentrate on meat, fresh fish, salads, fruits, vegetables, eggs and cheese. They should also drink plenty of water. People with dry hair should include vegetable oils in their diet.

SCALP MASSAGE

Massaging the scalp with the tips of the fingers increases the blood flow to the massaged area. This stimulates the follicles and can aid hair growth. It also means that the scalp is kept more healthy, with a greater supply of nutrients and speedier removal of waste products.

In scalp massage, it is important that the fingers do not slide over the scalp. This exerts pressure on the hair and can damage it. Massage can go from the neck up to the crown, and then again from the temples back to the crown, thus covering the whole scalp.

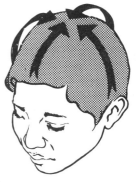

Direction of massage

©DIAGRAM

HAIR 2

LOSING YOUR HAIR

A number of hairs are lost from the scalp every day. These are usually replaced by new head hairs, but if they are replaced by fine, downy, hairs of the kind found on the face or arms, a thinning of the general growth of head hair results. This happens in varying degrees to nearly all men and women.

Only in very exceptional cases do women lose all their hair, though many notice a general thinning, particularly as they grow older. This condition is known as diffuse alopecia and is caused by an increase of male sex hormones in the body. If this hormonal imbalance is corrected, the full head of hair is usually restored. Stress can also cause hair loss because it interferes with the production of the hormones that stimulate hair growth. When the period of stress is over, normal hair growth is resumed.

After childbirth, many women notice an acute loss of hair. This too is a hormonal problem, but the hair soon returns to normal.

SUDDEN HAIR GROWTH IN WOMEN

This is usually due to a hormonal imbalance and can occur when the pill is first taken or is left off; during pregnancy; or during the menopause.

Tufts of hair on either side of the chin or a fine down on the upper lip may appear.

Very often these will disappear once hormonal balance restores itself. However, if the growth is unusually marked and distressing, other hormones can sometimes be given to help adjust the balance - although the treatment of hormonal imbalance is a very complex and delicate process.

BALDNESS

Hair can be lost for various reasons. Physical ailments that cause hair loss include:

scars or burns that destroy the hair follicles;

skin disturbances, such as dermatitis, psoriasis, or allergic reactions;

general bodily ailments, such as serious anemia;

and chemical pollution of the body (as in mercury poisoning).

Hair loss can also be caused by mental stress. This is because hair growth is linked with hormonal production, which in turn is closely linked to one's emotional state. The hair grows back when the period of stress is over.

In the above cases, by treating the illness, hair will usually be restored. But the commonest type of hair loss is male-pattern baldness. This is caused by a male hormone influenced by hereditary aging factors. Nothing, short of castration, can be done about this condition! Not only can it not be reversed but it cannot be stopped or even slowed down by any means so far discovered.

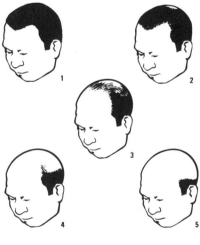

COPING WITH HAIR LOSS

There are several ways of coping with hair loss.

TRICHOLOGICAL TREATMENTS

Several clinics now offer a variety of different treatments for the scalp disorders that cause hair loss. These treatments can include: creams, lotions, massage, shampoos, and ultraviolet and infrared radiation.

Courses of these treatments are expensive; but there are no good medical grounds for them. In fact, there is a risk of wrong diagnosis, and inappropriate treatment being given for a condition that could be dealt with by a doctor.

TRANSPLANTS

This is a recent technique. Hair follicles are removed surgically from parts of the head where growth is abundant (often the nape of the neck) and implanted in the bald areas. This treatment is costly and takes considerable time, but there is no guarantee that the hairs will grow in their new location.

WIGS AND HAIRPIECES

The hairpiece is built from a base shaped to the bald area. Hair is attached to the base and cut to match the rest of the hair. The hairpiece is fixed to the scalp with strips of double-sided sticky tape.

HAIR WEAVING

Also known as hair linking or hair extension, hair weaving can take two different forms.

In the first, a hairpiece is made in the usual way and then attached to the scalp by stitching the side of it to the normal hair.

In the second, threads are strung across the bald area and pieces of hair, sewn together in clumps, woven directly into this.

HAIR REMOVAL FOR WOMEN

Superfluous hair can be removed in several ways. The different methods suitable for different parts of the body are shown below.

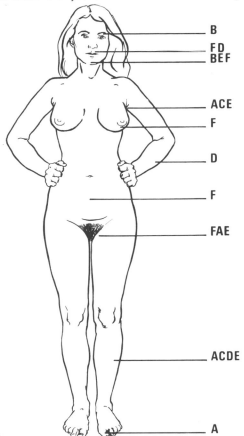

A Shaving:
armpits
pubic hair
legs
toes
B Plucking:
eyebrows
chin
C Waxing:
armpits
legs
D Bleaching:
upper lip
arms
legs
E Depilatory creams:
chin
armpits
pubic hair
legs
F Electrolysis:
upper lip
chin
breasts
abdomen
pubic hair

DANDRUFF

There are two kinds of dandruff. The first, affecting about 60% of the population to a mild degree, takes the form of fine, dry scales which fall from the scalp. The second kind, which is rarer, takes the form of thick, greasy scales adhering to the scalp.

The cause of both types is unknown and there is no real cure for dandruff, but there are things that can be done to control it.

Washing the hair with ordinary shampoo may not help, in which case a medicated shampoo can be used instead. These shampoos are designed to remove the scales and delay the recurrence of dandruff. Some have a simple antiseptic and others contain stronger chemicals. The most effective contain zinc pyrothionate, "ZP11."

However, if you do have difficulty in controlling dandruff, it may be caused by a skin disorder or some other condition that a doctor should treat.

PERMANENT WAVES

A "perm" or "permanent" is a 2-stage chemical process which causes each hair to alter the chain of the cells in its cortex. (The process is not literally permanent as the artificially created waves grow out as the hair grows.)

After washing, the first solution is applied to the wet hair. This is an alkaline-based solution designed to soften the hair by breaking the chain of the cells. The hair is then wound around small curlers and the second solution applied. This is an oxidizing lotion that halts the softening process and causes the cells to coalesce again, but this time under the stress of the roller which gives the hair its "permanent" wave. The hair is then rinsed and wound around larger rollers for drying.

Some people are allergic to the chemicals involved, so it is important to make a test curl first to check for hypersensitivity.

Dyed or bleached hair is particularly sensitive to the chemicals and it is essential to leave an interval of about 4 weeks between a "permanent" and a change of hair color.

HAIR LICE

Two species of lice affect humans: Phthirius pubis, found in the pubic hair; and Pediculus humanus, found in the hair on the head. The latter can be acquired not only by contact with an infected person, but also via objects such as combs and hats.

The infestation causes severe itching. It is most easily diagnosed by examining the scalp for the tiny eggs ("nits") attached to the hair shafts. The lice themselves are more difficult to find. Suitable treatment should be obtained from a doctor or pharmacist: it will include a special shampoo and often also a scalp emulsion.

TEETH 1

TEETH

Teeth are hard structures set in bony sockets in the upper and lower jaws. Their main function is to chew and prepare food for swallowing. They also help in the articulation of sounds in speech. In humans there are three main types of teeth.

Incisors are sharp, chisel-like teeth at the front of the mouth, used for cutting into food.

Canines are round pointed teeth at the corners of the mouth, used for tearing and gripping food.

Molars and premolars are square teeth with small cusps, which grind food at the sides of the mouth.

A tooth consists of two parts: the root, which is embedded in the jaw; and the crown, which projects out of the jaw. Where the root and crown meet is called the neck.

Each tooth is made up of enamel, dentine, pulp, and cementum. Enamel is the hardest tissue in the body, and it protects the sensitive crown of the tooth.

Dentine is a slightly elastic material which forms the bulk of the tooth under the enamel. It is sensitive to heat and chemicals.

Pulp is the soft tissue inside the dentine, and contains nerves and blood vessels, which enter the root of the tooth by a small canal.

Cementum is a thin layer of material which covers the root of the tooth and protects the underlying dentine. It also helps attach fibers from the gum to the tooth.

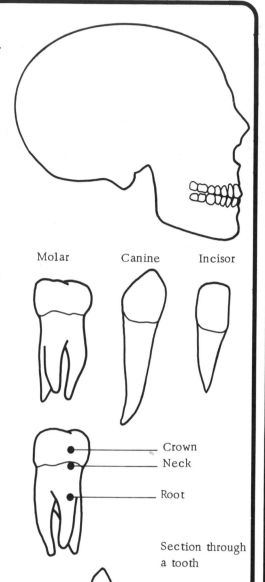

Molar Canine Incisor

Crown
Neck
Root

Section through a tooth

Enamel
Dentine
Pulp
Cementum
Periodontal membrane

THE TEETH WE'RE GIVEN

In humans there are two successive sets of teeth. The primary or "milk" set arrive 6 to 24 months after birth. Later they gradually fall out, from the age of 6 on, as the permanent teeth appear. Most of these are out by the age of 13, but the 3rd molar or "wisdom tooth" can erupt as late as the age of 25, or never.

Human teeth do not keep growing, but reach a certain size and then stop. Also, when the permanent teeth fall out, they are not replaced by a new set. But in some animals, such as the rabbit, the incisors keep growing, as they are worn down by use, while the shark grows set after set of teeth - to its great advantage!

AGE OF APPEARANCE

These are average figures only: actual dates vary greatly from child to child.

PRIMARY TEETH

Central incisors	6 to 8 months	1
Lateral incisors	9 to 11 months	2
Eye teeth	18 to 20 months	4
First molars	14 to 17 months	3
Second molars	24 to 26 months	5

ADULT TEETH

Central incisors	7 to 8 years	2
Lateral incisors	8 to 9 years	3
Canines	12 to 14 years	6
First premolars	10 to 12 years	4
Second premolars	10 to 12 years	5
First molars	6 to 7 years	1
Second molars	12 to 16 years	7
Third molars	17 to 21 years	8

The end numbers list the typical order of appearance.

The final number of adult teeth is between 28 and 32 - depending on how many of the wisdom teeth appear.

Primary teeth

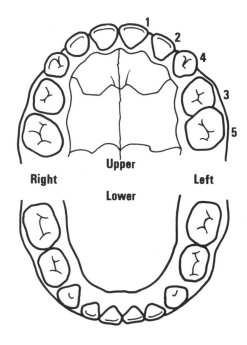

Upper

Right   Left

Lower

Looking into the mouth

Adult teeth

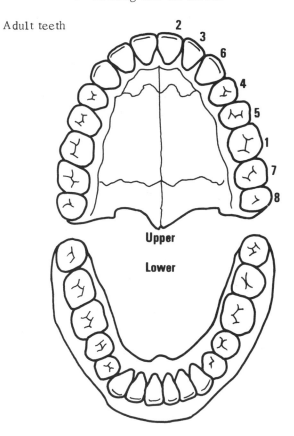

Upper

Lower

Sequence of appearance of adult teeth (upper jaw)

5 years

8 years

10 years

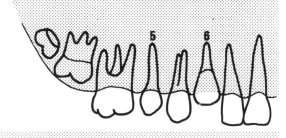

11 years

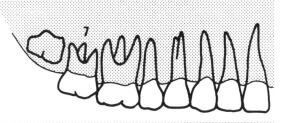

13 years

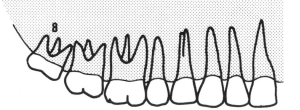

Adult

©DIAGRAM

DENTAL DISORDERS

Tooth decay is the most universal of human diseases. It especially afflicts those who eat a highly refined diet which is overcooked, soft, sweet, and sticky.

Bacteria in the mouth change carbohydrates in the food into acids strong enough to attack tooth enamel. Gradually the enamel is broken down and bacteria invade the dentine, forming a "cavity." The pulp reacts by forming secondary dentine to wall off the bacteria, but without treatment the pulp becomes inflamed and painful (toothache).

The infection may then pass down the root and cause an "abscess" - a painful collection of pus under pressure, affecting the gum and face tissues.

PERIODONTAL DISEASE

This is a general term for disorders in the supporting structures of teeth: the gums, cementum, and other tissues. The commonest cause is overconsumption of soft food, which cannot stimulate and harden the gums. Other causes include sharp food which scratches the gums, inefficient brushing, badly contoured fillings, ill-fitting dentures, irregular teeth, and teeth deposits. General factors such as vitamin deficiencies, blood disorders, and drug use may also be involved.

Periodontal diseases can be painless, but, if allowed to progress, the gum may become detached. from the tooth. The socket enlarges, securing fibers are destroyed, and the tooth loosens. Many teeth can be be lost in this way.

Painful periodontal disorders include abscesses in the gum and "periocoronitis." The latter is inflammation around an erupting tooth (usually the "wisdom tooth"), caused by irritation, food stagnation, pressure, or infection. It may accompany swollen lymph glands.

DENTAL TREATMENT

The dentist's intricate work has to be carried out in the confined, dark, wet, and sensitive environment of the mouth.

FILLING CAVITIES

Tooth decay is dealt with by drilling out the decayed matter and filling up the resulting cavity. All decayed and weakened areas must be removed, otherwise decay will continue beneath the filling. Also the cavity must be shaped so that the filling will stay in securely and withstand pressure from chewing.

High-speed electric drills are now usual, and so is the use of injected local anesthetic to make the procedure painless.

A lining of chemical cement is put in the prepared cavity to protect the pulp from heat and chemicals. The filling, placed on top of this, is usually an amalgam of silver, tin, copper, zinc alloy, and mercury. Alternatively, translucent silicate cement is used, for its natural appearance - but, since it can wear away, this cannot be used on grinding surfaces.

When the filling has hardened, it is shaped, and any excess trimmed off.

OTHER RESTORATIVE WORK

Some other replacement work can be prepared outside the mouth, and then cemented into place.

Inlays are cast gold fillings, shaped to fit a cavity in the crown of a tooth. A wax impression of the cavity is made, and the resulting mold filled with molten gold.

Crowns are extensive coverings to the crown of a tooth, made of porcelain or gold. The whole of the enamel of the tooth is removed, an impression made, and the crown made from a model.

PULP AND ROOT CANAL TREATMENT

If the pulp or root canal is decayed, normal fillings are complicated. Part or all of the pulp may have to be be removed. The root canal is sterilized and a silver pin sealed in place to fill it. The pulp cavity is then filled in.

EXTRACTION

Teeth need to be removed if they are irretrievably decayed, or so broken that they cannot be repaired, or if new teeth are erupting and have no room.

Forceps are used. They grip the tooth at the neck, while the blades of the forceps are inserted under the gum. The tooth is then moved repeatedly to enlarge the socket, and finally can be pulled out. Local or general anesthetic, by injection or gas, usually makes extraction painless.

TREATMENT OF GUM DISORDERS

Acute conditions are treated by pus drainage, antiseptic mouthwashes, antibiotics, and tooth extraction if necessary. Surgery may be needed to cut away the diseased gum. Long-term treatment aims at eliminating as many causative factors as possible, by improving oral hygiene, diet, and general health.

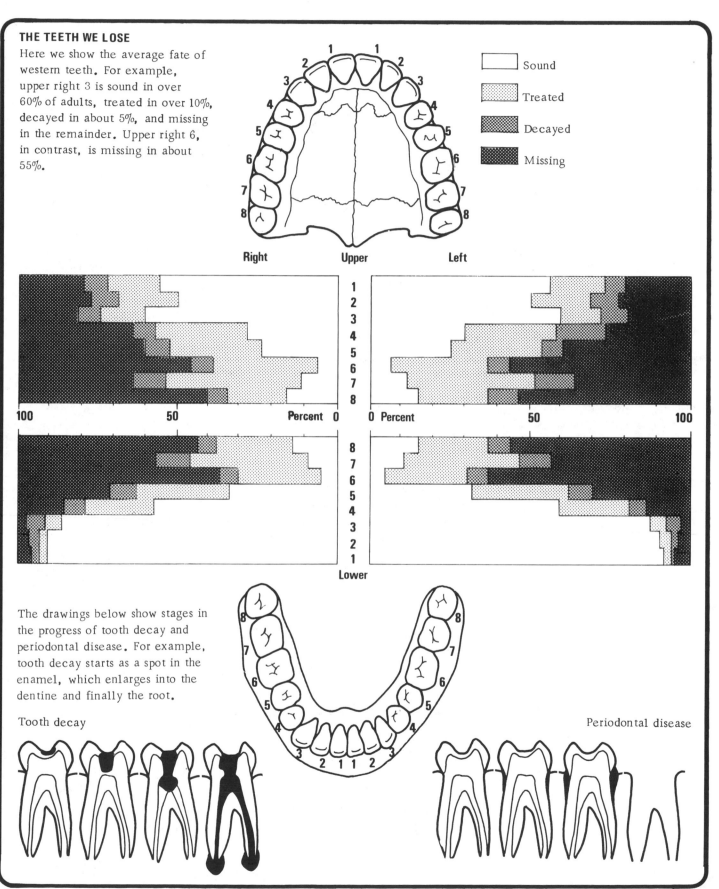

THE TEETH WE LOSE

Here we show the average fate of western teeth. For example, upper right 3 is sound in over 60% of adults, treated in over 10%, decayed in about 5%, and missing in the remainder. Upper right 6, in contrast, is missing in about 55%.

Sound

Treated

Decayed

Missing

Right   Upper   Left

100   50   Percent 0

0 Percent   50   100

Lower

The drawings below show stages in the progress of tooth decay and periodontal disease. For example, tooth decay starts as a spot in the enamel, which enlarges into the dentine and finally the root.

Tooth decay

Periodontal disease

©DIAGRAM

ORTHODONTICS

This is the branch of dentistry concerned with preventing and correcting irregularities of the teeth, eg variations in the number of teeth and abnormalities in their shape, size, position, and spacing. All these can cause defects in eating, swallowing, speech, and breathing. Malocclusion is the typical example. This means that the teeth are not in the normal position when the jaws are closed, relative to those in the opposite jaw. Teeth may stick out or in, or there may be spaces betweeen the biting surfaces due to uneven growth of teeth or jaws.

Irregularities may be caused by: bottle feeding and thumb sucking; loss of teeth, non-appearance of teeth, and appearance of extra teeth; birth injuries and heredity; and disease and poor health.

TREATMENT

Treatment may be long-term, but it is needed if the health, function, and esthetic appearance of the mouth are to be preserved. Methods include:
elimination of bad habits such as thumb-sucking;
practice of exercises to strengthen certain muscles and improve mouth movements;
relief of overcrowded teeth by extraction;
surgery on the soft tissues or bones to recontour the jaws.

But the commonest technique is to attach "braces" or similar appliances to the teeth, to apply continual pressure and so make them shift position. The braces, made of steel bands, wires, springs, or bands of elastic, may have to be worn for up to 2 years, or more. They are not attractive, and, if cemented in position, can make cleaning the teeth difficult. They are also more effective in the young than in older people.

FALSE TEETH

Ideally, false teeth ("dentures") should preserve normal chewing and biting, clear speech, and facial appearance.

TYPES OF DENTURES

These include full sets, partial dentures, and immediate dentures. For the construction of a full set, all the teeth are removed, and the healed bony ridge acts as a base. Impressions of both jaws are made in warm wax, and these give the basic patterns from which the dentures are made up.
With partial dentures, the new teeth are attached to surviving natural ones to keep them anchored. Where the anchoring teeth are not immediately alongside, they are linked to the false teeth by a bridge.
With immediate dentures, the false teeth are prepared before the teeth being lost have been removed. After extraction the empty sockets are immediately covered with the new dentures, and healing takes place beneath. A new set is then needed after about 6 months, as the ridge where teeth have been extracted shrinks.

USING DENTURES

Dentures can be uncomfortable. To avoid gum soreness, new dentures should at first be used only with soft food chewed in small amounts.
If soreness does occur, the dentist should be consulted. The dentures should not be left out of the mouth for more than a day or two, or the remaining natural teeth may begin to shift position.
False teeth should be brushed after every meal, and detachable dentures should be soaked overnight in water containing salt or a denture cleaner.
The wearer can regain her usual ease of speech by practicing reading aloud.

CLEANING THE TEETH

Most people use scrubbing motions, backward and forward and up and down.
Backward and forward strokes with the brush length are good for the tops of the molars (a), and the back of the front teeth (b).

But on the side teeth, use the brush sideways in a repeated stroke in one direction - upward on the bottom teeth (c), and downward on the top ones (d).

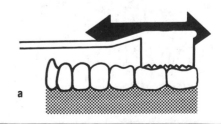

a

NEW TEETH FOR OLD

Crown

Bridge

Partial denture

PREVENTING DENTAL TROUBLE

DIET

At any age, the ideal diet for dental health should be: well balanced and adequate, so general health is maintained; chewable enough to stimulate the gums; and low in sugar content.

The balance and adequacy of the diet is especially important in the case of expectant mothers and growing children, so that strong teeth form.

ORAL HYGIENE

Teeth should be cleaned at least twice a day: after breakfast, and last thing at night. But it is better if they are cleaned after every meal. Cleaning polishes the teeth and removes stains and food debris.

Methods of cleaning vary from culture to culture. The toothbrush can be used ineffectively, and even cause damage (electric toothbrushes tend to be better). The value of toothpaste is doubtful, and it can give a misleading "clean feeling." Brushing with salt stimulates the gums and cleans just as effectively.

"Dental floss," or toothpicks of soft wood, are valuable for dislodging food between the teeth. Highly effective techniques used elsewhere include the fibrous chewing stick used in Africa, and the Moslem tradition of rubbing the teeth and gums with a towel.

DENTAL INSPECTIONS

Regular visits to the dentist about every six months catch disease in its early stages and so avoid drastic measures in the future.

FLUORIDE

Fluoride is a tasteless, odorless, colorless chemical, which, if added to drinking water in small amounts, reduces tooth decay in children by 60%. (Excessive amounts can cause the enamel to become mottled.) In the US, 60 million people now drink water with fluoride added, and 7 million others drink water that naturally contains fluoride. Some toothpastes also contain fluoride, and tablets can be bought to add to unfluoridated water. So far, no ill effects of the use of fluoride in these quantities have been established, but it only benefits the teeth of children under 14.

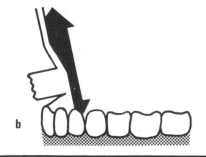

b

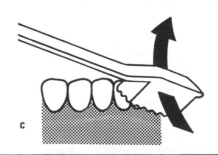

c

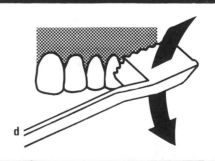

d

©DIAGRAM

THE EYEBALL

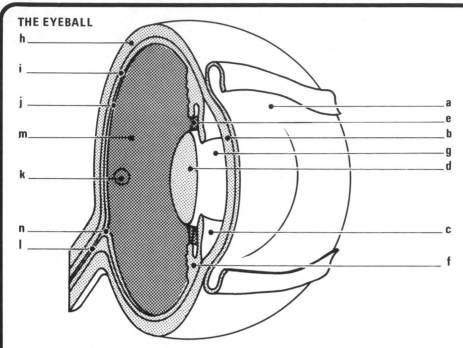

a	Conjunctiva
b	Cornea
c	Iris
d	Lens
e	Suspensory ligaments
f	Ciliary body
g	Anterior chamber
h	Sclera
i	Choroid
j	Retina
k	Fovea
l	Optic nerve
m	Vitreous body
n	Blind spot

THE CONJUNCTIVA is the membrane covering the front of the eyeball and the inside of the eyelids. It has a rich supply of blood vessels and is extremely sensitive.

THE CORNEA is the clear part of the eyeball which lets in the light.

THE IRIS controls the amount of light entering the eyeball. By contracting, it reduces the size of the pupil (the hole through which the light enters). It is the iris which gives the eye its "color."

THE LENS has a firm center, surrounded by a softer substance contained in a fibrous capsule. By being stretched or thickened, it focuses light on the back of the eyeball.

THE SUSPENSORY LIGAMENTS are attached at one end to the lens and at the other to the ciliary body. They hold the lens in place.

THE CILIARY BODY. The muscles of the ciliary body control the shape of the lens. If they contract, the lens is stretched and light rays from long distances are focused on the retina (are "accommodated"). If they relax the lens thickens, and close objects are accommodated. Both the lens and the iris are under the control of the autonomic nervous system, and cannot be controlled at will.

THE ANTERIOR CHAMBER lies in front of the lens and is filled with a watery fluid called the aqueous humor.

THE SCLERA or sclerotic coat is a layer of dense white tissue. It completely surrounds the eyeball, except where the optic nerve enters at the rear, and where it is modified at the front to form the transparent cornea. The sclera forms the "whites" of the eyes.

THE CHOROID tissue lies beneath more than two-thirds of the sclera. It is colored brown or black, and contains blood vessels. The ciliary body and the retina are formed from the choroid. Its color absorbs excess light within the eyeball, making for clearer vision.

THE RETINA is a thin layer of light-sensitive cells which lines the inside of the eyeball. It has a rich blood supply.

THE FOVEA lies on the visual axis of the eyeball. It is a small depression in the retina, at which vision is sharpest. It contains only "cone" cells.

THE OPTIC NERVE is a direct extension of the brain. It enters the eyeball at the rear. The head of the optic nerve is called the optic disk. It forms a blind spot in the vision, as there are no light-sensitive cells there. We are sometimes aware of this blind spot as a black dot at one corner of our vision.

THE VITREOUS BODY occupies the space behind the lens. It is a transparent jelly-like substance that fills out the eyeball, giving it its shape. It contains small specks which are often seen when looking at white surfaces.

PROTECTIVE STRUCTURES

The eyes - the organs of sight - lie in deep hollows in the skull, on either side of the nose, and are protected in various ways.

THE EYEBROWS prevent moisture and solid particles from running down into the eye from above.

THE EYELIDS are folds of skin which, when closed, cover and protect the eyes. The inner membrane of each eyelid is a continuation of the "conjunctiva" which covers the front of the eyeball.

THE EYELASHES are hairs that protrude from the eyelids. They prevent foreign bodies from entering the eye, and trigger off the protective blinking mechanism when touched unexpectedly.

THE LACRIMAL GLANDS produce a watery, salty fluid that cleans the front of the eyeball. It also lubricates the movement of the eyelid over the eyeball. When stimulated by strong emotion or irritants, the glands produce excess fluid.

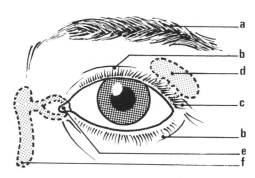

a Eyebrows
b Eyelids
c Eyelashes
d Lacrimal glands
e Lacrimal ducts
f Lacrimal sacs

THE LACRIMAL DUCTS drain the fluid from the eyeballs into the lacrimal sacs which lead into the nasal passage. When the ducts cannot clear the fluid fast enough it overflows and runs down the face as tears.

BLINKING is a protective action of the eyelids which spreads the lacrimal fluid over, and cleans, the front of the eyeball. Blinking is controlled by the brain. It occurs every 2 to 10 seconds, and the rate increases under stress, in dusty surroundings, or when tired, and decreases during periods of concentration.

EYE MOVEMENT

Movement of the eyeball is controlled by six muscles attached to the outside of the sclera.

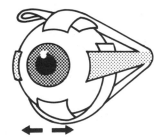
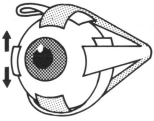

THE BLIND SPOT

How to find your blind spot. Hold the book at arm's length, and shut your left eye. Then look at the cross with your right eye, while slowly moving the book towards you. At one point the dot will disappear.

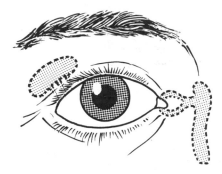

© DIAGRAM

EYES 2

SIGHT

When the light rays from an object enter the eye they are bent ("refracted") by the cornea and the lens (and to a lesser extent by the aqueous humor and vitreous body). Because of this refraction the rays are focused on the retina (though the image is upside down). The action of light on the cells of the retina triggers off an impulse which travels down the optic nerve to the visual centers of the brain. Here the impulses are interpreted and "seen" as colors and shapes the right way up.

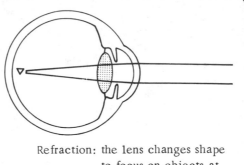

Refraction: the lens changes shape to focus on objects at different distances

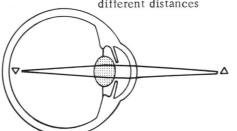

TRANSFER OF IMAGES

The nerves from the left sides of the two retinas travel to one side of the brain, those from the right sides to the other side of the brain.

a Eyeball
b Nerve
c Brain

THE RETINA

There are two types of light-sensitive cells in the retina. They are classified by shape: rods and cones. They are connected by nerve fibers to the optic nerve.

RODS

There are about 125 million rods in each eyeball. They are sensitive to low intensity light, and are used mainly in night vision. They are not sensitive to color, and therefore give only a monochrome image (black, white, and shades of gray). They are less than one four-hundredth of an inch in length and one-thousandth of an inch thick. The rods contain a purple pigment called rhodopsin. Light bleaches the rod as the pigment breaks down. This sets off electrical charges in the rods, which are transmitted down the optic nerve to the brain in the form of nervous impulses.

CONES

These are shorter and thicker, for most of their length, than the rods. They are used for high intensity light, such as daylight, and give color vision.

The actual process of color vision is not known, but it is thought that there are three different classes of cones, each containing a different pigment. Each pigment would be sensitive to a different color: blue, green, or red. Other colors would be combinations of these. It is thought that the nerve messages are produced by bleaching, as in the rods.

RESPONSE TO LIGHT

When a light-cell pigment has been broken down, and an impulse has been passed, the pigment must re-form before another impulse is possible. This takes about one-eighth of a second. The eye is therefore like a cinema screen. It does not give a continual picture, but successive "stills" at intervals of one-eighth of a second. These seem continuous because they run together.

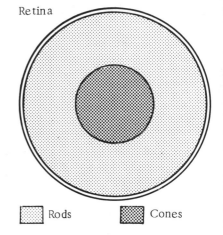

Retina

☐ Rods   ▦ Cones

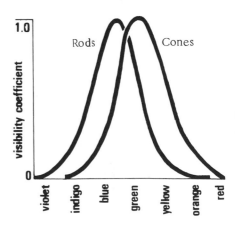

Sensitivity of rods and cones

COMPOSITE IMAGES

Each eye sees a slightly different view of the same object. The images received in the two visual centers (at the rear of the cerebral hemispheres) are composite images from both eyes, mixed 50:50. The further away the object is, the less the discrepancy between the two views. This, plus the amount of tension needed to focus and the amount of blurring, forms the basis of judgment of distance.

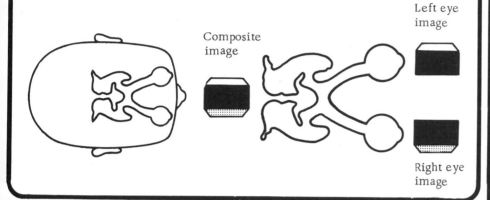

Composite image

Left eye image

Right eye image

VISUAL SCOPE

THE FIELD OF VISION is the area that can be seen by an eye without moving it. The size of the field varies with different colors. White has the largest, then yellow, blue, red, and green.

THE RANGE OF MOVEMENT of the eyeball, with the head still, is also limited. The human eyeball can tilt 35° up, 50° down, 50° in (i.e. toward the other eye), and 45° out. The greater angle available when turning in allows an eye to focus on an object that is just within the other eye's outer range.

THE AREA OF VISION is the total range through which a creature can see without moving head or body. It is determined by:
the position of the eyes in the head;
the shape of the head;
the eyes' range of movement;
and, at the edge, by the eyes' field of vision.

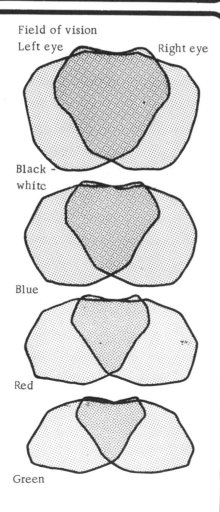

Field of vision

Left eye   Right eye

Black - white

Blue

Red

Green

PERCEPTION

The ability to perceive objects, colors, and distances is learned by experience. To the newborn child, the images received are meaningless and confused. It takes time to learn to use the eyes and correlate past with present information to bring about recognition.

This dependence of perception on the brain's judgments can be shown by presenting the eye with trick pictures; ones that allow alternative interpretations, or that give evidence that seems contradictory (a). Our perception will then shift or struggle between the alternative interpretations. The same process can be observed when waking up in unfamiliar surroundings - a series of alternative pictures flash through the brain, as it tries to make familiar sense of the data it is receiving.

In other cases, though, the brain accepts deceptive information unquestioningly (b).

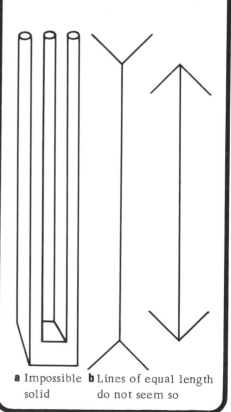

a Impossible solid   b Lines of equal length do not seem so

©DIAGRAM

EYE CARE

INJURIES to the eye and the immediately surrounding area should receive expert medical attention. Infection is a danger even if there is no significant damage. The eyes are tough but vital, and should be treated with care.

SMALL FOREIGN BODIES that get stuck in the eye can usually be removed by blinking. If this fails, then pull the upper lid outward and downward over the lower lid. When the upper lid is released the particle may be dislodged. A particle can also be removed with the corner of a clean handkerchief or by blowing it towards the edge. If none of this succeeds, get help and if necessary medical attention.

BLACK EYES are bruises of the eyelids and tissues around the eyes. They can be treated by applying a cold compress. If a black eye appears after a blow elsewhere on the head, see a doctor.

A STYE is an inflammation of the sebaceous gland around an eyelash, and is caused by bacterial infection. It is most often found in young people. A large part of the eyelid may become affected. To treat a stye, remove the relevant eyelash and bathe the eye with hot water. Antibiotics should only be used in extreme cases.

CONJUNCTIVITIS is inflammation of the conjunctiva. It can be caused by infection or irritation. If due to bacterial or viral infection, it needs the appropriate antibiotic eyedrops; if due to irritation, the irritant (eg an ingrown eyelash) is removed. Bathing the eye with warm water and lotions is soothing and is all that is needed in mild cases. Bandages or pads encourage the growth of bacteria, but dark glasses or eyeshades protect the eye from light and wind. Conjunctivitis is not very serious in itself (except for the trachoma form found in the tropics), but can sometimes cause serious complications such as ulceration of the cornea.

CONTACT LENSES

Contact lenses are thin round disks of plastic, that rest directly on the surface of the eye. They are increasingly used instead of spectacles, as they do not affect the appearance, often give better vision, and counteract many year-to-year changes in the eyesight. However, not everyone can wear contact lenses successfully, and some people find that they can wear them for only part of the day. They also require more care because of their smallness and fragility, and because of the effect a damaged, dirty lens can have on the eye. They need to be cleaned and stored in special fluid when out of the eye, and it is wise to insure them against destruction or loss.

TYPES OF LENS

Contact lenses can be "hard" or "soft." Hard lenses are either "scleral" lenses – covering the whole of the visible part of the eye – or "corneal" lenses, which rest on the center of the eye, floating on a film of tear fluid.

CORRECTIVE LENSES

Spectacles (or contact lenses) are used because of faulty focusing in the eye. The artificial lens corrects the work of the defective part of the eye.

NEARSIGHTEDNESS (myopia) is due to the refractive power of the eye being too strong (eg the lens may be too thick) or to the eyeball being too long. In both cases, the light rays are focused in front of the retina, giving a blurred image. Concave corrective lenses are needed to focus on distant objects.

FARSIGHTEDNESS (hypermetropia) is due to the eye's refractive power being too weak or the eyeball too short. The light rays are focused behind the retina, again giving a blurred image. Convex corrective lenses are needed for close work such as reading.

ASTIGMATISM means that the cornea does not curve correctly, and the person cannot focus on both vertical and horizontal objects at the same time. A special spectacle lens is needed, that only affects the light rays on one of these planes. Alternatively, a hard contact lens can be used, as the fluid layer between eye and lens compensates for the cornea.

PRESBYOPIA occurs in old age.

Nearsightedness

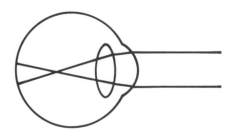

Farsightedness

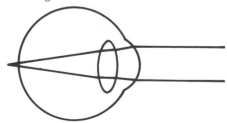

Corneal lenses are the most popular of all contact lenses, but scleral lenses are useful for very active sports. Soft lenses differ because they absorb water from the tear fluid.

COMPARING HARD AND SOFT LENSES

Soft lenses are immediately more comfortable, easier to get used to, can generally be worn for longer periods, can be left off for several days and then worn again without discomfort, can be alternated more easily with spectacles, and can be worn with less discomfort in dirty atmospheres.

Hard lenses are much less easy to damage, much cheaper, last perhaps 6 to 8 years (compared with 2 to 3 years for soft lenses with routine wear and tear, and often under a year as damage occurs), are more suitable for the majority of eye prescriptions, often give clearer vision, are easier to keep free from bacteria, and are much easier for the optician to adjust if difficulty arises.

Concave lens

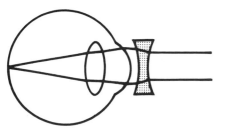

Convex lens

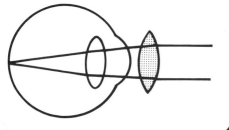

BLINDNESS
OBSTRUCTION OF LIGHT

When areas of the naturally transparent part of the eye become opaque, light rays are prevented from reaching the retina. Opacity of the cornea can be caused by corneal ulcers, or by keratitis i.e. inflammation of the cornea. Opacity of the lens is commonly caused by its becoming hard - "forming a cataract." Cataracts most often occur with aging, but can also be caused by wounds, heat, radiation, and electric shock.

DISEASES AFFECTING THE RETINA

These are often caused by diseases elsewhere in the body, especially those involving the blood supply.

a) Retinitis is inflammation of the retina with consequent loss of vision. It is associated with diabetes, leukemia, kidney disorders, and syphilis.

b) Retinopathy covers any disease of the retina that is not inflammatory. It is usually caused by degeneration of the blood vessels, impairing the retina's structure and function. It can be due to high blood pressure, diabetes, kidney disorders, and atherosclerosis.

c) Detachment of the retina. Primary detachment occurs if damage to the retina allows fluid from the vitreous body to leak through and lift the retina from the choroid. Treatment is possible. Secondary detachment occurs if the retina is pushed away from the choroid and damaged by underlying tumors, bleeding, or retinal disease. No treatment is possible.

d) Glaucoma can occur in age.

e) Choroiditis is inflammation of the choroid due to infection (especially syphilis) or allergy. The effects depend on the size and position of the inflammation: the nearer the fovea, the greater the vision loss. The inflammation can be treated, but damaged vision is seldom improved.

Declining field of vision in a case of progressive blindness in both eyes

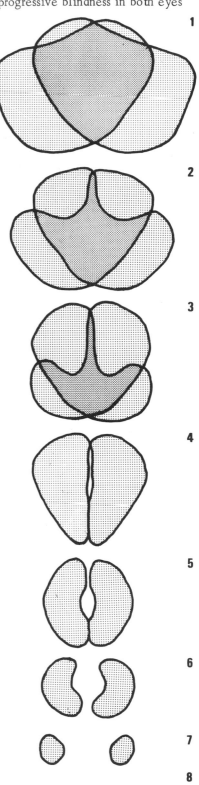

Left eye   Right eye
showing declining field of vision

©DIAGRAM

EARS 1

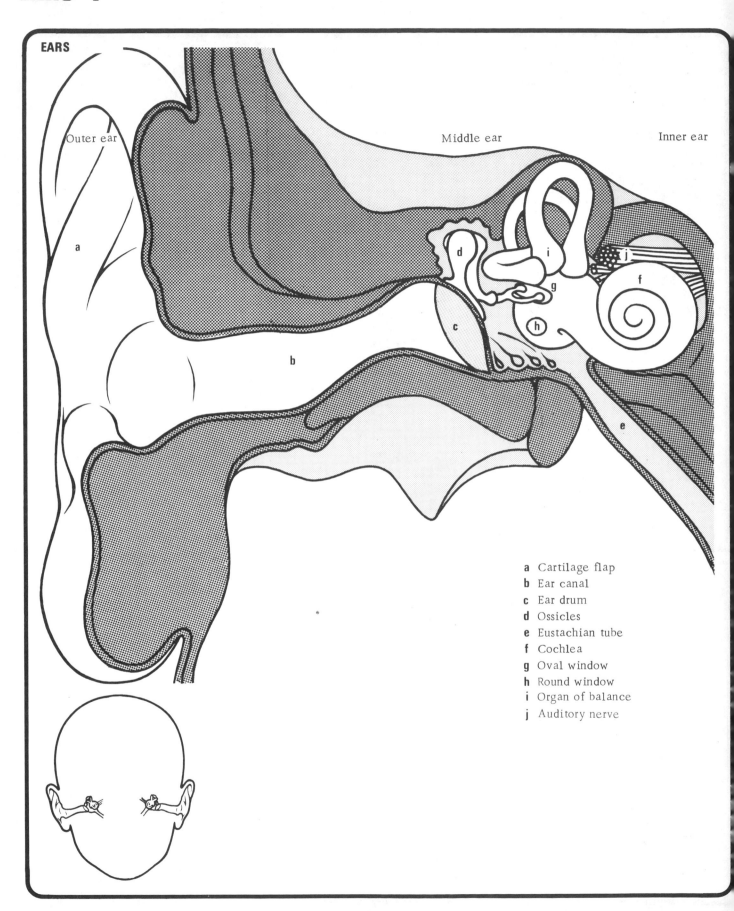

EARS

Outer ear   Middle ear   Inner ear

a Cartilage flap
b Ear canal
c Ear drum
d Ossicles
e Eustachian tube
f Cochlea
g Oval window
h Round window
i Organ of balance
j Auditory nerve

THE EAR

The structures of the ear fall into three groups.

THE OUTER EAR includes:
the external flap of cartilage (the "pinna" or "auricle");
and the ear canal (the "meatus").

THE MIDDLE EAR includes:
the eardrum (the "tympanic membrane");
three small bones called the "ossicles," and known individually as the hammer ("malleus"), anvil ("incus"), and stirrup ("stapes");
and the eustachian tube, which opens into the back of the throat, and keeps the air pressure in the middle ear equal to that outside.

THE INNER EAR includes:
the cochlea, a spiral filled with fluid and containing the "organ of corti;"
the oval window;
the round window;
and the organs of balance.

SOUND

When a solid object vibrates in air, it passes on this vibration to the surrounding air molecules. Sound waves are the vibration of air molecules.

Sound has three qualities:

PITCH, the highness or lowness of a sound, depends on the "frequency" of the sound waves, i.e. the number of vibrations per second. High pitched (piercing) sounds have a high frequency. Low pitched (deep) sounds have a low frequency.

INTENSITY is the loudness of a sound, and depends on the amount of energy in the sound waves, i.e. how widely they vibrate. Intensity is measured in "decibels."

TIMBRE is the quality of a sound. Sounds with the same pitch and intensity can be distinguished by their timbre. Timbre is created by the subordinate tones that accompany the pitch, or main sound.

HEARING

THE OUTER EAR Sound waves are collected by the pinna and funnelled into the ear canal.

THE MIDDLE EAR The eardrum vibrates in time with the sound waves. This vibration is passed on along the three ossicles to the oval window. The lever action of the ossicles increases the strength of the vibration. This allows the vibration to be passed from the air of the outer and middle ear to the fluid of the inner ear.

THE INNER EAR The vibration of the oval window makes the fluid in the cochlea vibrate. The pressure changes in the fluid are picked up by specialized cells in the organ of corti. This organ converts the vibrations into nerve impulses, which pass along the auditory nerve to the brain. Meanwhile, the vibrations pass on through the cochlea and back to the round window, where they are lost in the air of the middle ear and eustachian tube.

SENSITIVITY

LOUDNESS The human ear can hear sounds ranging in loudness from 10 decibels to 140 decibels (though the loudness becomes painful after 100 decibels). On the decibel scale, a ten unit increase means 10 times the loudness. Therefore the quietest sound the human ear can hear is one 10 million millionth the loudness of the loudest.

PITCH Different frequencies stimulate different parts of the organ of corti. That is why we can distinguish one sound from another. The human ear can hear sounds ranging in pitch from 20 cycles per second (low) to 20,000 cycles per second (high).

Frequencies above this are called ultrasounds, and can be heard by some animals but not humans.

DIRECTION The slight distance between the ears means that there are minute differences in their perception of a given sound. The brain interprets these differences to tell from which direction the sound came. But if a sound comes from directly behind or in front of the listener, both ears receive the same message, and the listener must turn his head before he can pinpoint the location.

DECLINE in hearing often progresses with age.

BALANCE

The organ of balance is in the inner ear next to the cochlea. It consists of three U-shaped tubes ("semicircular canals"), at right angles to each other. They are filled with fluid, which is set in motion when the person moves. Hairs at the base of each canal sense this movement and send messages to the brain, which are interpreted and used to maintain the person's balance. The organ also contains two other structures, the saccule and the utricle. These have specialized cells which are sensitive to gravity, and so keep a check on the body's position.

©DIAGRAM

EAR CARE

THE OUTER EAR should be kept clean at all times, to prevent wax and bacteria from collecting in the ear canal and damaging the eardrum.

To examine the outer ear, a beam of light from a flashlight is shone down the ear canal.

THE INNER EAR is tested by using a tuning fork. The fork should be heard clearly when it is held in front of the ear. If the tuning fork is heard more clearly when placed on the bone behind the ear, then: either the outer ear is blocked with wax; or, if not, the middle ear is faulty, since sound vibrations are being heard better through the skull. If hearing is still poor when the fork is placed on the bone behind the ear, it is the inner ear or the auditory nerves that are at fault.

SYRINGING of the outer ear cleans it, and washes out obstructions such as wax or foreign bodies. A large glass or metal syringe is used - one with a blunt point not more than 1in (2.5cm) long, so it cannot hurt the eardrum. The syringe is filled with warmish water, containing, if necessary, an antiseptic and/or wax dissolving agent. The fluid is directed along the upper wall of the canal, and flows out along the lower.

EAR DISORDERS

OTITIS EXTERNA

This is infection and inflammation of the outer ear, due to physical damage, allergy, boils, or spread of inflammation from the middle ear. There is itching and often a discharge, which may cause temporary deafness if it blocks the ear canal. Treatment is by antiseptic syringing and use of soothing lotions. Hot poultices and aspirin may relieve the pain.

OTITIS MEDIA

This is middle-ear infection, usually due to bacteria arriving via the eustachian tube. The eardrum becomes red and swollen, and may perforate. Pressure and pain increase as pus fills the middle ear. There is often temporary deafness and ringing, and sometimes fever. Treatment is with antibiotics.

A form of otitis in which a sticky substance is discharged in the middle ear is common in children. The ossicles cannot function, and in severe cases permanent deafness results.

MASTOIDITIS

Middle-ear infections can spread to the mastoid bone - the part of the skull just behind the ear. Infection swells the bone painfully, and the patient is feverish. Treatment is by antibiotics or the surgical removal of the infected bone (mastoidectomy).

MENIERE'S DISEASE

This affects the inner ear, and results in too much fluid in the labyrinths. Its cause is not known. It tends to occur in middle age, usually affecting more men than women. The symptoms are attacks of giddiness and sickness, followed by deafness with accompanying ringing in the ears.

Treatment is with drugs and control of fluid intake - not more than $2\frac{1}{2}$ pints (1.2 liters) a day. In extreme cases the labyrinths or their nervous connections are destroyed.

FUNGUS INFECTIONS

Fungus infections can occur in the outer ear. They are more common in tropical climates. There is persistent irritation and discharge, which is treated with antibiotics and antiseptic cleansing of the ear canal.

SYMPTOMS OF DISORDER

DEAFNESS can be temporary or permanent, caused by obstruction or disease.

EARACHE is usually caused by infection and inflammation in the ear.

In the outer ear this can occur through physical damage, boils, or eczema (a skin disorder). Large wax deposits can also cause earaches.

Germs from throat infections may spread up the eustachian tube and cause inflammation of the middle ear (especially in children). This is common after tonsilitis, measles, flu, or head colds, and can be very painful.

DEAFNESS

TYPES OF DEAFNESS

"Conductive deafness" refers to any failure in the parts of the ear which gather and pass on sound waves, eg blockage of the ear canal, eardrum damage, ossicle damage, etc.

"Perceptive deafness" refers to any failure in: that part of the ear which translates the sound waves into nerve impulses (the cochlea); or in the auditory nerves which transmit the impulses to the brain; or in the auditory centers of the brain which receive the message. Perceptive deafness may not mean that the person can perceive no sound. It may be that sound is received, but so scrambled as to be unintelligible.

CAUSES OF DEAFNESS

a) Disease. Some disorders can end in deafness.

b) Noise-induced. Any exposure to extremely loud noise, or continued exposure to moderately loud noise, can damage the eardrum and middle ear, causing hearing decline and eventually deafness. The main victims are those who work in very noisy surroundings,

Earache can also arise without any ear disorder, because of disturbances affecting the nerves it shares with other parts of the head. Tonsilitis, bad teeth, swollen glands, and neuralgia can all cause earache in this way.

RINGING IN THE EARS ("tinnitus") is usually associated with earache in the middle ear and/or high blood pressure. It is also caused by certain ear diseases.

GIDDINESS or vertigo can be caused by infections of the inner ear that affect the organs of balance.

DISCHARGES can come from boils or other infections.

SITES OF DISORDER

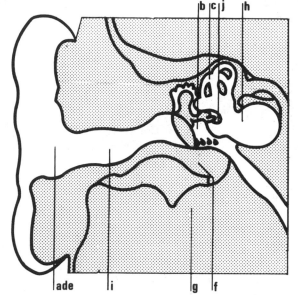

Possible sites of disorder in the ear
a Blockage
b Ringing
c Vertigo
d Discharge
e Otitis externa
f Otitis media
g Mastoiditis
h Ménière's disease
i Fungus
j Otosclerosis

and also the fans - and performers - of loud popular music.

(c) Congenital deafness. Deformities at birth range from complete absence of the ears, to minute mistakes in the internal structure. The latter can often be cured surgically. Congenital deafness can be due to heredity (genetic defects). It can also result from certain infections in the mother in the first few months of pregnancy, including German measles, flu, and syphilis. If there is anything in your child's response to sounds that gives rise to worry, consult your doctor.

(d) Otosclerosis. This is a condition in which the stirrup becomes fixed within the oval window, due to deposits of new bone. About one person in every 250 suffers from this, and it is more common in women than men. Surgical treatment may give improvement, but there is no way of halting the process responsible (though it may stop spontaneously).

HEARING AIDS

Hearing aids work by amplifying sound. If the amplification is loud enough, it can overcome the blockage or damage that causes conductive deafness, and allow the sound to reach the inner ear. Amplification also seems to help in many cases of perceptive deafness. However, sometimes the aid does not allow speech to be distinguished: it only makes the person more aware of unintelligible noise.

The performance of a hearing aid depends on:
(a) the frequency response. Normal speech usually lies between 500 and 2000 cycles per second;
(b) the degree of amplification;
(c) the maximum amount of sound that the aid can deliver. Too much sound can make speech unintelligible, and/or damage the ear mechanisms.

One problem with hearing aids is "acoustic feedback." This is the reamplification of sound vibrations that have already passed into the ear but have partly leaked out again.

INSERT RECEIVERS are the most common type of aid. They are molded to fit into the ear canal and form a perfect seal. No sound escapes, there is little or no acoustic feedback, and background noise is at a minimum. They can also be very small and, if transistorized, need no wires or attachments. A high degree of amplification is possible.

FLAT RECEIVERS fit against the external ear cartilage, and are kept in place by a metal band. They are usually used only if there is a continuous discharge from the ear, or if there has been a serious mastoid operation. Because of the bad contact, many sounds escape, and acoustic feedback produces much background noise.

BONE CONDUCTORS amplify the sound waves and send them through the bone of the head, not the air passages of the ear. They are uncomfortable and not very efficient, and are usually only used where some ear condition rules out an insert receiver.

© DIAGRAM

NOSE AND MOUTH

THE NOSE

The outer nose consists of tissue and cartilage supported at the top by the nasal bones. The nasal cavity beneath is divided into two (left and right) by a wall of bone and cartilage called the nasal septum. The nose has two functions - it is an opening to the respiratory system, and it houses the organs of the sense of smell (the olfactory system). The olfactory system consists of tiny hair-like nerve endings in the roof of the nasal cavity which detect odorous molecules in the air, and transfer signals to the brain via olfactory bulbs and a set of nerve fibers called the olfactory tract.

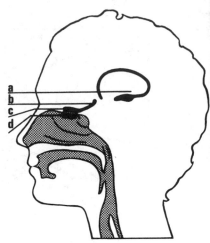

a Brain center for smell
b Olfactory tract
c Olfactory bulbs
d Olfactory hairs

SENSE OF SMELL

The sense of smell is one of the keenest of all the human senses. (It can for instance distinguish more odors than the ear can distinguish sounds.) It is a vital part of the taste process - people who have lost their sense of smell cannot taste the full flavor of foods.
The precise mechanism of smell remains a mystery. Similarly, attempts to classify different odors have mostly been unsuccessful. A recent theory proposes four categories - fragrant, burnt, acid, and rancid - and suggests that every smell is a blend of these four basic odors.

DISORDERS

COMMON COLD is an infectious disease of the respiratory system caused by viruses. It results in a running nose, reduced sense of smell, and sometimes a cough.
HAY FEVER is an allergic reaction which causes the mucus membranes in the nose and eyes to become swollen and irritated. It may also cause a watery discharge or sneezing. Antihistamine drugs are often prescribed to give relief.

NOSE BLEEDS are usually caused by the rupture of a blood vessel inside the nose. The bleeding may be the result of a blow, violent exercise, or exposure to high altitudes, or may be a sign of high blood pressure. It can usually be relieved by pinching together the nostrils; severe or persistent bleeding needs medical attention.
POLYPS are benign tumors in the nasal passages which cause a permanently stuffy nose. Often the result of frequent colds, they are easily removed by simple surgery
RHINITIS is inflammation of the nose's mucous membranes caused by colds or hay fever.

SINUSES

Sinuses are air cavities in the skull which open into the nasal passages. Inflammation of the sinuses is called sinusitis and symptoms include headaches and pain in the cheek bones. Drugs are used to treat most cases, and minor surgery can improve more severe ones.

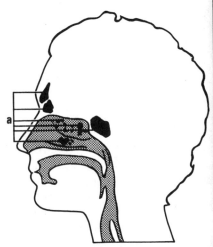

a Sinuses

COSMETIC SURGERY

Surgery can correct the bridge line, shorten the nose, build up a depressed nose, straighten a crooked one, or alter the shape of the tip. All involve shaving back or implanting extra bone or cartilage. The incisions are made inside the nose, and the new shape takes about 6 months to settle. Though cosmetic surgery of this kind is often performed at the whim of the patient, in many cases it relieves very real distress.

MOUTH

The mouth is the entrance to the digestive system and one opening of the respiratory system. It is completely surrounded by muscle except for the hard palate and lower jaw which are rigid. Behind the hard palate is the soft palate from which hangs the uvula, a projection of muscle tissue important in speech. On either side at the back of the mouth are the tonsils - oval masses of lymphoid tissue. The cavity is lined with mucous membrane. The mucus it secretes, along with saliva from the salivary glands, cleanses the mouth and keeps it lubricated.

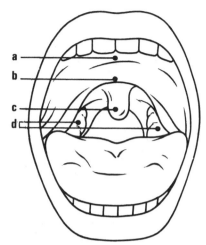

a Hard palate
b Soft palate
c Uvula
d Tonsils

DISORDERS

Very few serious disorders affect the mouth. A mixture of saliva, mucus, and secretion from the tonsils keep it moist and mostly germ-free. Any injuries to the mouth seem to clear up more quickly than they do elsewhere, and there is considerable resistance to infection. The most common minor disorders are ulcers, cold sores, and thrush.

ULCERS are inflamed sores in the mouth's mucous membrane usually caused by a scratch or similar injury. Most people suffer occasionally from small ulcers of this type (aphthous ulcers) which usually heal on their own, but mouth ulcers can also be signs of diseases such as diphtheria, leukemia, and cancer.

COLD SORES (HERPES SIMPLEX) are small inflamed blisters that appear around the mouth. They are usually the result of a virus which many people carry around in their bodies all their lives. An eruption can be triggered by another infection, or by exposure to very hot or very cold weather.

THRUSH (MONILIASIS) is an infection of the mucous membrane in the mouth, caused by a yeast-like fungus. It produces white patches inside the cheeks, but it can usually be treated by antifungicides.

CONGENITAL DEFECTS

CLEFT PALATE In the unborn baby the palate develops in two halves which fuse. In some babies the palate has not fused completely. This condition is known as cleft palate.

HARELIP This is the failure of the three parts of the upper lip to join - a congenital defect associated with cleft palate.

Both defects can usually be corrected by plastic surgery in a series of operations beginning soon after birth.

THE TASTE PROCESS

Before it can be tasted, a piece of dry food must be moistened and partly dissolved in the mouth by saliva from the salivary glands. The saliva, containing the particles of food, stimulates the taste buds on the tongue. Different areas of the tongue register different tastes (see below). The taste buds send signals to the brain which interprets these signals as tastes.

The sense of smell is also part of the taste process. The odors of food enter the nasal cavity and stimulate the olfactory system. This greatly heightens the sensation of taste.

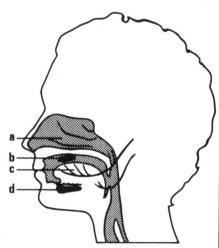

a Nasal cavity
b Food
c Nerves
d Salivary gland

Taste areas of the tongue

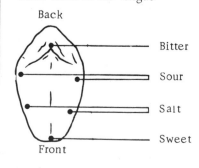

Back

Bitter
Sour
Salt
Sweet

Front

©DIAGRAM

BREASTS (FEMALE)

THE BREAST

The adult female breast, or mammary gland, consists of 15-25 lobes that are separated by fibrous tissue, rather like the segments of an orange. Each lobe resembles a tree and is embedded in fat.

After childbirth, milk produced in the alveoli of each lobe (the "leaves" of each "tree") travels along small ducts into the main "trunk" or milk duct. This duct is enlarged to form a reservoir just below the areola - the dark ring visible around the nipple. A narrow continuation of the duct links this reservoir with the nipple's surface. Each of the breast's 15-25 lobes has its own opening on the nipple.

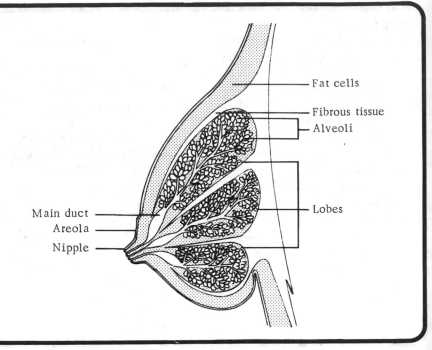

Fat cells
Fibrous tissue
Alveoli
Lobes
Main duct
Areola
Nipple

CHANGES IN THE BREAST

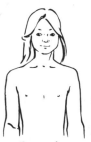 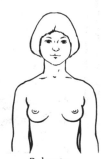 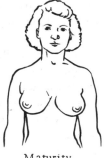 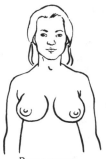 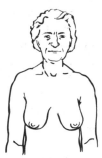

Pre-puberty   Puberty   Maturity   Pregnancy   Later life

The breasts undergo great changes during a woman's life.

BEFORE PUBERTY
The breast is simply a nipple projecting from a pink area - the areola.

PUBERTY
By the 11th year the areola bulges, the nipple still projecting from it. Secretion of the hormones estrogen and progesterone stimulates breast development. The milk ducts develop from the nipple inward and fat accumulates around them so that by the age of 16 or so, the breasts are prominent.

IN PREGNANCY
Early indicators of pregnancy often include swollen areolae, breast tenderness, and a marbled appearance produced by prominent veins in the breast. In the first 3 months, changes in blood supply and growth of milk ducts and alveoli enlarge the breasts by 20-25%. Toward the end of pregnancy, breasts are about $\frac{1}{3}$ larger than normal. Breast-feeding triggers further development but the breasts resume their former shape once breast-feeding ceases.

IN LATER LIFE
Around the menopause, breasts begin to droop and become less firm, as fibrous tissue slackens and milk ducts and alveoli shrink.

BREAST-FEEDING

Breast milk is sterile and better suited to most human babies than cow's milk. Breast-feeding also helps to establish the important physical contact between mother and baby.

MILK SECRETION by the breast begins just before childbirth as estrogen and progesterone output from the ovaries decreases. The reduction of the level of these hormones in the bloodstream affects the hypothalamus which then causes the pituitary to produce prolactin. It is this hormone that sparks off milk secretion in the breasts. The first substance secreted is not milk but colostrum - a thick, yellow liquid rich in antibodies (as the milk itself is). These antibodies give the newborn baby protection against disease and infection for up to 6 months.

MILK YIELD starts only about 3 days after childbirth. The flow is set off by the baby sucking the nipple (**1**). This sends nerve impulses (**2**) to the hypothalamus, which releases oxytocin that travels via nerve fibers (**3**) to the pituitary. From there, oxytocin flows through the bloodstream (**4**) to the breasts, causing the alveoli to contract and force liquid through the ducts to the nipples. Milk flow usually starts about 30 seconds after suckling begins.

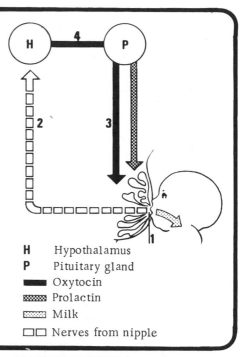

H Hypothalamus
P Pituitary gland
■ Oxytocin
▨ Prolactin
▥ Milk
▢▢ Nerves from nipple

SHAPES AND SIZES

The shape and size of breasts vary enormously. Though their size often corresponds to the overall body size, many slim women have large breasts and many obese women have comparatively small breasts.

LARGE BREASTS can occur because of fluid retention, obesity, or excessive hormone stimulation. Breasts tend to become larger during pregnancy and lactation, and many women find that their breasts enlarge after the age of 35. A firm supporting bra will help, and extreme enlargement can be corrected by hormone treatment, or by cosmetic surgery which involves the removal of some of the pads of fat which give the breasts their size.

SMALL BREASTS can occur because of too little hormone stimulation or just because the woman is slim. Underdevelopment at puberty can occasionally be helped by rubbing estrogen creams into the breasts, but these creams enlarge only the ducts and not the pads of fat. Cosmetic surgery can also help. Surgeons can insert silicone, or bags filled with fluid, between the breasts and the pectoral muscles - though the breasts do sometimes become infected later as a result. For most women who are anxious about their small breasts, however, a padded bra, upright posture, and perhaps exercises to strengthen the underlying pectoral muscles are all that is necessary.

IS A BRA NECESSARY? The need for a bra has been overemphasized but women who are in late pregnancy or breast-feeding are generally advised to wear one to avoid stretching supporting tissues.

ABNORMAL NIPPLES Naturally inverted nipples are a developmental fault that make breast-feeding difficult. (But inversion of previously normal nipples may be a sign of breast cancer). Extra nipples sometimes occur - usually in the armpits.

SEXUAL RESPONSE

Sexual stimulation affects the breasts in several ways. The nipples become erect and enlarged, then intermittently soft and pliable. Increased blood supply to the breasts causes their temporary enlargement by as much as 25%, especially in younger women who have not suckled. Also the areolae swell often, in younger women, sufficiently to engulf the base of the nipples. (In women over 50, this swelling is less marked, and may occur in one breast only.) Finally in many younger women, a pink flush mottles the breasts just before orgasm.

After orgasm, first the flush, then the areola swelling, soon vanish. But nipple erection may persist for hours, especially in older women or in those who have not had full release of sexual tension.

©DIAGRAM

HANDS AND NAILS

HANDS

THE HAND is remarkable for its flexibility. In particular, the thumb's ability to move in opposition to the fingers enables the hand to grasp objects and perform other delicate tasks.
The hand contains 3 important sets of bones: carpals in the wrist, metacarpals in the hand itself, and phalanges in the fingers.
Movements of the fingers are controlled by tendons attached to the muscles in the forearm. The hand is very strong - even a tiny baby can exert a very powerful grip.

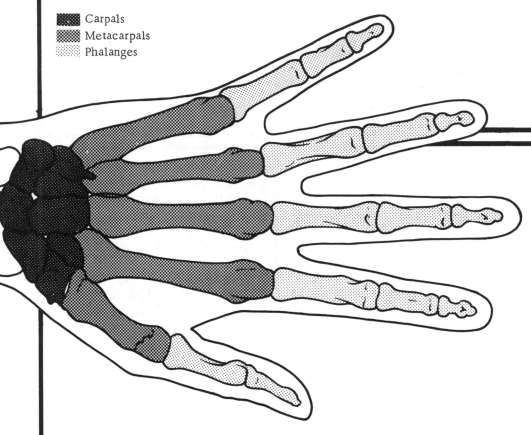

Carpals
Metacarpals
Phalanges

HAND CARE
The following points will help to keep the hands in good condition.
a) Do not wash hands more than necessary - soap removes some of the oils that keep the skin pliable.
b) Use hand cream when necessary to prevent dryness and redness, eg if hands have constantly been in water, and in cold weather.
c) Wear rubber gloves for all heavy jobs and for washing up.
d) Avoid direct contact with detergents and scouring products - an allergic reaction (contact dermatitis) may result.
e) Do not expose hands to extremes of temperature. Exposure to extreme cold, for example, causes chilblains.

NAIL CARE
Nails are easily damaged and it can be a long time before the damage grows out. To keep the nails in good condition, several points may be helpful.
a) Never use a steel file; it will tear the delicate keratin layers. Instead, use an emery board.
b) Always file from each side of the nail towards the center - never a sawing, back-and-forth motion.
c) Never file down the sides of the nail as this will weaken future growth.
d) Do not shape your nails to a point as this will encourage them to break.
e) Use a nail conditioning cream on the base of the nail if necessary, to strengthen new nail growth.
f) Clip away pieces of skin around the nail only if they are causing irritation. Regular use of hand cream will soften the skin and make it less likely to split.

NAILS
A NAIL consists of a small plate of dead cells. The horny, visible portion is made up of keratin - the substance found in skin and hair. The nail grows from a bed, or matrix, which is protected by a fold of skin at the base. The white crescent at the base of the nail is the visible part of the nail bed.
The nail rests on soft tissues which contain blood vessels to nourish the matrix.
Nails grow about $1\frac{1}{2}$in (3.8cm) year, though the rate varies with the individual.

HAND PROBLEMS

Only a few serious disorders affect the hands.

RHEUMATOID ARTHRITIS is probably the most severe of the common diseases affecting the hands. The lining of the finger joints becomes inflamed, causing the joints themselves to swell painfully.

A Rheumatoid arthritis
B Osteoarthritis
C Swollen fingers
D Warts
E Chapped skin
F Dermatitis
G Whitlows

NAIL PROBLEMS

WEAK AND BRITTLE NAILS can be the result of incorrect filing, dietary deficiencies, nail biting, too frequent immersion in water, or general ill health. The nails can be strengthened by:
a) the use of a nail hardening preparation;
b) a diet rich in calcium;
c) careful filing.
"Doses" of gelatin in the form of jello cubes, and courses of iodine tablets, may also help.

RIDGES across the nail are due to a deficiency caused by ill health. A course of vitamin A, iodine, and calcium may help improve their condition. Ridges down the nail are a feature of old age and rheumatism, but vitamins and special nail creams can alleviate this condition.

WHITE SPOTS are common on weak nails and are usually caused by injury. This causes the nail cells to separate and allows air to filter between them. Over acidity may also be a cause of white spots.

Treatment varies, but medical advice should be sought as early as possible.

SWOLLEN FINGERS are often a symptom of heart disease, and need immediate medical attention.

WARTS often appear on the hands and are a particular problem in childhood.

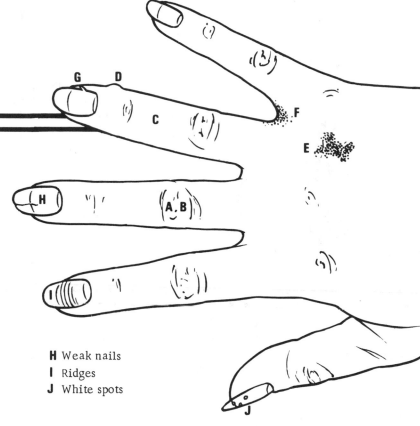

H Weak nails
I Ridges
J White spots

NAIL BITING

This common problem often begins in childhood, perhaps caused by stress of some kind, whether from a nervous disposition or a specific outside cause. It then develops into a habit difficult to break. Biting off the nail leaves it weak and rough, and the irritation caused by the ragged edge will encourage further biting.

WHITLOWS, or felons, are areas of inflamed tissue surrounding the nail. Pus often develops, and a poultice may be used to draw it out, and antiseptic creams applied to prevent the spread of infection. Alternatively a whitlow may be lanced by the doctor.

Use of evil-tasting chemical preparations painted on the nails is an effective means of discouraging nail biting; but it is also a good idea to try to pinpoint and remove any causes of stress. Adults anxious to break the habit may find it helpful to concentrate on allowing one nail at a time to grow longer.

©DIAGRAM

LEGS AND FEET

THE LEG

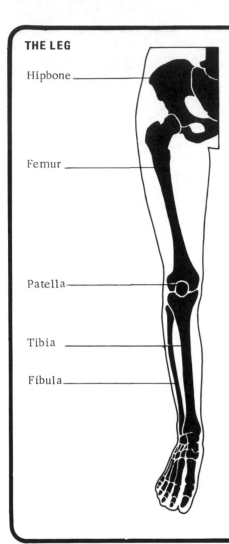

Hipbone

Femur

Patella

Tibia

Fibula

The leg contains 3 important bones - the femur (thighbone) in the upper leg, and the tibia (shinbone) and fibula in the lower leg.

The upper femur meets the hip in a ball and socket joint (which allows free movement). The lower femur and upper tibia meet at the knee joint, a hinge joint (allowing movement in one direction only). Here cartilage forms buffers between these bones, and the front and base of the lower femur abuts the synovial membrane - a sac filled with lubricating fluid. The patella (kneecap) covers and protects the knee joint.

The thigh muscles are used to bend the knee, while the lower leg muscles move the feet and toes. The sciatic nerve - the longest and thickest in the body - provides the nervous system for most of the leg. Blood feeds into the thigh through the femoral artery, and back toward the heart through two systems of veins, one deep, one superficial.

THE FOOT

Each foot comprises 26 bones with 33 joints, linked by more than 100 ligaments. Muscles, tendons, and ligaments keep the foot in different positions.

There are 3 sets of foot bones:
a 7 tarsal or ankle bones forming the ankle and the rear of the instep, and jointed for foot rotation;
b 5 metatarsal or instep bones forming the front of the instep; and
c 14 phalangeal or toe bones - 3 for each small toe and 2 for the big toe. (The ball of the foot is formed where the phalanges and undersides of the metatarsals meet.)

Bones in the foot

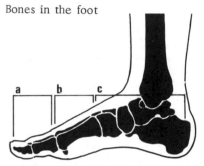

a Phalanges
b Metatarsals
c Tarsals

ARCHES

The normal foot has 2 important arches, one running lengthwise from the heel to the ball of the foot, the second running across the ball of the foot. These allow the spring needed for walking.

In flat feet, the arches have fallen so that weight is borne on the sole as well as the ball and heel.

Print of normal arch

Print of fallen arch

ATHLETE'S FOOT

This is a common disorder, best avoided by keeping the foot clean, dry, and cool, and avoiding contact with infected people and with changing-room floors. It is caused by a fungus infection, and first appears in the toe clefts. There may be splits and flaking, or pieces of dead white skin.

Treatment involves rubbing away the dead skin, applying a mixture of water and medicinal alcohol, and using a special dusting powder. A fungicidal ointment may be prescribed by the doctor. The feet should be exposed to the air as much as possible, and socks and pantihose should be washed after each wearing.

COMMON FOOT DISORDERS

Calluses are areas of skin hardened to form protection for parts that suffer pressure or friction. To clear them, soak the foot in warm water, and remove the callus by rubbing with an emery board.

CORNS are a type of callus usually caused by ill-fitting shoes. They have a cone-shaped core which causes pain when it presses on nerve endings. Corn pads may relieve pain but removal should be left to a chiropodist.

PLANTAR WARTS, or verrucas, are the result of a virus infection, and are often contracted at swimming pools. The warts grow into the skin and cause pain. Verrucas should be treated by a doctor.

DISORDERS OF THE LEG

DISLOCATED HIP Hip dislocation may be present at birth or occur later. It causes a lurching gait, backache in middle age, and sometimes osteoarthritis. Treatment varies with age, and may involve an operation.

DISLOCATED KNEE Slipping of the kneecap to the side may result from an injury, or the tendency may be present from birth. Doctors may recommend rest, or sometimes an operation.

HOUSEMAID'S KNEE Kneeling on hard surfaces for long periods may inflame the fluid-filled sac (bursa) that lies in front of the knee-cap. Tissues at the knee joint swell, become tender, and make knee-bending painful. Poulticing or minor surgery may be needed.

WATER ON THE KNEE This is a collection of fluid beneath the knee-cap. It may be due to infection, rheumatoid arthritis, or to a blow or strain. Rest usually brings recovery.

RHEUMATOID ARTHRITIS Hips, knees, ankles, and feet can be badly affected. Painful joint swelling is sometimes followed by joint erosion and dislocation. Cortisone treatment is used, and hip and knee joints are sometimes replaced with metal or plastic devices.

SCIATICA Inflammation of the sciatic nerve produces a form of neuritis called sciatica. A slipped disk in the spine may press on a nerve, producing intense pain in the leg and lower back. Bed rest, heat applied to the painful area, physiotherapy, and special exercises, may bring relief.

VARICOSE VEINS are swollen veins in the legs, which stand out above the surface and can be acutely painful. Their exposed position also makes them vulnerable to bleeding and ulceration. Varicose veins develop if the valves in the leg veins fail to prevent the back flow of blood. They are more likely in occupations involving long periods of standing - and also where there is a swelling of the abdomen, as in obesity, chronic constipation, and pregnancy This last is why 1 in 2 women over 40 suffer from varicose veins, but only 1 in 4 men of the same age. Possible treatments include: wearing pressure bandages and resting with the leg raised; courses of injections; and surgical tying or removal of the varicose veins.

THROMBOPHLEBITIS A blood clot in a deep vein produces deep pain and a swollen ankle. A blood clot in a superficial vein produces tenderness and a red, cord-like formation beneath the skin. The patient may feel ill and have a high temperature. Medical aid must be sought. Treatments include anticoagulants, supportive stockings or bandages, and rest for the leg.

BUNIONS AND HAMMER TOES

A BUNION is a hard swelling at the base of the big toe. Ill-fitting shoes cause the big toe to bend in, forcing out the base of the toe in a bony outgrowth. A fluid-filled sac (bursa) may develop between the outgrowth and the skin.

A HAMMER TOE is a toe bent up at the middle joint, where it presses on the shoe and causes a corn. Both these conditions can improve with exercise, manipulation, or well-fitting shoes, but severe cases may need surgery.

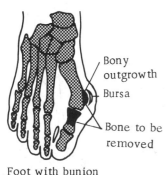

Foot with bunion

Bony outgrowth
Bursa
Bone to be removed

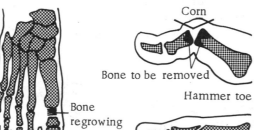

Bone regrowing

Foot after surgery

Corn
Bone to be removed
Hammer toe

Toe after surgery

FOOT CARE

a) Wash feet at least once a day. Soaking in cold, salt water or diluted cider vinegar refreshes the feet in hot weather.

b) Dust the feet daily with a special foot powder to avoid friction, to help dry out the feet, and to prevent odor.

c) Cut toe nails straight across and not in at the edges.

d) Wear clean socks or pantihose every day.

e) Keep the feet free of corns and calluses.

f) Wear comfortable, well-fitting shoes. Very high or very flat heels on shoes can force the body into an unnatural position, and cause considerable discomfort.

©DIAGRAM

BONES AND JOINTS

THE SKELETON

The skeleton gives structure to the body and affords protection to delicate internal organs. The human skeleton develops gradually from connective tissues that become first cartilage and then bone. In fact, at birth, some "bones" are still cartilage, and the process does not complete itself until about age 25. At the same time, many bones fuse together, so that the 330 in a baby's skeleton become 206 bones in an adult. Proportions also change. A newborn baby has a short neck, high shoulders, and a round chest. Between ages three and 10 the shoulders lower, the neck lengthens, and the chest broadens and flattens. Posture changes too: bow legs and knock knees, for example, are common to age five.

Skeletal differences between a newborn male (**1**) and an adult male (**2**)

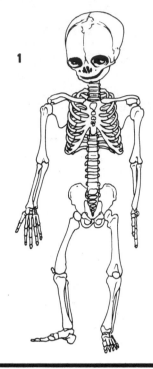 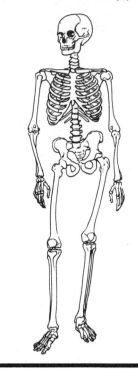

FONTANELS

The fontanels are softer areas of cartilage between the bones of a baby's skull. They make birth easier by allowing the head to distort. The largest fontanel – the "soft spot" – is on the top of the head toward the front. The fontanels gradually disappear as the skull bones fuse together. This process is usually completed by the age of 18 to 26 months.

Diagrammatic representation of a baby's fontanels – soft areas of the skull

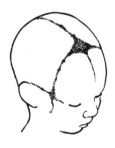 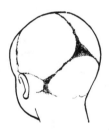

BONE GROWTH

A bone is first formed by the calcification of cartilage, and grows in the same way. Stages in the development of a bone are:
1 cartilage model;
2 formation of a small area of calcified cartilage (bone);
3 growth of calcified area as blood vessels bring in new bone cells;
4 formation of new bone is concentrated in growth areas at

each end of the bone;
5 fusion of growth areas with the bone itself to form the mature bone.

▦ Cartilage
▢ Bone
✎ Blood vessels

  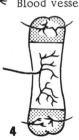 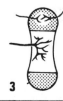 

BONES

Bones may be long, short, flat, or irregular. All bones have a hard outer casing, and a porous honeycomb interior where minerals are stored. Many bones also have a growth area at one or both ends, and a hollow center that contains marrow where blood cells are made.

Cross section of a bone

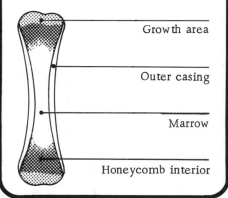

Growth area

Outer casing

Marrow

Honeycomb interior

JOINTS

Fully movable joints consist of a layer of cartilage around each bone end, and a joint capsule with a lining membrane that secretes lubricating fluid.
Different types include:
a hinge (eg elbow);
b pivot (eg neck);
c ball and socket (eg shoulder).

Cross section of a joint

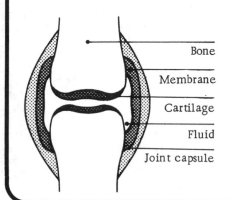

Bone

Membrane

Cartilage

Fluid

Joint capsule

Types of joint

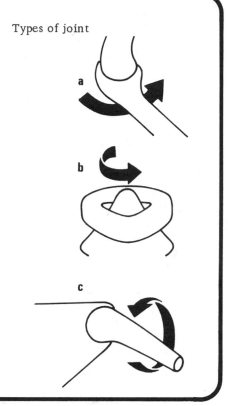

HIP DISLOCATION

Congenital hip dislocation is found in four to five live births per 1000, and is four times more common in girls than boys. In this condition the thigh bone is out of its hip socket (usually only one thigh is dislocated). It is usually checked for soon after birth, and can then be corrected by fitting a special splint. Surgery may be necessary if discovery is delayed.

Normal hip joint

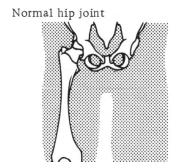

Congenital dislocation

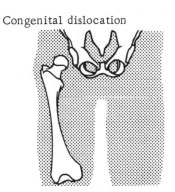

CLUB FOOT

This condition occurs in about one in 1000 live births, and one or both feet may be affected.
The foot is twisted out of position, with malformation of bones and abnormal stretching or shortening of muscles and tendons.
The condition is discovered at a post-birth check, and treatment must be started at once.
Mild cases can be cured by manipulating the foot into its correct position, putting it in plaster for about six months, and then possibly into a splint.
Special shoes may have to be worn for a time, and exercises will be needed to strengthen weak muscles and tendons.
More severe cases require surgery to correct the condition.

Types of club foot

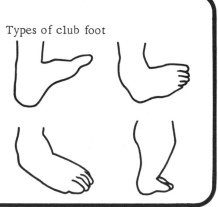

©DIAGRAM

NERVOUS SYSTEM

NERVOUS SYSTEM

Except for certain nerve sheaths that develop during infancy and childhood, the nervous system is fully developed at birth. It is made up of three different units: the central, the peripheral, and the autonomic nervous systems.

CENTRAL NERVOUS SYSTEM This consists of the brain and spinal cord, and is concerned with central control and coordination.

PERIPHERAL NERVOUS SYSTEM This consists of the nerve connections to all the different parts of the body; it receives information and sends out instructions.

AUTONOMIC NERVOUS SYSTEM Comprising nerve centers alongside the spinal column, this controls elementary functions such as heartbeat (and is involved in responding to the stress of feelings and emotions).

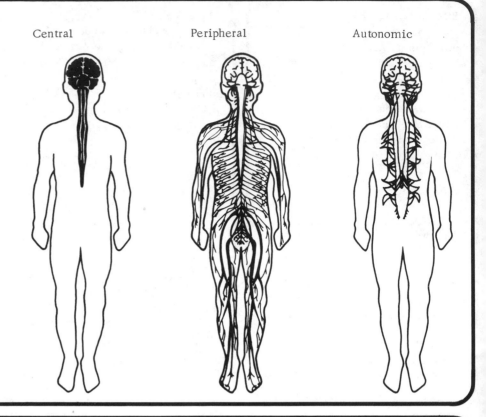

Central   Peripheral   Autonomic

THE BRAIN

The brain is a convoluted mass of nerve cells and fibers, protected by the skull and by a cushion of fluid. It is the coordinating center of the nervous system.

FOREBRAIN Including the cerebral hemispheres, the thalamus, and the hypothalamus, the forebrain is the center of memory, intelligence, and voluntary actions. The cerebral hemispheres, one on each side, are linked by bridges, of which the corpus callosum is the largest. The right hemisphere controls muscles on the left side of the body, and the left hemisphere those on the right.

MIDBRAIN This consists mostly of connecting nerve fibers. Its roof is formed by the optic lobes.

HINDBRAIN The hindbrain includes the cerebellum, which coordinates balance and movement, and the medulla oblongata, which directs involuntary actions such as heartbeat and breathing.

Section through the brain

Cerebral hemisphere

Thalamus

Hypothalamus

Medulla oblongata (brain stem)

Corpus callosum

Midbrain

Cerebellum

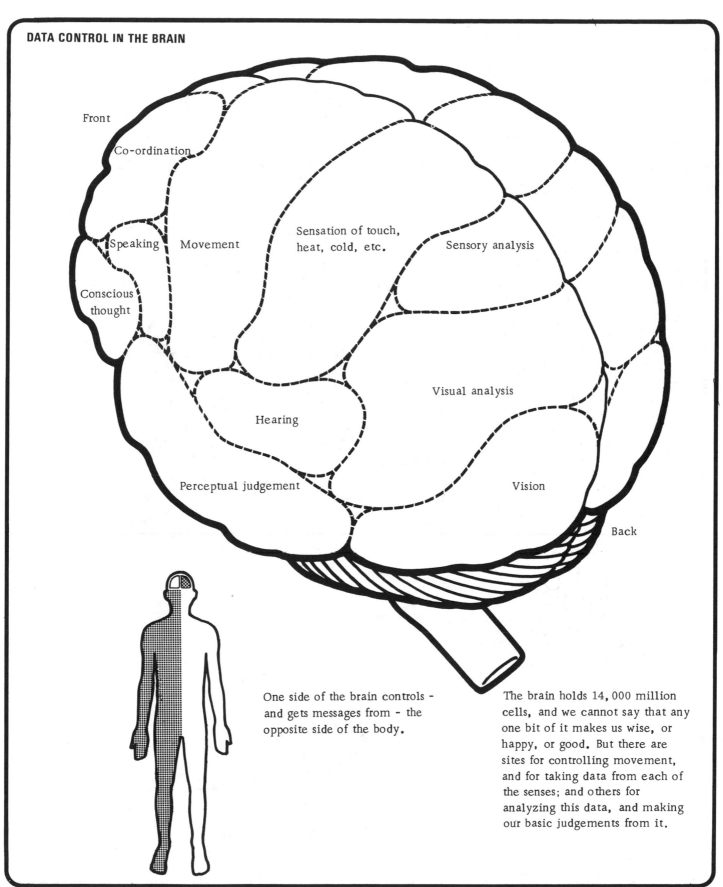

DATA CONTROL IN THE BRAIN

Front

Co-ordination

Speaking

Movement

Sensation of touch, heat, cold, etc.

Sensory analysis

Conscious thought

Visual analysis

Hearing

Perceptual judgement

Vision

Back

One side of the brain controls - and gets messages from - the opposite side of the body.

The brain holds 14,000 million cells, and we cannot say that any one bit of it makes us wise, or happy, or good. But there are sites for controlling movement, and for taking data from each of the senses; and others for analyzing this data, and making our basic judgements from it.

©DIAGRAM

79

RESPIRATORY SYSTEM

RESPIRATORY SYSTEM

Every organ in the body needs oxygen to keep it working. This oxygen is taken into the body during breathing, and carried around the body in the blood. Carbon dioxide, a waste product of body activity, leaves by the same route.

The nasal cavity and mouth lead by way of the pharynx, trachea, and bronchi to the lungs. (When swallowing, this route is closed off by a flap of cartilage.)

The lungs are pink, spongy organs lying in the chest cavity, bounded by the ribs, chest muscles, back muscles, and beneath by a muscular wall called the diaphragm.

Within the lungs the bronchi divide repeatedly into branches called bronchioles, each of which ends in an air sac surrounded by tiny blood vessels.

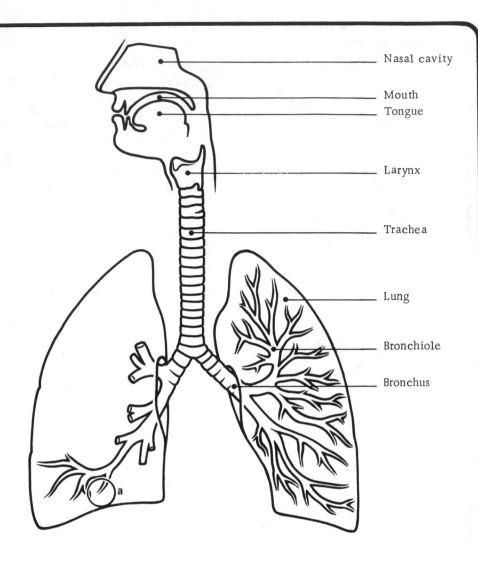

Nasal cavity

Mouth

Tongue

Larynx

Trachea

Lung

Bronchiole

Bronchus

Bronchiole

Alveolar duct

Alveoli

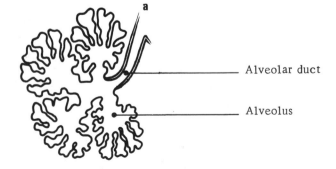

Alveolar duct

Alveolus

LUNGS AND BREATHING

The two lungs lie in the chest (pleural) cavity, bounded by the ribs and the chest muscles and back muscles, and beneath by a muscular wall called the diaphragm.

Breathing occurs because muscular effort enlarges the chest cavity: the diaphragm moves down, and the cavity skeleton forward and outward. This threatens to create a vacuum in the cavity, so air rushes into the lungs under atmospheric pressure, and the lungs enlarge and fill out the cavity. Breathing out also occurs through muscular effort: the chest cavity contracts, forcing air out of the lungs.

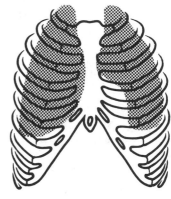 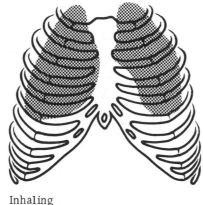

Inhaling   Exhaling

BREATHING MECHANISMS

During breathing in - inspiration - the diaphragm is pulled down and the ribs move forward and outward, enlarging the chest cavity. This threatens to create a vacuum in the cavity, so air rushes into the lungs under atmospheric pressure. During breathing out - expiration - the diaphragm rises, pressure in the chest cavity increases, and air is forced out of the lungs.

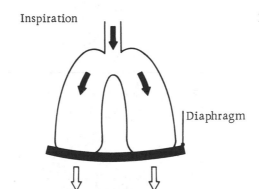 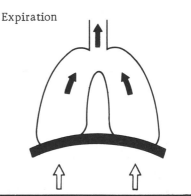

Inspiration   Expiration

Diaphragm

EXCHANGE OF GASES

Gases pass easily through the thin walls of the air sacs and tiny blood vessels in the lungs:

a oxygen enters the air sac during inspiration; carbon dioxide is brought to the lungs in the blood vessels;

b oxygen has entered the bloodstream to travel around the body; carbon dioxide has entered the air sac to be breathed out.

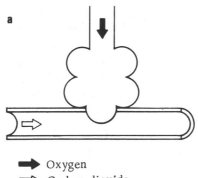 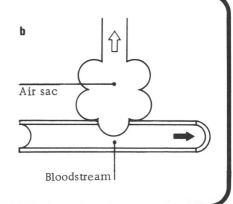

a   **b**

Air sac

Bloodstream

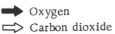

➡ Oxygen
⇨ Carbon dioxide

©DIAGRAM

81

CIRCULATORY SYSTEM 1

J51 HEART AND CIRCULATION

The blood is the transport system by which oxygen and nutrients reach the body's cells, and waste materials are carried away. The heart, a muscular organ positioned behind the rib cage and between the lungs, is the pump that keeps this transport system moving.

Before birth, oxygen and nutrients are obtained from the mother via the placenta, and most waste passes back the same way.

After birth, oxygen enters the blood via the lungs, and nutrients via digestion.

Carbon dioxide leaves via the lungs, while most other waste is filtered out of the blood by the kidneys.

These fundamental differences in supply and disposal require the circulatory system to be different before and after birth. (The fetal circulatory system is described in J52.)

Once a baby is born, circulation starts to function as it will throughout the rest of life.

In the lungs, blood takes in oxygen from the air sacs (J50), becoming "oxygenated." This blood is then transported to the left side of the heart for distribution around the body, entering the left atrium. When the atrium contracts blood is forced into the left ventricle. The left ventricle then contracts, forcing blood along blood vessels called arteries to all organs except the lungs. (Within the heart, backflow of blood is prevented by one-way valves.)

In the organs of the body, oxygen from the blood is used up and waste products collected - the blood becomes "deoxygenated." This blood travels back to the heart in blood vessels called veins. Deoxygenated blood is dark,

Diagram showing the flow of blood through the heart's chambers

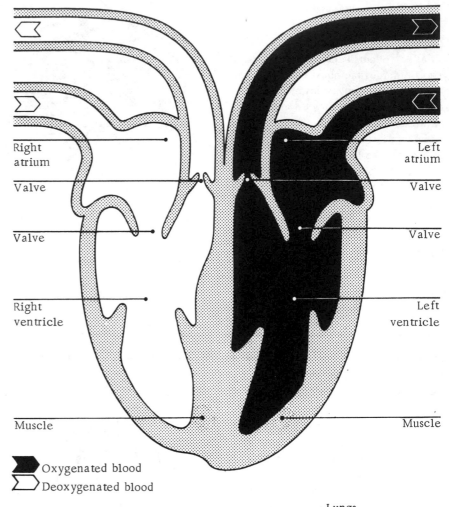

Right atrium

Valve

Valve

Right ventricle

Muscle

Left atrium

Valve

Valve

Left ventricle

Muscle

Oxygenated blood
Deoxygenated blood

bluish red compared with the bright red color of oxygenated blood. Blood from the veins enters the right side of the heart to be pumped to the lungs to be oxygenated. It passes from the right atrium to the right ventricle, and from there to the lungs. The blood has now completed its full circuit - lungs, heart, body, heart, lungs - as shown in the small diagram (right).

Normal circulation

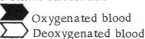

Oxygenated blood
Deoxygenated blood

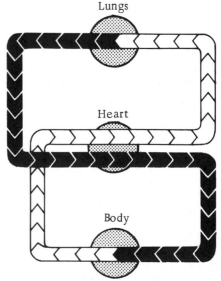

Lungs

Heart

Body

THE BLOOD SYSTEM

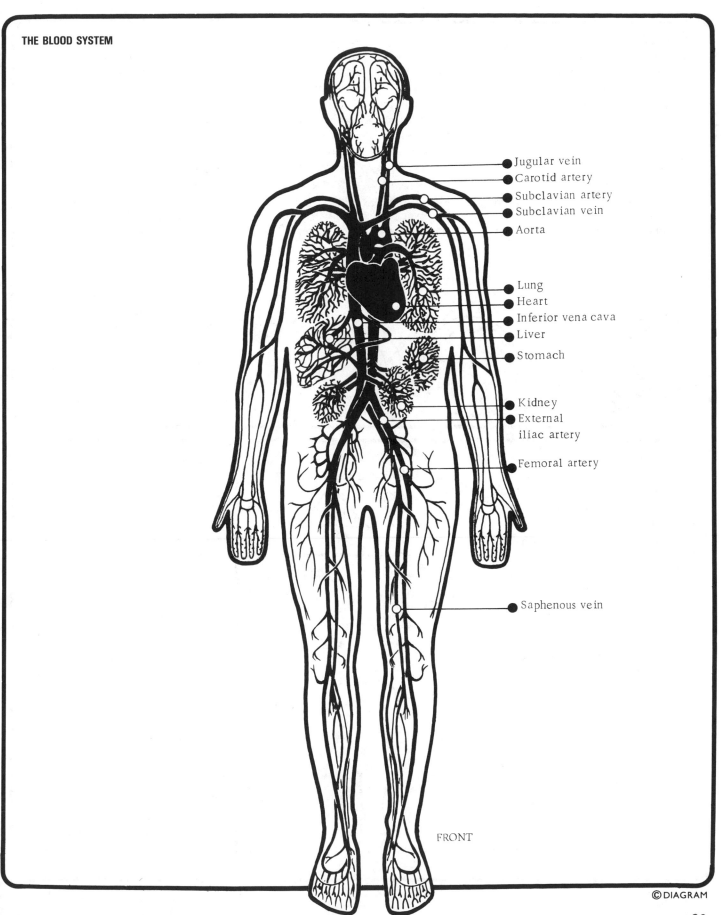

Jugular vein
Carotid artery
Subclavian artery
Subclavian vein
Aorta

Lung
Heart
Inferior vena cava
Liver
Stomach

Kidney
External
iliac artery

Femoral artery

Saphenous vein

FRONT

© DIAGRAM

CIRCULATORY SYSTEM 2

THE BLOOD

The blood transports oxygen, nutrients, and waste around the body, and is also vital in fighting infection. It is made up of red cells, white cells, platelets, and plasma.

RED BLOOD CELLS - also called red "corpuscles" - are rounded disks, dented on each side. They measure about three ten-thousandths of an inch in diameter, and have no nucleus. Hemoglobin, a pigment in the cells, gives them their color and, through a chemical reaction, allows oxygen and carbon dioxide to be transported.

WHITE BLOOD CELLS - or white corpuscles - are roughly spherical in shape. They are slightly bigger than red blood cells, but less numerous (about 500 red cells to each white one). There are three basic types - neutrophils, lymphocytes, and monocytes - with nuclei of different shapes. They devour bacteria.

PLATELETS - or "thrombocytes" are tiny colorless cells that can clump together and so make the blood clot.

PLASMA is an almost colorless liquid that is mostly water. It transports the blood cells and many dissolved substances. Proteins in the plasma include disease-fighting antibodies.

Red blood cells:
a front view **b** side view

Types of white blood cell:
c neutrophil **d** lymphocyte **e** monocyte

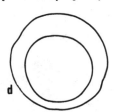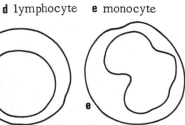

Platelets

BLOOD MANUFACTURE

Before birth, red blood cells are made in the blood vessels, liver, and bone marrow. After birth, they are made only in the bone marrow. In a young child blood cells are formed in the marrow of most bones, but in an adult they are usually formed only in the marrow of the spine, ribs, breastbone, pelvic bones, and part of the upper arm and leg bones. Lymphocytes are formed in the lymph nodes, spleen, and the thymus of a child. Other white cells are made in the bone marrow. Platelets, too, are made in the bone marrow. Most of the proteins in the blood are formed in the liver (J66).

A red blood cell typically lives for 120 days, a lymphocyte for over a year, other white cells for only 10 hours, and platelets for about 10 days.

Worn-out cells are broken down in the liver and spleen.

Sites of blood cell manufacture

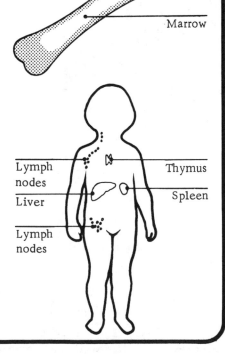

Marrow

Lymph nodes

Thymus

Liver

Spleen

Lymph nodes

BLOOD DISORDERS

Inherited disorders of the blood include the following.

HEMOPHILIA In this disorder, a deficiency of one of the factors that causes the blood to clot means that special treatment is needed even for minor injuries. Treatment includes transfusions of fresh plasma, and injections of concentrated clotting factor.

SICKLE-CELL ANEMIA is commonest among Negroid people. Red blood cells become sickle-shaped when deprived of oxygen. These sickle cells impair circulation. Over-exertion produces severe pains in the abdomen and joints. Blood transfusions may be needed.

THALASSEMIA, or Mediterranean anemia, is an inherited disorder in which red blood cells are broken down more quickly than the body can replace them. It is treated by blood transfusions, and sometimes by removal of the spleen.

VITAL STATISTICS

The adult heart beats 60 to 80 times a minute, and about 40 million times a year.

The smaller the heart, the faster it beats. An average male heart is about 10 ounces in weight, and a female heart 8 ounces. So a woman's heart makes about 6 to 8 more beats a minute than a man's. At each beat the heart takes in and discharges over $\frac{1}{4}$ pint (US) of blood (130cc). It pumps over 2000 gallons a day, and 50 million gallons in a lifetime. In a healthy man during exercise, the heart beat may increase to 180 beats a minute, pumping 40 pints a minute.

BLOOD VESSELS

Blood vessels are of three types: arteries, veins, and capillaries. Arteries carry oxygenated blood from the heart to all parts of the body. They have strong elastic walls that squeeze the blood along with waves of contraction.

After profuse branching the blood passes through the tissues of the body in tiny capillaries, where nutritive substances and oxygen are exchanged for waste products. The blood is then conveyed back to the heart along veins. These have thinner walls than the arteries, and contain valves to prevent a back flow of blood.

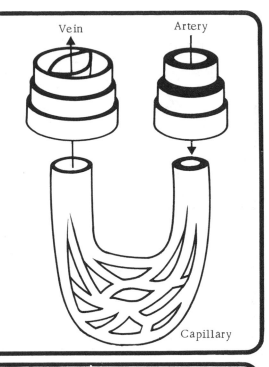

Vein   Artery

Capillary

BLOOD PRESSURE

Blood pressure is expressed in two figures that indicate the pressures in the aorta:

a) on contraction, called the systolic pressure;

b) on relaxation, called the diastolic pressure.

For a healthy young person at rest the systolic pressure is usually between 100 and 120mm Hg, and the diastolic between 70 and 80 mm Hg. These are expressed as 100/70 or 120/80.

Pressure in the right ventricle only rises to about 20 mm Hg. This is enough to send blood through the lungs and back to the left ventricle.

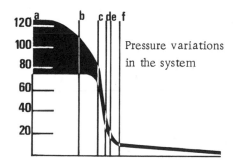

Pressure variations in the system

a Large arteries
b Small arteries
c Arterioles
d Capillaries
e Venules
f Veins

WHERE THE BLOOD GOES

This shows the relative distribution of blood to the various organs of the body, when at rest. During exertion, the distribution changes. At rest, the heart-lung-heart route takes 6 seconds, the heart-brain-heart route, 8 seconds, and the heart-toe-heart route, 16 seconds.

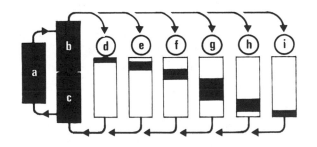

a Lungs
b Heart (right)
c Heart (left)
d Heart blood vessels 5%
e Brain 15%
f Muscles 15%
g Intestines 35%
h Kidneys 20%
i Skin, skeleton etc 10%

©DIAGRAM

DIGESTIVE SYSTEM 1

THE DIGESTIVE SYSTEM

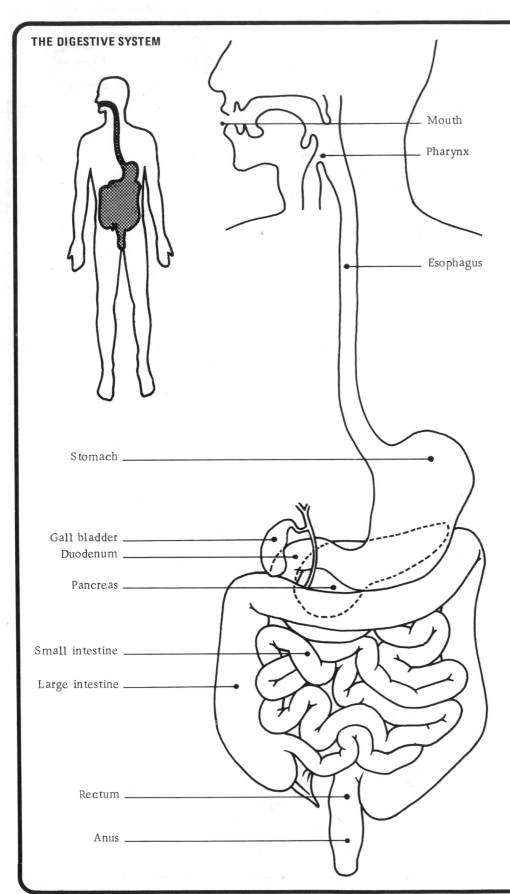

Mouth

Pharynx

Esophagus

Stomach

Gall bladder

Duodenum

Pancreas

Small intestine

Large intestine

Rectum

Anus

The digestive tract forms a tube over 30ft long, beginning in the mouth and ending in the anus. Between these it includes the esophagus (gullet), stomach, small intestine, and large intestine.

In the mouth, food is chewed into smaller pieces, mixed with saliva, and formed into a rounded ball ("bolus").

On swallowing, the bolus passes down the esophagus into the stomach.

The stomach varies in shape and size according to its contents. Its maximum capacity is about $2\frac{1}{2}$ pints. Here food is churned into even smaller pieces, mixed with gastric juice, and disinfected with hydrochloric acid. Fat is melted by the heat.

From the stomach food passes into the small intestine. In the first 12in of this (the duodenum), the food is mixed with pancreatic and intestinal juices and with bile from the gall bladder. Then here, and in the remaining 21ft of small intestine, most of the useful elements in food are absorbed through the intestinal walls into the blood and lymph streams.

In the 6ft long large intestine, water is absorbed into the body, turning the waste product into a soft solid (feces): a mixture of indigestible remnants, unabsorbed water, and millions of bacteria. Finally, the feces pass out of the body via the anus.

Food takes from 15 hours upward to pass through the whole system. It usually stays in the stomach 3 to 5 hours, the small intestine $4\frac{1}{2}$ hours, and the large intestine (where the sequence of meals may get jumbled) 5 to 25 hours or more.

THE DIGESTIVE PROCESS

Digestion breaks food down so it can be absorbed by the body. The process is partly mechanical: food eaten is chewed and churned to a mushy consistency. But it is mainly chemical: the complex chemicals taken in as food are worked on by enzymes and other chemicals made by the body, and so broken down to simple substances that the body can absorb. Absorption itself occurs almost entirely in the small intestine: the nutrients pass through the intestinal wall into thousands of tiny blood and lymph vessels. The remaining waste matter is a mixture of indigestible remnants, unabsorbed water, and millions of bacteria, and it passes out of the body via the anus.

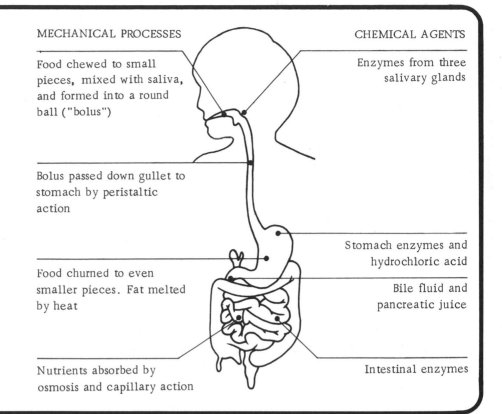

MECHANICAL PROCESSES

Food chewed to small pieces, mixed with saliva, and formed into a round ball ("bolus")

Bolus passed down gullet to stomach by peristaltic action

Food churned to even smaller pieces. Fat melted by heat

Nutrients absorbed by osmosis and capillary action

CHEMICAL AGENTS

Enzymes from three salivary glands

Stomach enzymes and hydrochloric acid

Bile fluid and pancreatic juice

Intestinal enzymes

CONGENITAL DISORDERS

ANATOMICAL ABNORMALITIES
Abnormalities of the digestive tract occur in one in 500 live births. The diagram gives examples. Most can now be corrected by surgery.
OTHER CONGENITAL DISORDERS
a) Celiac disease: fats and some other elements of food are not digested. There is diarrhea and weight loss, and stools passed are fatty. The cause is unknown, but recovery is usual if gluten is kept out of the diet. (Gluten is a protein found in wheat and oats.)
b) Cystic fibrosis: an inherited disorder found in one in 1,000 live births. All mucous glands are affected - including digestive ones, causing diarrhea and loss of weight (though the main danger is from bronchitis and pneumonia due to excessive lung mucus.)
Treatment is still being developed.

Diagram showing possible congenital disorders of the digestive tract

Gullet is a dead end: stomach links with windpipe

Fibrous lump blocks way to duodenum

Nerve fibers missing from rectum and anus - muscles cannot respond to pass stools

Bile duct missing

Part of intestine solid - no route through

Anus is a dead end, or opens into urinary tract or vagina

©DIAGRAM

DIGESTION AND ABSORPTION

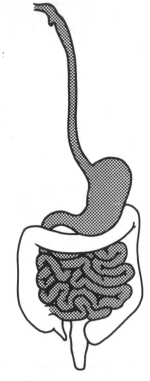

Carbohydrates

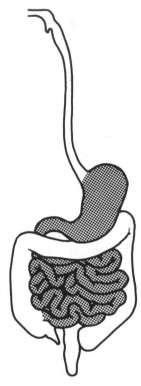

Fats and fat-soluble vitamins

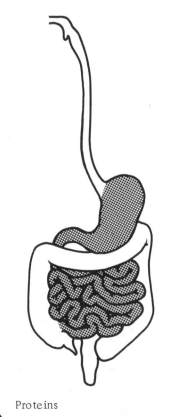

Proteins

Water and water-soluble vitamins

CARBOHYDRATES Digestion of starch begins in the mouth. It continues in the stomach, but the stomach usually empties itself before this is completed. In the duodenum, pancreatic juices break the carbohydrates down into monosaccharides, which are then absorbed into the blood stream. But some forms of carbohydrate (eg cellulose) cannot be digested, while some sugars begin to be absorbed even in the mouth.

FATS Digestion begins in the stomach, where naturally emulsified fats are converted into fatty acids and glycerol. (Unconverted fat causes food to be retained longer in the stomach.) In the small intestine, bile emulsifies the unemulsified fats, and pancreatic juice converts them into fatty acids. These are absorbed into the lymph vessels (70%) or the blood stream (30%). Fat-soluble vitamins are absorbed at the same time.

PROTEINS Digestion begins in the stomach, where proteins are broken down into peptones. In the small intestine, the pancreatic and intestinal juices break the peptones down into amino acids. The amino acids are absorbed into the blood stream.

WATER is absorbed in the large intestine, into the lymph vessels and blood stream.

URINARY SYSTEM

URINARY SYSTEM

The urinary system is responsible for removing waste products from the blood, and for regulating the body's salt and fluid levels.

The system comprises a pair of kidneys, two tubes called ureters linking the kidneys with the bladder, the bladder, and another tube called the urethra. In a male the urethra leads to the tip of the penis; in a female to an opening in front of the vagina.

Urine produced by the kidneys drips down the ureters to the bladder, which acts as a reservoir. A muscular ring surrounds the exit from the bladder into the urethra. When this is contracted, it prevents leakage of urine out of the bladder. Relaxation of the muscular ring causes urination as the urine passes into the urethra and out of the body.

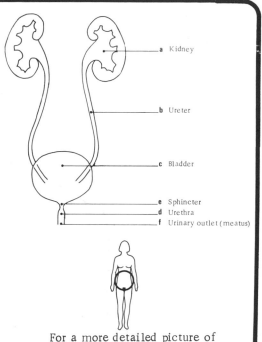

a Kidney
b Ureter
c Bladder
e Sphincter
d Urethra
f Urinary outlet (meatus)

For a more detailed picture of the male and female urinary systems, see p. 144.

URINE

Urine consists of 96% water and 4% dissolved solids. Only 60% of the water taken into the body is normally eliminated as urine. The rest passes out in sweat and feces, and through the lungs. Urine is normally straw- or amber-colored. In 24 hours an adult usually passes between $1\frac{3}{4}$ and 3 (US)pt (0.8 to 1.4 liters), spread over 4 to 6 occasions. Most do not find it necessary to get up to pass urine at night. However, all these characteristics vary normally with: the amount of fluid drunk, and when; the amount lost in sweat; the size of the bladder; etc.

KIDNEYS

The kidneys are dark red, bean-shaped organs that purify the blood and regulate salt and fluid levels in the body.

Each kidney is composed of tiny tubular filter units called nephrons, which are coiled in the outer cortex of the kidney and straight in the inner medulla. Nephrons open into pyramids that project into the expanded end of the ureter.

As the blood flows through the kidneys, water, salts, and urea (a waste product formed in the liver,) pass into the nephrons through the thin walls of the blood vessels and the nephrons. Farther along the nephron tubes, water and salts needed by the body pass back into the blood. Surplus water and waste products become urine, which passes via the pyramids into the ureter for elimination from the body.

Cross section of a kidney
a Cortex
b Medulla
c Renal artery
d Renal vein
e Pyramid
f Ureter

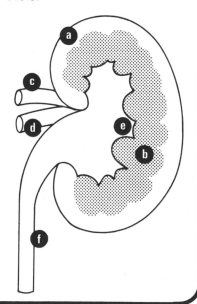

KIDNEY ABNORMALITIES

Abnormalities such as double kidneys or a single horseshoe kidney are quite common (2-3% of the population have double kidneys). Usually they cause no trouble, and are discovered in a general check-up. Surgery is rarely needed.

Double kidneys

Horseshoe kidney

©DIAGRAM

GLANDS AND HORMONES

GLANDS AND HORMONES

The word gland is sometimes used to refer to lymph nodes - as when we speak of "swollen glands" in the neck, armpits, or groin. But in medical terms the glands are those body organs that make and put out liquid chemical substances, and that through these have an effect elsewhere in the body. There are two types of gland.

EXOCRINE GLANDS

Glands of this type have tubes to carry their output to a nearby point where it will act. They include the liver, the salivary glands in the mouth, and the sweat glands in the skin.

ENDOCRINE GLANDS

These glands release their chemicals directly into the bloodstream. The chemicals they produce are called hormones and play a vital role in controlling many body processes.

The endocrine system is made up of the following glands: pituitary, thyroid, parathyroids (embedded in the thyroid), thymus, pancreas, adrenals (lying against the upper end of each kidney), and the sex glands or gonads (ovaries in a female and testes in a male).

HORMONAL RELATIONSHIPS

The action and interaction of hormones in the body is far from simple. In the first place, most endocrine glands consist of two or more separate parts, and each part may produce several different hormones with different roles in the body. Second, hormones from several different glands may play a part in a single role. For example, childhood growth depends on hormones from the pituitary, the thyroid, and the testes or ovaries. Third, and most important, pituitary hormones regulate the hormone production of other glands in the body.

Location of endocrine glands

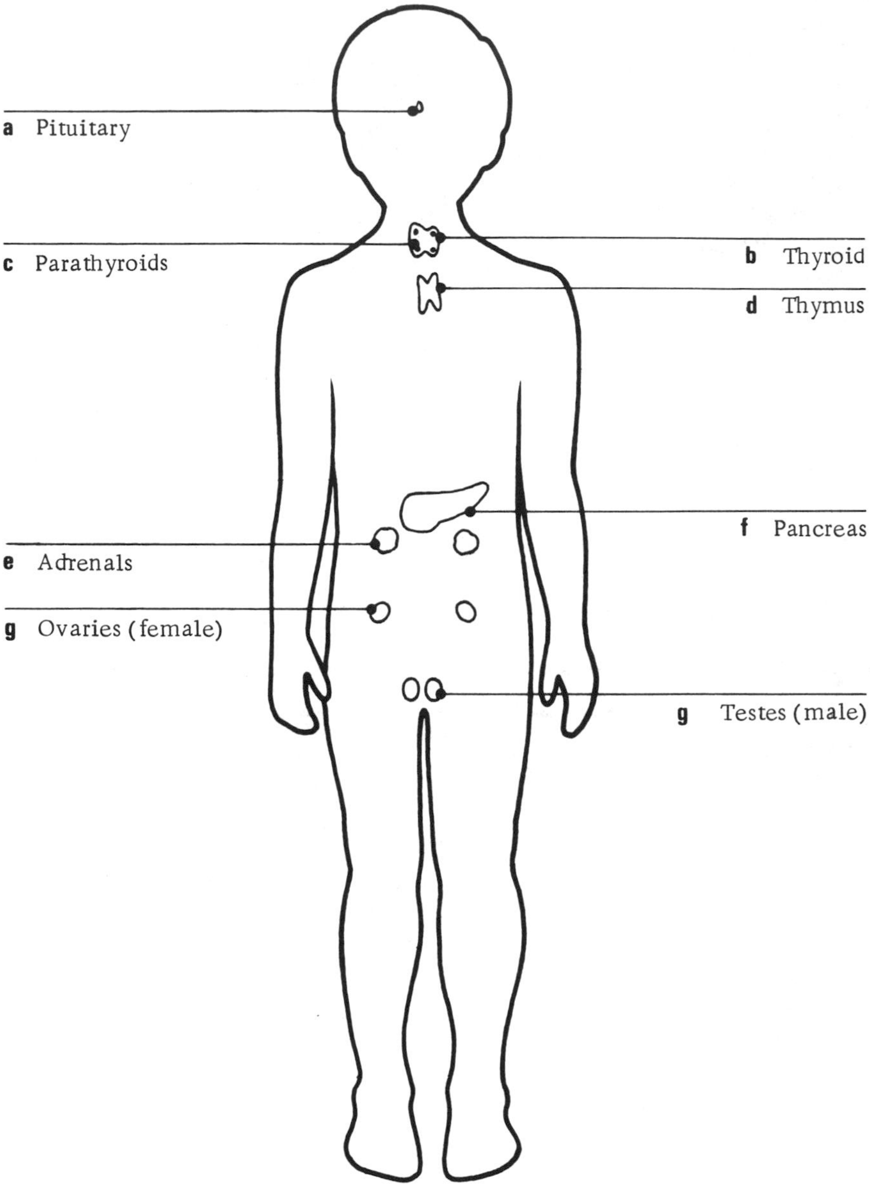

a Pituitary

c Parathyroids

b Thyroid

d Thymus

f Pancreas

e Adrenals

g Ovaries (female)

g Testes (male)

SUMMARY OF HORMONAL ACTION
Although the endocrine system is extremely complex, some of the processes affected by hormones from the various glands can be listed briefly, as follows:
a pituitary - growth;
b thyroid - energy level, and rate of growth and sexual development;
c parathyroids - use of calcium and phosphorus in the body;
d thymus - production of agents to combat infection early in life;
e adrenals - physical effort, use of glucose, inflammatory response to infection, and use of fat;
f pancreas - level of sugar in the body;
g testes or ovaries - production of sex hormones.

PITUITARY GLAND

The pituitary is a two-lobed gland connected by a thin stalk to the base of the brain. It is about the size of a pea.

Some of the hormones produced by the pituitary act directly on body processes. These include a hormone that regulates the growth of bones and other tissues, two hormones thought to affect body pigmentation, and a hormone that regulates the amount of water absorbed by the kidneys.

Other pituitary hormones affect body processes indirectly - by regulating the production of hormones in the thyroid gland, adrenals, and sex glands. Hormone production within the pituitary itself is regulated in two ways - directly by the brain, and by feedback mechanisms from elsewhere in the body.

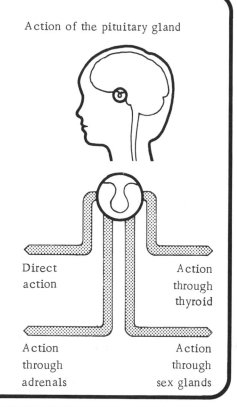

Action of the pituitary gland

Direct action

Action through thyroid

Action through adrenals

Action through sex glands

THE LIVER

The largest gland in the body, the liver is a wedge-shaped mass of tissue containing a great many blood vessels. Its glandular liquid is called bile, which is stored in the gall bladder until needed and then flows into the duodenum. Bile allows the body to digest fats and helps it take in vitamins A, D, E, and K.

Other important functions of the liver are the production of blood proteins, the production of heparin (a substance that in normal circumstances prevents the blood from clotting), the breaking down of worn-out red blood cells to be excreted in the bile, the storing of iron in a fetus, the regulation of the sugar content of the blood under the influence of the hormone insulin from the pancreas, and the conversion of waste into urea for excretion via the urinary system.

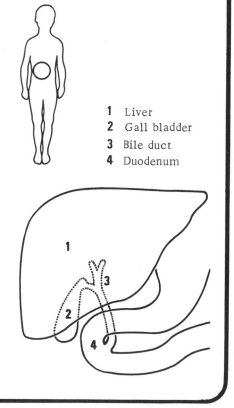

1 Liver
2 Gall bladder
3 Bile duct
4 Duodenum

HORMONAL DISORDERS

Various disorders result from too much or too little of a hormone. Examples include abnormal growth, and diabetes (diabetes mellitus).

DWARFISM

Deficiency of pituitary growth hormone produces a type of dwarfism in which proportions are normal but body size is extremely small. Mental ability is not affected. This type of dwarfism (unlike hereditary dwarfism) can sometimes be prevented if growth hormone is supplied before the normal growth period has ended.

GIGANTISM

Excessive length of the long bones of the arms and legs may be due to overproduction of pituitary growth hormone. Such overproduction is often caused by a non-cancerous pituitary tumor, which can be treated by irradiation or surgery.

DIABETES

This fairly common condition is due to a deficiency of the hormone insulin, made in the pancreas and vital to the body's use of sugar. Insulin deficiency, which may have a variety of causes, sets up a complex chain of reactions.

a) Unused glucose builds up in the bloodstream. It is filtered out by the kidneys but extra urine is needed to carry it away. This increases thirst.

b) Since the body cannot use glucose, it burns fat and protein instead. Weight is lost. Burning fat without glucose results in a build-up of waste products that can cause a coma.

c) Another chain of effects may gradually lead to degeneration of small blood vessels, especially in the eyes and kidneys.

Treatment checks these effects. It involves limiting carbohydrates (sugars) in the diet, and in some cases tablets that act like insulin, or daily insulin injections.

©DIAGRAM

LOCATION

The male sexual system is partly visible, and partly hidden inside the body.

The visible parts are the penis, and the scrotum containing the testes.

Inside the body are the prostate gland, the seminal vesicles, and the tubes that link different parts of the system.

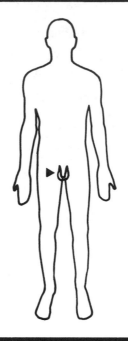

ERECTION

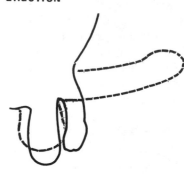

When a man is sexually aroused, the penis becomes swelled with blood. Instead of being "floppy" and hanging down, it becomes stiffer and longer, and juts out from the body. This is called "erection."

MALE SEX ORGANS

THE TWO TESTES (a) are the male reproductive glands. They hang in an external pouch (the scrotum), which is below and behind the penis. Each testis is a flattened oval in shape, about $1\frac{3}{4}$ inches long and 1 inch wide.

The scrotum is divided into two separate compartments (scrotal sacs), one for each testis. (Usually the left testis hangs lower than the right, and its scrotal sac is slightly larger.)

The testes make:

a male sex hormone, testosterone; and sperm cells, which are the male reproduction cells.

The sperm cells are needed to fertilize the egg in the female body, if new life is to be produced.

THE EPIDIDYMIDES (b) are found one alongside each testis.

A number of small tubes lead to each epididymis from its testis. In the epididymides the young sperm cells (spermatocytes) are stored and develop into mature sperm.

THE VAS DEFERENS (c) are the two tubes - one from each testis - that carry sperm from the testes to the prostate gland.

They are about 16 inches long, and wind upwards from the scrotum into the pelvic cavity.

They come together and join with the urethra tube just below the bladder.

THE PROSTATE GLAND (d) surrounds the junction of the vas deferens and urethra tubes.

Here the sperm cells are mixed with seminal fluid: the liquid in which the sperms are carried out of the body. The resulting mixture is semen: a thick whitish fluid.

THE SEMINAL VESICLES (e) make part of the seminal fluid that the prostate gland mixes with the sperm cells. More seminal fluid is made by the prostate gland itself.

THE URETHRA (f) is the tube that carries urine from the bladder to the penis. It is S-shaped and about 8 inches long.

In the prostate gland it is joined by the vas deferens - so it is also the route by which the semen reaches the penis from the prostate gland.

THE PENIS (g) is inserted into the female body during copulation. Most of the penis is made up of spongy tissue, loosely covered with skin.

The urethra tube enters the penis from the body and runs inside it to the tip of the penis.

The external opening in the tip (the meatus) is where semen or urine leaves the body.

In its natural state, the sides of the penis near its tip are covered by a fold of skin, called the foreskin. But this is often removed - usually because of religious or social custom, shortly after birth, but sometimes for medical reasons.

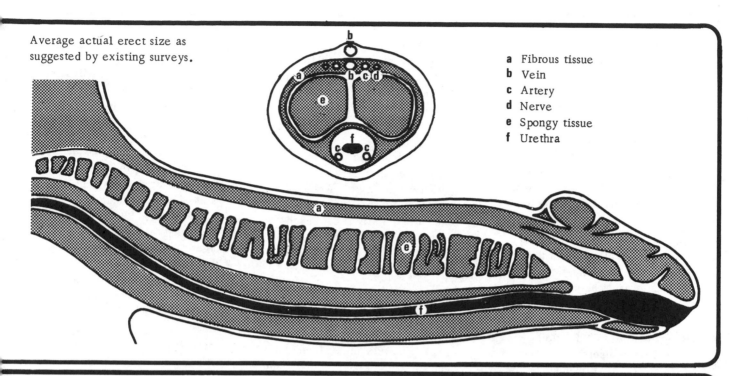

Average actual erect size as suggested by existing surveys.

a Fibrous tissue
b Vein
c Artery
d Nerve
e Spongy tissue
f Urethra

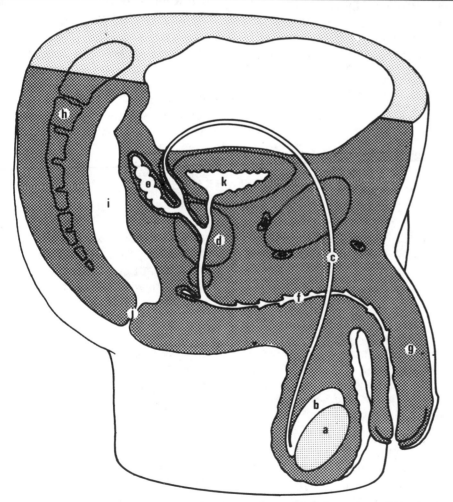

a Testes
b Epididymides
c Vas deferens
d Prostate gland
e Seminal vesicles
f Urethra
g Penis
h Pelvis
i Rectum
j Anus
k Bladder

©DIAGRAM

POTENCY

The characteristics of potency are sperm production, erection, and ejaculation.

SPERM PRODUCTION

Sperm cells are formed in tiny tubes inside the testes, at the rate, in maturity, of 10 to 30 billion a month. Each is about one 500th of an inch long, and each takes about 60 to 72 days to develop. They are stored during this time in the two epididymides (tubes which would be 20ft long if uncoiled), and remain there until ejaculated from the body by sexual orgasm - or until they disintegrate, being reabsorbed into the testes.

Sperm production can only occur at a temperature 3 or 4°F below normal body temperature. That is why the testes are suspended below the body in the scrotum. High temperatures not only prevent new sperms being formed, but also kill those in storage. The effects may be temporary (infertility) or - in extreme cases - permanent (sterility). Very low temperatures also halt sperm formation, but do not kill those in storage.

Otherwise sperm production is continuous, though there is some evidence of seasonal variations (the sperm concentration in semen seems to be lower in the warmer months).

ERECTION

Erection has been recorded at all ages - from baby boys a few minutes after birth, to old men in their late 80s. But the ability usually develops with puberty, and may be lost as old age approaches. Erection can occur gradually, or in as fast a time as 3 seconds, and can also be lost rapidly or slowly. The length of time for which erection can be maintained varies considerably, and depends on circumstances, but tends to decline with age.

EJACULATION

On orgasm, a man ejaculates on average about 3.5 millileter of semen - a small teaspoonful. But the amount varies greatly, even for a given person. 3.5ml is typical after three or more days without ejaculation; but the normal variation in the same circumstances is from 0.2 to 6.6ml. Also, the volume diminishes with repeated ejaculation, while it can reach 13ml after prolonged abstinence. Whatever the volume, it is almost entirely fluid rather than sperm. Of an average volume, 60% is fluid from the seminal vesicles, and 38% fluid from the prostate (the latter gives semen its characteristic smell). The remaining 2% includes other small fluid contributions, and the sperm themselves. All this is over 90% water. Nevertheless, the sperm count in a typical ejaculation totals between 150 and 400 million. Each sperm has lived between one and 21 days since it reached maturity, and most of them 7 to 14 days.

These different elements of semen appear in a fairly set order.

Ejaculation is preceded by fluid from the Cowper's gland: 1 or 2 drops of clear, colorless liquid, which neutralizes any acidity in the urethra left by urine. (This fluid can also be released after the plateau level has continued for some time, even if no orgasm occurs. It is not semen, but does contain a very large number of sperm cells. Hence pregnancy may follow intercourse even if no semen has been ejaculated.)

This is followed, on ejaculation, by:

first, a thin milky fluid from the prostate, which usually contains few sperms;

second, the fluid containing sperms;

third, the sticky yellow fluid from the seminal vesicles.

The overall appearance of semen is milky, opalescent, and opaque. The opalescence increases with the concentration of sperms.

A forceful ejaculation, after prolonged sexual abstinence, may shoot semen, if there is nothing in the way, 3ft or more; but 7 to 10in is the average distance.

REPEATED EJACULATION

The ability to have repeated orgasms with ejaculation in a short time varies enormously, and begins to decline almost immediately once puberty is complete. Kinsey records one man who had had 4 to 5 orgasms a day, with ejaculation, for 30 years; and another man who had one ejaculation in all that period.

Within a space of one or two hours, most men can manage one ejaculation, some a second, a few three or four. Kinsey records one achievement of about 6 to 8 ejaculations in a single session; but regular multiple ejaculation is typical of only a small number of men.

IMPOTENCE

Being impotent can mean two things: being unable to get an erection, or being unable to reach orgasm even if there is an erection. They can be dealt with together because they have similar causes.

Over 90% of impotence is caused psychologically: see pp.278-9. The few physical causes of long-term impotence can be categorized into: physical defects from birth; defects of physical development; and changes in the adult body state. Defects from birth concern, of course, the formation of the genitals. Defects of physical development concern the absence of puberty, due to hormonal failure. Neither of these can happen to someone who has had normal sexual functioning.

Long term impotence from physical causes can in his case only arise from changes in the adult body state, including possibly:

a) some diseases of the genitals;

b) some hormonal disorders;

c) some general disorders, such as diabetes, debilitating illnesses, and infectious damage to the spinal cord;

d) some surgery (eg for cancer of prostate or colon);

e) continual heavy drug or alcohol use; and

f) aging.

However, none of these is certain to cause impotence.

Physical causes of short-term impotence can include almost anything that lowers the body's vitality: immediate factors like great fatigue or heavy doses of alcohol or drugs, and more mild ones like poor health, poor nutrition, and perhaps even lack of exercise.

(These things are also likely to affect sperm production. But poor sperm production does not itself cause impotence. A man can produce few sperm, but still ejaculate normally.)

There are great variations in normal potency, and almost all men experience some failure of potency at some time in their lives.

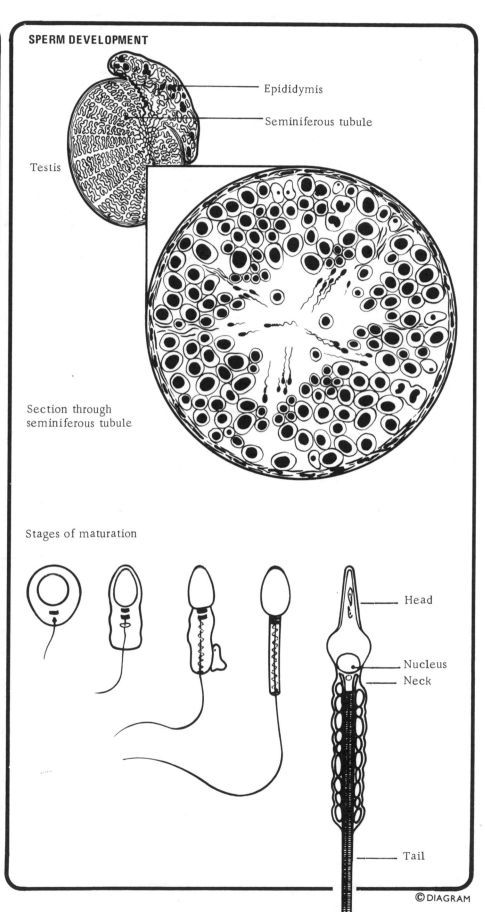

SPERM DEVELOPMENT

Epididymis

Seminiferous tubule

Testis

Section through seminiferous tubule

Stages of maturation

Head

Nucleus

Neck

Tail

©DIAGRAM

PENIS VARIATIONS

CIRCUMCISION

In its natural state, the sides of the penis near its tip are covered by a fold of skin, called the foreskin. In primitive circumstances, this probably served as some protection to the penis at its most sensitive part. When rolled forward, the foreskin is like a hood around the penis tip but it can also be pushed back along the shaft.

In many societies, the foreskin is often removed – it is cut away, in a minor surgical procedure known as circumcision. In rare cases there are medical reasons for this: sometimes the foreskin of an adult man can become very tight and difficult to move. But usually it is a religious or social custom. Male Jews and also Moslems are circumcised as a religious requirement; and in many hospitals in the USA, and some other countries, it is routine practice to circumcise all baby boys.

The value of routine circumcision is debatable. There is no clear evidence that presence or absence of a foreskin makes much difference to sexual sensitivity, or pleasure, or time taken to reach orgasm. One practical argument for circumcision is a hygienic one When the foreskin is intact, white secretions called "smegma" can accumulate underneath it. Unless these are regularly washed away, the foreskin can become smelly, dirty, and even inflamed. There is also a link between smegma and cancer of the penis in men; and a possible link between smegma and cervical cancer in women. However, adequate hygiene will cope with this just as well as circumcision.

Uncircumcised

Uncircumcised: rolled back

Circumcised

ERECT DIMENSIONS

The longest erect penis for which there is reasonable scientific evidence measured 12in. The smallest recorded with testes and normal functioning has been under $\frac{1}{2}$in in length. In other abnormal cases the penis can, of course, be totally absent.

Body size is no guide to penis size: the erect penis has a less constant relationship to body size than any other organ. Nor is flaccid size decisive: penises which hang longer when flaccid tend to gain less when erect. A short penis can gain as much as $3\frac{3}{4}$in, and a long one as little as 2in.

There is no relationship between penis size and sexual prowess or female satisfaction: the female vagina accommodates its size to that of the penis, and stimulation of the female clitoris does not depend on penis length.

FLACCID DIMENSIONS

In its normal state, the penis hangs down loosely. It averages about $3\frac{3}{4}$in long, and most examples are between $3\frac{1}{4}$in and $4\frac{1}{4}$in, though a few cases will fall well outside this range.

The penis gets temporarily smaller than usual in certain circumstances, eg in cold temperatures; through immersion in cold water; in extreme exhaustion; or after a failed attempt at sexual activity. These circumstances also pull the testes and scrotal sac closer to the body.

The penis may get permanently smaller in old age, or after longish periods of impotence.

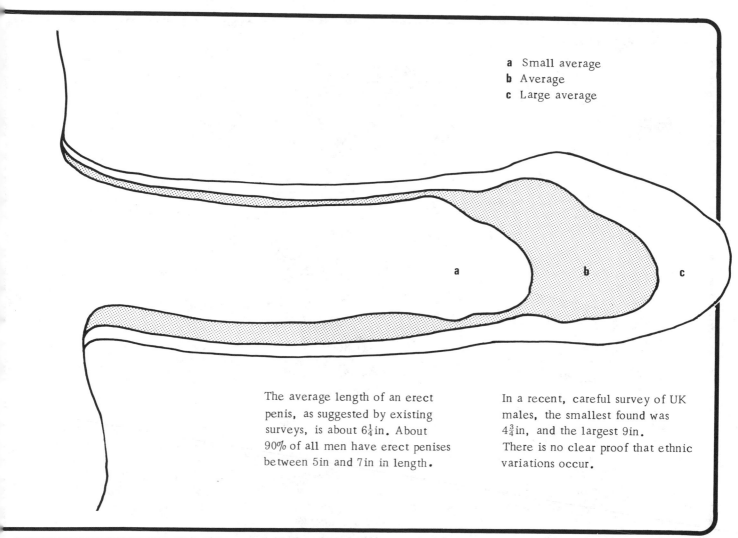

a Small average
b Average
c Large average

The average length of an erect penis, as suggested by existing surveys, is about $6\frac{1}{4}$in. About 90% of all men have erect penises between 5in and 7in in length.

In a recent, careful survey of UK males, the smallest found was $4\frac{3}{4}$in, and the largest 9in. There is no clear proof that ethnic variations occur.

ANGLE OF ERECTION

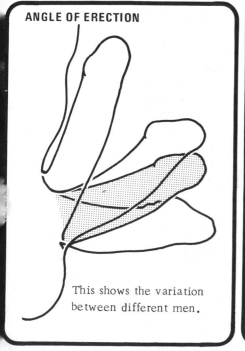

This shows the variation between different men.

APPEARANCE

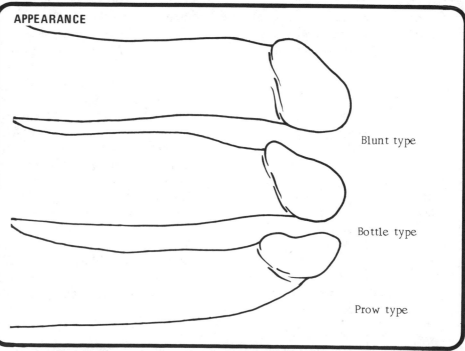

Blunt type

Bottle type

Prow type

© DIAGRAM

FEMALE SEXUAL AND REPRODUCTIVE ORGANS 1

EXTERNAL SEX ORGANS

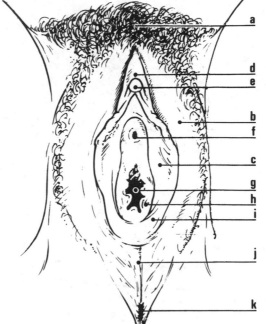

a Mons veneris

d Clitoral hood
e Clitoris

b Labia majora
f Urethra

c Labia minora

g Vaginal opening
h Hymen
i Bartholin's glands

j Perineum

k Anus

Most women are vague about the appearance and function of their sexual and reproductive organs. For unlike those of a man, those of a woman are almost entirely hidden, so that in a standing position the only obvious sign is the pubic hair. Collectively the female external sex organs or genitals are known as the vulva. At the front, as if one were looking between a woman's open legs, is:
a the mons veneris (mount of Venus) or mons pubis, a pad of fatty tissue over the pubic bone. From puberty this is covered with pubic hair. Extending downward and backward from the mons veneris are the
b labia majora (outer lips), two folds of fatty tissue which protect the reproductive and urinary openings lying between them. These outer lips change size during a woman's life and from puberty their outer surfaces are also covered

with hair. Between them lie the
c labia minora (inner lips). These are delicate, hairless folds of skin quite sensitive to touch. During sexual arousal they swell and darken in color (see B03). Below the mons area the labia minora splits into two folds to form
d a hood under which lies the
e clitoris. This is a small, bud-shaped organ and the most sensitive of the female genitals. The clitoris corresponds exactly to the male penis and like it is made up of erectile tissue. During sexual excitement the clitoris swells with blood and for most women is the center of orgasm. Just below the clitoris are the
f urethra - the external opening of the urinary passage which leads direct to the bladder - and the
g vaginal opening, the outside entrance to the vagina.
h The hymen, or maidenhead, is a thin membrane just inside the

vaginal opening. It varies greatly in shape and size, and in a virgin, it may be stretched or torn during the first experience of sexual intercourse, but quite often has already been stretched either by the use of tampons or during petting. If torn during intercourse there is usually some bleeding and possibly pain.
i The Bartholin's or vestibular glands lie either side of the vaginal opening. Contrary to previous belief, these glands play little part in vaginal lubrication. They may occasionally become infected (eg by gonorrhea).
j The perineum is the triangular area of skin lying between the end of the labia minora and the anus. Below its surface are muscles and fibrous tissue that are stretched during childbirth.
k The anus lies below the perineum and is the external opening through which feces pass from the rectum.

98

INTERNAL SEX ORGANS

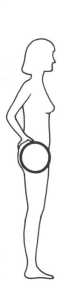

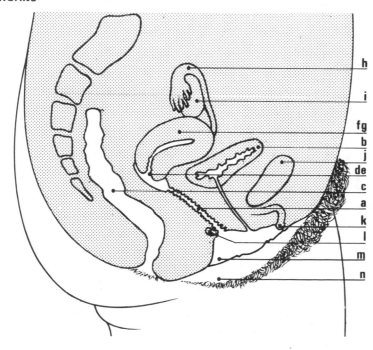

h	Fallopian tube
i	Ovary
fg	Uterus and endometrium
b	Bladder
j	Pubic bone
de	Cervix and os
c	Rectum
a	Vagina
k	Clitoris
l	Bartholin's glands
m	Labia minora
n	Labia majora

These are a woman's reproductive organs and consist of the vagina, uterus, Fallopian tubes, and ovaries.

a The vagina is a muscular passage, lying between **b** the bladder and **c** the rectum. It leads from the vulva upward, and at an angle, to the uterus. It is about 4-5in (10-12.5cm) long and capable of great distension. Normally the vaginal walls, which are lined with folds or ridges of skin, lie close together. During sexual intercourse they stretch easily to take the male penis and extend even more considerably during labor to allow a child to be born. The vagina is usually moist, though moistness increases with sexual excitement and may also vary at different times of the menstrual cycle. A continuous secretion from the cervix and vagina of dead cells mixed with fluid lubricates the vagina, keeping it clean and free from infection. It is this self-cleansing quality that makes vaginal douching unnecessary.

d The cervix is the neck or lower part of the uterus. It projects into the upper end of the vagina and can quite often be felt by sliding a finger as far back as possible into the vagina. This may not be possible at certain times during the menstrual cycle or during sexual excitement if the uterus changes position.

e The os, a tiny opening through the cervix, is the entrance to the uterus. It varies in shape and size depending on whether a woman has had children, but remains very small. It cannot be penetrated by a penis, finger, or tampon.

f The entire uterus (including the cervix) is a hollow, muscular, pear-shaped organ, about the size of a lemon in its non-pregnant state. Seen from the front the uterine cavity is triangular in shape and it is here that the fetus develops during pregnancy, pushing back the muscular walls in a surprising

manner. During labor the fetus moves from the uterine cavity through the cervix and vagina to be delivered through the vaginal opening.

g The endometrium is the mucous membrane lining the body of the uterus. Once a month it undergoes various changes as part of the menstrual cycle.

h The Fallopian tubes extend outward and back from either side of the upper end of the uterus. They are about 4in (10cm) in length and reach outward toward the ovaries.

i The ovaries are the female egg cells, equivalent to the male testes. They produce ova and also the female sex hormones, estrogen and progesterone. Once a month an ovum (egg) is released, which floats freely into the end of one of the Fallopian tubes.

©DIAGRAM

FEMALE SEXUAL AND REPRODUCTIVE ORGANS 2

MENSTRUATION

Around every 28th day, from about the age of 12 to about the age of 47, a woman has a discharge of blood and mucus from the vagina. The discharge lasts from 2 to 8 days (4 to 6 is most usual) and may be preceded or accompanied by various unpleasant symptoms such as headaches and nausea.

This, of course, is menstruation, or "the period" - the outward sign of the routine cycle of egg production and hormone change in a woman's body. It is a process that requires the wearing of pads or tampons (absorbent tubes placed in the vagina), if the menstruating woman is to avoid soiling her clothes.

EGG PRODUCTION

Each ovary contains groups of cells called follicles, which themselves contain immature eggs (ova). When the girl is about 12 these eggs begin to mature at the rate of one every 28 days or so - usually in alternate ovaries. (At birth, a female child's ovaries contain perhaps 350,000 immature eggs. Between puberty and menopause, only about 375 ever mature.) As each egg matures, it bursts from the ovary - a process called ovulation - and passes into the Fallopian tube leading down from that ovary to the uterus.

PROCESS OF MENSTRUATION

If the egg is not fertilized by a sperm, it begins to degenerate 24 to 48 hours after leaving the ovary, and eventually passes unnoticed out of the body in the normal flow of fluid from the vagina. But meanwhile the uterus has been preparing to receive a fertilized egg. Hormones have caused the lining of the uterus to thicken, and to excrete a fluid so that the fertilized egg could be nourished while implanting itself. When no fertilization occurs, further hormone stimulation causes the thickened lining to crumble, and to be discharged along with a little blood, through the vagina. This process is called menstruation.

PROBLEMS OF MENSTRUATION

PREMENSTRUAL DISCOMFORT
Symptoms (most noticeable in the 7 days before the start of the period) can include headaches, backache, nausea, breast tenderness, psychological tension, and depression. Hormone treatment is used in some extreme cases.
PAINFUL PERIODS (DYSMENORRHEA)
Two types of dysmenorrhea are distinguished - spasmodic and congestive. Spasmodic dysmenorrhea begins with the onset of the period and involves pain in the lower abdomen ("cramps"), thought to be caused by contractions of the uterine muscle. Congestive dysmenorrhea is felt as a dull ache just before the period. Spasmodic period pains often disappear after pregnancy, while congestive pains can persist until the menopause.
IRREGULARITY Menstruation is often irregular during adolescence. Some adult women also find, however, that the duration of their menstrual cycle varies from month to month. Regular cycles can vary from between 21 to 35 days.

ABSENCE OF PERIODS
(AMENORRHEA)
There are two types of amenorrhea - primary and secondary. If a girl reaches the age of 18 without experiencing menstruation she is said to be suffering from primary amenorrhea. This rare condition may be the result of an endocrine abnormality and must be investigated by a doctor.
Secondary amenorrhea is the term used to describe the absence of periods in a woman who has already begun to menstruate. This may be quite normal: menstruation does not occur in pregnant women, and sometimes does not recommence until some weeks after the birth, especially if the mother is breast feeding her child.
Secondary amenorrhea, however, can also be caused by emotional stress such as shock, fear, tension, or depression, and also by endocrine disorders, illness, drug-taking, traveling, and poor general health.

HEAVY PERIODS
During a menstrual period, a woman usually sheds between 2 and 4 tablespoons of blood. Some women, however, discharge considerably more.
If menstrual bleeding is heavy or prolonged (and this often happens to women fitted with IUDs), iron-deficiency anemia can result. This can often be remedied by an iron-rich diet or a course of medicinal iron.
Heavy bleeding (and bleeding between periods) can sometimes be symptomatic of problems such as hormone disorders, fibroid tumors, or cancer of the uterus. Also, occasionally heavy periods can be caused by psychological factors. Metropathia hemorrhagica is one of the conditions caused by hormonal imbalance which can result in very heavy bleeding. If the condition does not respond to curettage. a hysterectomy may be needed. In all cases of excessive bleeding a doctor should be consulted.

JOURNEY OF THE EGG

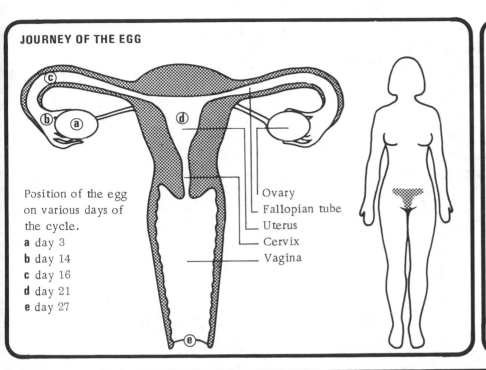

Position of the egg on various days of the cycle.

a day 3
b day 14
c day 16
d day 21
e day 27

Ovary
Fallopian tube
Uterus
Cervix
Vagina

MYTHS ABOUT MENSTRUATION

Throughout history, almost all societies have surrounded the menstrual process with myth and ritual. Even today, in some primitive cultures, the menstruating woman is thought to turn milk sour, turn food bad, damage crops, and even cause animals to abort! Elsewhere she may be completely isolated from the rest of the community in a special building. Modern Western society still preserves some old myths about menstruation – all of which can be ignored. It is perfectly safe for the menstruating woman to bathe, shower, swim, wash her hair, have intercourse, and take part in any other activity she wishes.

NORMAL MENSTRUAL CYCLE

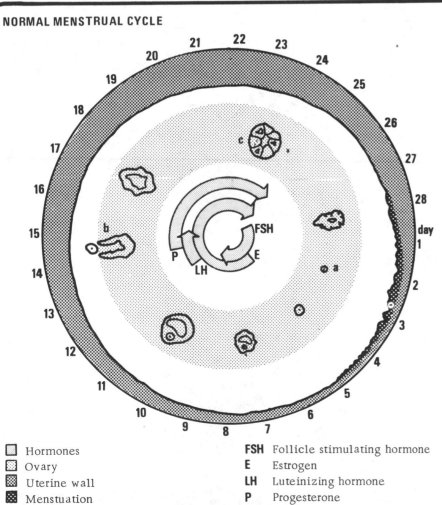

▨ Hormones
▨ Ovary
▨ Uterine wall
▨ Menstuation

FSH Follicle stimulating hormone
E Estrogen
LH Luteinizing hormone
P Progesterone

Day 1 onward: pituitary is already producing the follicle stimulating hormone (FSH) – a new egg begins to mature in one of the follicles (a).

Day 4 onward: the follicle produces estrogen (E). As this builds up, it stimulates growth of uterine wall and breasts; halts FSH production; and stimulates the pituitary to release luteinizing hormone (LH).

Day 12 onward: LH causes the follicle to burst (b), releasing the egg; and makes the follicle develop into a "corpus luteum," producing progesterone (P) as well as estrogen.

Day 14 onward: P makes the uterine wall prepare for a fertilized egg; and halts LH production. Without LH support, the corpus luteum degenerates (c), E and P levels fall, and eventually the uterine lining breaks up – menstruation (day 28).

E no longer inhibits FSH, and the cycle begins again.

©DIAGRAM

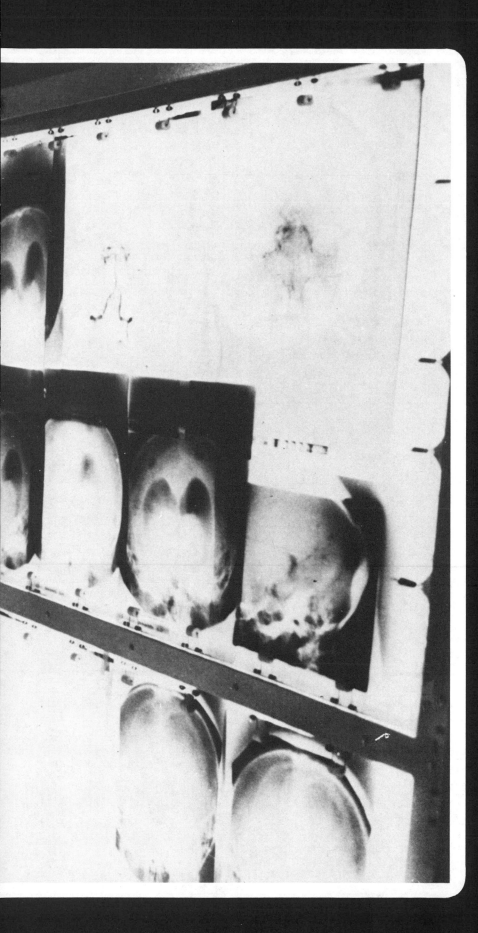

3 Illness and disease

The body is a complicated and fascinating machine, requiring careful maintenance and skilled diagnostic attention when something goes wrong.

Left: A team of diagnosticians examining skull x-rays (Camera Press Ltd.)

ILLNESS

ILLNESS

Every year, the population of the USA loses about 3,500 million days with normal activities restricted because of illness. Over a third of these are actually spent ill in bed. The number of entire days lost from work due to illness (i.e. days on which the person did no work at all), is almost 500 million. Almost 250 million entire days are lost from school. The average person has about 6½ days in bed ill every year, and has over 5 entire days off sick from either work or school. There is a slight difference between men and women. The average male (all ages) spends 5½ days a year ill in bed, the average female 7½.
In all countries, illness is by far the main cause of loss of working activity. In the UK, for example, about twice as many days are lost from injuries as from strikes, and about nine times as many from illness as from injury.
Days lost due to illness rise steadily with age. The average 40 year old loses almost twice as many as the average 20 year old, and the average 60 year old over 7 times as many. Yet the number of illnesses per person is higher in the lower age groups. In other words, a young man has several separate days off ill, scattered over the year; an old man has more days off, but they are more likely to be in one block, due to a single illness.

ACUTE

Frequency of different sicknesses changes with medical advances and social habits. This table shows some sicknesses for which frequencies have changed most in the last 25 years. Measles, whooping cough, poliomyelitis, and tuberculosis have been greatly reduced by immunization. Better drugs and careful control have kept syphilis in check. But hepatitis has risen with drug abuse, salmonella food poisoning with the rise of carry-out cooked food, and gonorrhea with changing sexual behavior.

■ 1950
▨ 1972

SICKNESS AND AGE

Showing the percentage with incidents of illness in a year in different age groups.
a Infective illnesses
b Upper respiratory illnesses
c Other respiratory illnesses
d Illnesses of the digestive tract
e Injuries

CHRONIC

There are over 25 million people in the USA who suffer badly enough, from some long-lasting disability, for their activities to be restricted.
Of these people, 73% are restricted in their major activity (whether work, school attendance, or housework) - including 18% who cannot carry it out at all.
There are slightly more disabled men than women, but much more dramatic is the way in which disablement increases with age. About a third of all disabled are 65 or over. This age rise applies both to those restricted in their major activity, and to those suffering from less significant limitation.
Causes of disability include: the standard causes of death (heart conditions, cancer, respiratory disorders, etc); sensory disorders (eyesight and hearing); impairments of movement (paralysis, arthritis, back disorders, etc); mental and nervous conditions; isolated "trouble spots" (ulcers, hernias, varicose veins, hemorrhoids, etc); and disorders of certain body systems (ranging from diabetes to digestive trouble and asthma).

The main causes of disability are (listing the most important first): heart conditions; arthritis and rheumatism; visual impairments; high blood pressure without any heart condition; and mental and nervous conditions. These are fairly equally distributed between the sexes, except that women are twice as likely to suffer from arthritis and rheumatism and (though it is far less important in both sexes) from high blood pressure without heart condition. (Men are rather more likely to suffer from visual impairment.)

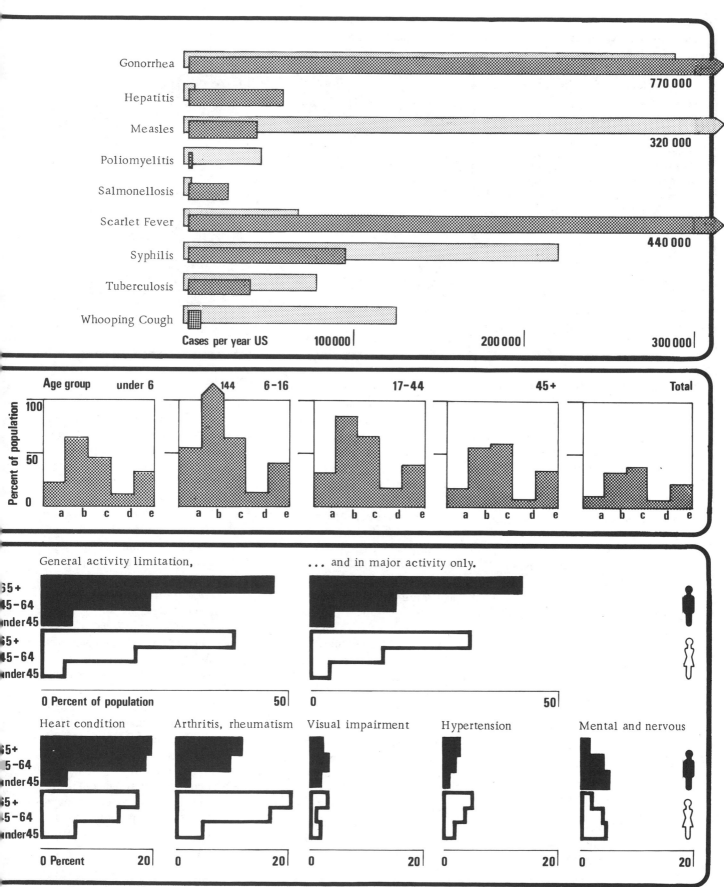

Gonorrhea **770 000**

Hepatitis

Measles **320 000**

Poliomyelitis

Salmonellosis

Scarlet Fever **440 000**

Syphilis

Tuberculosis

Whooping Cough

Cases per year US   100 000   200 000   300 000

Age group   under 6   144   6-16   17-44   45+   Total

Percent of population
100
50
0
a b c d e   a b c d e   a b c d e   a b c d e   a b c d e

General activity limitation,   ... and in major activity only.

65+
45-64
under 45

65+
45-64
under 45

0 Percent of population   50   0   50

Heart condition   Arthritis, rheumatism   Visual impairment   Hypertension   Mental and nervous

65+
45-64
under 45

65+
45-64
under 45

0 Percent   20   0   20   0   20   0   20   0   20

©DIAGRAM

105

BODY AND INFECTION 1

INFECTIOUS ILLNESSES

Infection is not the only cause of illness. Other causes include: inborn defects; metabolic disorders; developmental changes; degenerative processes and malignant growths; nervous conditions; poisons; nutritional disorders; and irritation by external sources, whether mechanical, chemical, thermal, or from radiation. All these can also interact to support each other. However, the illnesses that we most expect to have to deal with, in day to day living, are the infectious ones; and they are still among the major causes of death, in the less developed parts of the world.

ATTACK AND DEFENSE

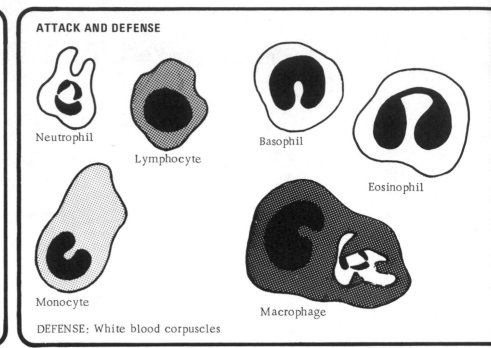

Neutrophil

Lymphocyte

Basophil

Eosinophil

Monocyte

Macrophage

DEFENSE: White blood corpuscles

AGENTS OF INFECTION

Infectious illnesses occur when certain microscopic living organisms gain access to the body. The symptoms of illness arise from their effect on the body, and from the body's attempts to cope with them. The creatures involved are usually single-celled rather than many-celled, and mostly either bacteria or viruses. In fact, bacterial infection is the source of most infectious illness.

BACTERIA

These are a form of tiny single-celled plant life, round or rod shaped, and each from one to 20 thousandth of a millimeter in diameter. They consist simply of an outer wall inside which there is protoplasm and DNA. Most are incapable of any independent movement. They almost always reproduce simply by dividing in two; and many can form themselves into spores - a seed-like inactive state, in which they can survive adverse conditions. The conditions they prefer vary greatly between the different types, but mostly they

do not like too great heat or cold, and like moisture which is not too acidic.

Bacteria commonly occur in vast numbers in almost every corner of life - including on and in man's body. Most are utterly harmless to man, some he is dependent on, and some no life forms could exist without. Bacteria, for example, play a vital part in the body (eg in digestion, vitamin manufacture, and destruction of dangerous substances), while all life depends on bacteria in the air and soil, without which dead matter would not decay and return into the cycle of existence.

Illnesses from bacteria arise in two ways. Firstly, because bacteria that normally exist on, or in, the body - and may be very useful - get into the wrong part of the body. Examples of this include acne, pimples, and boils; some meningitis; and many urinary infections (especially in women). The first group are caused by normal skin surface bacteria

gaining entry into a sweat duct or skin wound; the second by throat bacteria gaining access to the brain; and the third by bacteria from the rectum finding their way into the urinary tract.

Secondly, illnesses can occur because bacteria that are always harmful gain access to the body. Examples of illnesses that are always carried by one specific organism include scarlet fever, tuberculosis, whooping cough, typhoid, syphilis, and gonorrhea. Examples that can be caused by a range of bacteria include tonsilitis, dysentery, most pneumonia, and "food poisoning". In fact, certain types of bacteria have so far evolved, from being free living, that they are dependent on other living cells for their very existence. Some of these are useful or harmless - but others are among the most dangerous to man. Outside the living cell they either die immediately (eg syphilis bacteria) or have to form spores (eg tetanus).

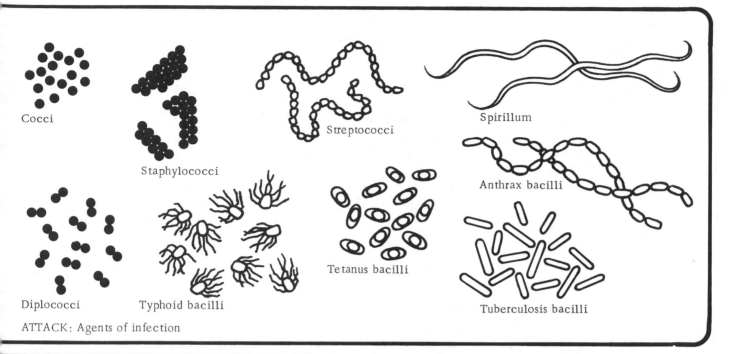

Cocci

Staphylococci

Diplococci

Typhoid bacilli

Streptococci

Tetanus bacilli

Spirillum

Anthrax bacilli

Tuberculosis bacilli

ATTACK: Agents of infection

VIRUSES

If bacteria seem small enough, they are giants compared with viruses. These are the most primitive form of life we know: each a minute quantity of nucleic acid, wrapped up in a protein sheath. They are also the ultimate parasites. They can survive well enough in many conditions, but can only become active and reproduce inside another living cell, where there are enzymes which they lack. Having gained entry to a cell (sometimes by a syringe-like injection process), a virus takes over the cell's chemical processes, and uses it to produce hundred of new viruses. Finally the cell breaks open and dies, spilling out the new viruses to enter into other cells. Sometimes the cells attacked are easily replaced - as in the nose lining, for example. But usually the viral attack at least interferes with the body processes, and may cause irreparable damage - for example, if nerve cells are destroyed.

Illnesses caused by viruses fall into two groups: those that attack particular organs (eg influenza and the respiratory system, mumps and the salivary gland, polio and the nervous system); and those that cause general symptoms, often with a skin rash (eg measles, rubella, chicken pox, smallpox, and yellow fever).

RICKETTSIAE

These are intermediate between viruses and bacteria. Most live in the intestines of insects, and can infect a human being if the insect is a parasitic blood-sucker. Typhus fever and Rocky Mountain spotted fever are both caused in this way.

TOXINS

Toxins are not organisms in themselves - they are immensely powerful chemical poisons, produced by certain bacteria when active in the human body. For example, it has been estimated that one 6,000th of an ounce of pure botulin (the cause of botulism food poisoning) would be enough to kill the entire population of the world. Other toxic infections

include tetanus and diphtheria. In many toxic illnesses, the bacteria themselves are not harmful to the body - only the substance they produce, if it is not neutralized.

OTHER ORGANISMS

Other single-celled organisms, much larger than bacteria, can cause infection, but such illnesses are mostly found in tropical or subtropical climates, where they are often fatal.

Examples include amoebic dysentery, malaria, and sleeping sickness. Some of these are caused by parasites, which have developed a life cycle that depends on passing part of their existence in another creature, and part in man.

Multi-celled organisms may also infect man, eg pinworms, tapeworms, and flukes. More common are infections by fungi - a type of plant that has no chlorophyll, and so must obtain its food from organic material. One example is athlete's foot (ringworm).

©DIAGRAM

BODY AND INFECTION 2

HOW INFECTIONS OCCUR

ROUTES INTO THE BODY

Infective agencies have four main routes into the body:

a) through breaks in the skin, or in the mucous membrane that lines the mouth, nose, etc (the breaks may be wounds, that germs happen to enter, or bites inflicted by the insect that brings the infection);

b) down the respiratory tract into the lungs;

c) down the digestive tract, into the stomach and bowels; and

d) up the reproductive and urinary systems, via the genitals.

In all cases, the infective agent may remain localized at its point of entry, or may enter the blood or lymph system and be distributed through the body. For example, infection of a wound may result in an abscess filled with pus; and/or may spread to the surface, possibly infecting other flaws in the skin; and/or may travel up the lymph canals to the regional lymph nodes - perhaps being trapped there and causing a further abscess. Serious blood-borne infections are rare, but blood-borne bacteria do often attack already damaged heart valves.

SOURCES OF INFECTION

The most frequent source of common infections is inhalation of water droplets carried in the air. Breathing, speaking, coughing, and sneezing, all spread droplets of saliva, sputum, or secretion into the air. These can bear bacteria, and can be breathed in (or taken in with food) by other people. Ordinary breathing can spread droplets over a range of 4ft, and loud speaking over about 6ft, while a sneeze can spread 20,000 droplets over a distance of up to 15ft. Infections spread in this way include colds, influenza, sore throat, scarlet fever, diphtheria, measles, mumps, whooping cough, meningitis, and tuberculosis.

Other sources of infection include:

a) inhalation of dried bacterial spores, in dust-carrying air (anthrax);

THE BODY'S DEFENSES

The body has three levels of natural defense to infection. It tries to prevent foreign organisms from entering the body's tissue. It tries to kill the ones that it cannot prevent. It tries to render harmless the ones it cannot immediately kill.

BARRIERS TO INFECTION

The main barrier to infection is the skin surface - a physical barrier that cell repair is constantly trying to maintain, and that few foreign organisms can penetrate when it is unbroken. The skin also secretes antiseptic substances in its sweat, so that not many infectious agents are able to survive on it for long. (Tear fluid does the same job for the crevices around the eyelid.) Where the orifices of the body form necessary openings, their lining of mucous membrane also presents a physical barrier and a trap, coated as it is with antiseptic substances in a layer of mucus. The digestive and respiratory tracts are also guarded by a ring of lymph tissue, around the pharynx, at the back of the mouth. Finally, if outside organisms do get beyond this, they are usually either caught in the layer of mucus that coats the respiratory tract, or destroyed by hydrochloric acid in the stomach juice.

PHAGOCYTIC CELLS

Certain cells in the body are "phagocytic" - they recognize and eat intruders. If infective agents do penetrate the body tissue, they may be of a kind that these cells can deal with. The white corpuscles in the blood are mobile phagocytes. Others are fixed at points throughout the body - especially in the spleen and liver, where they can filter infection out of the blood, and the lymph nodes, where they can filter the lymph circulation.

ANTIBODIES

Some foreign organisms the body cannot destroy immediately. But it may be able to neutralize them - the process is a chemical one. Bacteria and viruses have an outer sheath of protein. To neutralize them, the body manufactures another protein, called an

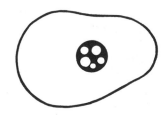

1　　　2　　　3

Stages in the phagocytic digestion of a bacterium

108

b) direct physical contact with an infected person (venereal disease, some skin conditions);
c) contact with "fomites", i.e. intermediate objects, such as clothing and eating utensils, that someone else has infected;
d) eating or drinking infected food and liquid (dysentery, typhus, cholera, brucellosis);
e) entry of soil or dust into a wound (tetanus, gas gangrene);
f) the bites of parasitic insects, such as the mosquito (malaria), tsetse fly (sleeping sickness), rat flea (bubonic plague), and louse (typhus fever);
g) the bites of infected animals (rabies), or contact with infected animals (brucellosis);
h) insufficiently sterile medical procedures, as in surgery and hypodermic injection (hepatitis);
i) infection carried by the mother in her blood stream and passed on to the fetus during pregnancy.

Infection from a human origin can come from someone who has no symptoms. With many infections, someone who has had, or is about to have, the infection, can infect others. With some, a person who never shows signs of the illness can infect others: the person is immune, but "carries" the infection (typhoid, diphtheria, cholera, dysentery).

The final cause of infection is self-infection: the transference of bacteria from a part of the body where they are harmless (such as skin surface or rectum) to a part where they can cause infection (such as a wound). This is mostly caused by lack of hygiene (especially of the hands), and is often compounded by scratching. Resulting illnesses are not usually passed on to others.

With all sources of infection, it is very important how long the organisms have to multiply, before they enter the body. The taking in of a few isolated bacteria is not normally going to cause illness.

antibody. Antibodies are formed in the lymphoid tissue, and released into circulation. The molecules of the antibody interlock with those of the protein sheath, as pieces of a jigsaw interlock. This is then coated with other proteins, called "complement", always present in the blood, and the whole mass can be eaten in the usual way by phagocytic cells. However, antibodies are "specific" - i.e. they can usually only cope with one particular organism. For another organism is almost always sheathed in another protein; and another protein requires another antibody, for the correct chemical neutralizing reaction. Again, it is like a jigsaw - only certain pieces

interlock; and complement, the general immunizer, cannot act until this interlocking has been done.

This means that the body has to learn anew how to neutralize each new threat; and if the threat comes in overwhelming numbers, the body has no time to learn. Sometimes the response of producing the correct antibody has been acquired from the mother. But otherwise the body needs mild contact with the threat first, so it has a chance to set up the right process. Then, in the face of a serious threat, it can immediately put that protein into mass

production. For the body does not have to learn anew how to neutralize old threats as they recur. It recognizes them, and remembers what was done before; and sometimes the memory lasts for life.

Incidentally, allergies are over-violent antibody reactions to organisms that are not dangerous but that the body regards as foreign.

CELL DEFENSES

Antibodies are carried in the body fluids, not in the cells. But some cells can produce substances that will attack organisms that enter them. Chemicals for destroying the cold virus, for example, are produced in the mucous membrane of the nose.

4   5

6

© DIAGRAM

BODY AND INFECTION 3

DEFENSE CELLS IN ACTION

Neutrophils, formed in the bone marrow, are the first white cells to arrive (a). They are followed by monocytes (b), and then by lymphocytes (c). The lymphocytes increase in size as they come into contact with the bacteria, while the monocytes concentrate on consuming the dying neutrophils, and so turn into ever larger macrophages (d). Both monocytes and lymphocytes also play a part in antibody formation.

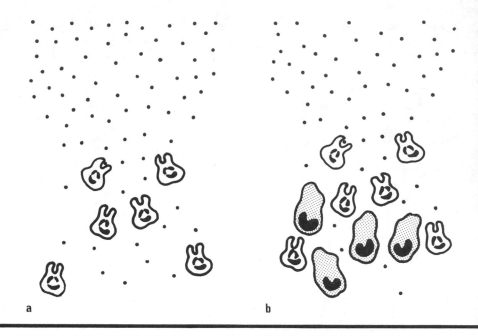

a   b

DEFENSE PROCESS

If the infective agent is not in the blood stream, the white corpuscles and antibodies must get to the point of infection and fight their battle there. The body brings this about by what is called the "inflammatory response". Any damage to body cells causes them to release "histamine" - a substance that automatically increases blood flow to the area, by widening the small arterial branches and the capillary openings. Also gaps appear, between the cells that form the capillary walls; and, through these gaps, blood plasma, containing white corpuscles and antibodies, escapes into the neighboring tissue, until the fluid pressure between tissues and capillaries is equalized.

The fluid dilutes the infectious organisms; the white corpuscles eat the organisms and damaged body cells; antibodies neutralize what organisms they can; and the mixture of cell debris and white blood cells forms the familiar fluid called pus. (If the amount of pus is large enough, and not drained away by the body, it forms a massive accumulation called an abscess.) Meanwhile the area is walled off by a blood clot, which gradually forms into connective tissue, completing the isolation of the area. Any organisms that escape into the blood or lymph vessels are killed off in the liver, spleen, and lymphatic nodes. Finally, after the infection is overcome, healing begins. The fibrous wall is digested away, and body tissue grows inwards to repair the damage.

It is the process of the inflammatory response that causes the standard symptoms of a local infection: redness, swelling, heat, and pain. The redness and heat arise from the increased blood supply, the swelling from the build-up of fluid, the pain from the swelling. Throbbing may also be felt, from the pulse of blood in the neighboring arteries.

All this is familiar for surface infections. But even in an infective invasion deep within the body, the same basic principles are involved.

More general effects may also be felt throughout the body, especially if the infection is serious. The numbers of a type of white blood corpuscle may increase rapidly from 3,000 or 4,000 up to 20,000 per cubic millimeter; the body temperature rises; and the pulse rate rises by about 10 beats a minute for every degree Fahrenheit rise in temperature. Together with flushing, as surface blood vessels enlarge, these give the familiar symptoms of illness.

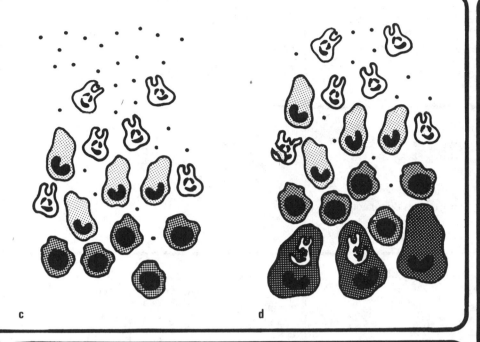

c

d

COURSE OF INFECTION

An infectious illness may be either acute or chronic. Acute illnesses are brief, with suddenly developing symptoms. They soon end with the illness defeated or the patient dead. Chronic illnesses are long drawn out. Most specific diseases always produce one type of illness rather than the other.

As all infections involve the same basic process - invasion by a foreign organism, and the body's response - so all tend to show a common pattern (though the stages are more obvious in an acute illness).

a) Incubation, beginning with the moment of infection. The organisms multiply inside the body.

b) Prodrome, a short interval of generalized symptoms (eg headache, fever, nasal discharge, irritability, and general feelings of illness). The infection at this stage is usually very communicable to others - but diagnosis is difficult.

c) Peak, when the illness is at its height. Each infection shows its own characteristic symptoms. Temperature is at its highest, and the rash appears where relevant.

d) Termination, the end of the illness. This is marked either by a "crisis" - a period of 12 to 48 hours in which the symptoms rapidly disappear - or by a "lysis", a more gradual termination.

e) Convalescence, after the illness. The patient regains health and strength.

Note that, from the micro-organism's point of view, a successful infection is one strong enough not to be killed by the body's defences, yet not strong enough to kill off its host. Chronic infections, such as syphilis and tuberculosis, show this pattern: the organism is just able to hold its own against the body's defences. Acute infections often spread rapidly through a population and then die out, because everyone is either dead or recovered.

MEDICINE AND INFECTION

ARTIFICIAL IMMUNIZATION

The antibody process shows how our defences against an infection are strengthened if we have already met the infection in mild form. This principle underlies artificial immunization. A vaccine is given to the patient, usually by injection. It gives him only slight symptoms, if any, but makes him immune to a specific disease.

Vaccines may use:

a) a related, weak infection (eg using cowpox to immunize against smallpox);

b) the dead organism of the disease (eg whooping cough, influenza, and Salk polio vaccines);

c) a weakened organism of the disease (eg BCG tuberculosis, measles, plague, and Sabin polio vaccines); or

d) (where the threat is from toxin rather than infection) a "toxoid", i.e. a toxin that has been made non-poisonous, but can still stimulate anti-toxins in the body.

The protection of many vaccines lasts a lifetime. Immunization can also be achieved less efficiently and for a shorter period by injecting a ready-made antibody or anti-toxin (eg against diphtheria, or tetanus).

ANTIBIOTICS

Immunization protects against future infection; antibiotics fight an existing one. Micro-organisms have to compete with each other for scarce food supplies, so some have developed the ability to produce chemicals that kill off their rivals. These chemicals form the bases of antibiotics. However, resistant strains of bacteria often develop, helped by their rapid reproduction rate; and against viruses, less progress has been made, because their similarity to the human cell makes it hard to destroy one without the other.

©DIAGRAM

INFECTIOUS DISEASES

INFECTIOUS DISEASES

	INCUBATION	SYMPTOMS	TREATMENT	QUARANTINE
Gastro-enteritis	0-24 hours	Continual vomiting, diarrhea, irritability, dehydration.	In babies: reduce feeding, or give clear liquids, eg dextrose in water.	Isolation from infants
Bacterial meningitis	1-10 days	In young babies: lassitude, irritability, poor feeding, fits, fever, vomiting, possible increase in head size. In older children: headache, fever, vomiting, convulsions, neck rigidity, rash.	Immediate antibiotics.	Yes, until recovered.
Viral meningitis; Encephalitis	0-7 days	As for bacterial meningitis.	Supportive treatment. Antibiotics to prevent recurrence.	Yes, until recovered.
Coughs and colds	0-48 hours	Stuffy or runny nose, raised temperature, possibly vomiting.	In babies: supportive treatment and in severe cases antibiotics in liquid form to prevent pneumonia.	Isolation from infants
Rheumatic fever	1-6 weeks	Sore throat, temperature, diffuse rash, pains in limbs and large joints, may be small rheumatic nodules beneath the skin.	Bed rest, aspirin, prolonged use of penicillin in large doses, and occasionally steroid hormones.	Yes, for 2 weeks after start of treatment.
Diphtheria	2-5 days	Fever, sore throat, swollen neck glands, may be a dark, offensive-smelling membrane at the back of the throat.	Heavy doses of antitoxin, and sometimes antibiotics.	Yes, until free of symptoms.
Poliomyelitis (Polio)	10-12 days	Minor: fever, headache, diarrhea, vomiting (this stage lasts 1-2 days). Major (beginning 7 days later): fever, signs of meningitis, tender muscles, severe restlessness. May go on to paralytic stage: weak muscles, asymmetrical paralysis possibly affecting breathing. Improves after 7 days but may leave residual disability.	Bed rest and sedatives. In paralytic stage, massage of affected muscles and possibly use of a respirator.	Sufferers: 6 weeks isolation. Contacts: 3 weeks isolation.

	INCUBATION	SYMPTOMS	TREATMENT	QUARANTINE
Chicken pox	4-7 days	Chill, fever, headache, malaise. Red spots - on face, chest, back - that later contain clear fluid, burst, and develop brown crusts.	For symptoms only: rest, and lotion for itching.	Yes, until appearance of crusts over lesions.
German measles	14-21 days	Malaise, fever, headache, inflamed mucous membrane. Fine pink spots first on face and neck, and then elsewhere.	Little or none. Lotion for itching.	Yes, until rash disappears.
Measles	8-13 days	Fever, cough, conjunctivitis. All-over rash appears later - white spots with red perimeter, along with inflamed background skin.	For symptoms: rest, cough syrup, soft diet, protection from cold, damp, and bright light. Treatment also for any complications.	Isolation from infants.
Mononucleosis (Glandular fever)	Not known	Headache, fever, sore throat, swollen lymph nodes. Loss of appetite.	For symptoms: rest, mouth wash, aspirin. Treatment for complications, if any.	Yes, until temperature has been normal for a week.
Mumps	12-20 days	Chill and fever, headache, temperature, swollen salivary glands (pain on chewing). Other glands may be swollen.	For symptoms: rest, soft diet, aspirin, perhaps sedatives. Also treatment for complications, if any.	Isolation from men and youths who have no immunity.
Roseola	4-7 days	High fever 3-4 days. Convulsions. Enlarged spleen. Later, purple-brown spots on chest, abdomen, face, and extremities.	For symptoms only: aspirin, water sponging to lower temperature.	No
Scarlet fever	1-3 days	Chills, fever, vomiting. Rash 24 hours after fever: small red spots join to form redness on whole body. Strawberry tongue. Sore throat.	Penicillin, rest, soft diet, water sponging to lower temperature, lotions for itching.	Yes, for not less than 7 days after onset.
Whooping cough	7-14 days	Sneezing, listlessness, and cough becoming convulsive with typical whooping breathing. Vomiting. May expel thick mucus.	Rest, fresh air, small meals, refeeding after vomiting. Mild sedatives and antibiotics. Hospital for serious cases.	Yes, for 21 days after onset of cough.

©DIAGRAM

MALE INHERITED DISORDERS

INHERITED DISORDERS

A few disorders exist, that usually occur only in men, but that they have always inherited through their mothers. These include hemophilia, red-green color deficiency, and two forms of muscular dystrophy.

Of course, many defects can be inherited. Each chromosome inherited from a parent carries many thousands of "genes" or units of genetic information. If any one of these genes is faulty, it will not pass on the correct instructions, and a defect can occur.

However, these few disorders such as hemophilia are passed on in the odd way described because they are linked with the X chromosome. This is one of the chromosomes that determine sex; but other genes on the same chromosome have other jobs - including helping to ensure normal color vision, blood clotting, and so on.

Both men and women have X chromosomes, and both men and women can have X chromosomes in which one of these genes is faulty. But whenever the defective chromosome is matched by another, normal X chromosome, the defect will not appear: the correct function (eg color vision) is guaranteed by the normal chromosome i.e. the normal gene is "dominant".

So in any woman the defective chromosome is normally masked by a healthy one, from the other parent. And a father with the defective gene cannot pass it on to his sons at all, because to them he contributes only his Y chromosome.

But a mother can pass it on to her sons, because to them she contributes their X chromosome, which may be defective, and their other, Y, chromosome, will not "mask" it, because it does not have a gene responsible for the defective function.

The only way in which a woman can show the signs of one of these defects is if she has inherited defective X chromosomes from both sides of the family. This is very unlikely, but does happen

rather more often in the case of color-vision deficiency.

A woman who does not herself show signs of the disorder, but can pass it on, is called a "carrier". Only chance decides whether or not any one of her children inherits the defective X chromosome: the child can equally inherit the healthy one. So if a carrier becomes pregnant, there is: a one in four chance of her having a normal son; the same chance of her having a normal daughter; the same of her having an affected son; and the same of her having a carrier daughter.

If a woman is found to be a carrier, there is risk not only to her own subsequent children, but also to those of her female relatives on the maternal side - because they may also have inherited the defective gene.

If no previous family history is discovered after careful check, it is likely to be an isolated mutation in either mother or child. If in the mother, she can still pass it on to subsequent children.

HEMOPHILIA

This disorder is characterized by uncontrollable bleeding, even after slight wounds. It is caused by a deficiency of one of the elements needed to make the blood clot. There are about 8 cases per 10,000 of the population, and about 3 to 4 severe cases per 100,000.

SYMPTOMS

In mild cases, the disorder may remain undiscovered until revealed by some incident (eg loss of a tooth). In severe cases, it will be obvious soon after birth. There is persistent blood flow from any cut, or in any bruise. Without special treatment this may

continue for hours or even days, despite normal attempts to stop it. Even when the flow does finally stop, it may recur soon after. The real danger, however, is from internal bleeding. Superficial cuts and scratches are not threatening unless the mucous membrane is involved, and deep cuts do not kill unless a major vein or artery is involved. But bleeding in soft tissues, such as the kidney, is serious, and bleeding in large joints can eventually cripple them. Both these can occur spontaneously in severe cases.

The symptoms may decline with age, and at any age there may be

periods free from trouble. Laboratory tests are necessary for a definite diagnosis.

TREATMENT

The hemophiliac has to take special care in all he does, and in severe cases his sphere of activity is drastically curtailed. Any sports involving bodily contact or danger of injury must be avoided (though he can still swim, run, etc.)

Hemophiliacs often wear warning tags, in case they are involved in accidents. If bleeding does occur the missing factor is injected intravenously to help clotting. This is also done if an operation or tooth extraction is necessary.

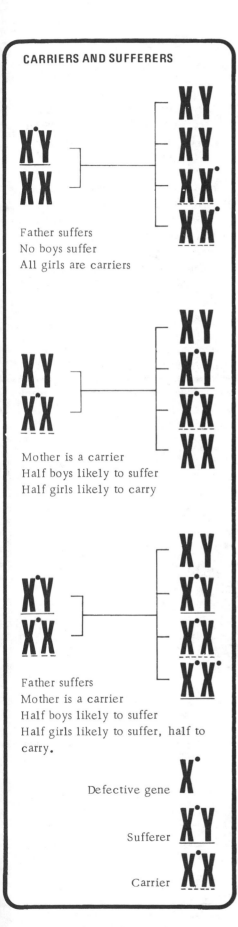

CARRIERS AND SUFFERERS

Father suffers
No boys suffer
All girls are carriers

Mother is a carrier
Half boys likely to suffer
Half girls likely to carry

Father suffers
Mother is a carrier
Half boys likely to suffer
Half girls likely to suffer, half to carry.

Defective gene

Sufferer

Carrier

MUSCULAR DYSTROPHY

This is a disease in which the muscles waste away. Muscle tissue does not replace itself, and slowly gives way to fibrous tissue and fat. It may be due to absence or excess of protein, or presence of abnormal protein. There are several forms of the disease. All are usually inherited, but two are sex linked - inherited almost only by men, and through female carriers. The carriers themselves may have slight muscle weakness, but are usually apparently normal.

DUCHENNE TYPE
This is the most common and most severe form. Half those with muscular dystrophy are boys with the Duchenne type. Most are dead by the time they are 25. It is invariably fatal.
The first symptoms develop between the age of 2 and 5. Walking is clumsy, running poor, falls frequent. Later, climbing stairs and getting up after falls become difficult. Weakness begins with certain muscles of the shoulders, upper arms, and thighs. Diagnosis is by measuring enzymes in the blood serum, and examining small muscle samples under a microscope. (If muscular dystrophy is suspected in a family, these tests can also diagnose it within a few days of a child's birth).

There is at present no cure or effective drug treatment.
Exercise and muscle stretching can slightly slow the progress of the disease, but eventually (usually between 8 and 11) the child has to take to a wheelchair. The spine curves, muscle weakness spreads, eventually affecting even eating and drinking, and muscle contraction distorts limb positions. Finally respiratory and heart muscles are involved, and death occurs, usually between 16 and 25.
There are tests that prove a woman is a carrier (though none can prove she is not).

BECKER TYPE
This is similar in the muscles affected and pattern of inheritance, but rarer, later, slower, and milder. Patients can usually still walk in their thirties and often into middle age.

COLOR VISION DEFICIENCY

Color vision deficiency is much more common than is usually realized, and ten times more so in men than in women. Estimates range from 2 to 8% of all men, so millions suffer without realizing it. It usually results from a hereditary defect in the structure of the eye. There are various forms, but more than half of all cases have "red-green" color deficiency, and this is one of the types that are linked with sexual inheritance. In this condition, all three normal pigment responses occur in the eye, but there is difficulty in telling red, green, and yellow apart - all are likely to appear grayish. This can occur in varying degrees, ranging from slight graying from a distance to identical response to vivid colors close at hand.

An affected person will not normally realize for himself that he has this problem, but special tests can reveal the defect. Charts covered in colored dots are used. These show up a pattern or number to a person seeing them with normal vision, but no pattern (or a different one) to those with color defects.

The condition cannot be cured, but the newly developed X-Chrom contact lenses can give sufferers near-normal color perception.

© DIAGRAM

CARDIOVASCULAR DISORDERS 1

HEART DISORDERS

ANGINA PECTORIS

Angina is any pain caused by muscular spasm. Angina pectoris is a pain in the chest caused by an increased demand for blood that the heart muscle is unable to meet. It can be caused by any activity that requires an increased heart rate. For example, those prone to it may suffer an attack after eating a heavy meal. It is not an illness in itself, but the possible symptom of other cardiovascular malfunctions.

HEART FAILURE

This occurs when the heart cannot keep pace with the demands made upon it. It can be caused by: heart attacks; intense fevers; chronic lung disease; high blood pressure; faulty heart valves; or strain on the heart muscle in extreme physical exertion.

The symptoms are breathlessness and chest pain. Usually one side of the heart fails first, and then pressure builds up as the other side continues to function. Legs and ankles swell and the lungs flood with fluid.

The rhythm of the heartbeat is affected, becoming abnormally fast (tachycardia), slow (bradycardia) or irregular (fibrillation).

Unless the process is arrested and cured, death results.

Medical advice should be obtained as quickly as possible. If the heart has stopped completely, external cardiac massage should be applied within three minutes - but only if it is certain that the heart has stopped.

Rest is the most important factor in recuperation: the amount and intensity of rest depend on the extent of heart damage. Drugs are given for sedation; also to increase kidney output, to clear the lungs and tissues of excess fluid. Where the rhythm of the heartbeat is permanently disrupted, an electronic pacemaker is fitted. This co-ordinates the contractions of the atria and ventricles.

BLOCKAGE AND TISSUE DEATH

THROMBOSIS is blockage of a blood vessel by a blood clot. It occurs when the blood flow is too slow, due to prolonged inactivity; or when the blood flow is disturbed by a change in the smooth lining of a blood vessel, and this sets off the clotting process.

EMBOLISM is blockage of a small blood vessel by a mass of foreign material, eg a mass of bacteria, or fragments from a thrombosis elsewhere.

INFARCTION refers to the death of living tissue when its blood supply is cut off by a thrombosis or embolism.

Thrombosis   Embolism

Three pulmonary embolisms: their areas of effect

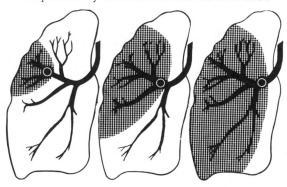

How leg thrombosis at (a) may lead to embolism at (b)

ARTERIAL DEGENERATION

ARTERIOSCLEROSIS is thickening and hardening of the arterial walls. It particularly occurs in old age.

ATHEROMA is a fatty deposit on the arterial walls. Such deposits contain large amounts of cholesterol. With age they are usually accompanied by the deposition of calcium salts.

ATHEROSCLEROSIS is the condition in which atheroma, accompanied by arteriosclerosis, interferes with the blood supply. It may also encourage thrombosis.

ANEURYSM is a bulge in the wall of a vein or artery at a weak point. Under the pressure of blood, the aneurism may balloon out, and finally break, spilling out blood.

HEART ATTACK

A "coronary thrombosis" is a thrombosis in the arteries that supply the heart muscle itself. It may result in a cardiac or "myocardial" (muscle of the heart) infarction. When these two disorders have occurred, the person has suffered a heart attack, which may have resulted in a cardiac arrest (stoppage of the heart).

The severity of the attack depends on:

a) whether or not the blood can find an alternative route;

b) the size and position of the area affected; and

c) whether or not the nerves that regulate the heartbeat are affected.

The dead tissue of the infarction is gradually replaced by scar tissue. This cannot contract, and weakens the heart.

By the age of 40, most men will probably have suffered minor coronary infarctions, most of which will have passed unnoticed. The chance of serious infarctions increases with age. They happen suddenly and are accompanied by severe pain, cold sweating, breathlessness and faintness. As with all heart disorders, rest, to allow repair, is necessary.

RHEUMATIC HEART DISEASE

This affects about 1% of the US population. It results from rheumatic fever, a bacterial infection which causes inflammation of certain joints. In rheumatic heart disease the heart and the heart cavity also become inflamed. Heart failure can occur as a direct result, but more usually the valves of the heart are left scarred, thickened, and deformed, which threatens heart failure in later life. The fever itself is often recurrent, with serious attacks following mild ones. In the USA, it is the cause of 90% of all heart trouble in children.

HIGH BLOOD PRESSURE

High blood pressure (hypertension) is associated with arteriosclerosis, kidney diseases, glandular disorders, obesity and anxiety. It can also occur without such factors. The cause is obscure at present, but is linked with contraction of the small arteries

Hypertension often starts in the early 30s, but no age is exempt, and it is not caused by aging. Heredity seems to be one factor. Hypertension is usually controlled effectively by drugs.

Atheroma is deposited. Clots of blood adhere to it. Finally a large thrombus blocks the artery. But blood may find ways through, or take alternative routes.

HEART SURGERY

The main reason for heart surgery is the heart's importance to the body - and that is also the main difficulty. Till a few years ago, surgeons could only carry out operations that did not interfere with the heart's pumping action. For if the flow of oxygenated blood to the body stops, brain damage and death follow within 4 minutes due to lack of oxygen.

But the development of the heart-lung machine has changed the situation. For a short time, this machine can take over the functions of the heart and respiratory system, while the heart is being operated on. Usually, the two large veins that lead into the right atrium are connected to the machine instead. Deoxygenated blood flows into the machine, where it is mixed with oxygen, in a way similar to the lungs' processes, and the blood flows back into the body through a connection to one of the branches of the aorta.

This allows sufficient time for certain procedures, especially:

a) closure of atrial or ventricular septal defects; or

b) replacement of diseased valves, such as the aortic or mitral. (The mitral valve is that between the left atrium and left ventricle, and is often damaged in cases of rheumatic fever.)

Difficulties that can arise afterwards, especially when the machine is used for longer than normal, include:

a) kidney damage; and

b) psychological psychosis (from which recovery usually occurs).

Both are possibly due to the pressure output of the machine being too low for pumped blood to reach some tissues.

Other surgical techniques include:

a) replacement of segments of coronary artery or aorta;

b) artificial pumping aids for the left ventricle; and

c) use of artificial "pacemakers", to stimulate the heartbeat.

© DIAGRAM

CARDIOVASCULAR DISORDERS 2

BRAIN DISORDERS

ARTERIOSCLEROSIS PSYCHOSIS

This results from a generally impaired blood supply to the brain. It is usually associated with old age, but may occur earlier. The person feels restless and emotional, is inclined to wander at night, and complains of headaches, giddiness, and momentary blackouts. Memory may fail and strong personality traits become exaggerated. These symptoms fluctuate.

As the physical change is permanent, treatment concentrates mainly on creating a relaxing environment. Sedative drugs may be used if the person is overactive or worried, anti-depressants if he is depressed. Drugs are also used to try to increase cerebral circulation (but this usually has little effect).

"STROKE"

Technically known as a cerebrovascular accident, a stroke results from failure of the blood supply to a part of the brain. This may be due to a thrombosis or embolism, or to a hemorrhage in the brain from a ruptured blood vessel. Infarction (death) of part of the brain may occur.

A stroke is more common in men than women. It can vary in severity from a minor disturbance, forgotten in a few minutes, to a major attack causing unconsciousness and death. The severity depends on the position and extent of the damage. In a severe attack, the patient loses consciousness almost immediately. Death may then follow in a matter of hours; alternatively, consciousness is regained, but there is usually lasting damage. An attack may also show itself in sudden paralysis of one side or part of the body, without loss of consciousness. Or, again, it may develop over several hours, with persistent throbbing headache, vomiting, dizziness, and numbness of the limbs.

Recovery depends on age, general health, and the site and size of damage. Even if recovery is possible, it may take years; but in other cases, control of the body has suffered permanent damage, with muscles paralyzed or very weak. The effects occur in the opposite side to the side of the brain affected (because one side of the body is controlled by the opposite side of the brain). A stroke in the side of the brain that is "dominant" may also affect speech. Mentally, concentration may be impaired, but judgement and basic personality need not be. Treatment consists mainly of rest and prolonged convalescence, with careful nursing, physiotherapy and (if needed) speech therapy. Drugs are sometimes used to lower the blood pressure, if it is high, and so help prevent further damage.

CONGENITAL HEART DEFECTS

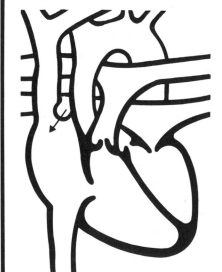

Atrial septal defect

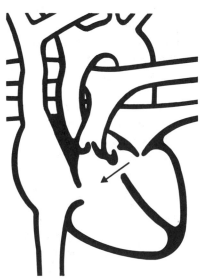

Ventricular defect

Two common types of congenital heart defect are shown here:
a) atrial septal defect, which is the presence at birth of a large hole between right and left atria; and
b) ventricular septal defect, which similarly is the presence of a hole between right and left ventricles. In each case, because pressure in the left side of the heart is higher, the result is that oxygenated blood from the left passes through and mixes with deoxygenated in the right.

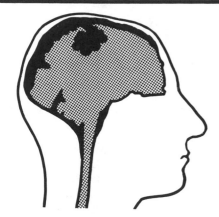

Hemorrhage

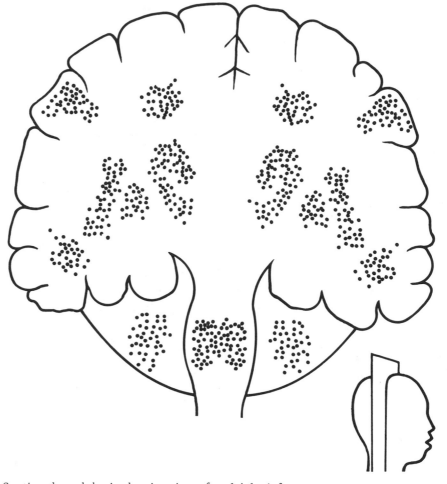

Section through brain showing sites of multiple infarctions

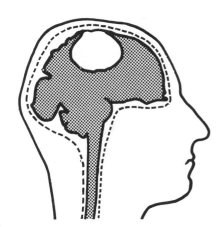

Thrombosis or embolism

OTHER DISORDERS

THROMBOSIS can occur in arteries in the pelvis or legs, causing the leg to swell.

THROMBOPHLEBITIS is inflammation of a vein due to damage or infection, and resulting in a clot.

PULMONARY EMBOLISM is blockage of one of the arteries of the lungs, caused by fragments of a thrombosis elsewhere. It is a serious condition sometimes requiring a major operation.

DEEP VEIN THROMBOSIS is a thrombosis of the veins deep in the legs. The superficial veins take over, as the main route for blood returning to the body; but the condition is dangerous because of the chance of a pulmonary embolism.

Because of inactivity and tissue damage, both deep vein thrombosis and pulmonary embolism are more likely after an operation. (Any pain deep in the calf muscle should be reported at once.)

VARICOSE VEINS

Varicose veins are swollen veins in the legs, that stand out above the surface and can be acutely painful. Their exposed position also makes them vulnerable to bleeding and ulceration.

Varicose veins develop if the valves in the leg veins fail to prevent the back flow of blood.

They are more likely in occupations involving long periods of standing - and also where there is a swelling of the abdomen, as in obesity, chronic constipation, and pregnancy. This last is why 1 in 2 women over 40 suffer from varicose veins, but only 1 in 4 men of the same age. Treatment ranges from wearing pressure bandages and resting with the legs in an elevated position as often as possible, to surgical tying or removal of the vein, making the blood find alternative routes.

©DIAGRAM

CARDIOVASCULAR DISORDERS 3

RISK FACTORS

The two factors most directly associated with heart disorders are high blood pressure and high cholesterol level.

HIGH CHOLESTEROL LEVEL Cholesterol is a main part of the substances deposited on the arterial walls in atherosclerosis. It has been established that heart disease is more common in those with a high blood level of cholesterol than in those with a low level, though no reason for this has been demonstrated to everyone's satisfaction.

HIGH BLOOD PRESSURE It is known that the higher the blood pressure, the higher the risk of atherosclerosis. This may be because high blood pressure forces cholesterol into the arterial walls. Alternatively, it may be that high blood pressure is more of a symptom than a cause. But high blood pressure certainly means that excessive strain is being exerted on the heart muscle, the coronary arteries, and the whole arterial system. The heart has to pump with greater force, and eventually the heart muscle enlarges. In men between 30 and 60, a systolic blood pressure over 150 doubles the likelihood of a heart attack (and increases that of a stroke four times).

CAUSES OF THESE Overeating and obesity, smoking, lack of exercise, and stress and anxiety all tend to produce high blood pressure and high cholesterol levels, as well as predisposing the person to cardiovascular disorders in other ways.

OTHER FACTORS A history of cardiovascular trouble in past generations sometimes means that a person has inherited a disposition to heart disease.

In addition, some factors work without raising blood pressure or cholesterol level; eg excessive alcoholic consumption can damage many tissues, including the heart.

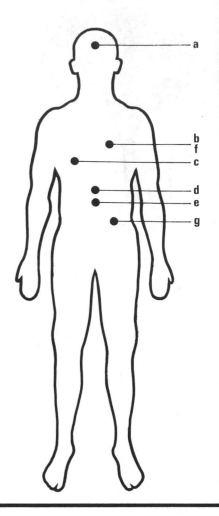

LIKELIHOOD OF DISEASE

These diagrams show the likelihood of a person getting heart disease (i.e. angina, heart attack, or death). They compare the likelihood at different ages, and the effect of the three biggest and most measurable risk factors: high blood pressure, high blood cholesterol level, and cigarette smoking.

First, they show how the likelihood changes with age. For example, with no risk factor, a 35 year old man is in little danger - but his likelihood has risen considerably by age 50, and again by 65. Similarly, for a man with all three risk factors, likelihood rises between 35 and 50. But then it falls slightly by 65, because so many of the easiest victims are by then dead.

Second, they show the relative importance and the interaction of the risk factors at different ages.

At 35, high cholesterol is the worst single risk factor; by 65, it is high blood pressure. Similarly, at 35, by far the best single thing to do, if you have all three risk factors, is to get your cholesterol level down. At 65, it is to get your blood pressure down.

The final bar shows that the likelihood for a woman of 65, with all the risk factors, is only $\frac{2}{3}$ that of a 65 year old man.

a Stress
b Cholesterol
c Smoking
d Drinking
e Overeating
f High blood pressure
g Obesity

CHOLESTEROL

Cholesterol is a waxy substance produced in the liver, and also acquired from outside in certain foods, especially animal fats, milk products, and eggs. It is found throughout the body, but especially in the brain, nervous tissue, and adrenal glands. It is important in the repair of ruptured membranes, and in the production of sex hormones and bile acids. Blood cholesterol level varies with age, sex, race, hormone production, climate, and occupation, and is thought to depend mainly on the amount manufactured in the body. Surplus adrenalin, due to stress situations, may be one cause of excess cholesterol.

However, nutrition does affect the level to some extent - though in this the cholesterol content of the diet seems less significant than the fat content. A diet rich in "polyunsaturated fatty acids" (i.e. those of most vegetable oils) can lower the blood cholesterol level. if saturated fat consumption is reduced.

HIGH BLOOD PRESSURE

Blood pressure depends on the resistance of the arteries to the heart's pumping efforts. It falls if the heart pumps less and/or if the arteries are dilated. It rises if the heart pumps more and/or if the arteries are constricted.

Heart output can be raised temporarily by exercise , but this is only dangerous in some cases of existing heart disorder - it does not cause disorder. Other temporary high blood pressure is caused by excitement, stress, or apprehension, and whether these are dangerous probably depends on their duration and frequency and how high they go.

Consistent high blood pressure is certainly dangerous. Its causes are discussed on p. 117. In the absence of a specific causative disorder, blood pressure can be kept low by exercise, elimination of body fat, and limiting salt intake.

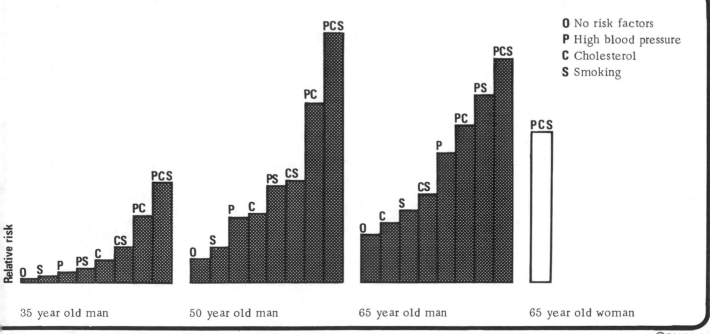

O No risk factors
P High blood pressure
C Cholesterol
S Smoking

Relative risk

35 year old man   50 year old man   65 year old man   65 year old woman

©DIAGRAM

CANCER 1

WHAT IS CANCER?

Cancer is one of several disorders which can result when the process of cell division in a person's body gets out of control. Such disorders produce tissue growths called "tumors". A cancer is a certain kind of tumor.

Cancer attacks one in every four people.

NORMAL CELL DIVISION

The body is constantly producing new cells for the purposes of growth and repair - about 500,000 million daily. It does this by cell division - one parent cell divides to form two new cells. When this process is going correctly, the new cells show the same characteristics as the tissue in which they originate. They are capable of carrying out the functions that the body requires that tissue to perform. They do not migrate to parts of the body where they do not belong; and if they were placed in such a part artificially they might not survive.

TUMORS

In a tumor, the process of cell division has gone wrong. Cells multiply in an unco-ordinated way, independent of the normal control mechanisms. They produce a new growth in the body, that does not fulfill a useful function. This is a tumor, or "neoplasm". A tumor is often felt as a hard lump, because its cells are more closely packed than normal.

Tumors may be "benign" or "malignant". A cancer is a malignant tumor. That is, it may continue growing until it threatens the continued existence of the body.

BENIGN TUMORS

In a benign tumor:
the cells reproduce in a way that is still fairly orderly;
they are only slightly different from the cells of the surrounding tissue;
their growth is slow and may stop spontaneously;
the tumor is surrounded by a capsule of fibrous tissue, and does not invade the normal tissue;
and its cells do not spread through the body.

A wart is a benign tumor. Benign tumors are not fatal unless the space they take up exerts pressure on nearby organs which proves fatal. This usually only happens with some benign tumors in the skull.

MALIGNANT TUMORS

In malignant tumors, the cells reproduce in a completely disorderly fashion.
The cells differ considerably from those of the surrounding tissue. (Generally, they show less specialization.)
The tumor's growth is rapid, compared with the surrounding tissue.
The tumor has no surrounding capsule, and can therefore invade and destroy adjacent tissue.
The original tumor is able to spread to other parts of the body by metastasis, and produce secondary growths there.
A malignant tumor is usually fatal if untreated, because of its destructive action on normal tissue.

BIOPSY

A biopsy is the most certain way of distinguishing between benign and malignant tumors. A piece of the tumor is surgically removed, and then studied under a microscope.

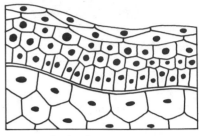

Normal body tissue

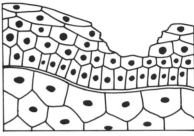

Damaged body tissue

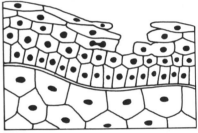

Normal cell replacement

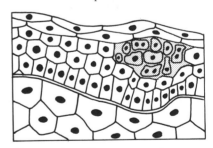

Abnormal malignant growth

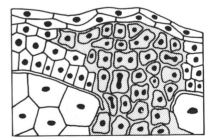

Loss of basement membrane integrity

CANCER GROWTH
CELL DIVISION
Cancerous cells cannot divide faster than normal cells. But normal cell division reaches its maximum rate only in times of injury and repair. Cancerous growths are continually producing cells at this maximum rate without check. They are less successful than normal tissue could be, because many of the faulty cancerous cells die. Nevertheless, the result is that cancerous growths grow faster than normal tissue.

METASTASIS
Metastasis is the process by which cancerous cells travel from the original (primary) cancer site to other parts of the body. It occurs when cancerous cells get caught up in the flow of blood or lymph. The cells are carried along in the vessels, until they lodge in another part of the body. If they succeed in establishing themselves there, this becomes a new (secondary) cancer site. If a secondary site gets large enough, it can also metastasize in turn.

Cancer that has metastasized along the lymph vessels normally sets up its secondary sites in the glands. Cancer that has metastasized in the blood stream sets up secondary sites in the bones, lungs, and liver. Cancers in the brain do not metastasize, but cancers elsewhere can metastasize to the brain. Some sites are more receptive than others. Common locations for secondary growths include the lungs, bones, brain, kidney and bladder, and larynx, the testes in men, and the breasts in women. A cancer can also spread through the body simply by the process of growth.

CANCER BY COUNTRY

In all the sample, with the exception of Iceland, more men than women die of cancer.

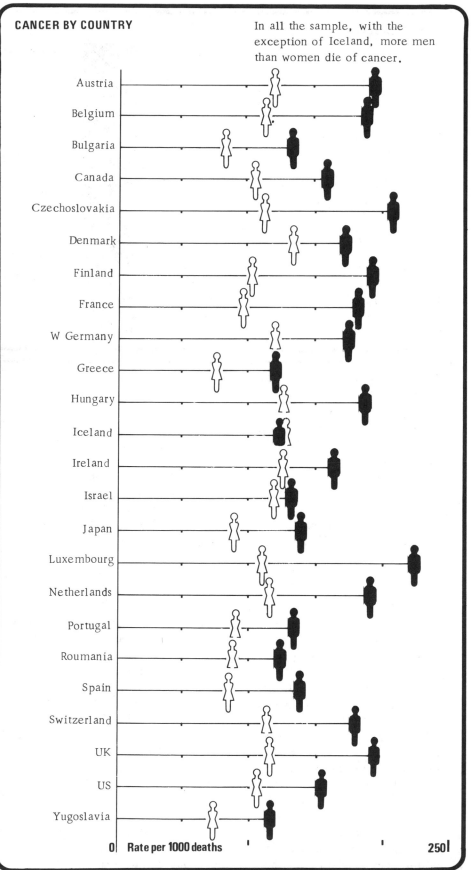

Austria
Belgium
Bulgaria
Canada
Czechoslovakia
Denmark
Finland
France
W Germany
Greece
Hungary
Iceland
Ireland
Israel
Japan
Luxembourg
Netherlands
Portugal
Roumania
Spain
Switzerland
UK
US
Yugoslavia

0 **Rate per 1000 deaths** 250

© DIAGRAM

123

CANCER 2

SITES OF CANCER

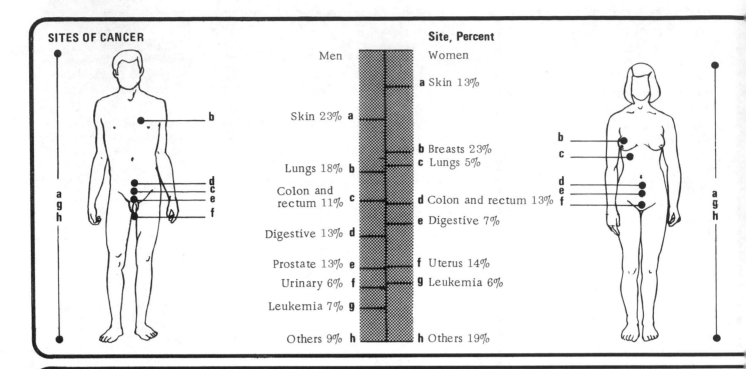

Site, Percent

Men		Women
		a Skin 13%
Skin 23% **a**		
		b Breasts 23%
		c Lungs 5%
Lungs 18% **b**		
Colon and rectum 11% **c**		**d** Colon and rectum 13%
		e Digestive 7%
Digestive 13% **d**		
Prostate 13% **e**		**f** Uterus 14%
Urinary 6% **f**		**g** Leukemia 6%
Leukemia 7% **g**		
Others 9% **h**		**h** Others 19%

CAUSES OF CANCER

CHROMOSOME DAMAGE

The characteristics of a cell are inherited from its parent cell. They are passed on in the DNA in the chromosomes. This forms a set of coded instructions, which controls the cell's structure and function. In cancerous cells, the characteristics of malignant growth are passed on from one generation to another. This means that the genetic code must have been damaged.

This, in fact, is seen, if the chromosomes of cancerous cells are examined. Normal cells have 46 chromosomes arranged in 23 pairs. Almost all cancer cells are abnormal in the number and/or structure of these chromosomes.

NORMAL DEVIANCY

Cells with genetic defects appear in the body every day: so many millions of cells are being made, that some mistakes are inevitable. But most die almost immediately, because they are too faulty to survive, or because they are recognized as abnormal and eaten by white blood corpuscles. Others are only slightly defective, and not malignant. Only very rarely do malignant cells survive and reproduce successfully.

Appearance of cancer in a person may simply be due to this unlucky chance. Alternatively, it may be that the body has "immunity" to such malignant cells, and that this sometimes breaks down. This would explain why cancer can remain "dormant" in a person for many years.

SPECIAL FACTORS

A few factors have been recognized, that do make genetic damage in cells more likely. But they can only explain a tiny proportion of the cancer that occurs.

a) Certain chemicals can cause cancer to form, if they are repeatedly in contact with the body over a period of time. Such chemicals are called carcinogens, and include some hydrocarbons. Apart from tobacco smoke, these carcinogens usually only affect workers whose job brings them into regular contact with them. (However, atmospheric pollution may also be slightly carcinogenic.)

b) Certain viruses can pass malignant tumors from one animal to another, and the same may occur in man. But so far only one rare form of cancer is thought to be caused this way.

Apart from this, human cancer seems not to be virus induced – and therefore not infectious.

c) Ionising radiation. Without correct protection, X rays can cause skin cancer, and radiation can cause leukemia. Also ultraviolet rays (as in sunlight) may cause skin cancer in some circumstances.

d) Continued physical irritation. There is disagreement over this, but some experts believe that continued physical disturbance of the skin or mucous membrane can cause cancer. If so, the sharp edges of a broken tooth, for example, could eventually cause a cancer in the mouth.

Others argue that such irritation can only accelerate an existing cancerous growth.

Cancers can grow almost anywhere in the body, but the most common sites are shown here.

Cancers are classified by the kind of tissue in which the primary growth occured. Tumors originating in the "epithelial" cells (eg skin, mucous membrane, and glands) are called carcinomas; those in connective tissue (eg muscle and bone), sarcomas. Secondary sites are classified by the kind of primary tumor that they came from. This is possible because metastasized growths still show some of the characteristics of the tissue from which they originally came.

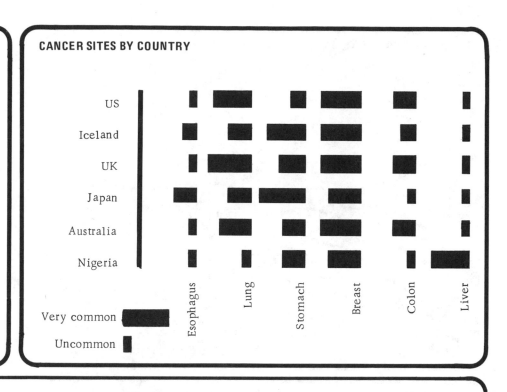

CANCER SITES BY COUNTRY

US
Iceland
UK
Japan
Australia
Nigeria

Esophagus
Lung
Stomach
Breast
Colon
Liver

Very common
Uncommon

CORRELATIVE FACTORS

Some individuals are more likely to develop cancer than others.

a) Heredity. Actual cancerous growths are not inherited. But a predisposition for cancer can be passed on. It may be that some inherited characteristics make a person's cells more likely to become malignant.

b) Age. Most cancers occur in the 50 to 60 age group. However, children and adolescents are susceptible to leukemia, brain tumors, and sarcomas of the bone.

c) Sex. In almost all countries, cancer occurs more frequently in men than in women.

d) Geographical location. Eg, for some unknown reason, gastric cancer is most frequent in coastal countries with cold climates.

e) Cultural habits. Eg, cancer of the penis is less common in societies where circumcision is usual.

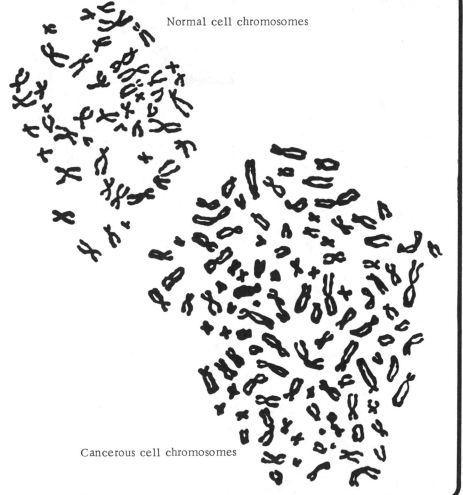

Normal cell chromosomes

Cancerous cell chromosomes

© DIAGRAM

CANCER 3

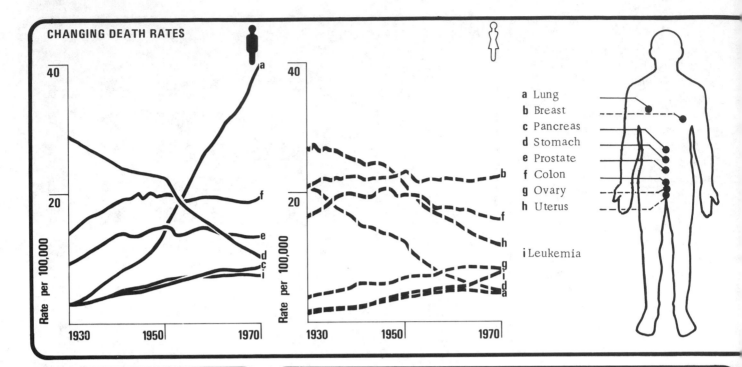

CHANGING DEATH RATES

40 — 20 — Rate per 100,000

1930 1950 1970

a

f
e
d
c
i

40 — 20 — Rate per 100,000

1930 1950 1970

b
f
h
g
i
d
a

a Lung
b Breast
c Pancreas
d Stomach
e Prostate
f Colon
g Ovary
h Uterus

i Leukemia

SYMPTOMS

a) Any unusual bleeding or discharge from mouth, genitals, or anus (including, in women, bleeding from the breast and menstrual bleeding between periods).
b) Any lump or thickening or swelling on the body surface, or any swelling of one limb.
c) Any increase in size or change in color or appearance in a mole or wart.
d) A sore that will not heal normally.
e) Persistent constipation, diarrhea or indigestion that is unusual for the person.
f) Hoarseness or dry cough that lasts more than three weeks.
g) Difficulty in swallowing or urinating.
h) Sudden unexplained loss in weight.
If you develop any of these symptoms, you should visit your doctor. Nearly always, the cause will be something else, not cancer. But do not delay. If it is cancer, quick diagnosis is essential.

TREATMENT

Treatments for cancer have a good chance of success only if the tumor is still localized. Early diagnosis is vital. Once a tumor has metastasized, successful treatment is almost impossible.
SURGERY Surgical removal of localized malignant tumors at an early stage is the only completely successful form of treatment known at present.
In later stages, surgery may be attempted in conjunction with other techniques.
RADIOTHERAPY Cancer cells are killed by radiation more easily than normal cells. Radiotherapy seeks to destroy cancerous tissue by focussing a stream of radiation on it. This can be done only if the cancer is still localized, and can be destroyed without causing radiation damage to the rest of the body.
The rays used are either X-rays or those of radioactive materials such as radium or cobalt.
CHEMOTHERAPY This is treatment by the administration of chemicals. Again, the major

difficulty is finding drugs that will destroy cancer cells without harming normal cells. Three main types of chemical are used:
those that interfere with the cancer cells' reproductive processes;
those that interfere with the cells' metabolic processes;
those that increase the natural resistance of the body to the tumor cells.
These chemicals can affect the whole of the body, specific regions, or the tumors themselves, depending on how they are applied.
HORMONE THERAPY is used mainly for tumors of the endocrine glands and related organs. It is also useful in the treatment of metastases originating from these areas (eg in women, against disseminated breast cancer).
Success depends on whether the cancerous cells still have the specialized relationship with the hormone that the original tissue had.

Cancer is not a modern disease - it has been found in dinosaur fossils, and in the remains of Java man who lived about 500,000 years ago. But in the present century it has become vastly more prevalent. In 1900 it was the seventh main cause of death in the USA. Today it is the second. Some experts, though, believe this is simply because people are living longer - for likelihood of cancerous growth increases with age.

However, some types of cancer have shown a dramatic fall in recent years, eg stomach cancer. (This particular example may be linked with changes in techniques of food preservation.)

LUNG CANCER

This is one of the most deadly forms of cancer, for it is not usually diagnosed until too late. The tumor begins in the walls of the bronchial passages or sometimes in the body of of the lung, and usually produces no symptoms until it has become firmly entrenched in the lung tissue and even metastasized. Only one in twenty cases live for more than two years after lung cancer has been diagnosed.

Lung cancer is very much associated with cigarette smoking. Over 95% of all lung cancer patients are smokers.

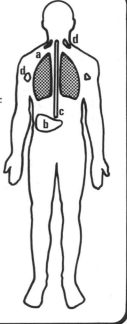

Lung cancer metastasis

Primary site:
a Lung

Secondary sites:
b Liver
c Spine
d Lymph nodes

LEUKEMIA

Leukemia is a cancerous disease of bone marrow and of tissues that produce blood corpuscles. It results in an abnormally large number of white corpuscles at the production site, or in the bloodstream, or both. There may be up to 60 times the normal number, and also many immature forms that never appear in healthy blood.

TYPES OF LEUKEMIA

There are two main types: acute, in which the onset is sudden, the duration usually only a few weeks, and the termination fatal; and chronic, in which the onset is gradual, and the duration up to 20 years before death ensues. Each type subdivides according to the type of corpuscle that multiplies.

a) Acute lymphoblastic leukemia is the most common form. It is most frequent under the age of five, and rare after 25, but appears again in the old.

b) Acute myeloblastic leukemia can occur at any age, but it is most common in the middle aged.

c) Chronic lymphatic leukemia occurs most often after the age of 50. It is almost three times more common in men than in women.

d) Chronic myeloid leukemia is most frequent between 20 and 40, and more common in men.

CHARACTERISTICS

Shortage of red blood corpuscles make the patient pale, tired, and anemic. Shortage of normal white corpuscles means that their protective work is undone, and the body is open to infection. Shortage of blood platelets reduces blood clotting ability. Swellings of feet and legs are common, and there may also be diarrhoea.

In chronic forms there are enlarged spleen, liver, and lymph nodes, and often a high temperature.

TREATMENT

There is no known cure for leukemia. Drugs and radiotherapy are used to try to hold it in check. In acute leukemia, periods without signs of illness can sometimes be obtained.

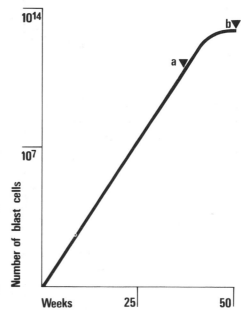

The above graph shows a typical growth rate for leukemia cells in the bone marrow.

The leukemia cell population doubles 40 times before the disorder can be properly diagnosed.

Diagnosis (a) occurs after 37 weeks, death (b) after about 50.

©DIAGRAM

BREAST CANCER 1

BREAST CANCER

Almost a quarter of cancer in women is breast cancer, and 5% of women contract it. If it is caught before metastasis, 9 out of 10 survive. But metastasis can occur within a month of the tumor appearing. Despite medical advance, the death rate from breast cancer has been steady for over 40 years.

WHO GETS BREAST CANCER? The disease is commonest in women aged 40 to 60. Heredity, diet, and estrogen levels in the body are all under investigation as possible significant factors. Research suggests that early first childbirth, and breast-feeding, may both make breast cancer less likely in the mother.

SELF EXAMINATION It is vital to check the breasts once every month for any changes: see p.130 for the procedure and what to look for. Anything you find - lumps or other changes - will usually be due to some other cause, not cancer; but do check with a doctor immediately. Early detection can save your life.

CLINICAL EXAMINATION If a lump is confirmed, it is not necessarily cancerous. A needle may be inserted into the lump, to see if it collapses indicating a cyst). If not, a biopsy will be taken, to distinguish between cancer and a benign tumor. Again, sometimes a needle may be used for this, but usually a surgical incision is necessary.

SURGICAL EXAMINATION As a result it is common, for speed of treatment, for a surgeon to obtain and examine the suspicious tissue while the patient remains under anesthetic - and then go on to operate at once if the tissue is judged cancerous.

NON-CANCEROUS DISORDERS

Other disorders have symptoms similar to breast cancer.

FIBROADENOSIS (CHRONIC MASTITIS) features a permanent increase in the breast's glandular content, due to hormonal imbalance The breast may feel lumpy or rubbery, all over or in patches, and there may be some pain, particularly before menstruation or after heavy lifting. Most common between the ages of 40 and 55, fibroadenosis is a normal bodily change for many women. Treatment may not be needed - but always check with a doctor.

CYSTS AND BENIGN TUMORS Perhaps 75% of breast lumps are non-malignant. Breast cysts are small sacs in the breast tissue, filled with liquid, and usually harmless. Most common in women aged 35 to 45, they do not always need treatment. Benign tumors may swell until pressure on nerves or neighboring tissues causes pain, and requires their removal; but they cannot invade or destroy other tissue, as cancer can.

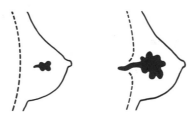

Growth of malignant tumor

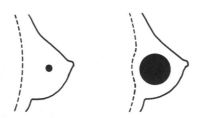

Growth of benign tumor

SCREENING TECHNIQUES

MAMMOGRAPHY This X ray technique is often used as an aid in the diagnosis of breast cancer. It involves placing the breasts in direct contact with the X ray film. Usually two views of each breast are taken. Any malignancy shows up as an irregular opaque patch in the breast. One run of 2000 mammograms detected 92% of cancers present. Mammography alone is not usually thought sufficient for certain diagnosis, but in combination with clinical examination and biopsy, around 97% of lumps examined can be correctly diagnosed.

THERMOGRAPHY This is another method helpful in the detection of breast cancer. In a normal person, 45% of the heat given off by the skin is infrared radiation. Thermography is the technique used to record in a photograph the way in which this heat is given off. A malignant growth emits more heat than the surrounding tissue, and so shows up as a light patch on the photograph. Thermography, however, is less successful than mammography and clinical examination.

Typical thermographs

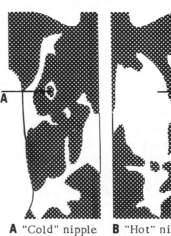

A "Cold" nipple (normal) B "Hot" nipple (cancer?)

TREATMENT

Breast cancer usually spreads in the lymph system, beginning with the armpit nodes (also those of the chest and spine). This determines the surgical possibilities:
a) removal of the lump alone (lumpectomy);
b) removal of the breast (simple mastectomy);
c) removal of breast and some armpit nodes (modified radical mastectomy);
d) removal of breast, some armpit nodes, and some chest wall muscles (radical mastectomy); and, occasionally
e) removal of breast, all armpit nodes, some chest wall muscles, and some chest nodes (superradical mastectomy).
Which is chosen depends on:
a) the decision which to minimize - deformity or the risk of recurrence;
b) surgical opinion: eg does simple mastectomy with radiotherapy give results just as good as a "radical"? - and especially
c) information about the tumor's size, type position, and spread. This last must mainly come from actual surgical observation; and as a result a patient can go under anesthetic not knowing how much of her body she will lose. Preliminary diagnostic surgery, though, may be impossible to arrange, and take up vital lifesaving time.
Surgery may, as noted, be accompanied by drug and/or radiotherapy. But again there are difficult issues: of effectiveness versus side effects, and of timing.
The more extensive operations are followed by physiotherapy, to minimize the effects of muscle loss on arm movement and breathing abilities.

Lymph nodes near the breast, and likely routes of spread in breast cancer

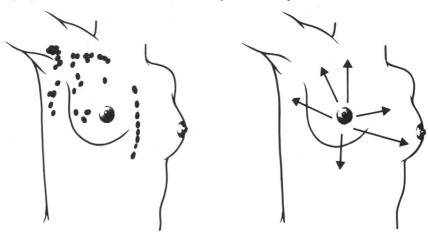

Simple mastectomy: area removed and stitching

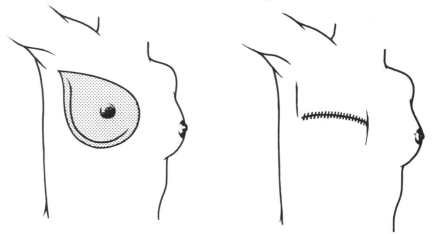

Radical mastectomy: area removed and stitching

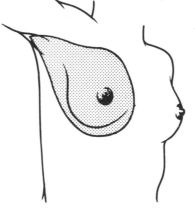
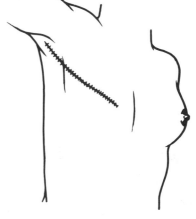

Area removed

©DIAGRAM

BREAST CANCER 2

BREAST EXAMINATION

Examine the breasts regularly for lumps. First stand before a mirror arms by side, undressed to the waist. Look for irregularities in outline of the breasts, or any puckering or dimpling of the skin. Then check nipples for discharge or bleeding.

The next part of the examination is performed on 5 separate areas in turn - each quarter of the breast, and lastly the armpit.

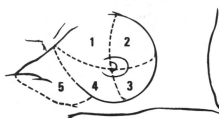

1 Lie on bed, folded towel under right shoulder, right arm behind head. With flat of fingers, feel upper, outer quarter of right breast
2 Repeat on lower, inner quarter
3 Bring right arm to side, and examine lower, outer quarter
4 Repeat on upper, outer quarter, and area between breasts and armpit
5 Examine armpit
6 Move towel under left shoulder and repeat the examination on the left breast.

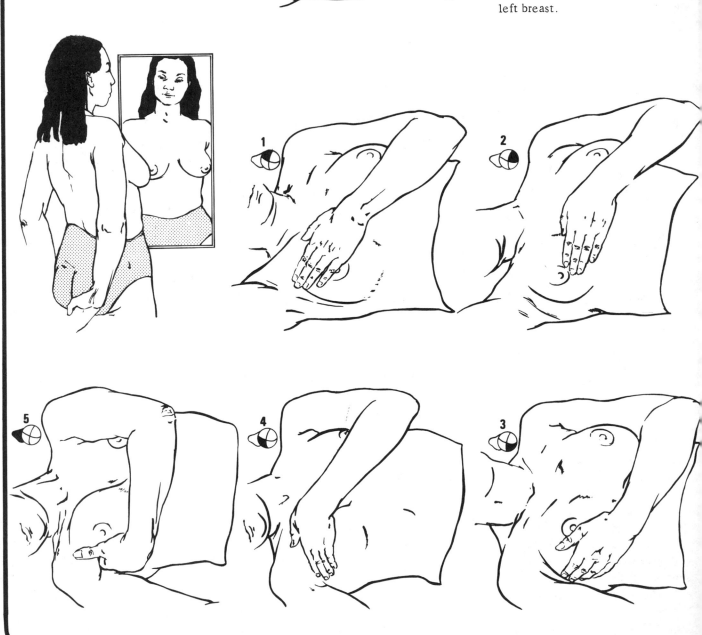

CANCER OF THE FEMALE REPRODUCTIVE TRACT

CANCER OF THE FEMALE REPRODUCTIVE TRACT

CERVIX

Cervical cancer is on the decline. Nevertheless, about 2 women in 100 get it, and one of these dies from it. It can occur at any age, but 45-50 is most common. A possible symptom is unusual vaginal bleeding, eg between periods, after intercourse, or more than 6 months after the menopause. But analysis of the cervical tissue is the only sure evidence. The well-known "Pap" smear test involves the painless gathering of a few sample cells on the end of a wooden spatula (the physician may well do a pelvic examination to check the uterus and ovaries at the same time). The sample is sent to a specialist laboratory, fixed in alcohol, stained in solution, and examined for abnormal cells. In 1000 smears, 20 might show some abnormality, and perhaps 3 the early signs of cancer. (The other abnormalities will include signs of vaginal infection, etc.)

A woman should have had such a test by the time she reaches 25 (or when she first becomes pregnant, if this is earlier). It should **be**
repeated twice in the first year, and thereafter every year till she is 65. (Occasionally cancer can appear after a recent negative smear; but this is rare.)

If possible signs of cancer are found, a repeat smear may be taken, followed by a larger specimen using curettage or a tiny punch. If cancer is confirmed, the alternatives are:
a) conization, in which cervical tissue is cut away (the cervix is stitched, and rapidly heals with little pain and usually no after-effects); or
b) hysterectomy.

Which is used depends on the state of the tumor. Hysterectomy is necessary once malignant growth has begun. It gives almost 100% success, but a few very advanced cases may need radiation therapy or further surgery.

UTERUS

Cancer here is less common; it usually only occurs in older women (typically 50-60). The diabetic and obese are susceptible, and it can run in a family. Tumor growth
is slow; bleeding symptoms are significant. Diagnosis is by curettage of the uterus under anesthetic; treatment by hysterectomy and X ray therapy. If treated early enough, 80% of patients survive more than 5 years.

OVARIES

Ovarian cancer accounts for 5% of cancers in women. It is more frequent after 40, and especially after the menopause. Its slow growth is hard to detect. The first sign is enlargement of the ovary, showing up on pelvic examination - but only 5% of ovary enlargements are cancerous. Pelvic examinations every 5 years should catch them in time.

VULVA

Cancer of the vulva is rare, and usually only found in old women. It is typically preceded by long-standing vulval itching, and sometimes an ulcer (but in 99% of cases these symptoms do not signify cancer). Diagnosis is by examination of tissue samples taken under anesthetic.

Cancer of cervix and uterus
Possible sites

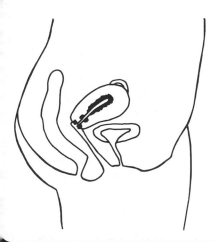

Possible routes of spread

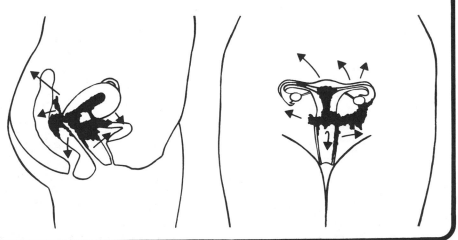

©DIAGRAM

DISORDERS OF THE BRAIN AND CENTRAL NERVOUS SYSTEM

MALDEVELOPMENT AND DAMAGE

Maldevelopment of the brain may be due to spontaneous chromosome abnormality, as in Down's syndrome (mongolism), or to an infection such as German measles during the mother's pregnancy.

Mental retardation or problems of nervous control may also result from damage to a normal brain, because damaged brain cells, unlike other cells in the body, can never be replaced. Possible causes of damage include the following.

The brain may sometimes be damaged during birth, if the head strikes hard against the mother's perineum (the muscular outlet of the pelvis) during a too rapid birth, or if the brain is starved of oxygen through constriction of the umbilical cord. Modern obstetric techniques have greatly reduced such risks.

In a baby that is very premature, physical immaturity can result in glucose deficiency and subsequent brain damage.

Thyroid deficiency is responsible for cretinism, in which brain damage is accompanied by lack of activity, chronic constipation, slow growth, and distinctive coarsening of the features. Prompt treatment is essential to prevent severe mental retardation.

In hydrocephalus, often associated with spina bifida, the head becomes swollen because of inadequate drainage of the fluid around the brain. Untreated, the condition produces brain damage as the fluid presses on the brain. Treatment is by the insertion of plastic drainage tubes.

Other causes of brain damage include severe head injury, and poisoning, for example with lead.

EPILEPSY

Epilepsy is a disorder of the nervous system. It is usually due to brain damage but may also be inherited. Mental ability is not affected. There are two main types: petit mal and grand mal. Both can be controlled with drugs.

PETIT MAL is characterized by momentary lapses of attention, sometimes with blinking or slight twitching. Attacks last only a few seconds, but may occur several times in a day. In many cases the sufferer is unaware that anything has happened.

GRAND MAL is a more severe form of epilepsy in which actual fits occur: the sufferer loses consciousness and has convulsions in which he thrashes about and may foam at the mouth. A fit usually lasts one or two minutes, followed by sleep (see pp. 518-19 for first aid).

SPINA BIFIDA

In a normal backbone (a), each bone (vertebra) forms a closed ring around the spinal cord. In spina bifida (literally, split spine) one or more vertebrae are open at the back. Genetic and other factors may be responsible. Approximately 5% of the population have a mild form of the disorder, which causes no problems and usually remains unnoticed.

In other cases, spinal fluid passes through the gap to form a bulging bag at the back (b) although the spinal cord is not affected. Such cases are usually cured by an operation soon after birth.

In the most severe form of spina bifida the spinal cord is deformed or damaged before birth, producing varying degrees of paralysis and deformity in the lower body. An operation only sometimes succeeds, and physiotherapy and special treatment will be needed.

CEREBRAL PALSY

Cerebral palsy is a disorder in which brain damage sustained before or around the time of birth interferes with muscular control. There are three main forms:
a) spasticity, characterized by stiff and difficult movements;
b) athetosis, in which pronounced involuntary movements interfere with normal body movements;
c) ataxia, in which a disturbed sense of balance and depth results in an unsteady gait.

The degree of disability varies considerably. Associated defects include impaired speech and hearing, and - in about 50% of cases - reduced intelligence. Treatment involves training healthy muscles to take over the work of affected ones as far as possible. Drugs, surgery, and special equipment may help in this.

Section through trunk

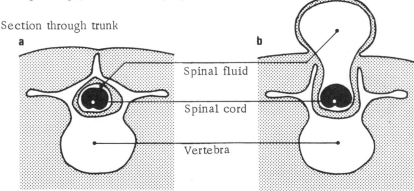

a   b

Spinal fluid

Spinal cord

Vertebra

RESPIRATORY DISORDERS 1

BRONCHITIS

Bronchitis is an inflammation of the bronchi - though the bronchioles and smaller passages are often involved too. Its severity varies. Mild cases may seem like a severe chest cold. Severe cases lead to pneumonia and death. Bronchitis is most prevalent in the UK, where it kills over 30,000 people a year. There are two forms, acute and chronic.

ACUTE BRONCHITIS

This results from:
viral and bacterial infection following colds and flu;
exposure to damp cold air;
inhaling irritating dust or vapors;
or from combinations of these.
The attack begins with a short, dry, painful cough, and a general feeling of acute illness. There may also be slight fever. After two or three days the cough begins to bring up sputum (mucus from the lungs) in increasing quantities. The symptoms then begin to subside. The cough may last for three or four weeks, but the more distressing symptoms pass off in about ten days.
The condition is most dangerous in the old, especially if emphysema is also present. Treatment consists of antibiotics, bedrest, warmth, hot drinks, inhalation of steam preparations, and abstention from smoking.

CHRONIC BRONCHITIS

Chronic bronchitis mainly affects the middle aged and the old. It can lead to emphysema and heart failure. It usually develops after repeated respiratory infections. There are several major contributing factors:
excessive cigarette smoking;
exposure to a cold, damp climate;
damp living conditions;
exposure to irritating environmental dust and fumes, eg from industrial pollution;
obesity;
and, probably, constitutional predisposition.

The disease produces a constant cough which is worse during the night and in the mornings. The mucous membranes of the bronchial tubes become thickened, and the nutritional blood supply to the lungs may be impaired. Emphysema and other complications may lead to constant breathlessness. Treatment depends on the patient's age, the severity of the illness, and whether there are complications. It may include:
expectorants to loosen the mucus in the air passages;
steam inhalations;
and antibiotics if there is any bacterial or viral infection.
To prevent recurrence, a sufferer should avoid cold, dusty, or polluted air, and should not smoke. Care must be taken to prevent colds from developing into bronchitis.

BRONCHITIS AND EMPHYSEMA

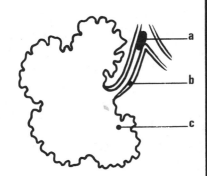

a Duct blocked with sputum
b Mucous membrane swollen
c Alveolus swollen

EMPHYSEMA

Emphysema is linked with bronchitis, and with cigarette smoking. It is mainly seen in older people. Due to infection, inflammation, and obstruction of the air passages, the lungs lose their elasticity. The small alveolar air spaces become enlarged: the dividing walls are stretched thin and break down, and large air sacs are formed.
This greatly reduces the surface area available for gas exchange, so the blood and body get less and less oxygen for each breath. Breathing becomes increasingly labored, as more breaths are needed to take in the necessary oxygen.
The lung's deterioration often also hinders the passage of blood through the arterioles. This puts a strain on the right side of the heart. It becomes weakened and dilated, and death from heart failure can result.

©DIAGRAM

RESPIRATORY DISORDERS 2

PNEUMONIA

Pneumonia is a disease in which large parts of the lungs become inflamed and filled with fluid.

TYPES OF PNEUMONIA

Bacterial pneumonia results from infection by bacteria. When one or more lobes of one lung are infected, it is called lobar pneumonia. If both lungs are involved it is called bilateral pneumonia.

Broncho-pneumonia occurs in patches of the lung tissue, not in whole lobes. It often comes about as a complication of bronchitis and other illnesses.

Hypostatic pneumonia occurs in bed-ridden people, especially the elderly. Fluid collects in the lungs because of lack of movement.

Primary atypical pneumonia is caused by viral infection.

Predisposing factors for pneumonia include the common cold, chronic alcoholism, malnutrition, and bodily weakness.

SYMPTOMS

In lobar pneumonia the illness begins with chest pains, vomiting, and shivering, closely followed by a rapid rise in temperature to 104°F. Breathing is difficult. A harsh dry cough brings up rust colored sputum which may contain blood in untreated cases. The temperature stays high for about a week. It then falls within 24 hours to normal, and pulse and breathing become regular. The patient recovers quickly (but may be fatigued for many weeks).

Broncho-pneumonia and other forms have similar symptoms, but do not end suddenly. The temperature tends to fall and rise, gradually returning to normal over a number of weeks.

TREATMENT

Treatment includes antibiotics and measures similar to those for severe bronchitis. An oxygen tent is used in extreme cases. Convalescence should last for a month or two.

Pneumonia can be fatal in weak or aged people; in cases where the extent of inflammation prevents respiration; and in those whose resistance is low for other reasons (eg because of other illness, or alcoholism). Because of this it is often quoted as a cause of death for old people who could not withstand the illness or the accompanying fever and fatigue. But apart from these and other extreme cases, it is not normally fatal.

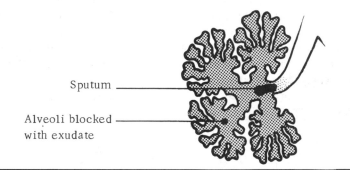

Sputum

Alveoli blocked
with exudate

PLEURISY AND EMPYEMA

Pleurisy is inflammation of the pleura - the membrane which lines the chest cavity and covers the lungs. It nearly always accompanies pneumonia and other lung inflammation. In dry pleurisy the inflamed membranes rub against each other as the patient breathes, causing acute pain. In wet pleurisy the pleural cavity fills with fluid - there is no pain, but breathing is impaired.

Pleurisy is seldom fatal in itself, but may increase the risk of fatality in the diseases it accompanies.

Empyema is any condition in which there is pus in the pleural cavity.

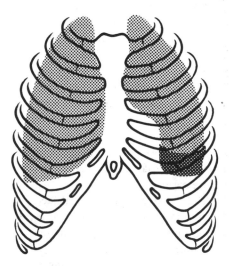

Pleurisy

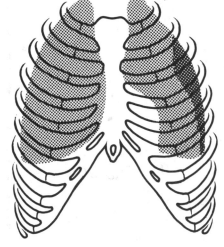

Empyema

TUBERCULOSIS

Tuberculosis (TB) is not just a respiratory disorder. It is a general term for diseases caused by the bacterium "myobacterium tuberculosis". These are contagious: the bacteria are carried in the sputum of the patient, and are spread when he sneezes or coughs. They enter other bodies by being breathed in or swallowed. Since the bacteria are also very hardy, and can survive for a long time in dried sputum and dust, everything within the vicinity of a patient soon becomes infected.

The most common form of tuberculosis is pulmonary tuberculosis, because the lungs, in adults, are the most common point of entry into the body. But tuberculosis can also affect bones, joints, skin, lymph nodes, larynx, intestines, kidneys, testes, prostate gland, and nervous system.

PROCESS OF THE DISEASE

The bacteria enter the body through the lungs or through the intestines (most common in children). They are carried around in the lymph or blood vessels, settle in an organ, multiply, and produce small greyish nodules (or "tubercles") around themselves - big enough to be almost visible to the naked eye. When adjacent tubercles touch they fuse, forming a larger, yellow tubercle. This has a soft, yellow, cheesy substance inside.

As fusion spreads, the healthy tissue is broken down, to be replaced by the diseased substance of the yellow tubercles. In pulmonary tuberculosis this infected substance will eventually burst into a bronchial tube and be coughed up, leaving a hole in its place.

Often areas of fibrous scar tissue are built up as the body tries to surround and contain the infection.

SYMPTOMS

These depend on the organ attacked. In all cases the bacteria disrupt and then destroy the organ and its functions.

In pulmonary tuberculosis the first infection is often unrecognized, and thought of as a bad cold or flu. There is a cough, fever, and possibly chest pain. This often clears up, leaving a hardened, scarred area called the primary complex. Many people have signs of this primary complex with no further trouble.

Secondary infection occurs when the bacteria spread to the rest of the body. The patient spits blood, has a chronic cough, and experiences pain when inhaling. He loses appetite and weight, is constantly tired, and sweats profusely.

TREATMENT

Since the discovery and use of streptomycin and other drugs, the dangers of tuberculosis have been greatly reduced. Surgical treatment is now seldom needed, though fresh air, a healthy diet, and plenty of rest are still essential to full recovery.

PNEUMOTHORAX

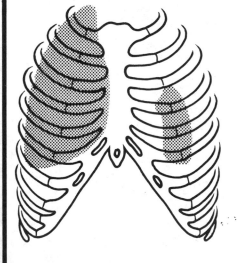

This is air in the chest cavity. It may have come from outside through a wound, or from inside through a hole in the lung. The lung on that side collapses as a result, causing sudden severe pain and breathing difficulties. But it expands again in a short time, because air in the chest cavity is quickly absorbed.

Pneumothorax may be induced surgically, in cases of tuberculosis for example, to allow the collapsed lung to rest and heal.

SOUNDS OF DISORDER

As air passes in and out of the lungs, it makes sounds that can be heard through a stethoscope. In a healthy person these have a sighing or rustling character. But in an unhealthy person, unusual sounds and their location can tell of lung disorders. Tubes that are constricted but dry cause whistling and "snoring"; those narrowed by mucus, sibilant sounds; those filled with fluid, bubbling noises. In each case the coarseness and loudness of the sound suggests the size of the tube involved.

©DIAGRAM

BOWEL DISORDERS

THE INTESTINES

The intestines are the long tube by which food leaves the stomach and is eventually excreted from the body. The tube is made up of sheaths of muscle, coated on the inside with mucous membrane.

The small intestine leads directly from the stomach. It is about 22 feet long, and up to $1\frac{1}{2}$ inches wide. It continues the process, begun in the stomach, of absorbing nutrients from the food.

The large intestine (the "colon") follows on from this. It is about 6 feet long and up to $2\frac{1}{2}$ inches wide. Its main function is the absorption of water from the waste products ("feces").

Many physiological disorders may affect the small intestine, eg bacterial infection, or fever. It can also be a site for ulcers and cancer.

However, the term "bowels" refers mainly to the large intestine - and often simply to the last 6 to 8 inches (the "rectum") and the surface opening (the "anus") through which the waste products are excreted, usually in a fairly solid form known as "stools." This is another potential cancer site. But it is also affected by certain well known disorders, linked with the physical process of waste evacuation, and dealt with on these pages.

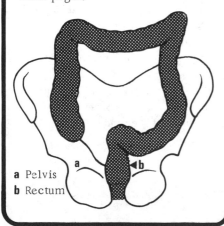

a Pelvis
b Rectum

NORMAL BOWELS

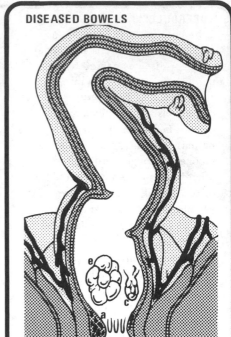

a Rectum
b Veins
c Mucous membrane
d Skin

DISEASED BOWELS

a Internal hemorrhoid
b External hemorrhoid
c Polyp
d Perianal abscess
e Rectal carcinoma

DIARRHEA AND CONSTIPATION

These common complaints are both usually caused by the failure of the colon to carry out its job of controlling the level of water in the feces.

This may be due to any one of many causes: a change of eating habits; gastritis (inflammation of the stomach); gastro-enteritis (inflammation of stomach and intestine); or bacterial or viral infection of the intestine.

DIARRHEA

Diarrhea is the excessive discharge of watery feces. The primary danger in serious cases is body dehydration, and this can be combated by an increased intake of fluid.

CONSTIPATION

Constipation is infrequent or absent defecation. It is usually caused by a poor diet, expecially one lacking in roughage. But it may follow diarrhea in the course of an infection- and is also sometimes caused by intestinal obstruction. However, much imagined constipation is only the consequence of judging bowel habits by an excessive norm of "regularity." In fact, "normal" bowel motions may occur as often as three times a day, or as infrequently as once every three or four days, depending on the individual.

HEMORRHOIDS (PILES)

Hemorrhoids (piles) come about through the enlargement of veins in the wall of the rectum or in the anus.

This may be due to acute constipation, or overstraining during excretion. It can also result from tumors.

The swellings cause the mucous membrane to press against passing feces, causing discomfort, pain, and sometimes bleeding.

Internal hemorrhoids occur at or before the rectum's junction with the anus. If they protrude beyond the anal opening the pressure of the anal muscle (the "sphincter") often causes great and constant pain - this is known as "strangulation".

External hemorrhoids occur under the skin just outside the anus. In addition to the usual causes, they can also result from a ruptured vein, leading to a hemorrhage.

Internal hemorrhoids may eventually develop "polyps". This is a condition in which the hemorrhoidal protrusion becomes fibrous and elongated.

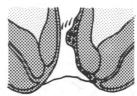

Internal hemorrhoid

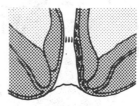

External hemorrhoid

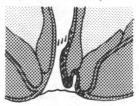

Enlarged hemorrhoid

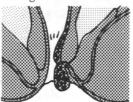

Strangulated hemorrhoid

COLITIS

Colitis is inflammation of the colon - often with an associated ulcer. The symptoms are abdominal discomfort, diarrhea, blood in the feces, and fever. Anemia and even emaciation result. The first (acute) phase can be fatal if untreated. More usually, a prolonged (chronic) phase develops.

The causes are unknown, but may be linked in different individuals with: infection; allergy; deficiency of vitamin B and certain proteins; or simply nervous stress. Sometimes several causes occur together.

Treatment involves bed rest until the fever has passed; and also careful dietary control, excluding milk and all products derived from milk. Steroids may be used. Relapses are frequent. In extreme cases surgery is needed.

RECTAL PROLAPSE

This is the collapse of the rectal wall. It occurs mostly in young babies and the aged. It is caused by excessive straining during excretion, and (in the old) by weak rectal and anal muscles. In severe cases an entire area of the rectal wall passes through the anal sphincter. Extreme pain from strangulation results.

ABSCESSES

An abscess is caused by bacterial infection. In order to combat the bacteria, body fluid and white blood corpuscles collect in the tissue spaces, and form pus. A painful swelling results that continues to grow until it bursts and discharges its fluid.

To avoid discomfort and the possibility of further complications abscesses are usually drained surgically. Anorectal and perianal abscesses are extremely painful, because of the pressure of the anal sphincter, the passage of feces, and the constant irritation due to their anatomical positioning.

FISSURE-IN-ANO

This is splitting of the walls of the anus. It is usually due to the passing of an exceptionally large stool. An "acute" fissure involves only the outer surface of the wall (the mucous membrane). If it does not disappear after a few days, it develops into a "chronic" fissure, which is deeper. This causes great pain and needs intensive treatment.

© DIAGRAM

HERNIAS

HERNIAS

A hernia has occured if a body organ protrudes through the wall of the body cavity in which it is sited. This happens most often in the abdomen: part of the stomach or intestine is pushed through the abdominal wall.

Hernias occur where the cavity wall is weak, either because of a natural gap where a blood vessel or digestive tube passes, or because of scar tissue. They are often called "ruptures", but this really means any tearing or breaking of tissue, eg ruptured blood vessels.

INGUINAL AND FEMORAL HERNIAS

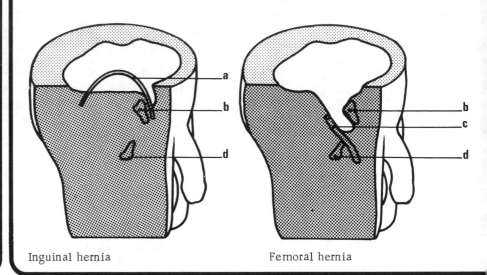

Inguinal hernia   Femoral hernia

TYPES OF HERNIA

INGUINAL HERNIAS are by far the most common. In men, the inguinal canal is the pathway down which the testes descend just before birth. In later life it contains the spermatic cord and blood vessels. In an inguinal hernia, part of the intestine protrudes down this canal, into the scrotum. Since the inguinal canal is much smaller in women (containing only a fibrous cord), inguinal hernias are much more common in men.

FEMORAL HERNIAS are more often found in women. The femoral canal is the route through which the main blood vessels to the leg pass from the abdomen. In a femoral hernia, part of the intestine passes down the canal and protrudes at the top of the thigh.

UMBILICAL HERNIAS occur where the abdominal wall has been weakened at the navel by the umbilical cord. They are found mostly in young children.

VENTRAL HERNIAS occur where the abdominal wall has been weakened by the scar of a wound. (When the scar is due to an operation, it is called an incisional hernia.)

EPIGASTRIC HERNIAS are protrusions of fat and sometimes intestine through the abdominal wall between the navel and the breastbone.

OBTURATOR HERNIAS occur when part of the intestine passes through a gap between the bones of the front of the pelvis.

HIATUS HERNIA occurs when the upper part of the stomach protrudes upwards through the hole in the diaphragm occupied by the esophagus.

Hernias can also be classified in other ways.

CONGENITAL OR ACQUIRED Congenital hernias exist at birth. All others are acquired hernias. The only congenital hernias are umbilical or inguinal, but congenital weakness in the abdominal wall may give rise to hernias later on.

REDUCIBLE OR IRREDUCIBLE Reducible hernias can be pushed back into place in the abdomen. Irreducible hernias cannot - the opening is too small.

STRANGULATED OR UNSTRANGULATED Strangulated hernias are those in which the tightness of the opening has cut off the blood supply. This is a very serious condition, leading rapidly to tissue death, gangrene, and death. In unstrangulated hernias, the protruding tissue still has its blood supply.

a Spermatic cord
b Pubic bone
c Femoral canal
d Ischium

e Site of inguinal hernia
f Site of femoral hernia

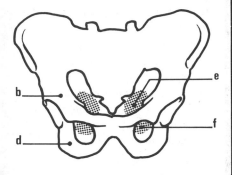

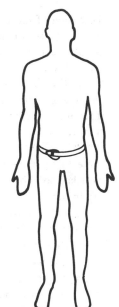

Truss for right
inguinal hernia

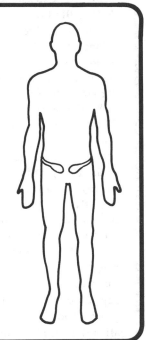

Double truss

CAUSES, SYMPTOMS, AND TREATMENT

CAUSES

Congenital hernias are caused by the failure of some channels to close properly during fetal development. The intestine is either displaced at birth or easily becomes so.

Acquired hernias are caused by any form of straining or exertion that increases pressure in the abdomen, and forces it through a weak spot in the abdominal wall, eg physical work, straining at the bowels, violent coughing, etc. Strain and exertion equally act as predisposing factors, i.e. they weaken the abdominal wall, as also does any large, sudden gain or loss in weight (including pregnancy). Because of the use of men in heavy manual work, the male is much more likely to suffer from hernias.

SYMPTOMS

These depend on the type and condition of the hernia, the size and tightness of the opening, and the amount of the organ involved. Also the onset of the hernia may be gradual, with the symptoms increasing till they become noticeable; or sudden (perhaps whilst lifting a heavy weight), in which case the person is often aware of something having "given way", perhaps with varying degrees of pain.

In general there is a feeling of weakness and pressure in the area, occasional pain or a continual ache, and a gurgling feeling in the organ under strain. A swelling may be present all the time or may appear only under pressure. Swellings that are continually present may increase in size.

Digestion is disrupted, usually causing constipation.

Strangulated hernias produce special acute symptoms. When the blood supply is cut off the protruding tissue dies and swells, increasing the pressure in the opening. The hernia becomes inflamed and acutely painful and the skin over the area may redden. (With intestinal hernias, forward movement in the intestine ceases, and there may be vomiting.) The dead tissue in the hernia quickly becomes gangrenous, often within five or six hours, and this in turn causes peritonitis - inflammation of the abdominal lining and its contents. If untreated, death occurs within a few days.

TREATMENT

Reducible hernias are sometimes held in place by a truss - a belt with a pad which is fitted over the hernia. But as long as the hernia exists, the risk of future strangulation remains. Most hernias are therefore treated surgically. Any damaged tissue is removed, the protruding organ replaced in the abdomen, and the opening stitched up again.

Strangulated hernias require immediate operation.

©DIAGRAM

ULCERS AND GOUT

PEPTIC ULCER

An ulcer is a breach in the surface of the skin or in the membranes inside the body. The breach does not heal, and it spreads across, and through, the tissue.

PEPTIC ULCERS

These are ulcers of the stomach and duodenum (the first part of the small intestine). They occur if the lining of the stomach or duodenum fails to stand up to the digestive properties of the gastric juices, i.e. the stomach and intestine begin to digest themselves. Peptic ulcers rarely exceed $\frac{3}{4}$ in across.

Duodenal ulcers form 80 to 90% of peptic ulcers, "gastric" (stomach) ulcers 10 to 20%. Men form 90% of all sufferers. (In women, gastric ulcers are slightly more common than duodenal ones.) Duodenal ulcers can occur at any time after the age of 20, gastric ulcers usually occur after 40.

CAUSES

The exact cause of peptic ulcers is not understood. It is thought that any small cut or tear in the lining is eroded and deepened by the action of the digestive juices. But why these ulcers do not occur more often (when such a cut or tear is likely to happen to all of us at one time or another), and why the erosion works through but not right across the surface, is just not known.

ASSOCIATED FACTORS

Some things are known to increase the likelihood of developing a peptic ulcer:
living under considerable stress;
drinking large amounts of alcohol;
eating rich food;
having excess acidity of the stomach;
suffering from frequent stomach or intestinal infection;
being of blood group "O";
having a family history of ulcers;
and being of "personality type A"*. The fact that women are much more likely to develop gastic ulcers after the menopause suggests that the female hormone, estrogen, may have some preventative value.

SYMPTOMS AND TREATMENT

SYMPTOMS

There is pain in the upper abdomen, which gets worse when the stomach is empty and can often be relieved by taking more food. There is also tenderness in the area of pain. Indigestion, nausea, and vomiting may occur. If the cause is an ulcer, it can be seen on X rays.

TREATMENT

Bed rest is needed and use of antacids to neutralize the stomach juices. Diet should be controlled: rich, strong foods, alcohol, tea, and coffee must be avoided. Frequent snacks of soft, bland food

---

When does pain occur?

Is it made better by food?

Vomiting?

Appetite?

SITES OF ULCERS

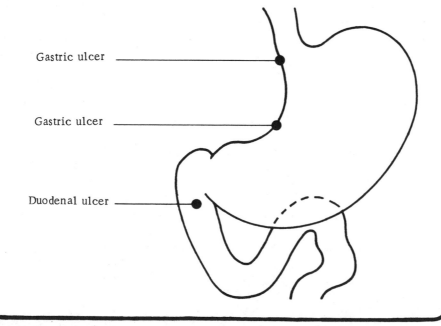

Gastric ulcer

Gastric ulcer

Duodenal ulcer

DANGERS

PERFORATION occurs when the ulcer eats right through the stomach or duodenal wall. This need not happen because there is a continual laying down of scar tissue during the process of erosion. But if it does happen it is extremely serious, because digestive juices are released into the abdominal cavity, threatening fatal peritonitis. Immediate surgery is needed: by 8 to 10 hours after perforation, the patient's situation is very grave.

Perforations are 8 to 10 times more likely with duodenal ulcers than with stomach ulcers.

are taken, so the patient is eating about every two hours. This also helps to reduce stomach acidity. If the ulcer does not improve, surgery may be needed. With a stomach ulcer, the part of the stomach containing the ulcer is removed. With duodenal ulcers, the amount of gastric juice reaching the ulcer is reduced. This may be done by:

cutting some of the nerves that trigger gastric juice production;
or removing a part of the stomach where production occurs;
or diverting the outflow past the ulcer.

DUODENAL	GASTRIC
Before meals or 2 to 2½ hours after	½ to 2 hours after meals
Yes	Sometimes
Rare	Common
Good	Fair

OBSTRUCTION Peptic ulcers may block the passage of food through the stomach and/or duodenum, by causing swellings or muscular spasms. This is treated by administering special foods intravenously, in the hope that the obstruction dies down when irritation is removed. If this fails, surgery is needed.
HEMORRHAGE occurs when a blood vessel is ruptured by the ulcer. Blood is vomited, or passed in the feces. If the bleeding cannot be controlled within 24 hours, surgery is needed.

GOUT

Gout is a recurrent - and extremely painful - inflammation of certain joints. It is caused by overproduction of uric acid in the body. (Uric acid is produced during the breakdown of proteins, and is usually excreted from the body in the urine.) Excess uric acid is carried round in the bloodstream, and crystals of the acid and its salts (urates) are laid down in the cartilage of the joints - most often those of the feet (especially where the big toe joins the foot) and hands. The exact cause of the increase in uric acid is not known, but gout is associated with overconsumption of rich and high protein food, and of alcohol, and with sedentary living. It seldom occurs before the age of 45. In 80% of cases there seems to be a family history, so it may be partly hereditary. Men form 95% of all sufferers: women only develop gout after their menopause, so there is thought to be a connection with sex hormones.

ACUTE GOUT

Attacks of gout usually begin at night, with acute pain in the big toe or thumb. The joint becomes red, swollen, shiny and very tender, and the patient is feverish and irritable. He passes less urine than normal, but of a much thicker color. If untreated, the attacks can last from four to ten days. With drugs, the pain can be relieved within 24 hours. Large amounts of fluid - at least 5 pints a day - should be drunk, to stimulate the kidneys.

CHRONIC GOUT

The first attack of gout is very seldom the last. In the first subsequent attacks, the same site is usually affected, but later more joints become involved.
In chronic gout the acute attacks occur more often. They tend to be less painful, but the symptoms do not clear up completely in between. The joints affected become arthritic, as the deposits of crystal become permanent. The deposits gradually form stones ("tophi"), and the joints become swollen, disfigured, and fixed. Crystals are also deposited in the kidneys (which may lead to kidney failure), under the skin, in the eye, along the tendons, and in the cartilage of the ear.
In cases of long standing, gout is accompanied by degenerative changes in the heart, arteries, and liver.
In treatment, drugs are used to combat the pain and to reduce the level of uric acid in the blood. Diet is also controlled: kidneys, liver, brains, fish roes, sardines, spinach, strawberries, and rhubarb should not be eaten. Also alcohol should not be drunk, and the weight is best kept below average.

©DIAGRAM

URINARY DISORDERS 1

L20 SYMPTOMS OF DISORDER

For a doctor, the urine and urination are among the most useful signs of disorder - relating sometimes not just to the urinary system, but to the general health of the body.

Characteristics of urination that may interest a doctor include: changes in quantity and frequency (including rising at night); slow and weak, or unusually forceful flow; stopping and starting, and dribbling; difficulty in beginning or continuing; inability to restrain (incontinence); sudden stopping; and, of course, pain or other unusual sensation on urinating, or inability to urinate at all.

Characteristics of the urine that may be of interest include unusual color, odor, cloudiness, frothiness, and content. Abnormal chemical content can include albumen (which may indicate kidney disorder) or sugar (diabetes). Chemical testing can be carried out very easily, using a treated paper that changes color when moistened with urine. Other abnormal contents can include bacteria, parasites, kidney tube casts, bile pigment, and especially blood or pus.

However, many unusual characteristics of the urine or urination will more usually be due to insignificant causes than to disorder. For example, having to get up from bed to urinate is often due to drinking tea or coffee last thing at night. Strikingly unusual colors can be produced just by certain medicines and foods.

OTHER SYMPTOMS

Other symptoms of disorder include: itching, redness, or stickiness at the urethral opening; any discharge of fluid from the urethra; pain or swelling in the area of the kidneys, and shivering, temperature, or fever.

L21 TYPES OF DISORDER

INFECTION

This can reach the urinary system in two ways: "downward," via the bloodstream and then the kidneys; or "upward," via the urethral opening in the genitals. An example of the first can be tuberculosis. But the second is much more common, and especially common in women:
a) because in women the closeness of anus and genitals helps bacteria pass between them; and
b) because the shortness of the female urethra allows bacteria to reach the higher parts of the tract more easily.

Most bacteria entering the tract from outside are killed by the urine; but 5% of women (both adults and children) do have active bacteria in the bladder. Often there are no symptoms. If there are, frequency of urination and pain on urinating are typical. Diagnosis is by bacteriological examination of a urine sample. Treatment is with an antibiotic.

INFLAMMATION

Inflammation of the tract is mostly caused by infection, but also by: dietary irritation (eg alcohol, and perhaps food allergy); use of chemicals (vaginal deodorants, contraceptive foams, etc); and tissue damage during sexual activity, childbirth, or surgery. Even when not caused by infection, it can offer a favorable site for infection. Inflammation of the urethra is called "urethritis," that of the bladder "cystitis" - but see p.146. The symptoms are: discharge of pus from the penis, pain or urinating, tenderness of the urethra, and possibly inflammation of other organs such as testes, bladder and even kidneys. In women it often leads to cystitis. Treatment depends on the cause, but drinking large quantities of fluid usually helps.

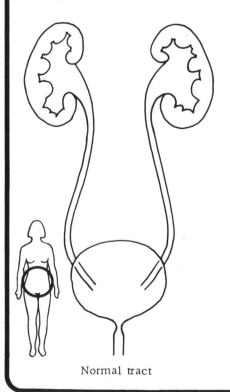

Normal tract

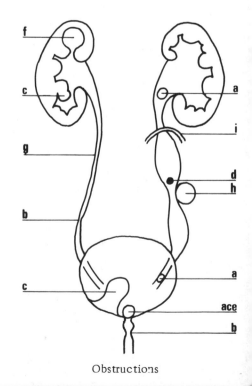

Obstructions

FLOW ABNORMALITY

This includes obstruction of flow, complete or incomplete; also apparently normal flow that nevertheless leaves stagnant pools of urine in the tract. Causes include:

a) blockage by extraneous objects (eg stones, blood clots, etc);
b) malfunction of the tract itself (eg through congenital controlling malformation, tumors and other growths or tissue changes, and temporary spasm); and
c) outside pressure on the tract (eg from fibroids, displaced uterus, or pregnancy).

Stagnant urine is always a likely site for infection. Where there is flow blockage as well, pressure builds up behind the obstruction, and that section of the tract may be stretched and dilated. Eventually the pressure and dilation may reach back up the ureters towards the kidneys. Kidney infection may result, and rapid surgical treatment is needed, before the kidneys suffer permanent damage.

INCONTINENCE

This is inability to control urination. For incontinence in the old, see M16; but it also occurs in younger women. Causes include: psychological stress (eg severe fright); disorders of the bladder; congenital defects; tissue damage occurring in childbirth or surgery; and impairment of the nerves due to injury or disease.

Two types are fairly common:

a) urgency incontinence, where there is a shortened time gap between the desire to urinate and uncontrollable urination - it occurs quite often in women over 40; and
b) stress incontinence, typified by small amounts of urine escaping when the person strains, coughs, or laughs - whether the bladder is full or virtually empty. This is usually only seen in post-menopausal women; special exercises, or sometimes surgery, are needed.

STRICTURES

A stricture is an abrupt narrowing of the ureters or urethra. In the ureters, it may be congenital, or caused by physical irritation such as the passage of kidney stones or surgical instruments. Treatment is difficult, and surgery is usually necessary.

Strictures of the urethra are more common. They may also be congenital, but are usually "spasmodic" or "organic".

Spasmodic strictures are temporary, and due to irritation by cold, excess alcohol, or physical objects. They last only a few hours or days, and cause no permanent discomfort. Organic strictures follow prolonged inflammation or laceration, and if untreated may cause distension and inflammation of bladder and kidneys. Both forms are treated by stretching the urethra with special instruments. In organic strictures this must be repeated regularly.

Sites of infection

OBSTRUCTIONS include:

a stones,
b strictures,
c tumors,
d blood clots,
e foreign bodies, and
f TB fibrosis.

Also external pressure from:

g pregnancy,
h tumors, and
i congenitally displaced arteries.

INFECTIONS include:

j TB, and
k kidney infections.

Also especially, bacteria:

l in the bladder,
m in stagnant or obstructed urine,
n in stones,
o in foreign bodies, and
p in structural inflammations.

©DIAGRAM

URINARY DISORDERS 2

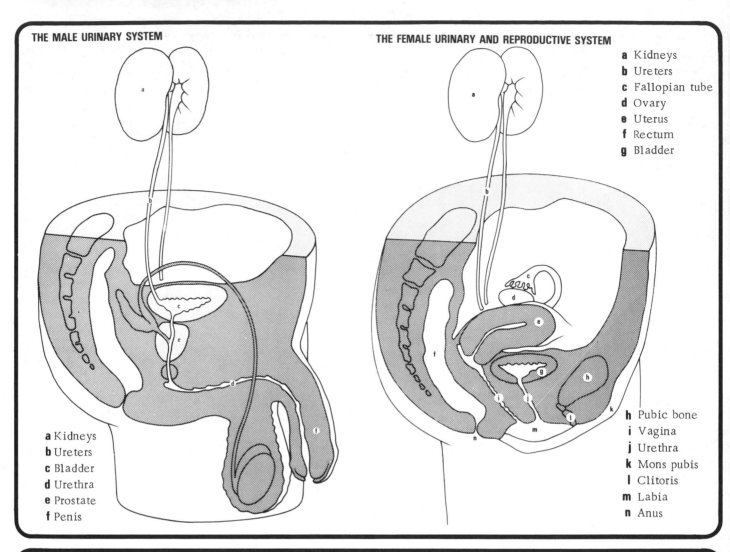

THE MALE URINARY SYSTEM

THE FEMALE URINARY AND REPRODUCTIVE SYSTEM

a Kidneys
b Ureters
c Fallopian tube
d Ovary
e Uterus
f Rectum
g Bladder

a Kidneys
b Ureters
c Bladder
d Urethra
e Prostate
f Penis

h Pubic bone
i Vagina
j Urethra
k Mons pubis
l Clitoris
m Labia
n Anus

KIDNEY DISORDERS

These include: congenital defects; tumors; stones; damage through injury; inflammation without infection; and infection. INFECTION is especially common in women. It can arrive via the bloodstream or the urinary system. In acute attacks, bacterial infection via the urinary tract is typical. Symptoms are shivering and fever, acute pain in loin or under ribs at back, and frequent urination. Qualified medical attention is vital: prescribed antibiotics, bed rest, and plenty of fluids. Long-term infection may follow acute infection, or arise from urinary obstruction or blood-borne infection. (Stones are frequent sites.)

Symptoms include dull back pain, painful and frequent urination, tiredness, headache, nausea, loss of appetite, and fever. Treatment depends on causes. In neglected cases, kidney damage may result, with possibly high blood pressure and blood poisoning.

Kidney damaged by back pressure

Normal kidney

Kidney damaged by infection

©DIAGRAM

144

PROSTATIC DISORDERS – MALE

The prostate gland* surrounds the junction of the bladder and urethra. It is normally about 1in by $\frac{3}{4}$in by $1\frac{1}{2}$in, and during ejaculation it supplies a fluid without which the sperm is sterile.

However, the prostate can become enlarged, especially in the elderly. There are three possible processes:

a) Benign enlargement, which is the usual form, and occurs when fibrous cells multiply inside the gland. The cause is unknown, but may be due to hormonal imbalance.

b) Cancer of the prostate, which is a common form of cancer in men.

c) Prostatitis, or inflammation of the prostate, which is usually due to infection in the urinary tract (such as gonorrhea, cystitis or urethritis). Such inflammation may be temporary or chronic.

SYMPTOMS

Enlargement eventually causes retention. In this case the following symptoms occur:
urination is slow to begin, lacks force, and is often interrupted by pauses;
the desire to urinate occurs with increasing frequency, especially at night;
there are further dribbles of urine after urination has stopped;
and, if the stagnant urine retained in the bladder sets up infection, or if infection passes up from the urethra, pain on urination occurs.
Finally, the urethra may be completely blocked, so no urine can pass.

TREATMENT

Prostatitis may be dealt with by antibiotics and prostatic massage, but more serious prostate enlargements require surgical removal of the gland itself. After this operation, a man is usually sterile, and ejaculates his semen backwards into the bladder, rather than externally. However, erection and the experience of orgasm remain as before.

In the case of cancer, hormonal treatment may be an alternative to surgery. Cancer of the prostate may, of course, metastasize elsewhere.

Other possible disorders of the prostate include tuberculosis and the formation of stones.

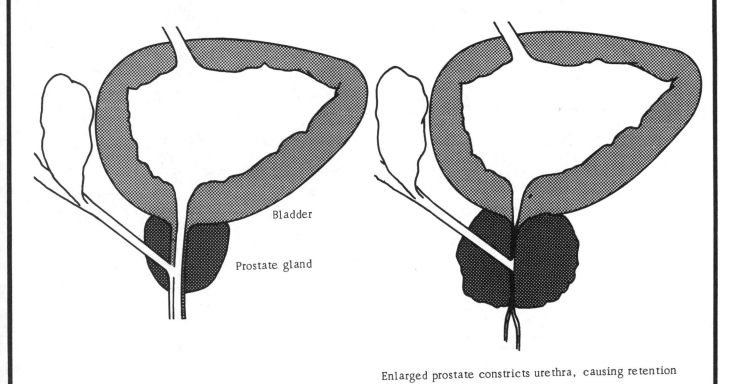

Bladder

Prostate gland

Enlarged prostate constricts urethra, causing retention

©DIAGRAM

CYSTITIS

CYSTITIS

Strictly, this means inflammation of the bladder. However, it is now generally used for a certain collection of symptoms, usually in women, which can arise in a variety of ways. The main symptoms of an attack (acute cystitis) are:

a) great frequency of urination (perhaps every few minutes);
b) pain on urination - often extreme; and yet
c) a recurrent or even continuous desire to urinate, even when there is no urine to pass.

As the attack continues, there may also be increasing incontinence, and often blood in the urine. Other associated symptoms can include: pain just above the pubic bone, or in the loin; and a foul smell from, and perhaps debris in, the urine. Also extreme pain may be felt if sexual intercourse is attempted. This syndrome is very common: perhaps 80% of women suffer from it at some time in their lives, and it is often recurrent (chronic cystitis) and hard to eradicate. In its extreme forms it can bring depression, disrupted career and home life, and even (since it can both derive from sexual intercourse and interfere with it) broken emotional and marital relationships.

CAUSES OF CYSTITIS

There are two main alternative causes:
a) infection;
b) inflammation without infection (though infection may also set in later).

INFECTION

This is usually by Escherichia coli (E. coli) bacteria from the rectum finding their way into the urethral opening. E. coli are often found on the perineum (the skin between anus and genitals). Their progress towards the vulva is often helped mechanically by careless use of toilet paper or by sexual activity (petting, or

INVESTIGATION

The sufferer should always see a doctor - and always try to get proper tests made to pinpoint the cause. First step should be laboratory testing of a urine sample for infection and (if present) responsiveness to drugs. The patient should drink before going to the doctor, so as to be ready to pass urine for this. A clean sample is important: the vulva should be swabbed, and only a small midstream sample (i.e. from halfway through urination) taken. During menstruation, a tube (catheter) inserted into the urethra should be used. The patient may also be able to give useful information, eg the amount of time between the attack and the last previous intercourse. (Cystitis due to inflammation alone will follow intercourse sooner than that due to infection, since bacteria need time to multiply. Unfortunately, estimates vary - from "very soon after" for inflammation and 12-24 hours after for infection, to 24 hours after for inflammation and 36 for infection.)

If no infection is found, or if it fails to clear after a course of drugs, hospital investigations may be needed, such as:

a) physical examination by a specialist;
b) taking of bacteria samples from vagina and perineum;
c) early morning urine samples;
d) blood samples;
e) X rays of the urinary tract, often using injections of dye into the bloodstream to show up obstructions, or introducing dye into the bladder to show its action; and
f) cystoscopy, which is the surgical inspection of the inside of the urethra and bladder, using a "periscope tube" inserted into the urethra under general anesthetic.

The diagrams show "intravenous pyelograms" (IVPs): X rays of the urinary tract taken after iodine dye has been injected into the bloodstream. The iodine passes out through the urinary system.

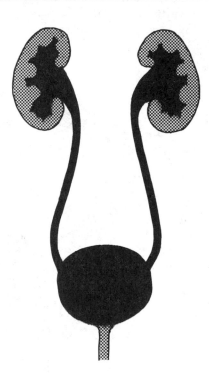

IVP showing normal functioning of the kidneys and urinary tract

just the movement of the penis).
Other sources of infection are:

a) similar cross-infection from the vagina (eg candidiasis, trichomoniasis, or gonorrhea);
b) lack of male hygiene (eg when uncircumcised); and
c) infections from the kidneys that pass downward (eg tuberculosis).
Infection may be aided by: stones; stagnant pools of urine due to retention; lowered resistance, as in anemia; and (for bacteria preferring non-acidic urine) diabetes.

INFLAMMATION
For general causes, see p.142.

In cystitis, the normal cause (apart from infection) is bruising or skin cracking through sexual activity. Relevant here are: frequency of intercourse (hence "honeymoon cystitis"); insufficient lubrication; use of certain positions (depending on the individuals); and over-forceful petting. Other relevant causes are:

a) tissue irritation through use of vaginal deodorants, foam contraceptives, unsuitable lubricants, etc;
b) strain on the bladder due to prolapse of the uterus;
c) damage through childbirth or

surgery; and possibly
d) allergic reaction of the urinary tract to certain foods.
Inflammation can, in turn, provide a breeding ground for infection (and an entry for infection into the bloodstream).

CHRONIC CYSTITIS
This is usually a case of repeated attacks of acute cystitis, but there may be long-term tissue changes also involved, including changes in the urethra due to the menopause, and changes in the bladder lining from bacterial or other infection. Occasionally there may be psychological factors.

IVP showing blockage in one ureter, distention above the blockage, and a growth in the bladder

TREATMENT

Depending on the cause of trouble, treatment may include:
a) antibiotics and similar drugs, to combat urinary and/or kidney infection;
b) increase of the patient's fluid intake;
c) drugs to relax the muscles of the bladder;
d) drugs to combat vaginal infection;
e) hormone therapy to restore mucus and tissue characteristics;
f) surgery for urinary blockages;
g) surgery to deal with other causes of inflammation, eg repair for a prolapsed uterus.
When on a course of drugs:
a) the symptoms may vanish soon after starting the course;
b) there may be side effects, eg nausea or depression.
But it is very important to finish the whole course.
Drinking a vitamin C source may be requested, to bring urinal acidity into a range where the drug works best.

SELF-HELP IN AN ATTACK

Cystitis attacks still have to be dealt with, despite medical help and preventative precautions.
At the first hint of trouble:
a) pass a urine specimen into a clean, closed container, for the doctor;
b) drink 1pt ($\frac{1}{2}$ liter) of cold water;
c) take a mild painkiller;
d) lie or sit down with two hot water bottles, one against the back, one (wrapped in a towel) high between the legs;
e) drink $\frac{1}{2}$pt ($\frac{1}{4}$ liter) of water, diluted fruit juice, or barley water, every 20 minutes;
f) (but not if a heart patient) take a teaspoon of bicarbonate of soda in a little water, and repeat each hour for 3 hours;
g) use diuretic pills if prescribed; and
h) after every urination, wash the skin between anus and vulva and dab it dry.
After half an hour, the attack should begin to ease.

© DIAGRAM

DISORDERS OF THE UTERUS

PROLAPSE

Prolapse of the womb is a not uncommon condition, in which the uterus sags down into the vagina, and may even protrude out between the legs. The symptoms include frequent and difficult urination; incontinence; vaginal discharge; low backache; a feeling that something is coming out of the vagina; and, especially, that all the above symptoms immediately disappear on lying down.

The condition is produced by weakening of muscles that support the uterus. The cause is usually damage done in childbirth; 99% of women with prolapsed wombs have given birth. But aging and heavy physical activity also contribute, and the symptoms often appear only after the menopause, when the affected muscles may lose tone and ligaments atrophy. Mild cases require no treatment, but more serious or troublesome ones need a pessary inserted by a doctor, or sometimes surgery.

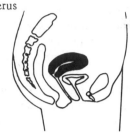

Normal uterus

Prolapsed uterus

With pessary

RETROVERSION

In most women, from puberty on, the upper end of the uterus is tilted forward in the body, and moves backward only as the bladder fills or when the woman lies on her back. But in about 10% of women the uterus is always retroverted (tilted backward). Once blamed for many ailments, in fact this may be troublesome only in pregnancy, when the enlarging uterus may fail to rise into the abdomen. Urine retention in the bladder results. A doctor can usually correct the situation by hand. Untreated, it could cause cystitis, and even miscarriage. Retroversion can also start after childbirth. Doctors disagree whether this can cause backache, etc. If it seems very troublesome, surgery is needed. Other causes of displacement can include: pelvic tumors (such as ovarian cysts); and connective tissue joining to other structures. Surgery can deal with these if necessary.

"CERVICAL EROSION"

The cells lining the cervical canal sometimes extend down till they show as a reddened area at the head of the vagina. This happens naturally in puberty and first pregnancy, and needs no treatment unless it persists over 6 months after childbirth and causes much vaginal discharge. It then needs electric cauterization, which produces a heavy, discolored vaginal discharge for 4 to 6 weeks till healing is complete. It may also occur in women on the pill or using IUDs. Again it usually disappears without symptoms or treatment, but needs regular "Pap tests" against cancer.

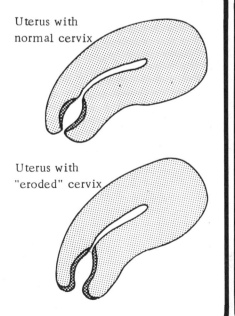

Uterus with normal cervix

Uterus with "eroded" cervix

ENDOMETRIOSIS

Endometrial cells can grow in the wrong place, forming cysts in the uterus muscle, the ovaries, or other parts of the pelvis. This is commonest in unmarried or infertile women in their thirties. On menstruation these cells bleed a little, so the cyst swells, causing pain in the lower abdomen, especially before or at the end of menstruation, and sometimes pain on intercourse. Where ovaries or Fallopian tubes are blocked, infertility results. Treatment involves hormones or surgery (eg removal of the cyst, or part of an organ, or in severe cases hysterectomy).

Retroverted uterus

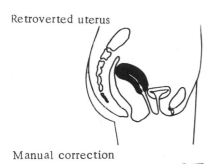

Manual correction

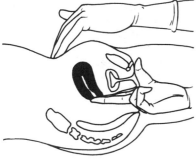

Retroversion due to a fibroid

FIBROIDS

These are lumps of fibrous tissue, growing in the muscle wall of the uterus, sometimes singly, sometimes in large groups, usually pea-sized, occasionally as large as grapefruits. They occur in about 20% of women over 30, especially the infertile, the sexually inactive, and those who only bear children late in life (also, for some unknown reason, in black women more than white). Their cause is unknown, but may be hormonal. Most give no trouble and need no treatment. Large ones can cause pain, heavy and irregular menstrual bleeding, womb enlargement that interferes with urination and bowel action, and infertility through spontaneous abortion. They can usually be removed by surgery, but in extreme cases hysterectomy is necessary. Long-term emotional stress can also enlarge the uterus and give heavy periods. This has occasionally resulted in unnecessary surgery.

Polyps are another type of lump found, also usually harmless, but developing from mucus tissue and forming dangling shapes.

Normal uterus

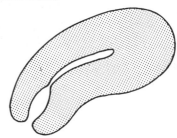

Uterus with fibroids

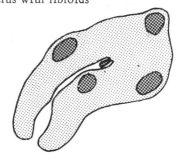

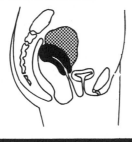 Typical endometriosis sites

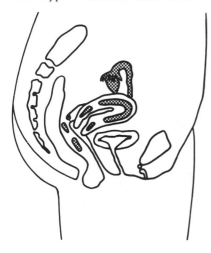

DILATION AND CURETTAGE

Also called "D&C," this involves:
a enlargement of the cervical opening, using dilators; and
b gentle scraping of the uterine wall with a metal curette.
It is used:
to diagnose cancer, pregnancy outside the uterus, or causes of abnormal bleeding or discharge;
to clear waste from incomplete delivery or abortion;
to help fertility;
to cause abortion;
and, sometimes, as routine preparation for gynecological surgery.
Anesthetic is needed. Recovery takes 6 hours to 2 days.

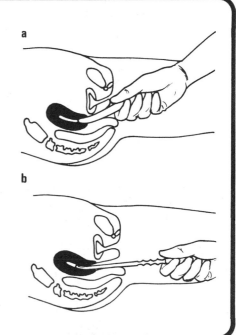

a

b

© DIAGRAM

HYSTERECTOMY

HYSTERECTOMY

This is surgical removal of the uterus (womb). It can be:

a subtotal (removal of the uterus except for the cervix);

b total (removal of uterus and cervix); or

c radical (removal of uterus, surrounding tissue, and part of the vagina).

With any of these, ovaries and Fallopian tubes may also be removed.

If the uterus is not enlarged by disease, removal can be via the vagina. Otherwise, an incision is necessary: a vertical one below the navel or a horizontal one just above the pubic region. (The scar left is almost invisible.) The operation takes less than an hour.

Area removed

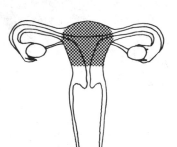
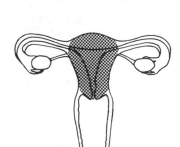
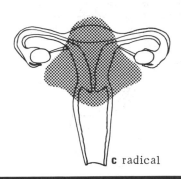

a subtotal   **b** total   **c** radical

REASONS FOR HYSTERECTOMY

In the USA, 25% of all women aged 50 and over have had hysterectomies. The diagram shows some valid reasons for the operation; but very often it occurs for no good reason (eg for removal of small fibroids). Some US doctors even favor routine hysterectomy once childbearing is over, to forestall any risk of cancer; but many others view this as a surgical racket. Some American doctors have also urged hysterectomy as combined abortion and sterilization, for poor women with large families. But again, most doctors are against this, because:

a) hysterectomy within 13 weeks of conception can endanger the mother's life; and

b) the patient cannot afford the hormone therapy necessary to combat the resulting severe menopausal depression if the ovaries also are removed.

Possible reasons for hysterectomy

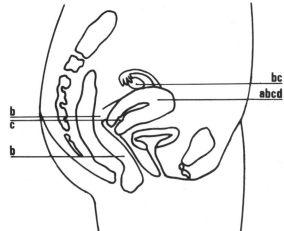

bc
abcd

a Fibrosis
b Endometriosis
c Cancer or pre-cancer
d Metropathia hemorrhagica

PHYSICAL AFTEREFFECTS

The diagram shows how the immediate effects of the operation wear off. Other considerations are long-term:

a) Menstruation ends immediately, and with it fertility and the need for contraception.

b) Subtotal hysterectomy usually has no sexual effect. With other forms, some women claim loss of sexual pleasure, after cervix removal. Some also complain of shortened vaginas.

c) If the ovaries remain, female hormone production continues. If they have been removed, severe menopausal symptoms result. Hormone therapy may be given to combat these.

d) Obesity does not follow hysterectomy unless the patient eats too much and exercises too little (but psychological factors may encourage this).

Weeks

① Patient usually out of bed in 1-2 days

② Patient usually home in 7-10 days

⑦ Most normal activity possible (including sexual intercourse)

⑦-⑧ Sedentary work becomes possible

① Perhaps temporary pain and inability to urinate (both treatable)

① Sometimes intravenous feeding for 2-3 days; brief gas pains on return to solid food

⑤ Moderate activity safe

⑩ Bending and lifting not safe until now

REMOVAL OF OVARIES

This is usually used to deal with or prevent cancer of the ovaries, which eventually affects 1% of US women over 40. It often accompanies hysterectomy, especially if this is performed for cancer of the uterus (about 10% of US women with the uterus only removed do later get ovarian cancer). Removal of large cysts can be another reason for the operation.

Where one ovary needs removal, the other may also be taken out to prevent it being a future site of disease (eg in cancer cases, or post-menopausal patients).

Post-menopausal women who lose both ovaries experience no special effects. Younger women have a menopause, often with severe symptoms, but treatable by hormone therapy.

Removal of the ovaries

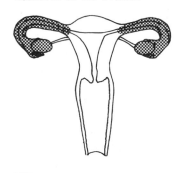

▓ Area removed

PSYCHOLOGICAL AFTEREFFECTS

Hysterectomy affects different women differently. Some enjoy the relief given from heavy bleeding or threat of disease, feel more active and healthy, and are happy that they can no longer conceive accidentally. But younger women often resent the loss of fertility, and many become depressed. Research shows that hysterectomy patients in the USA:

a) grow more dissatisfied with the operation as time passes;

b) are more likely to be dissatisfied if the ovaries are also removed (blaming the operation for hot flashes, lethargy, and obesity);

c) are four times likelier to become depressed in the 3 years after the operation than other women;

d) are likely to remain depressed for twice as long (2 years on average);

e) are especially liable to depression if under 40 when operated on; and

f) are five times more likely to make a subsequent first visit to a psychiatrist than other women - with the peak period 2 years after surgery, and the bulk of cases from those operated on for other than a serious physical condition.

In fact, all such statistics probably reflect a situation in which many hysterectomies have been performed unnecessarily; and even so, 41% of a typical sample were still satisfied with the operation, 4 years afterward. Where the operation is genuinely necessary, serious or long-term psychological disturbance is much less likely.

©DIAGRAM

VAGINAL DISORDERS

VAGINAL DISCHARGE

Apart from the contribution of the uterus at menstruation, normal vaginal discharge consists of:

a) clear watery mucus from the cervix (especially midway between periods);

b) clear fluid that has "sweated" through the walls of the vagina (usually only a small amount, but more in pregnancy or emotional upset, and a great deal during sexual excitement);

c) dead cells from the vaginal wall; and

d) a small contribution from the Bartholin glands at the vaginal entrance during sexual excitement. The resulting discharge is transparent or slightly milky, with little or no odor, slippery in feel, and perhaps yellowish when dry. It keeps the vagina moist and clean, and may be more noticeable at certain points in the menstrual cycle than at others.

SIGNS OF DISORDER

What is significant is not increased amount, but irritation, unpleasant odor, or unusual color. Irritation includes itching, chafing, soreness, or burning, of vagina, vulva, or thighs.

CAUSES

For abnormal menstrual discharge, see p. 100. There are several possible causes of other abnormal discharge.

a) Forgotten foreign bodies, eg tampons, or contraceptive caps. These can cause a very thick, odorous discharge, which clears up when the cause is removed and the vagina washed out.

b) Chemical irritation. Disinfectants in bathing water can cause soreness; vaginal deodorants, contraceptive foams, and even some soaps, can cause soreness and discharge. Symptoms may take time to clear after the cause is eliminated.

c) Post-menopausal atrophy.

d) "Cervical erosion".

e) Infection, including candidiasis, trichomoniasis, gonorrhea, and NSU.

INFECTION

Many bacteria live harmlessly in a normal healthy vagina. Some help keep its surface a little acidic, and this restricts the development of other, harmful organisms. Factors favoring infection include:

a) generally lowered resistance (due to lack of sleep, bad diet, other illness, etc);

b) cuts, abrasions, etc (eg from childbirth or intercourse without sufficient lubrication); and

c) potentially, all factors which affect the quantity and acidity of the vaginal mucus, including: menstruation and pregnancy; taking birth control pills, other hormones, or antibiotics; excessive douching; diabetes or pre-diabetes; and the menopause.

TRICHOMONIASIS ("TRICH")

Trichomoniasis vaginalis is a one-celled animal parasite, and the most common infectious cause of vaginal discharge. Perhaps 50% of women carry it at some time; about 15% develop symptoms at least once. The discharge is often greenish-yellow or grayish, thin and foamy, but may be thicker and whiter if other infection is also present. Other symptoms are: itching and soreness of vagina and vulva; clusters of raised red spots on cervix and vaginal walls; and an unpleasant odor. If it spreads to the urinary tract, it can cause

cystitis; if to the Fallopian tubes infertility. Transmission can occur sexually (men carrying it generally have no symptoms), and also very occasionally via moist objects such as towels, washcloths, and toilet seats (the parasite can live briefly outside the body). It cannot be passed on to a baby in childbirth. Qualified medical treatment is vital, especially as it often occurs in conjunction with gonorrhea (this should be checked for once the symptoms of trichomoniasis clear up). Both partners should be treated

But the usual treatment is with oral metronidazole (Flagyl) or tinidazole, both thought suspect drugs by some. (Prescribed vaginal suppositories or gels may be adequate alternatives, though infection may recur.) Especially avoid oral metronidazole if you are pregnant, have peptic ulcers or another infection, or have a history of blood or central nervous system disease. Also do not take alcohol with it. Avoid intercourse till tests show clear.

CANDIDIASIS ("THRUSH")

This is caused by a yeast organism (a type of fungus). It can be passed on sexually, men usually having no symptoms. But it often occurs in the vagina anyway, kept at bay by the acidic conditions, and only thriving if these get milder. Itching and soreness of vagina and vulva then result, especially when the body is warm (eg in bed at night). There may also be a thick creamy discharge that smells of yeast and looks like cottage cheese.

Self-treatment may help (eg one or two yoghurt treatments; or vinegar douches twice a day for 3 days), but will not usually clear up an established infection. Normal treatment is with nystatin suppositories, inserted to the top of the vagina - one a night for one or two weeks. (These have fewer side effects than oral doses, and can be used during pregnancy.) Tampons cannot be worn, as they will soak up the medication; but a pad and old pants are needed, as nystatin gives a yellow stain. Other suppositories have also begun to be used recently. Also creams are still sometimes prescribed for direct application to the vagina, cervix, and vulva; but these destroy all vaginal bacteria, so when treatment ends it is important to re-create acidic conditions in the vagina, to encourage normal bacterial growth.

Very recurrent infection suggests: lack of hygiene by the partner (eg thrush can live under the male foreskin); or developing sugar diabetes; or the effects on the vaginal mucus of taking oral contraceptives and/or antibiotics.

If a woman with thrush gives birth, the baby may have the infection in its digestive tract, and should be treated with nystatin drops.

OTHER INFECTIONS

NON-SPECIFIC VAGINITIS This is the name for any unidentified vaginal infection. Cystitis- like symptoms may be the first sign of disorder, followed by a discharge that is often white or yellow and may be streaked with blood. The vaginal walls may be puffy and coated with pus, and there may also be lower back pains, cramps, and swollen glands in abdomen and thighs. The usual treatment is with sulfa creams or suppositories (eg Vagitrol, Sultrin, or AVC cream).

HEMOPHILIS VAGINALIS This bacterium has now been identified, and in some areas is found to be more common than trichomoniasis or candidiasis. Symptoms are similar to trichomoniasis, and it is often misdiagnosed; but the discharge is usually white or grayish, creamy in consistency, and smells especially unpleasant after intercourse. It is often transmitted sexually, and treatment of both partners is necessary: for women, with nitrofurazone (Furacin) in suppository or cream form, or with sulfa suppositories; for men, with tetracycline or ampicillin.

L19 VAGINAL ORGANISMS

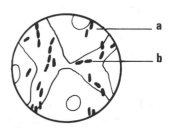

1) Normal vaginal organisms:
a dead mucus tissue cells;
b lactobacillus bacteria.

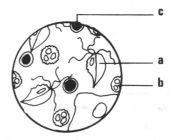

2) Trichomoniasis:
a one-celled animal parasites;
bc dead white defensive blood cells (pus).

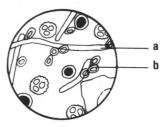

3) Candidiasis:
a yeast masses;
b yeast buds;
with other abnormal organisms.

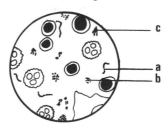

4) Non-specific vaginitis:
a streptococci;
b staphylococci
c bacilli;
and other organisms.

©DIAGRAM

DISORDERS OF THE PENIS AND SCROTUM

DEFECTS AT BIRTH
DISPLACED OUTLET
The outlet of the urethra should be at the tip of the penis. Epispadias is the condition in which the urethra comes out on the upper surface of the penis, instead of at the tip. Hypospadias is the condition in which it comes out on the under surface. Both cause difficulties. During urination, the man may have to sit, or tilt his penis. During intercourse, he may be effectively infertile, because too little semen finds its way to the uterus. Both conditions can be dealt with by surgery. If the foreskin is still present, it can be used to form a new passage for the urethra; if not, skin grafts are needed.

UNDESCENDED TESTES
The testes normally move down from the fetus' abdominal cavity to his scrotum during the eighth month of pregnancy. If they are still in the abdominal cavity at birth, the condition is called cryptorchidism. It can be corrected surgically. If this is not done before puberty, sterility results.

INTERSEX
Otherwise normal people can be born with genitals that are intermediate between those of the two sexes: for example, an unusually short penis, perhaps surrounded with folds of skin, and a half or fully formed vagina. In such cases, the dominant hormonal activity that begins at puberty may well be different from that of the sex they have been brought up as.

BLOCKED DUCTS
These can be birth defects, but are usually due to infection: see J19.

DEFECTS OF DEVELOPMENT
GENETIC ABNORMALITIES
Some people, from the moment of their conception, do not have the normal sex chromosomes of either a male (XY) or a female (XX). This occurs because of errors in cell division or fertilization, and the consequences generally appear at puberty. Among such people, those that appear to be male fall into two groups. The first group have an extra Y chromosome (XYY): they have normal male functioning, though they often have other, non-sexual difficulties. The second group have one or more extra X chromosomes, sometimes with extra Y ones as well: XXY, XXXY, XXXXY, XXYY, XXXYY. All these are male, but usually their genitals fail to develop at puberty, and they may also show some female secondary sexual characteristics. The first three types listed (with the single Y chromosome) are always sterile. The fertility of XXYY and XXXYY cases is not yet clear.

HORMONAL ABNORMALITIES
People with normal sex chromosomes can nevertheless have hormonal defects. At puberty, normal sexual growth and functioning may be late or never develop, even though the genitals in childhood were the normal shape and size. Hormone treatment may be needed, with occasional pauses to see if the body hormones have started up yet.

SCROTUM
VARICOCELE
This is a condition in which there is a collection of varicose veins around the scrotum. The blood carried in the dilated veins makes the testes warmer than they should be, and infertility can result. Regular bathing of the testes in cold water may be enough to counteract the temperature change. Also weight reduction will help prevent the veins getting worse, if obesity has been one cause of the condition. If the problem continues, surgical removal or tying of the swollen veins may be needed. This does not interfere with the general blood supply to the testes.

SCROTAL FLUID
Hydrocele is accumulation of fluid in the layers of cells around the testes. It can cause overheating

SITES OF DISORDERS

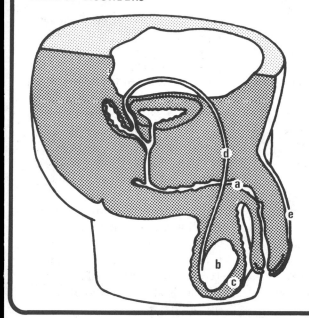

a Urethra
b Testes
c Scrotum
d Vas deferens
e Penis

and infertility, and may require surgical drainage. Hematocele is a similar condition resulting from injury; blood is mixed in the fluid. Spermatocele is accumulation of seminal fluid.

SCROTAL SWELLINGS
These can be caused not only by fluid accumulation, but also by cysts, tumors, inflammations, and hernias in the area, or by an attack of mumps.

BLOCKED DUCTS
Blockage can occur in the vas deferens tubes, that carry sperm from testes to urethra. Infertility results. Infection, such as venereal disease, is the usual cause. Corrective surgery bypasses the blockage, joining the unaffected part of the tube directly to the end of the urethra.

PENIS
PAINS
Pains in the penis can be caused by trouble either there or elsewhere, due to inflammations, stones, or growths.

SKIN DISORDERS
Growths on the penis surface can include warts, ulcers, sores connected with sexual infections, and cold sores. None are especially hard to treat.

Cancer of the penis may appear as irritation and discharge from beneath the foreskin, or simply as a pimple on the penis surface that does not heal. In most western countries it accounts for only 2% of all male cancers, and usually responds well to radium treatment.

PRIAPISM
This is non-sexual erection of the penis; it may be painful also. It is commonest in the elderly, when

causes can include prostate enlargement, inflammation, piles, etc. In children, it may be due to penis inflammation, circumcision, over-tight foreskin, or worms; in adults, to drug abuse, gonorrhea, epilepsy, leukemia, back injury, or just convalescence from an acute illness.

Continual priapism can be due to severe spinal injury, or to a clot in the prostatic veins.

CROOKED ERECTION
This can be due to short term inflammation of the urethra, especially from gonorrhea. Otherwise, usually in older men, it is linked with the spontaneous formation of scar tissue along one side of the penis. In this case, it gradually stops being painful, and the condition itself may eventually disappear without treatment.

©DIAGRAM

SEXUAL INFECTIONS 1

VENEREAL DISEASE

The term venereal disease (VD) is used for certain infections which are almost always passed on by sexual contact. This happens because:

a) the micro-organisms that cause them usually live in the infected person's genitals - or in some other place (such as mouth or anus) where they have been put by sexual activity; and

b) to infect another person, they usually have to enter his body through an orifice (such as the genital opening, anus, or mouth), and sexual activity gives them this chance. The first symptoms of disorder appear on the part of the body that has been in contact with the infected part of the infected person.

Otherwise, these disorders have little in common. Some are caused by bacteria, some by viruses, some by other micro-organisms. Some are rare in our society, others epidemic. And some can be only painful or troublesome, others, if untreated, crippling or fatal. Some kinds of venereal disease have been known since medicine began. Syphilis, the most notorious, may have been brought back to Europe from America as a result of Columbus' expedition in 1492. Cases of VD increased rapidly during World War II, and in the last 20 years cases in many countries have multiplied 3 or 4 times. The frequency of gonorrhea, for example, is now second only to that of the common cold.

SYPHILIS

Syphilis is sometimes nicknamed "the pox" or "scab". It is the most serious of sexual infections. Its prevalence varies. In the USA it is the third most common reportable disease. In the UK, for example, it is comparatively rare. Worldwide, there are about 50 million cases.

CAUSATION

Syphilis is caused by tiny bacteria shaped like corkscrews: "spirochetes". These thrive in the warm, moist linings of the genital passages, rectum, and mouth, but die almost immediately outside the human body. So the infection almost always spreads by sexual contact. Whether the probing organ is penis, tongue, or (perhaps) finger, and whether the receiving organ is mouth, genitals, or rectum, a syphilitic site on either one can infect the other. Very occasionally syphilis does occur from close non-sexual contact (and cases have occurred in doctors and dentists from their professional work); but it cannot be spread by physical objects such as lavatory seats, towels, or cups. It can, however, be inherited from an infected mother, resulting sometimes in stillbirth or deformity, and in other cases in hidden infection that causes trouble later.

INCUBATION

There is an "incubation period", between catching syphilis and showing the first signs - always between 9 days and 3 months, and usually 3 weeks or more. About 1000 germs are typically picked up on infection. After 3 weeks these have multiplied to 100 to 200 million. If the disorder is untreated, they can invade the whole body, eventually causing death.

Syphilis has four stages. Each has typical symptoms, but these can vary or be absent.

PRIMARY STAGE

The first symptom is in the part that has been in contact with the infected person: genitals, rectum, or mouth. A spot appears and grows into a sore that oozes a colorless fluid (but no blood). The sore feels like a button: round or oval, firm, and just under $\frac{1}{2}$in across. A week or so later, the glands in the groin may swell - but they do not usually become tender, so it may not be noticed. There is no feeling of illness, and the sore heals in a few weeks without treatment.

SECONDARY STAGE

This occurs when the bacteria have spread through the body. It can follow the primary stage straight away, but usually there is a gap of several weeks. The person feels generally unwell. There may be headaches, loss of appetite, general aches and pains, sickness, and perhaps fever. Also there are breaks in the skin, and sometimes a dark red rash, lasting for weeks or even months. The rash appears on the back of the legs and front of the arms, and often too on the body, face, hands, and feet. It may be flat or raised, does not itch, and looks like many other skin complaints.

Other symptoms can include: hair falling out in patches; sores in the mouth, nose, throat, or genitals, or in soft folds of skin; and swollen glands throughout the body.

All these symptoms eventually disappear without treatment, after anything from 3 weeks to 9 months.

LATENT STAGE

This may last for anything from a few months to 50 years. There are no symptoms. After about two years, the person ceases to be infectious (though a woman can still sometimes give the disease to a baby she bears). But presence of syphilis can still be shown by blood tests.

TERTIARY STAGE

This occurs in about $\frac{1}{3}$ of those who have not been treated earlier. The disease now shows itself in concentrated and often permanent damage in one part of the body. Common are ulcers in the skin, and lesions on ligaments, joints, or on bones. These are painful, but tertiary syphilis is more serious if it attacks heart, blood vessels, or nervous system. It can then kill, blind, paralyze, cripple, or render insane.

TESTS

Syphilis is not easy to diagnose. Its symptoms are often mild or indistinctive. Testing sores for bacteria or blood for antibodies is necessary. Neither always works, so repeat tests are important.

TREATMENT

This involves antibiotics - usually penicillin. Given in primary or secondary stages, it completely cures most cases. Tests and examination often last more than two years afterwards, to make sure the cure is complete.

In the latent and even tertiary stages, syphilis can still be eradicated and further damage halted; but existing tertiary stage damage often cannot be repaired.

Primary sore

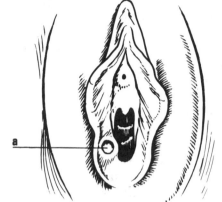

Secondary stage syphilis

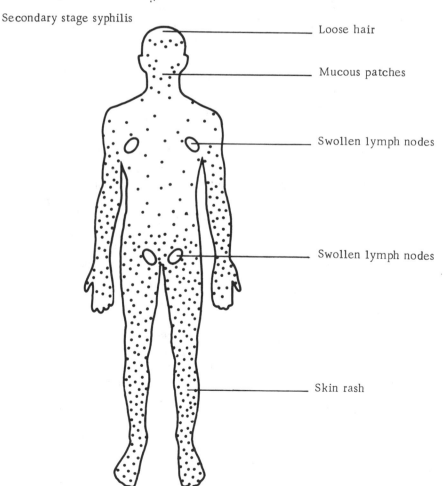

Loose hair

Mucous patches

Swollen lymph nodes

Swollen lymph nodes

Skin rash

Left

a The primary stage syphilitic sore is hard with a clearly defined edge.

Right

The primary sore may develop in or on the:

b genitals,

c anus,

d mouth, or sometimes

e hand.

Secondary stage symptoms include:

f skin rash,

g patches of loose hair,

h swollen lymph nodes, and

i secondary sores in the mouth, nose, throat, genitals, or skin.

Tertiary syphilis can attack almost any part of the body.

©DIAGRAM

SEXUAL INFECTIONS 2

GONORRHEA

Gonorrhea (sometimes nicknamed "the clap") has spread very rapidly among young people in recent years - partly because it is so easy for a woman to have it without knowing it. There are over half a million reported cases in the USA every year, and the true figure is probably many times that number. Several infections in one person in a single year are not too uncommon. Worldwide, there are about 150 million cases.

CAUSATION

Like syphilis, gonorrhea:

a) is caused by a bacterium that thrives in the warm moist lining of urethra, vagina, rectum, or mouth;

b) is normally only passed on by sexual contact, but may be sometimes by close body contact or by inheritance from an infected mother; and

c) cannot be picked up from objects (though it has been suggested that gonorrhea can be carried by pubic lice, and these can be picked up from objects such as lavatory seats).

Unlike syphilis, the form of sexual contact involved is normally only genital or anal intercourse. Oral contact does not often pass on gonorrhea; if it does, it is usually fellatio, rather than cunnilingus that is responsible. (But some scientists even allow the possibility of infection through kissing.)

SYMPTOMS IN MEN

After incubation (usually under a week, but sometimes up to a month), gonorrhea in a man shows itself in marked symptoms:

a) discomfort inside the penis;

b) a thick discharge, usually yellow-green, from the tip of the penis; and

c) pain or a burning sensation on urinating.

Later it may spread to near points, such as glands leading off the urethra (eg the prostate, seminal vesicles, and testes) and the bladder. A resulting abscess may obstruct the urethra, causing difficulty in urinating. Infection of the testes can also cause hard tender swelling: each may become as large as a baseball. If both are involved, and the infection is untreated, sterility may result.

Homosexual men can be infected in the rectum during anal intercourse. This may result in soreness, itching, and/or anal discharge, and sometimes severe pain, espcially when defecating. But often there are no symptoms, or only a feeling of moistness in the rectum. In either case, though, the infection can be passed on again during subsequent anal intercourse.

If oral contact results in infection, it is mainly as a throat disorder that is often not recognized as gonorrhea. It is also unlikely to infect others, because the lymph tissue where the bacteria can survive are deep in the tonsil area. Unlike syphilis, gonorrhea usually remains fairly localized, but if untreated can finally spread to the blood stream and infect bone joints, causing arthritis.

SYMPTOMS IN WOMEN

In women the incubation period is longer, and the eventual symptoms, if any, much less severe or identifiable. There may be discomfort on urinating, more frequent urination, and a vaginal discharge. The discharge is distinctively yellow, and unpleasant in smell - but this may be unnoticed due to the typically small quantities involved. Seventy to 90% of cases in women occur without the woman being aware of the disease; but she is still just as infectious, even where there are no symptoms.

If untreated, the infection may spread to:

a) glands around the vaginal entrance, making them swell, sometimes as large as a golf-ball;

b) the rectum (because of the closeness of the two openings), causing inflammation and perhaps a discharge; and/or

c) the cervix, uterus, and fallopian tubes. Fallopian infection can result in fever, abdominal pain, backache, sickness, painful or excessive periods, and pain during intercourse. If not treated quickly, sterility can result. It can also kill mother and fetus, by causing any pregnancy to be ectopic.

Even where gonorrhea does not affect the fallopian tubes, it can result in premature birth, umbilical cord inflammation, maternal fever, and blindness in the child.

TEST AND TREATMENT

Gonorrhea is diagnosed by laboratory analysis of any discharge or of a smear from the affected part. Treatment is with antibiotics - usually penicillin, though many forms of gonorrhea are becoming more resistant to it. Apart from avoiding infecting others through intercourse, the person being treated should also avoid masturbation and alcohol, since both can irritate the urethra and interfere with cure.

OTHER VENEREAL DISEASES

The other officially designated venereal diseases are much less common in temperate climates.
SOFT CHANCRE (CHANCROID) is caused by a bacillus (a rod-shaped bacterium) and is contracted sexually (usually by intercourse). After 3 to 5 days' incubation, it generally produces an ulcer on the genitals and painful swollen glands (but either sex can carry the infection without symptoms). Treatment is with antibiotics and other drugs.
LYMPHOGRANULOMA VENEREUM This is caused by a very small bacterium, and can be contracted from infected bedding and clothing as well as (more usually) from sexual intercourse. After 5 to 21 days' incubation, it produces a small genital blister or ulcer. Later there can be internal complications. Treatment is with antibiotics.
GRANULOMA INGUINALE This is caused by a bacillus, and is contracted sexually (usually by intercourse). After 1 to 3 weeks' incubation, it produces bright red painless genital sores. Treatment is with antibiotics.
In the USA, soft chancre cases do turn up in northern city clinics, but the others are seldom found outside the south.

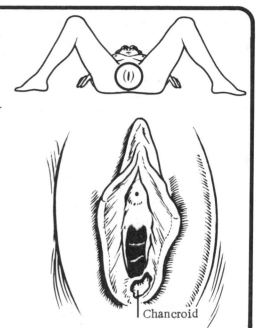

Chancroid

N.S.U.

This is not a venereal disease, in the strict sense, since it does not require intercourse with an infected person. But it is the most common of all sexual infections in men. (In women the symptoms and effects are often insignificant.)
CAUSATION
The cause of NSU is uncertain. Any urethritis with symptoms like gonorrhea but a different cause is called non-gonorrheal urethritis (NGU). If no specific cause is found it is then called NSU: non-specific urethritis. Two possible causes are:
a) an unidentified micro-organism; or
b) reaction of the penis to the chemistry of the vagina. The latter is suggested in the many cases where NSU occurs spontaneously i.e. in two people who have never had intercourse with anyone else. NSU often develops just because sex habits change: eg, when a man first starts having sex regularly, or has intercourse with someone new, or returns to a partner after an absence.

SYMPTOMS IN MEN
If NSU develops after a specific intercourse, there is usually 1 to 4 weeks incubation. Symptoms may then include:
a) greenish-yellow discharge from the penis;
b) blockage of the penis tip with dry pus; and
c) discomfort when urinating, and sometimes increased frequency.
Untreated NSU may spread to: the bladder (causing pain on urinating and perhaps blood in the urine); the testes (causing swelling); and/or especially the prostate (causing pains in groin or back or between the legs).

One per cent of men with NSU develop Reiter's syndrome:
a) inflamed urethra, and discharge;
b) inflamed eyes;
c) inflamed joints (arthritis); and perhaps also
d) skin rash and mouth and penis ulcers.
The arthritis may cripple, and the eye inflammation affect the sight.
TEST AND TREATMENT
Diagnosis is negative - when tests reveal no other cause of the symptoms. Treatment includes antibiotics, and abstinence from alcohol and intercourse. If NSU has spread, treatment takes longer. Sometimes NSU is hard to cure and keeps on reappearing months or years later. Reiter's disease may require hospitalization.

© DIAGRAM

SEXUAL INFECTIONS 3

THE PENIS AND SEXUAL DISORDERS

Most sexual disorders produce typical symptoms in the penis area, if contracted genitally.

a) Primary syphilis: a single, hard, painless sore; glands often swollen.

b) Gonorrhea: discharge from tip.

c) Chancroid: small painful pimples, breaking down to soft painful ragged ulcers that bleed easily; widely swollen glands.

d) Granuloma: red pimple growing into painless raised area, bright beefy red in color.

e) Pubic lice: the lice appear as bluish-grey dots about the size of a pinhead; the "nits" (eggs) can also be seen, attached to the hairs.

Syphilis

Chancroid

Gonorrhea

Granuloma

Pediculosis pubis

OTHER SEXUAL INFECTIONS AND INFESTATIONS

These include:

a) trichomoniasis and candidiasis, which usually develop only in women;

b) genital versions of two common skin disorders - warts and cold sores; and

c) infestation by certain minute insect parasites, such as pubic lice or the scabies mite.

GENITAL WARTS are fairly common and very contagious. They are spread by sexual contact, perhaps caused by a virus, and appear, after 1 to 6 months' incubation, on, in, or around genitals or anus. They are usually cured by repeated use of a resin application. If this fails, they may have to be burnt off with chemicals or electricity.

GENITAL HERPES is fairly contagious and increasingly common.

The virus responsible is thought to lie dormant in the skin for long periods. When activated, it causes a genital or anal sore that weeps colorless fluid and forms a scab. There is no sure treatment (though bathing in salt solution may help), but it usually disappears temporarily after about 10 days.

INFESTATIONS are passed on by sexual or sometimes other close body contact, and are not especially common.

Scabies, or "the itch", is caused by a tiny mite, which mainly lives on and around the genitals. The female mite burrows beneath the skin to lay her eggs. The symptoms

- itchy lumps and tracks - become noticeable after 4 to 6 weeks' incubation. They can occur between the fingers, on buttocks and wrists, and in the armpits, as well as on the genitals. The itching is worse in warm conditions (eg in bed).

Pubic lice, or "crabs", are genital versions of the lice that can also occur in other hairy parts of the body. They feed on blood, and cause itching that can be severe.

Treatment of both parasites involves painting the entire body with appropriate chemicals.

Pubic lice are called by the French "papillons d'amour": the butterflies of love.

a Scabies mite, very highly magnified

b Pubic louse, highly magnified

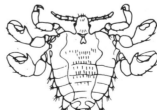

SYMPTOMS

Following is a list of possible signs of venereal disease in women. All these symptoms usually have some other cause - not venereal disease. But do not delay in getting proper medical advice. If symptoms disappear, it may just mean that the infection has progressed naturally to its next stage. You may still have a venereal disease; and you may still be able to infect others. Possible symptoms in the genital area include:

a a sore, rash, or ulcer, on, in, or around the genitals;

b similarly, a sore, rash, or ulcer, on, in, or around the anus;

c unusual vaginal discharge;

d pain or a burning feeling on urinating;

e increased frequency of urination;

f itching or soreness of vagina or vulva; and

g swollen glands in the groin.

Possible symptoms on the head and body include:

h a sore, ulcer, or rash on or in the mouth (or sometimes the nose);

i an eye infection;

j loss of patches of hair;

k persistant sore throat after fellatio;

l a rash on the body;

m sores in soft folds of skin;

n swollen glands in the armpits; and

o a sore, ulcer, or rash on the fingers or hand.

Possible symptoms if infection spreads up the reproductive tract include:

p nausea;

q backache;

r abdominal pain;

s pain during intercourse;

t painful or excessive periods; and fever.

PREVENTION

Worrying too much about getting a sexual infection is not a healthy attitude. You may quite sensibly avoid sex with people whose sexual experience seems likely to have been casual and widely indiscriminate. But since NSU can occur spontaneously in two virgins, the only way to be sure of not developing a sexual infection is not to have sex.

There is no immunity to VD or vaccine against it. But various measures can help reduce the chances of infection:

a) use of some contraceptive foams (Delfen, Emko), some contraceptive creams (Cooper, Ortho), and some contraceptive jellies (Cortane, Ortho-Gynol, Milex Crescent, Koromex AII, suppositories, Progonasyl antiseptic);

b) use of some non-contraceptive vaginal products (Lorophyn suppositories, Progonaryl antiseptic);

c) use of a condom by the male;

d) inspection of the male penis, for an ulcer or sore, or for infectious discharge from the penis tip;

e) use of a "morning after" antibiotic dose, under prescription from a doctor or clinic (but this may be very difficult to obtain);

f) urinating immediately after intercourse; and possibly

g) washing the genitals before and after intercourse.

In practice the first two methods are probably the best.

Note that:

a) As an anti-VD measure, a condom needs to be put on before any sex play begins. It then guards fairly effectively against gonorrhea and NSU but not against syphilis.

b) To wash out the vagina after intercourse, a low-pressure douche can be used. But no vaginal washing should be carried out if the contraceptive method used involves a foam, jelly, or cream, whether alone or with diaphragm or condom.

c) To check a penis for discharge, roll back the foreskin, if necessary, and squeeze the penis firmly - preferably before it becomes erect. One or two drops of thick white, gray, or colored fluid, appearing at the tip, may indicate infection. Clear liquid is usually just urine or semen.

Perhaps the most sensible attitude is not so much to guard against getting these infections, as to act carefully to stop them spreading if you do. It is important:

a) to get cured properly, following qualified medical instructions, and returning for prescribed checks and tests even if they seem unnecessary;

b) to avoid sexual contact with anyone until you are sure you are cured; and

c) to make sure that all your recent sexual contacts know what has happened, and that they all get themselves thoroughly tested and, if necessary, treated.

4 Food and fitness

Exercise and a proper diet go hand in hand in helping us to maintain our physical and emotional well-being.

Left: Outdoor exercise for two
(Photo: Huegelman/Petra - Camera Press Ltd.)

WHAT FOOD PROVIDES 1

WHAT FOOD IS

Food is anything that has a chemical composition which can provide the body with:
a) material from which it can produce heat, activity, and other forms of energy;
b) material that can be used in the growth, maintenance, repair, and reproduction of the body; and/or
c) substances to regulate these processes of energy production, growth, repair, reproduction, etc.
Not everything we eat or drink is food. Flavorings such as saccharin and pepper are not utilized by the body. Tea affects the nervous system and is a drug but not a food. (But alcohol, though it is a drug, also provides energy, and so falls within the definition of food.) Bran performs a useful function as a laxative, but again is not food as it is not absorbed by the body. The constituents of food that are of value to the body are: proteins, carbohydrates, fats, vitamins, minerals, and water. Energy uses carbohydrates, fats, and proteins. Growth and repair uses proteins, minerals, and water. Control of body processes uses proteins, minerals, vitamins, and water.

CALORIES

The constituents in food that help growth and repair cannot all be measured on one scale: different constituents do different jobs, which are not interchangeable. The same applies to the constituents that control body processes. It is no good, for example, trying to add together Vitamin A units and calcium units to get so many "control units." That is like trying to add cows and washing machines. But the constituents that provide energy can be measured on a single scale, and added together. For in the end all of them can be measured in terms of the amount of heat they produce in the body.
The basic unit for measuring any energy (including heat output) is the scientist's calorie. This is defined as the amount of energy needed to raise the temperature of 1 cc of water by 1° Centigrade. The measure used in talking about food and human energy needs is a thousand times larger than this: the kilocalorie, or Calorie (which should be - but is not always - written with a capital "C").
For example, a typical number of Calories for a woman to use up in a day is about 2500. So this is the amount of energy her food must supply, unless she is to run down her stored reserves. Protein, fat, and carbohydrate are all sources of energy (though protein is more vital as a source of other things). One ounce of protein produces over 113 Calories in the human body, one ounce of carbohydrate the same, and one ounce of fat 225 Calories. (An ounce of alcohol - rich in the carbohydrates that we call sugars - produces 180 Calories.) Actual foods range in calorific value from, for example, 105 Calories to the pound (tomatoes) to 4200 Calories to the pound (lard).

PROTEINS

Protein is the basic ingredient of the living organism. There is no substitute for protein, for it is the only constituent of food which contains nitrogen - essential for the growth and repair of the body. Proteins, in fact, provide raw materials for the body's tissues and fluids. They also have certain specialized functions. They help to maintain the chemical fluid balance in the brain, spine, and intestine, and they aid the transport of food and drugs.
Proteins are very complex substances made from a number of chemicals called amino acids. About 20 different kinds of amino acid are found in protein food, and the thousands of different ways these can be linked up produce the many types of protein that exist in food. A single protein molecule can contain as many as 500 amino acid units linked together.
The two main sources of protein are from animals, in the form of meat, fish, eggs, and dairy produce, and from plants in the form of nuts, peas and beans, grains and grain products (such as bread, especially wholemeal), and in small quantities in many tubers and vegetables. Most animal proteins contain all the essential amino acids that humans need, and so are called complete proteins. But vegetable proteins are all, individually, more or less incomplete. They can carry out some individual jobs in the body, but cannot fulfill the vital task of cell repair and growth unless combined together, or with animal protein.
Most proteins are insoluble in water. Some are soluble - such as casein (milk), and albumen (egg white) - but become insoluble when heated or beaten.

CARBOHYDRATES

Carbohydrates provide our main source of energy for immediate use. Energy is used up even during sleep, to keep body organs functioning. Carbohydrates play a vital role in the proper functioning of the internal organs and the central nervous system, and in heart and muscle contraction. Our bodies cannot manufacture carbohydrates so we get them from plants, or from animals that feed on plants. Plants synthesize carbohydrates out of the reaction of sunlight on water and carbon dioxide. This is called photosynthesis and it occurs in the green leaves of plants.

All carbohydrates are made up from carbon, hydrogen, and oxygen. The end product of these three elements is first sugar and then starch, which is stored in plants for future use.

There are several different kinds of carbohydrates.

SUGAR

This is one kind, of which there are various types:

Glucose is the form in which fuel is transported in the body (though, eaten, it gives energy no more quickly than other sugars).

Fructose comes from most fruits, and from honey. It can also be made out of sugar cane. Glucose converts to fructose during the process of the release of energy in the body.

Sucrose is a chemical combination of fructose and glucose and occurs naturally in sugar beet and sugar cane. It is also present in fruit and in carrots. Sucrose forms the common household sugar, which is available in various grades and crystal sizes due to different refinement processes.

Lactose occurs naturally in human and cow's milk and is not as sweet as sucrose. It is a combination of glucose and galactose.

Maltose is derived from malt and is also produced naturally from starch when grain germinates.

STARCH

This forms the largest part of the carbohydrate in our food. It is the stored food in plant seeds, intended for use in maintaining the growing plant until it is able to feed itself by photosynthesis. Unripe fruit contains starch which converts to sugar as the fruit ripens. Starch is composed of complex chains linked together with glucose units. Starch is indigestible unless it is cooked, when the starch granules swell and burst.

GLYCOGEN

Glycogen is similar to starch, and it serves the same purpose in animals as starch does in plants: i.e. it stores fuel - in this case in the liver and muscles. It is not found in most meat as it breaks down into glucose after the animal is killed, but horse meat and oysters do retain it.

OTHER FORMS

Cellulose is the compound produced in plants to give themselves rigidity and strength. It is fibrous, and indigestible to most animals except some insects (but does function as roughage).

Pectin is present in apples, other fruits, and turnips. It has no direct food value but has the property of making jam set.

Sources of useful carbohydrates include bread, potatoes, rice, wheat, sugar, honey, vegetables, fruit, jam, liver, milk, eggs, and cheese.

FATS

Fat is the most concentrated source of energy. Also, when stored in the body as a layer of fat beneath the skin and around organs, it provides insulation and protection for body structures. Finally, certain fats carry the fat-soluble vitamins (A, D, E, and K).

Fat contains the same three elements as carbohydrates - carbon, hydrogen, and oxygen, but combined in a different way. Chemically, fat is a combination of fatty acids and glycerine. At normal temperatures, fat can be solid as in animals or liquid as in vegetable and fish oil. But all fat can be made liquid by heating and solid by cooling.

Fat is not soluble in water, though it is in alcohol, ether, and chloroform. But by chemical treatment with alkalis, fat can be broken down into its separate units, and then can be mixed with water. This is the process by which fats are digested in the body. Mineral oils such as Vaseline and paraffin cannot be broken down in this way and are therefore not digestible by the body and of no value as food.

Fat in the diet falls into three categories. Sources such as butter, lard, margarine, and oils are added to recipes in a recognizable and measurable form. Other sources, such as the fat found in meat, fish, eggs, etc, are not so readily measurable, and vary with the quality of the source, the time of year, and so on. In addition, when fat is added as a cooking medium, it finds its way into the outer layer of the food, increasing its fat content.

©DIAGRAM

WHAT FOOD PROVIDES 2

VITAMINS

Vitamins are certain substances found in food in minute amounts. They are needed for the regulation of chemical processes inside the body, and through this have an important role in growth and development and in protection against illness and disease. The presence of vitamins in the diet is essential, as most of them cannot be made by the body.

The role of vitamins in nutrition was only discovered in the present century, but there are now known to be about 40, of which 12 or more are essential in the diet. Because of the haphazard process of their discovery, they originally formed a jumbled list of alphabetic names (A, B_1, B_6, etc). But now their chemical structures have been identified, chemical names are often used for many of them. Identification has also meant that some can now be made artificially.

Chemically, in fact, they are proving to be a mixed bag - only sharing the characteristic of being complex substances needed by the body in tiny amounts. For example, the body only needs one ounce of thiamin in its lifetime - despite the vital importance of that ounce. Above an average day-to-day requirement, increased amounts of a vitamin do no further good, and in some cases are actually harmful.

Vitamins in the diet can be divided into two classes: those soluble in fat (vitamins A, D, E, and K), and those soluble in water (vitamins C and the B vitamin complex).

VITAMIN A is found in halibut and cod liver oil, milk, butter, and eggs. It is destroyed by cooking and sunlight.

It plays a role in the formation of bone and of the enamel and dentine in teeth. It is also responsible for the ability to see in dim light.

VITAMIN D is found in eggs, milk, butter, and fish liver oils. It is also synthesized in the skin during exposure to sunlight.

It plays a part in the digestive absorption of some minerals, such as calcium, and phosphorus. It is also necessary for retaining calcium in bones.

VITAMIN E is found in wheatgerm, oil, lettuce, spinach, watercress, etc.

There is no definite evidence that it is essential to humans, but it does help in the healing of skin wounds, and may also be connected with fertility.

VITAMIN K is found mainly in green plants such as spinach, cabbage, and kale. But it is also synthesized in the gut by the action of bacteria.

It is a necessary factor in the blood-clotting mechanism, as it is needed for the production of prothrombin.

VITAMIN C is found in fresh fruit and vegetables, especially lemons, oranges, blackcurrants, tomatoes, and watercress. Human milk also contains vitamin C. This vitamin is easily destroyed by cooking, especially if the food has been chopped up.

One of its most important functions in the body is to control the formation of dentine, cartilage, and bone. It also helps the formation of red blood cells, and the correct healing of wounds and broken bones.

There is no conclusive evidence that vitamin C prevents colds.

VITAMIN B is in fact a complex of fifteen different substances, but they are classed together because they occur together in the same types of food, such as yeast and wheat germ. Unlike the other vitamins, at least some vitamins of the B group are found in all living plants and animals.

The following are the most important B vitamins.

THIAMIN forms the part of the

WATER

Water is not really a food, but it is an essential part of all tissues. Our bodies are composed of about $\frac{2}{3}$ water. It acts as a form of transport: the blood, which is mainly water, carries food in its basic forms to the tissues and takes away waste products to be excreted. Chemically, water is a simple compound of oxygen and hydrogen, but is never found pure as it contains traces of minerals, dissolved gases, and solids. The amount of these depends on the water's source.

As well as in liquid form, water is also found in most solid food. Since it is constantly being lost in sweat, urine, and breathing enzyme system essential for the breakdown of carbohydrates and the nutrition of nerve cells.

RIBOFLAVIN acts with thiamin and nicotinic acid in the oxidation of carbohydrates. It is also important for the growth of the fetus, and is thought to play a part in the mechanism of vision.

PYRIDOXINE (B_6) helps the breakdown of protein into amino acids and is necessary for the formation of blood cells. However, sufficient pyridoxine is produced in the intestine.

PANTOTHENIC ACID probably plays a part in the detoxification of drugs and the formation of chemicals that pass nerve impulses along the nerves.

NIACIN is needed for healthy skin and nerves, and food digestion.

FOLIC ACID is an anti-anemic factor found in green leaves and in liver and kidneys. It is especially important during pregnancy, to prevent anemia.

COBALAMIN (B_{12}) is the only vitamin containing a metal, cobalt. It is found in a high concentration in the liver and is essential for the formation of red blood cells. Unlike the other B complex vitamins, it has no vegetable source.

out, it must be replaced every day or dehydration of the body will occur. However, the body's need for water at any time is very accurately registered by the degree of thirst.

MINERALS

Minerals, like vitamins, do not supply any heat or energy, but play a vital role in the regulation of body fluids and the balance of chemicals.

MACRONUTRIENTS
These are the minerals needed by the body in comparatively large quantities.

CALCIUM is found in milk, cheese, fish, some green vegetables, and in "hard" drinking water. It is necessary for the proper formation of bones and teeth; also for the functioning of muscles and clotting of the blood. During growth, calcium is constantly being laid down in bones and simultaneously withdrawn into the bloodstream for use elsewhere. The body of an adult normally contains 2-3½lb of calcium of which at least 99% is present in the bones.

PHOSPHORUS is found in meat, such as brains, kidneys, and liver. Dairy produce such as cheese is also rich in phosphorus. It is important for energy transfer. Its function in the body is closely linked with that of calcium.

SODIUM AND CHLORINE occur together in the familiar form of common salt, and also in animal protein. Both are vital for life: they maintain water balance and distribution, osmotic pressure, acid-base balance, and muscular functioning. The amount taken in a normal diet is usually more than enough, but in hot weather much may be lost in sweat.

POTASSIUM is related in function to sodium and chlorine. It is found mainly in meat, fish, vegetables, chocolate, and dried fruit.

SULFUR occurs in certain amino acids, especially in animal proteins.

Sulfur in the body is found especially in insulin, which regulates the level of sugar in the blood and in the human hair.

MAGNESIUM occurs in nuts, beans, cereals, dark green vegetables, seafood, and chocoate. Its function is similar to calcium.

MICRONUTRIENTS
These are the minerals needed by the body in much smaller quantities.

IRON is found in fish, liver, eggs, "black pudding," beans, green vegetables, and oatmeal. The body of a healthy adult contains about 4gm of iron - roughly the amount of a 3in nail.

Iron is an essential part of red (hemoglobin) blood cells, which enable the blood to take up oxygen from the lungs and carry it to all cells in the body.

IODINE is important for the healthy functioning of the thyroid gland. It occurs in seafish, shellfish, iodized table salt, and vegetables grown on soil naturally containing iodine.

FLUORINE is found naturally in seafish, some "hard" drinking water, and china tea. It is also added to the water artificially in some localities.

Traces of fluorine are present in bones, teeth, skin, and thyroid gland. One known function is that it helps prevent tooth decay.

OTHER MICRONUTRIENTS are zinc, selenium, manganese, copper, molybdenum, cobalt, and chromium.

TRACE ELEMENTS
These are found in the body in tiny amounts, but their function, if any, is not yet known. They include strontium, bromine, vanadium, gold, silver, nickel, tin, aluminum, bismuth, arsenic, and boron.

©DIAGRAM

WHAT WE EAT

FOOD INTAKE

What people eat varies enormously. For many people in tropical countries, a typical day's food is based on rice and a little vegetable - totaling perhaps 1600 Calories, and containing only tiny amounts of necessary proteins, vitamins, and minerals. Yet in industrial countries, the daily diet of a food lover may total over 3500 Calories, and supply in all respects about twice the body's needs.

Even taking national averages, great differences remain. In Ghana, on a diet mainly of roots, cereals, and vegetables, the average total daily intake is perhaps 2000 Calories, including 1.7oz (47gm) of protein of which only ¼ is of animal origin. In Denmark, on a diet of meat, dairy produce, cereal products, vegetables, and fruits, the average intake is perhaps 3300 Calories including 3.3oz (95gm) of protein (2.2oz - 62gm - of animal origin).

Diets also vary greatly in their variety, their range of geographical source, and their handling and processing before consumption. Perhaps 75% of the world lives on a basic diet of one food, usually a cereal (typically rice), usually grown by themselves, and usually eaten in a simple boiled form.

Average individual grain intake on such a diet in a poor country totals perhaps 400 lb (180kg) a year. In contrast, about 1700 lb (770kg) of grain enters the food chain of a North American each year - but only 30% is ever eaten as cereal products. The rest goes to feed livestock for meat and dairy produce. People in industrial societies buy widely from restaurants, carry outs, and vending machines, as well as from an average supermarket stock of 7000 different food items that have been stored, transported (perhaps imported), usually processed and preserved, and wrapped for sale.

THE HAVES AND HAVE-NOTS

The maps on the opposite page show average daily protein and Calorie intake in different countries. The patterns are very similar: most countries with high total Calorie consumption are also high in protein consumption. But a few national diets have adequate protein, though total Calories are low, while rather more have ample Calories but deficient protein.

On this page we summarize the situation, but by inhabited regions rather than countries. Taking continental areas, on average:
(a) Europe, North America, and Oceania (Australia etc) have a sizable excess Calorie intake, and some excess protein intake;
(b) in South America the situation varies from country to country;
(c) the Middle East has slight Calorie and protein deficiencies; and
(d) Asia, Central America, and parts of Africa have sizable Calorie deficiencies and large protein deficiencies.

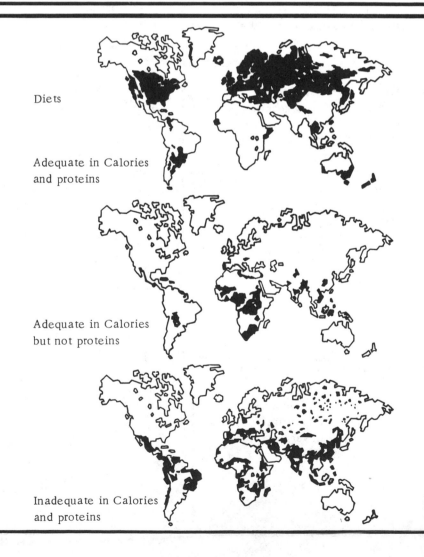

Diets

Adequate in Calories and proteins

Adequate in Calories but not proteins

Inadequate in Calories and proteins

USA
 1 Green leafy vegetables
 2 Citrus fruits and vegetables
 3 Other fruits and vegetables
 4 Milk and milk products
 5 Meat
 6 Eggs
 7 Oil and fat
 8 Sweetened products
 9 Potatoes
 10 Cereals

India
 1 Citrus fruits
 2 Vegetables
 3 Legumes (peas, beans, etc)
 4 Other fruits, sugar, meat, fish, eggs, milk, oil, and fat
 5 Rice

Average food consumption

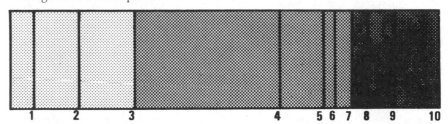

Fruit and vegetables   Animal products   Carbohydrates

Average food consumption

Here we show comparative Calorie consumption by food source for an average person in the USA (above) and in India (left). Calorie consumption in the USA is both far higher and far more varied in source.

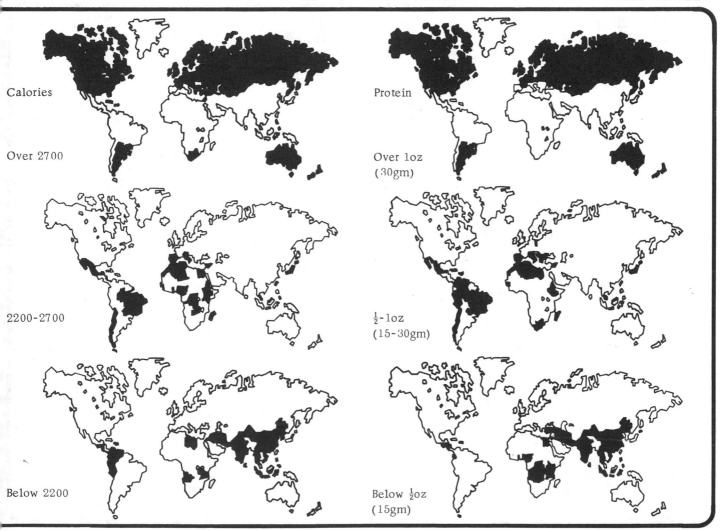

Calories

Over 2700

2200-2700

Below 2200

Protein

Over 1oz
(30gm)

½-1oz
(15-30gm)

Below ½oz
(15gm)

©DIAGRAM

NUTRITION 1

NATURE AND NUTRITION

Carbohydrates, proteins, fats, minerals, vitamins, and water are all ingredients of a healthy diet. Some aspects of diet planning require special attention, but generally speaking nature is an excellent nutritionist - letting us know what our bodies require. Moreover, in developed countries, where different nutrients are readily available, most people eat more than enough different types of food to prevent any serious nutritional deficiencies.

FATS AND DIET

Fats form an important part of a balanced diet. But medical research into heart disease suggests that many people should take greater care over what type of fats they eat.

There are two basic types of fats - saturated and polyunsaturated. Saturated fats harden at room temperature. They occur in meat and dairy products; many solid shortening products; and also coconut oil, cocoa butter, and palm oil, which are used in many bought cookies and pastries. Medical research has shown a relationship between consumption of saturated fats and high levels of cholesterol (itself a special kind of fat) in the blood. High blood-cholesterol levels are in turn associated with certain types of heart disease.

Polyunsaturated fats are most commonly consumed in the form of liquid vegetable oils (such as corn and soybean oils). Fats of this type apparently do not result in raised blood cholesterol levels. In fact, some authorities think that eating polyunsaturated fats can counteract the effect of eating other foods that are high in cholesterol.

PROTEIN AND DIET

Protein is vital to growth and cell repair - yet estimates of protein needs vary enormously. An average estimate for good nutrition is about 2½oz (70g) a day in adulthood (with protein supplying about 7% of Calorie intake). But people have been found to adapt in a healthy way to intakes under half to over twice this amount.

All this is dependent, of course, on all essential amino acids being eaten, and in the right proportions. Lack of protein can be a problem in old people with little money, and in those following unusual diets.

WATER REQUIREMENT

The normal requirement is the equivalent of about 6 or 7 glasses of fluid a day. Thirst usually provides very accurate indication of need, but in very hot conditions may not keep up with the intake needed to replace perspiration loss.

CARBOHYDRATE AND DIET

Too much carbohydrate in the diet shows itself in unhealthy weight gain. It is not, however, only the quantity of carbohydrate intake that is important. The type of carbohydrate eaten is also significant.

Traditional cereal products, such as bread and potatoes, contain other nutrients as well as carbohydrates and so can make an important contribution to total diet planning.

The current widespread replacement of traditional carbohydrate foods by highly refined and sweetened carbohydrate products has serious implications. Pastries, cakes, chocolate, ice cream, and alcohol all have high carbohydrate counts but contain little of nutritional value apart from their energy content. They provide little or no roughage.

Most important, however, are the harmful effects of too much sugar in the diet. Some scientists believe:

a) that the rush of sugar into the bloodstream causes the body to overreact - withdrawing too much sugar from the blood and so leaving us feeling tired and irritable;

b) that eventually excessive sugar intake can cause diabetes in people who would not otherwise suffer from it; and

c) that sugar has a role in producing heart disease.

Certainly sugar promotes tooth decay, and destroys the appetite for more nutritious foods.

Such criticisms apply not only to white sugar, but also to brown sugar, raw sugar, honey, and molasses. But white sugar is the main culprit, simply because the amounts of sugar added in the cooking or processing of foods usually dwarf the amounts of sweetening added at mealtime.

MINERALS AND DIET

Most of the body's mineral requirements are met without special diet planning. Some care, however, is needed in the following cases.

a) Sodium intake is usually far higher than necessary, but may be insufficient for very heavy work in hot conditions.

b) Calcium intake depends on ordinary consumption of milk and cheese.

c) Iron is needed for the hemoglobin in red blood cells. It is found in meat and eggs, brewer's yeast, and wheat germ. But it is only absorbed in tiny quantities, and hardly at all if vitamin C in the body is low. Women, with their regular, menstrual blood loss, often develop an iron shortage (anemia), with resulting fatigue and breathlessness.

d) Iodine shortage occurs if the diet contains no seafood and only vegetables grown in iodine-free soil. A lack of iodine causes thyroid deficiencies and thyroid gland enlargement. This is less common with modern food transport and availability of iodized salt.

VITAMINS AND DIET

Vitamins B_1 and C can be lost by bad cooking, but both have plenty of uncooked sources.

Vitamin A may be deficient if dairy produce, margarine, or green or yellow vegetables are not eaten.

Vitamin D is only much needed in the diets of children and nursing mothers. Butter, margarine, and liver are sources. Vitamin D deficiency has been found in children in poor urban areas - especially those with pigmented skin, which impedes the vitamin's formation in the body from sunlight. It can also occur in the aged and housebound poor.

EXCESS AND DEFICIENCY

In diet planning it is important to look at the diet as a whole. An excess of one nutrient will not compensate for a deficiency of another.

ENOUGH IS ENOUGH
Enough is enough; more is not better. You probably know that too many Calories are not good for you. But other excesses are harmful, or just useless, too.

It is no use eating protein above your needs: it cannot be stored. Too much of certain minerals or vitamins can cause deficiencies in others, by upsetting their absorption or storage. For example, too much B_1 can cause deficiencies in other B vitamins. And some nutrients taken in excess are positively harmful. People have killed themselves, for example, trying to get enough vitamin A and D and taking far too much. Both these vitamins are insoluble in water, and so the excess cannot just be excreted as can an excess of vitamin C. (In fact vitamin C is the only nutrient that some scientists think may be useful to us in enormous doses - and there is far from agreement on this point).

INTERACTION OF NUTRIENTS
Nutrients do not act totally independently of one another. The body is too complex for that. A deficiency of one nutrient can lead to a deficiency of another by affecting the body's ability to make use of the second nutrient, even if it is present in the diet. For example, vitamin A deficiency can lead to vitamin C deficiency; vitamin C deficiency to iron deficiency.

Interactions occur not only among vitamins, and between vitamins and minerals, but also between vitamins and proteins, vitamins and carbohydrates, vitamins and fats; and there are many multiple relationships as well.

FOOD NOT NUTRIENTS

Some people set out to eat quantities of "nutrients" and build up the "perfect diet." But you cannot buy every dietary ingredient in an individual package. You have to buy food; and food is a jumble of hard-to-measure ingredients. Even drinking a glass of milk becomes a nightmare of protein, calcium, fat, carbohydrate, vitamin A, vitamin D, riboflavin, and phosphorus - together with a few other things. And if that seems too simple - what was the fat content of that hot dog? or the protein content of that lobster thermidor? Start eating for nutrients, and you will probably eat twice as much as you need to. Start buying for nutrients, and you will be trying to get, for example, your daily vitamin C out of a handful of rosehips, rather than from a morning glass of orange juice, some lunchtime potatoes, and a helping of cabbage in the evening. The search for nutrients is an excellent way of wasting time and money.

©DIAGRAM

NUTRITION 2

A HEALTHY DIET

There is no one ideal diet. First, needs differ (and so does the impact of availability, cost, taste, habit, and cooking facilities and skills). But, more important, there are a million different ways of satisfying those needs, in healthy eating. It is possible to live healthily on a diet of milk, wholewheat bread, and green vegetables. It would not be very interesting, though. Variety is the spice of food.

PROCESSING AND COOKING

It is hard to generalize, but the more processed a food is, the less desirable it is likely to be as a regular part of a healthy diet. Canned and precooked foods, mass-produced breakfast cereals, cookies, and pastries, and ready meals, all tend to be open to criticism. Defects include:
a) lower nutritional value;
b) added sugar and saturated fats;
c) added preservative chemicals and, often, untested colorings and flavorings; and sometimes
d) unhygienic production.
In general, the "whole food" movement is a sensible one (though, incidentally, there is no agreed evidence that "organic" vegetables have higher food value than chemically fertilized ones - even though they may taste better and contain fewer pollutants). However, nutrients can be lost in home cooking as well as in processing, and undesirable ones added. "Boiling" of vegetables should always be done by steaming in a very shallow amount of water, if vitamin C is to be preserved. (Salt should only be added at the last moment.) "Frying" should be in a tiny amount of unsaturated oil, not hard fat. (Grilling is better, where applicable.)

DAILY FOOD GUIDE

Included here is a daily food guide devised by a dietitian to provide a balanced diet. (Those suffering from certain illnesses, such as diabetes, may need a more carefully planned diet, about which a physician should be consulted.)
The guide divides foods into four groups, and recommended daily servings per person are given for each group.

NOTES
1) Group A uses 1 cup whole or skimmed milk as the basic measure. Alternatives are: 1 cup buttermilk; $\frac{1}{2}$ cup evaporated milk; $\frac{1}{4}$ cup non-fat milk powder; 1oz cheddar cheese; $1\frac{1}{2}$ cups cottage cheese.
2) If amounts in group A are doubled in the course of the day, not more than one serving of group C is needed.
3) Whole milk (not skimmed) and butter or margarine should be used during childhood, pregnancy, and lactation.

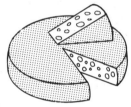

A: MILK AND CHEESE
Age 0-9 years: 2-3 cups
9-12 years: 3 or more cups
13-19 years: 4 or more cups
Adult: 2 or more cups
Pregnancy : 3 or more cups
Lactation: 4 or more cups

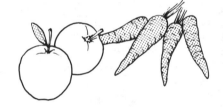

B: FRUIT AND VEGETABLES
Four or more servings.
Serving size examples:
a) $\frac{1}{2}$ cup dark green or deep yellow vegetable (served at least every other day);
b) $\frac{1}{2}$ cup or 1 medium-sized raw fruit or vegetable rich in vitamin C;
c) 1 medium potato.

C: MEAT AND PULSES
Two or more servings.
Serving size examples:
a) 2-3oz cooked meat, poultry, or fish (excluding bone and fat);
b) 2 eggs;
c) 1 cup beans, peas, or lentils;
d) 4 tablespoons of peanut butter.

D: BREAD AND CEREALS
Four or more servings.
Serving size examples:
a) 1 slice bread;
b) 1 cup ready-to-eat cereal;
c) $\frac{1}{2}$ to $\frac{3}{4}$ cup cooked cereal, macaroni, spaghetti, hominy grits, rice, noodles, or bulgur.

VEGETARIANISM

A vegetarian is a person who does not eat the meat of any mammal, bird, or fish. There are two main types of vegetarian:

a) vegans, who eat nothing at all of animal origins, and

b) lacto-ovo-vegetarians, who do allow themselves animal products such as milk, cheese, eggs, and honey.

There are also people who call themselves vegetarians but do eat fish.

REASONS FOR VEGETARIANISM

Reasons for vegetarianism vary from society to society and individual to individual. It has been advocated for religious, philosophical, moral, economic, and health reasons. It has also been adopted as a necessity. Many primitive peoples have lived on a diet of fruit, nuts, and berries, with meat only when it could be obtained.

Perhaps the most powerful arguments for vegetarianism in modern society are:

1) the inefficiency of the animal food production chain in a largely underfed world;

b) the relative cheapness of the ingredients of vegetarian diet; and

c) the possible unhealthiness of eating meat that contains crop pesticides and antibiotics and hormones given to the animals, and that has been processed in many ways that are not necessarily hygienic or beneficial.

Also many people feel that the slaughter of animals is cruel and debasing, and that vegetarianism is part of a more peaceful and harmonious way of life.

VEGETARIAN DIET

Despite the claims of vegetarians, there is no established evidence that eating meat is unhealthy in itself. But it is certainly as possible for a vegetarian to be healthy, strong, and long-lived as it is for a meat-eater.

A person who chooses to give up meat must be careful that his diet still provides enough of the right nutrients.

There are no problems with:

a) healthy carbohydrates (grains, cereal products, potatoes, fruits);

b) fats (vegetable oils, dairy products, nuts, margarine); and

c) minerals and most vitamins (vegetables and fruits).

Obtaining an adequate supply of protein and certain vitamins can, however, be more problematic for vegetarians than for meat-eaters.

DIET PLANNING

Vegetarians must take particular care that their diet provides them with adequate supplies of the following nutrients.

PROTEIN is readily available from eggs and dairy produce, nuts, soybeans, raisins, grains, and pulses. But a vegetarian should be sure to get a good selection of essential amino acids at each meal. This is not difficult where eggs or dairy produce are eaten: cereal and milk, bread and milk, and bread and eggs are all good amino acid combinations. But vegans must depend on soybeans, or on carefully planned vegetable combinations. These include: lentil soup and hard wholewheat bread; and beans and rice.

VITAMINS requiring particular attention in a vegetarian diet are:

1) cobalamin (vitamin B_{12}) - available from dairy produce and yeast, and, particularly useful for vegans, in synthetic form;

2) vitamin D - also needed in synthetic form by vegans where sunlight is unsufficient.

IRON AND CALCIUM are also worth mentioning, as they are sometimes lacking even in the diets of meat-eaters. In fact there are many excellent vegetarian sources.

Iron is found in raisins, lentils, wheatgerm, prunes, spinach and other leafy vegetables, and in bread, eggs, and yeast.

Calcium occurs in dairy produce, dried fruit, soybeans, sesame seeds, and in leafy vegetables.

1½ cups beans + 4 cups rice   =   protein equivalent of 12oz (340g) steak

©DIAGRAM

EXCESS WEIGHT

ARE YOU OVERWEIGHT?

One way of learning whether you are among the overweight is to check your weight against a desirable weight table - such as the one given below (H39) (Note that "desirable" weight tables give lower figures than "average" weight tables - in a society where more people are overweight than underweight the average will be higher than is healthy.)

Even without weighing yourself it is possible to do a quick check for overweight. Start by asking yourself the following questions. Do you have any telltale bulges? Do you look much fatter than you used to? Have your measurements increased appreciably? If you pinch your upper arm, thigh, or midriff, is there more than one inch (2.5cm) of flesh between your thumb and forefinger?

WHY PEOPLE PUT ON WEIGHT

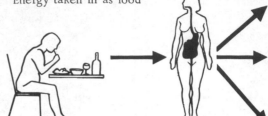

Energy taken in as food

Digestion

a Energy needed by the body

b Surplus energy burned up - "thermogenesis"

c Surplus energy stored as fat

Overweight is always caused by taking in more food energy than the body uses up. The bulk of food energy is taken in in the form of carbohydrates and fats. Both these supply Calories (the measure of energy); and both are converted to fat deposits if the Calories they supply are more than the body uses.

a Most of it is used to supply body energy needs - to maintain basic life processes and for all physical activity (H19-H20)

b It is still the subject of scientific controversy, but it does seem that some people get rid of surplus

input because their bodies automatically speed up their metabolism and burn up the surplus rather than store it.

Also there is a rise in the body's metabolism after every meal. So two people may eat exactly the same, but one will burn up more than the other if the food is taken in several small meals rather than two or three large ones.

c Food energy that is neither needed nor burned up is stored by the body in the form of fat. In overweight people the store far exceeds any normal future demand.

DESIRABLE WEIGHTS

Given here are desirable weights for men and women of different heights and body sizes at age 25. The table is based on statistics collected by US insurance companies. Older people can expect to exceed these weights, perhaps by 10-12lb (4.5-9kg), but should avoid gains beyond this. Weights given are without clothes; heights without shoes.

	Height ft in (cm)	Small frame lb (kg)	Medium frame lb (kg)	Large frame lb (kg)
MEN	5 3 (160)	122 (55.3)	130 (58.9)	140 (63.5)
	5 4 (162)	125 (56.7)	133 (60.3)	143 (64.9)
	5 5 (165)	128 (58.0)	136 (61.7)	147 (66.6)
	5 6 (167)	132 (59.9)	140 (63.5)	151 (68.5)
	5 7 (170)	136 (61.7)	145 (65.8)	156 (70.8)
	5 8 (172)	140 (63.5)	149 (67.6)	160 (72.6)
	5 9 (175)	145 (65.8)	153 (69.4)	164 (74.4)
	5 10 (177)	149 (67.6)	157 (71.2)	169 (76.7)
	5 11 (180)	153 (69.4)	162 (73.5)	174 (78.9)
	6 0 (183)	157 (71.2)	166 (75.3)	178 (80.7)
	6 1 (186)	161 (73.0)	171 (77.6)	183 (83.0)
	6 2 (188)	165 (74.9)	176 (79.8)	188 (85.2)
WOMEN	4 11 (149)	103 (46.7)	110 (49.9)	120 (54.4)
	5 0 (152)	106 (48.8)	113 (51.3)	123 (55.8)
	5 1 (154)	109 (49.4)	116 (52.6)	126 (57.1)
	5 2 (157)	112 (50.8)	120 (54.4)	130 (58.9)
	5 3 (160)	115 (52.1)	124 (56.2)	134 (60.8)
	5 4 (162)	119 (53.9)	128 (58.0)	138 (62.6)
	5 5 (165)	123 (56.0)	132 (59.9)	142 (64.4)
	5 6 (167)	127 (57.6)	136 (61.7)	146 (66.2)
	5 7 (170)	131 (59.4)	140 (63.5)	150 (68.0)
	5 8 (172)	135 (61.2)	144 (65.3)	154 (69.9)
	5 9 (175)	139 (63.1)	148 (67.1)	159 (72.1)
	5 10 (177)	143 (64.9)	152 (68.9)	164 (74.4)

APPETITE CONTROL

Most people have an effective appetite control - or "appestat" - which prevents them from putting on too much weight.

Generally, the appestat is remarkable for its precision. For example, eating an extra half slice of bread a day (30 Calories) above energy output, would bring a weight gain of 110 lb over a 40 year period. It is the appestat that normally protects people from this type of weight gain.

Some people, however, ignore the messages from their appestats. Typical reasons are:

a) social habit or custom;
b) excessive love of food in general or of certain foods in particular;
c) habits of overeating acquired during childhood;
d) lack of exercise; and
e) eating for psychological support, whether as a general addiction or as a response to shock or stress.

APPETITE AND EXERCISE

Some people put on weight because their appestat is put out of action by an excessively sedentary existence.

When physical activity falls below moderate levels, research has shown that appetite may actually increase - even though the body has no need for the extra food. Increasing the amount of exercise in such cases not only increases Calorie output but also appears to put the appestat back into good working order.

Of course, exercise above moderate levels will increase the appetite.

WHEN PEOPLE PUT ON WEIGHT

Childhood - overweight children often become overweight adults

Puberty - female sex hormones increase the deposition of fat

Taking the pill - can make some people put on weight

Social life - eating large meals bought out or made to impress

Pregnancy - excess weight gain may be hard to lose later

Feeding a family - eating up the family's leftovers

Later life - cutting down exercise but not food

Being depressed or worried - and finding consolation in food

Giving up smoking - and nibbling food instead

EXCUSES FOR OVERWEIGHT

Many overweight people like to blame something outside their control - their heavy bones, heavy family, hormones, even their body water level. But:

a) variations in bone density cannot account for more than about 7 lb weight difference;
b) though overweight does "run in families," it may be due more to acquired eating habits than genetic factors;
c) hormonal malfunctions can cause obesity, in very rare cases, but these show themselves clearly in other bodily symptoms; and
d) the body water level is very well regulated except in very hot weather and in some illnesses.

EFFECTS OF OVERWEIGHT

Overweight people are not just more tired, short of breath, and physically and mentally lethargic, with aching joints and poor digestion. They are also more likely to suffer from high blood pressure, heart disease, diabetes, kidney disorders, cirrhosis of the liver, pneumonia, inflammation of the gall bladder, arthritis, hernias, and varicose veins. They have more accidents, are more likely to die during operations, and have higher rates of mortality in general (including 3 times the mortality from heart and circulatory disease).

Some of these effects arise from mechanical causes: the burden of extra weight and its particular location as fat deposits. Others arise chemically, from the need to supply more body tissue than normal. For example, the spread of hormones over increased body tissue is sometimes a cause of infertility. It can also cause serious problems in pregnancy (eg toxemia).

In many cases, reduction to desirable weight removes all the symptoms of disorder and results in increased life expectancy.

©DIAGRAM

LOSING WEIGHT

LOSING WEIGHT

Losing weight is not easy. It demands controlled eating habits, discipline, patience, and a change of attitudes. Before you start:
a) do not be tempted by any promise of easy weight loss – there are no miracles;
b) adopt a definite diet plan and stick to it;
c) if you need advice, get it from your doctor.
It is best to aim to lose weight steadily over a long period. Constant yo-yo weight changes are as bad for you as being overweight. Once weight is lost, keep a constant check, and deal with small gains as they occur.

CHOOSING A DIET PLAN

All genuine diets restrict Calorie intake. If a person's intake of food does not contain sufficient Calories to meet energy requirements, the body makes up the deficiency by burning up its stores of fat.
Three basic types of diet plan are popular at the present time. Each of of them can work if you are sufficiently determined.
a) Low-Calorie plans set a numerical limit to daily Calorie intake (usually 1000 to 1500 Calories). Constant reference must be made to Calorie tables.
b) Low-carbohydrate plans also cut down Calorie intake, but only by reducing consumption of carbohydrates. Tables are simpler than for low-Calorie plans. Fat is unrestricted, making the diet more palatable and socially acceptable. Weight maintenance through excessive consumption of fats does not appear to occur in practice. (A good example of a low-carbohydrate plan is described below).
c) No-count plans, simplified versions of the low-carbohydrate system, divide food into three categories: high-carbohydrate food that must be avoided; high-Calorie, non-carbohydrate food that can be eaten in moderation; and unrestricted food.

CARBOHYDRATE UNIT DIET

The Carbohydrate Unit diet was devised by Professor John Yudkin, MD, and is described in detail in his book "Lose weight, feel great!" (published by Larchmont Press, NY, 1974). The following tables form the basis of the diet. Meat, poultry, fish, eggs, cheese, tea, coffee, butter, margarine, and fat, are all zero units: eat as much as you like. Try a limit of 15 CUs a day. Lower to 10 if necessary, or raise to 20 or even 30 if weight loss is too rapid. Each day get two helpings from each of these groups:
a) milk and cheese;
b) meat, fish, eggs;
c) fruit and vegetables;
d) butter, margarine.
Drink at least ½pt of milk a day. Note that sorbitol is not allowed but saccharin is.
The plan forms a very sensible basis for future healthy eating. Typical portions are indicated for each item.

DAIRY		CUs
Rice pudding	4oz	5
Fruit yoghurt	6oz	2
Milk	½pt	3
Custard	4oz	2
Plain yoghurt	5oz	2
Cottage cheese	2oz	½
Cream	1oz	0

CEREALS		CUs
Macaroni	6oz	10
Spaghetti	6oz	10
Noodles	6oz	9
Vermicelli	6oz	9
Roll	2oz	6
Breakfast cereals	¾oz	4
Buckwheat	3oz	4
Wheat flour	1oz	4
Rice	4oz	4
Semolina	1oz	4
Bread, 1 slice	1oz	3
Wheatgerm	½oz	1½
Roll, starch-reduced	1oz	½

VEGETABLES		CUs
Lentils	4oz	6
Sweet potato	3oz	5
Butter beans	4oz	4
Corn, 1 cob	4oz	4
Yam	4oz	4
Potatoes	3oz	3
Fried potatoes	1oz	3
Parsnip	4oz	2
Peas	4oz	2
Artichoke	5oz	1
Asparagus	4oz	1
Green beans	4oz	1
Beetroot	2oz	1
Carrot	3oz	1
Kohlrabi	4oz	1
Leeks	4oz	1
Swede	4oz	1
Turnip	3oz	1
Cauliflower	4oz	½
Cabbage, celery, chicory, cucumber, marrow, mushrooms, onion		0

SLIMMING AIDS

A considerable variety of slimming aids is widely available - but not all of them are effective or recommended.

a) Substitute meals (wafers, chocolate bars, packaged foods, etc) have a stated Calorie content and sometimes contain cellulose to give a fuller feeling in the stomach Some slimmers find them useful, but they do nothing to encourage the eating habits needed to stay slim.

b) Low-Calorie substitute foods and drinks (eg saccharin, skim milk, slimmers' bread, crispbread) can help slimmers reduce total Calorie intake.

c) Prescribed drugs (Apisate, Tennate) can reduce appetite. These new drugs do not seem to be addictive, but do nothing to encourage good eating habits.

d) Proprietary slimming pills usually contain cellulose and are meant to suppress appetite. Amounts are so small that their effectiveness is probably more psychological than real.

e) Saunas, Turkish baths, and reducing garments cause loss of body water through sweating. This can reduce measurements and weight, but the effect is rapidly canceled out by the drinking needed to replace the fluid loss.

f) Vibrator belts and other massagers are meant to break down fat deposits. There appears to be little evidence to support claims made for them.

g) Machines using electric impulses to relax and contract muscles are recommended by some slimmers.

h) Exercise will not on its own make much difference to your weight. It would, for example, take 12 hours of tennis to lose 1 lb of fat. Exercise does, however, increase the sense of well-being that dieting brings;

i) Attending a slimming clinic can be an effective - though expensive - way of getting slim.

j) Many slimmers find that joining a slimming club gives a valuable psychological boost.

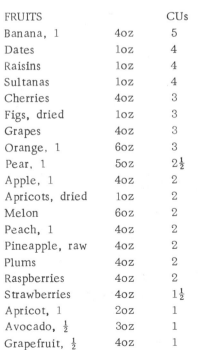

FRUITS		CUs
Banana, 1	4oz	5
Dates	1oz	4
Raisins	1oz	4
Sultanas	1oz	4
Cherries	4oz	3
Figs, dried	1oz	3
Grapes	4oz	3
Orange, 1	6oz	3
Pear, 1	5oz	$2\frac{1}{2}$
Apple, 1	4oz	2
Apricots, dried	1oz	2
Melon	6oz	2
Peach, 1	4oz	2
Pineapple, raw	4oz	2
Plums	4oz	2
Raspberries	4oz	2
Strawberries	4oz	$1\frac{1}{2}$
Apricot, 1	2oz	1
Avocado, $\frac{1}{2}$	3oz	1
Grapefruit, $\frac{1}{2}$	4oz	1
Prunes	1oz	1
Tomato	1oz	1
Rhubarb	4oz	0

SWEETS, SAUCES, SOUPS		CUs
Cake, fruit, iced	2oz	8
Apple pie	4oz	7
Cake	2oz	7
Doughnut	2oz	6
Glucose	2oz	6
Chocolate	2oz	5
Honey	1oz	5
Mince pie	2oz	5
Fruit in syrup	4oz	5
Molasses	1oz	4
Pancakes	2oz	4
Candies	1oz	4
Cookies	2 small	3
Ice cream	1oz	3
Sugar, white, brown	$\frac{1}{2}$oz	3
Cranberry sauce	1oz	$2\frac{1}{2}$
Syrup	$\frac{1}{2}$oz	$2\frac{1}{2}$
Jam	$\frac{1}{2}$oz	2
Soups, various	8oz	0-2
Peanut butter	1oz	1
Cocoa	1tsp	1
Mayonnaise	$\frac{1}{2}$oz	0
Salad dressing	$\frac{1}{2}$oz	0

DRINKS		CUs
Cider	$\frac{1}{2}$pt	7
Beer, heavy	$\frac{1}{2}$pt	7
Chocolate	8oz	6
Beer, light	$\frac{1}{2}$pt	5
Lemonade	8oz	5
Liqueurs	1oz	5
Vermouth, sweet	2oz	5
Port	2oz	$4\frac{1}{2}$
Apple juice	8oz	4
Bitter lemon	6oz	4
Brandy	1oz	4
Gin	1oz	4
Lager	$\frac{1}{2}$pt	4
Sherry	2oz	4
Wine, sweet	3oz	4
Whiskey	1oz	4
Coca Cola	6oz	3
Fruit juices, sugared	5oz	3
Vermouth, dry	2oz	3
Wine, dry	3oz	3
Tomato juice	5oz	1
Coffee, black	1 cup	0
Tea, clear	1 cup	0

©DIAGRAM

COMPULSIVE EATING

WHO BECOMES A COMPULSIVE EATER?

Compulsive eating is a common problem among women - and with the resulting obesity can cause considerable distress. Many women indulge in occasional bouts of "stuffing" but these are rarely significant. The compulsive eater, however, is addicted to food. She may use it to relieve feelings of loneliness, isolation, frustration, dissatisfaction, or boredom. She may use it to comfort herself if she feels guilty, depressed, or unattractive. During adolescence she may overeat to stifle her emerging sexuality, and later in life she may use her obesity to avoid contact with the opposite sex. Overeating - a secret and solitary activity - needs handling with sensitivity and understanding. To help overcome the problem, a woman should avoid being alone for longer than necessary, ensure that only low-Calorie snacks are kept in the house, and divert herself with physical activity when the craving for food starts. Severe cases often need clinical help.

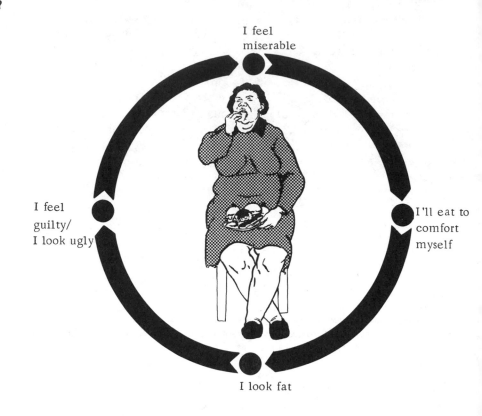

I feel miserable

I feel guilty/ I look ugly

I'll eat to comfort myself

I look fat

Reasons for compulsive eating vary, but the vicious circle illustrated above is common to many women.

TYPICAL PATTERN OF WEIGHT GAIN

A case history of a typical compulsive eater is described here.
1 Age 12: 135 lb (61kg)
2 Age 13: 165 lb (75kg) after distress caused by deaths in the family
3 Age 14: 145 lb (66kg) following treatment at a reducing salon
4 Age 15: 110 lb (50kg) after a

strict diet
5 Age 16½: 200 lb (91kg). Tired of dieting and of "watchful" parents the girl stopped paying attention to what she ate
6 Age 17: 220 lb (100kg) - highest weight despite psychiatric treatment
7 Age 18: 180 lb (82kg) after

another reducing treatment and diet
8 Age 19: 130 lb (59kg) after treatment involving amphetamines to which the girl became addicted. The addiction was cured and the girl's weight increased slightly over the next 20 years
9 Age 40: 140 lb (64kg)

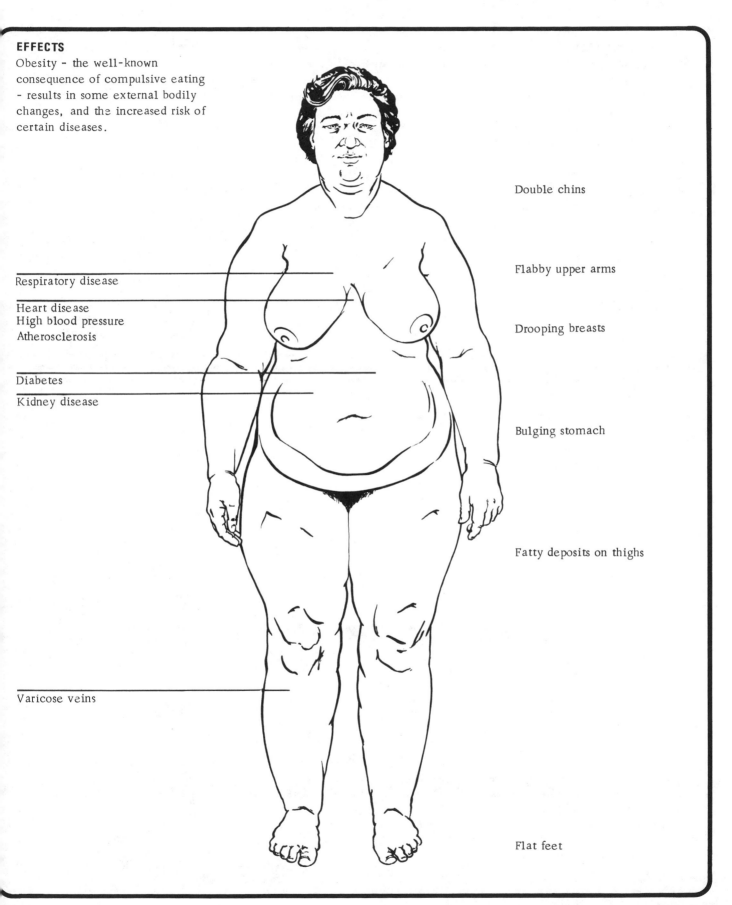

EFFECTS

Obesity - the well-known
consequence of compulsive eating
- results in some external bodily
changes, and the increased risk of
certain diseases.

Double chins

Flabby upper arms

Respiratory disease

Drooping breasts

Heart disease
High blood pressure
Atherosclerosis

Diabetes

Bulging stomach

Kidney disease

Fatty deposits on thighs

Varicose veins

Flat feet

©DIAGRAM

ANOREXIA NERVOSA

ANOREXIA NERVOSA

True anorexia nervosa has been described as the "willful pursuit of thinness through self-starvation." It is a serious disorder mainly affecting adolescent women.

SYMPTOMS AND BEHAVIOR

Dieting begins because the anorectic either is or believes herself to be overweight. It develops into a morbid fear of fatness and continues to the point of extreme emaciation (rarely recognized by the anorectic). Loss of weight is accompanied by loss of menstruation, constipation, discoloration of skin, and the growth of a fine body hair. Despite the extreme loss of weight most anorectics are hyperactive. True loss of appetite is rare and starvation is interspersed with secret eating binges followed by self-induced purging.

BACKGROUND AND CAUSES

Anorexia nervosa and the reasons behind it are highly complex. It has been noticed that many anorectics suffer from an overwhelming sense of ineffectiveness. Continuous starvation and refusal of food represents a gesture of independence, possibly from an overdominant mother, while thinness is seen as a desirable achievement in itself.

AGE OF ONSET

The onset of anorexia nervosa typically occurs during adolescence. The diagram below shows the percentages of a sample group of anorectics who became ill at different ages.

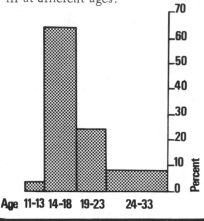

PATTERN OF DEVELOPMENT

The development of a typical case of adolescent anorexia nervosa is described in the diagram below - from initial carbohydrate starvation until medical diagnosis. Response to treatment varies considerably. At the two extremes are obesity and starvation. More typically a patient responds to treatment and attains normal weight.

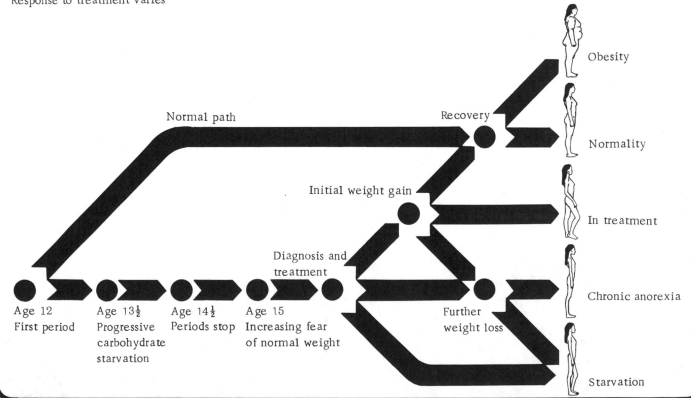

SELF-IMAGE

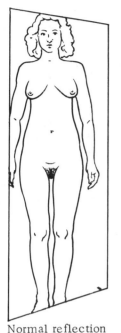
Normal reflection

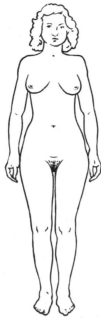
Average-sized girl

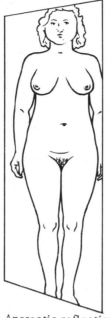
Anorectic reflection

All people have a distorted view of their own body proportions, but whereas the normal person underestimates face, chest, and hip size while slightly overestimating waist size, the anorectic grossly overestimates each of these sizes even when severely emaciated.

Degree of distortion in the perception of body size:

	Normal %	Anorectic %
Face	94.7	157.6
Chest	95	134.2
Waist	100.2	146.6
Hips	96.6	128.8

(100% equals actual size)

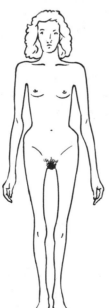
Chronic anorectic

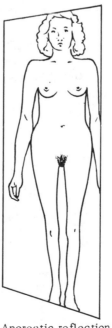
Anorectic reflection

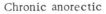

TREATMENT

Treatments for anorexia nervosa include the following:

DRUG THERAPY AND BED REST
The patient is hospitalized and kept in bed. Drug therapy is combined with encouragement to eat a high-Calorie diet. The patient is allowed up only after reaching a near normal weight (usually in 1 to 3 months).

BEHAVIOR MODIFICATION REGIMES
Similar to the treatment already described, but with a system of rewards to meeting daily and weekly weight gain targets.

PSYCHOTHERAPY
The patient and her family receive psychotherapy before and after the patient's release from hospital.

Follow-up of anorectic patients

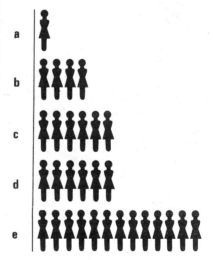

The diagram summarizes follow-up information on a group of anorectics. Follow-up material was available on 30 of the original sample of 45 patients.

a Obese (1)
b Dead (4)
c Anorectic (6)
d In treatment (6)
e Recovered (13)

©DIAGRAM

181

FITNESS

WHAT IS FITNESS?

In the most general sense, a person's fitness is his ability to cope with his environment and the pressure it puts on his mental and physical system.

But in a more usual, and useful, sense, fitness refers to the body's physical capabilities, as measured by tests of strength, speed, and endurance.

Notice that, though fitness is usually associated with health, it is not the same thing. An olympic athlete can be ill, and someone free of diagnosable illness may still be extremely unfit.

In the absence of planned exercise, work and transport effort (such as walking or climbing stairs) are the main determinants of a person's actual physical capabilities.

In modern society, as the element of physical activity in work and transport declines for most of us, we are increasingly dependent on planned exercise if we are to hold on to fitness and its benefits.

COMPONENTS OF FITNESS

There are four basic components of physical fitness:
a) general work capacity;
b) muscular strength;
c) muscular endurance; and
d) joint flexibility.

GENERAL WORK CAPACITY

This concerns the body's ability to supply itself with the oxygen and energy it needs to keep going during general physical activity. It depends on the efficiency of the cardio-vascular and respiratory systems, and is therefore usually called "circulo-respiratory fitness" (CR fitness). In general, CR fitness is called on in those activities that involve a good proportion of the body's muscles over an extended period of time, eg hard walking, running, jogging, swimming, cycling. The limit of CR endurance is marked by labored breathing and a pounding heart, rather than by failure of a particular muscle group to respond any more.

MUSCULAR STRENGTH

This concerns the maximum force a particular muscle group can apply in one action. There are two types.
a) Isometric strength is force applied against a fixed resistance.
b) Isotonic strength is force applied through the full range of movement available to a certain muscle or muscle group (as set by the joint or joints acted upon).

An example involving both types of strength is arm-wrestling. The beginning, when the participants' arms first lock motionless against each other, involves isometric strength. The latter part, in which one participant's arm forces the other's down to the table, involves isotonic strength. The two types are at least partly independent of each other.

MUSCULAR ENDURANCE

This concerns the ability of particular muscle groups to go on functioning over a period of time.

FACTORS AFFECTING FITNESS

POTENTIAL FITNESS

Even if all people were as fit as possible, their physical abilities in terms of strength, speed, and endurance would not be equal. Three main factors limit someone's potential fitness.

a) Age. The natural atrophy of age affects the efficiency of the whole body. But different aspects of fitness reach peak potential at different times. Speed is at its best at the beginning of adulthood, strength in the late 20s, while endurance can improve up to middle age.

b) Sex. Women are constitutionally more fit than men. For example, they are better able to withstand extremes of temperature, and have a longer life expectancy. Men are specifically more fit - they have a greater potential for strength and speed.

c) Somatotype. The shape of a person's body limits the degree of fitness that can be obtained. Most athletes have a high mesomorphic rating, and many also have a high ectomorphic one. People with a high endomorphic rating do not have the same capabilities.

ACTUAL FITNESS

Similarly, several factors determine actual fitness.

a) Medical health. A person cannot become or remain fit if his body is not in good health.

b) Nutrition. A healthy diet is essential in attaining and maintaining fitness and health.

c) Weight. If someone is above his desirable weight , his body is always functioning under the burden of an extra load. If someone is too far below optimum weight, his tissues will lack the ability to function at maximum efficiency.

d) Physical activity. With lack of activity, the body atrophies. This is shown by the muscular weakness that follows confinement to bed,

Again, there is both isometric and isotonic endurance. Isometric endurance involves the ability to maintain force as long as possible against a fixed resistance or in a fixed position (as when the opening lock of arm wrestling continues over a period of time). Isotonic endurance involves the ability to repeat a muscular movement against resistance as many times as possible (as with push-ups or repetitive weight-lifting).

In both cases, the limit is marked by inability of the muscle group to respond any more.

FLEXIBILITY

This concerns the range through which a joint will move. Except in the case of certain bone disorders, it depends more on the nearby muscles than on the structure of the joint itself.

INTER-RELATION

a) Localization. Levels of muscular strength, muscular endurance, and

joint flexibility are all localized i.e. development of one part of the body does not necessarily imply the development of any other part. CR fitness, in contrast, usually develops as a single entity.

b) Inter-dependence. Muscular strength and endurance in any one muscle or muscle group are inter-related. For example (taking isotonic strength), a muscle capable of a maximum 200 lb force through its whole range of movement will be able to go on moving 50 lb through that range longer than one with an 80 lb maximum. Otherwise the components of fitness are fairly independent of each other.

MOTOR ABILITIES

An individual's physical capabilities are limited not only by his fitness, but also by his motor abilities: coordination, balance, agility, reaction time, speed, movement time (i.e. speed of moving a part of the body), and power (i.e. ability for explosive movement).

HUMAN EFFICIENCY

The efficiency of any system is measured by how much energy output (work) it gives, for a set amount of energy input (fuel). Its efficiency will vary, depending on how near to its limits it is working. The nearer it is to maximum energy output, the more units input are needed for each unit gain in output. But all systems are more or less inefficient over all their output range i.e. all give less than 100% return.

The average human body is between 16 and 27% efficient - which compares badly with several products of the human mind. But by regular exercise the body's efficiency can be raised to 56%, which is better than many machines.

 Normal efficiency

 Normal variation

☐ Efficiency with fitness

a Electric motor
b Steam turbine
c Petrol motor
d Steam engine
e Human body

and by the way limbs change shape when encased in plaster casts.
e) Sleep. In order to function at optimum efficiency, the body and mind must have adequate rest.

This graph shows how strength varies with age in men and in women. The dotted line shows actual average strengths for women. The other line for women shows what their strength would be if they were the same size as men.

Muscle strength, by sex and age

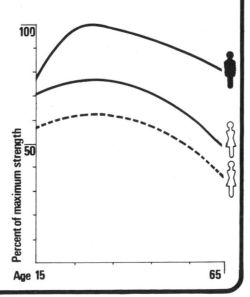

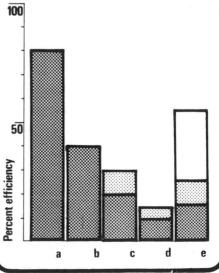

© DIAGRAM

BODY AND ACTIVITY 1

BODY DURING ACTIVITY

Any muscular effort requires some energy, and puts some demand on the rest of the body to supply that energy. When a large proportion of the body's muscles are involved over a period of time, the effects of this become noticeable.

CARDIO-VASCULAR SYSTEM

The blood network supplies the cells of the body with the oxygen and other nutrients (glucose, fat, etc) that they need for activity, and takes away the resulting waste products (carbon dioxide, lactic acid, etc). The heart is the pump that drives blood around the system. During exertion, more nutrient and waste transportation is needed. This is mainly supplied by increased heart activity. The total rate of blood flow may rise by up to 600% due to increase both in the rate of heart beat and in the amount pumped at each stroke. Systolic blood pressure may rise up to 70%. At the same time, levels of nutrients build up in the blood, as physical need increases, and levels of waste products as usage exceeds disposal.

Constriction of capillaries in uninvolved areas (such as stomach and skin) reduces their blood supply, while dilation of others increases the supply to involved areas (involved muscles, lungs, and heart). Later, as body heat builds up, dilation of capillaries near the skin surface allows blood flow to give radiation heat loss.

RESPIRATORY SYSTEM

This steps up activity in response to the cells demand for oxygen and the build up of carbon dioxide in the blood. The depth and rate of breathing increases, so ventilation may rise up to 12 times the resting rate. Overall oxygen consumption may rise from $\frac{3}{4}$pt (US) a minute to $10\frac{1}{2}$pt (US).

ANAEROBIC ENERGY SUPPLY

If oxygen supply does not equal need, the muscles can still go on working for a time, because they have some ability to function "anaerobically", i.e. without oxygen. Some forms of exercise can be completely anaerobic, because they occur in short concentrated bursts of up to 10 seconds (eg sprinting). But an "oxygen debt" is built up - i.e. large quantities of oxygen are then needed to clear the accumulated waste products.

Other forms must be chiefly aerobic, because of the length of time they last (eg long-distance running). So they have to be at a lower level of exertion, so oxygen supply can more nearly keep up.

LIVER

During physical exertion, the liver is important for its ability to convert waste lactic acid into useful glycogen or glucose (providing oxygen is present). This maintains glucose levels, slows the build up of oxygen debt, and increases the blood's ability to carry carbon dioxide.

ENDOCRINE SYSTEM

The main short-term hormonal response to physical exertion is release of adrenalin by the adrenal glands, stimulating the body to the "fight or flight" response. Other hormonal activity controls, for example, body water, and aids increased energy mobilization.

OTHER RESPONSES

a) It is possible that during exertion the spleen releases reserve red blood cells into the blood, increasing its gas carrying capacity.
b) The skin's heat loss by sweat and radiation is determined not only by body mechanisms but also by external conditions. High environmental temperatures limit radiation loss; high humidity limits sweat loss.

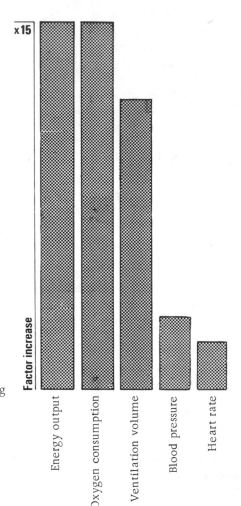

Response to exercise

Factor increase

x15

Energy output · Oxygen consumption · Ventilation volume · Blood pressure · Heart rate

INACTIVITY

In a mechanical system, use produces wear and deterioration. In the human body, the changes during strenuous activity are drastic - but it is chiefly disuse that causes deterioration. Once growth has ceased, there is no physical improvement without increased physical demand. This is true not only of muscular strength and cardio-vascular and respiratory response, but also of motor abilities such as co-ordination and reaction time. Moreover, inactivity not only fails to develop latent capacity, but also results in a general deterioration of many body systems, through disuse.

The effects on the muscular system are the most obvious. There are 639 muscles in the body, accounting on average for 45% of body weight. Loss of muscle tone results, for example, in:
a) general physical weakness and tiredness;
b) a weak and sagging abdomen;
c) back pain due to weak back muscles; and
d) a weak and lethargic heart.
Degeneration of the abdomen and heart have, in particular, general effects throughout the system. Abdominal weakness favours digestive trouble, while sluggish blood flow encourages blockage of arteries and capillaries.

In general terms, the effects of inactivity show themselves in reduced vitality, lowered resistance to infection, and perhaps in such mental conditions as lack of enthusiasm, inability to concentrate, nervousness, irritability, and insomnia. Physical inactivity also increases food intake (see G27), often producing excess weight with all its harmful consequences. When inactive man does have to undergo physical stress, his heart cannot greatly increase its stroke volume. A pounding heart results, as it tries to keep up with demand just by beating faster.

HARMFUL EFFECTS?

Despite the benefits of activity for the body, some harmful effects can occur from ill-judged exercise.
HEALTH AND AGE
Sudden violent exertion after a period of inactivity or ill health is to be avoided. Any fast or vigorous physical activity should never be attempted without a gradual development of fitness first - especially over the age of 30. The older one is, the more unhealthy, or the lower the existing level of fitness, the more careful one should be about the rate of progress - and the more sensible it is to check with a physician first, especially if more than one of these factors is involved. In fact, any inactive person aged 35 or over should check with a physician before starting a serious exercise program. And no one should suddenly begin doing activities that exhaust them or make their heart pound.
THE HEART
Appropriate exercise will usually benefit even a seriously diseased heart (see F08). Severely

inappropriate exercise can damage even a healthy one. Anyone, of any age, who suspects he has something wrong with his heart, or his blood pressure, should, again, check with a physician. A heart already weakened by disease or congenital deformity will suffer rather than gain from any sudden extra burden.
BODY HEAT
a) Always precede vigorous localized muscle activities with general warm up movements - especially when exercising in cold conditions. Sudden activity in a few muscles when the rest of the body is cold can send the blood pressure soaring. Isometrics, in particular, should be preceded by warming up movements.
b) Never wear sweat clothing or rubber suits during exercise, in the belief that this will help you "lose weight". The resulting dehydration could even kill you.
c) Don't stop suddenly after violent exertion, especially if suddenly changing to a warm environment. Slow down gradually.

SORENESS AND PAIN
Normal soreness or stiffness after exercise is simply due to the build-up of waste products, and will go as these are cleared. But during exercise sudden unaccustomed movements can wrench muscles and joints. You should get accustomed to new movements gradually, so your body can learn what positions and efforts it can safely allow. Take any feelings of pain or strain seriously, and ease off. Within a session of exercise (and within each exercise, where appropriate), work up gradually from gentle movements to forceful. Where some jolting against the ground in inevitable. use shock absorbing footwear or a suitable surface. Treat any muscle injuries that occur (whether impact or wrenching) with ice or cold water, elastic pressure dressing, elevation, and rest, rather than with heat.

©DIAGRAM

BODY AND ACTIVITY 2

THE GAIN FROM ACTIVITY

MOTOR ABILITIES When the same activity is repeated, learning allows control of movement to become involuntary and so freed from the retarding influence of thought. Co-ordination and skill improve, and wasted motions become fewer.

MUSCLES Depending on the degree of exertion and the type of exercise, there may be increase in the size, strength, hardness, endurance, and flexibility of muscles used.

a) Size changes with growth in muscle mass due to increased size of individual muscle fibers.

b) Strength changes with muscle size: muscle fiber has a contractile force roughly proportional to its cross-section area.

c) Short-term strength changes occur too rapidly to be due to muscle size, and are probably due to better nervous organization. Inhibitory impulses (designed to prevent muscle being torn away from the bone) may be allowed to weaken with learning, so that positive impulses reach the muscle sooner and produce a stronger response. Improvement in motor coordination may also help.

d) Changes in hardness are due to tighter contraction, and perhaps to muscle fiber replacing fat.

e) Development of endurance may be due to increased capillarization (see below), or, as with strength, to improved nervous organization.

f) The processes by which flexibility is increased are not yet understood. Gains in flexibility reduce the likelihood of joint injury and muscular stress.

THE HEART gains from appropriate exercise, as other muscles do. Its strength, coordination, and endurance improve: More blood is pumped per stroke, so the rate of heartbeat can be lower. In a normal adult at rest, the heart rate is perhaps 70 a minute, and 80 or 90 is not uncommon. After endurance training, 55 to 60 beats a minute is possible, increase during exertion is smaller, and return to normal more rapid. The heart may beat several thousand times fewer every day, and so suffer less from wear.

CARDIO-VASCULAR SYSTEM Regular exertion has several effects.

a) It increases the number of capillaries in active tissue, improving the blood supply so that the body cells are capable of using more oxygen and nutrients per minute. The heart especially benefits, and here the capillaries also provide alternative routes that aid recovery from heart attacks.

b) It also improves the speed of return of blood from the extremities. The action of improved muscle tone against vein walls aids flow and helps prevent thromboses and varicose veins.

c) It may also reduce some forms of high blood pressure, by relaxing the arterioles.

d) Arterial deposits are kept in check by the increased blood flow during exertion and by the effect exertion has of reducing the blood content of fats such as cholesterol and triglyceride (perhaps because exercise metabolism uses them up). Exertion may also lower the blood's clotting tendencies. If so, thromboses are less likely.

In general, appropriate exercise can reduce the likelihood of cardio-vascular disorder and make recovery more likely if it does occur. This can be seen by comparing the rates of coronary heart disease and death in active and sedentary occupations. Exercise, in mild form, may also aid in recovery from such disorders.

RESPIRATORY SYSTEM This is strengthened and improved by exertion. Air intake increases, both in the amount that can be breathed in at one time (vital capacity), and in the amount taken in over a period (ventilation). The efficiency of gas exchange in the lung alveoli also improves, while an increase in the number of red blood cells aids blood gas transport.

BODY POSTURE Better muscle tone improves both skeletal posture and organ position. Stronger back muscles lower the likelihood of spine disorder and back pain. Stronger abdominal muscles prevent stomach sag, with widespread effects on the efficiency of the internal organs. Appetite and digestion improve, and better bowel action discourages flatulence, constipation, and piles. Also the muscle tone of abdomen and hamstrings controls the tilt of the pelvis - so here again improvement reduces the likelihood of lower back pain.

NERVOUS SYSTEM Motor responses are improved by exertion that requires quick interpretation, decision, and action. The system becomes more coordinated, and better able to judge and respond to exact requirements. In addition, there seems to be a relationship in some people between physical fitness on the one hand, and, on the other, mental and perceptual alertness, absence of nervous tension, and even prevention of emotional illness. Stress and inturned emotion may also be more specifically dissipated by exertion that is especially rhythmic, or that involves enjoyable competition.

RESISTANCE TO DISEASE The physical fitness that results from activity should make the body better able to fight off infection and to recover quickly from any illness (or injury) that does occur.

WEIGHT CONTROL

In general, regular activity can use up excess Calories, stimulate the body to waste more Calories in thermogenesis, and in some cases lower the appetite. Exercise alone does not effectively reduce body fat, but it allows effective dietary measures to be - and feel - less severe.

ENERGY Someone used to physical activity usually has more energy than an inactive person.

a) His usually lower metabolic rate conserves energy resources.

b) The efficiency of his respiratory, cardio-vascular, and digestive systems makes oxygen and nutrients more quickly available during exertion, and speeds recovery afterwards.

c) He has increased oxygen debt tolerance i.e. his muscles can carry on without oxygen a little longer.

d) Any given physical demand is usually less near the limit of his acquired abilities, and so can be dealt with more efficiently. For example, he has lower oxygen needs for any specified task.

In general, a physically fit person is able to withstand fatigue for longer periods, and, after a similar work effort, he is probably more likely to have energy left for leisure activities.

WELL-BEING The physically active person is more likely to sleep well, look well, and feel the exhilaration of a healthy body.

LIFE EXPECTANCY The overall consequence of all the factors involved is that those taking regular exercise have a lower death rate for each age group than those who do not, and that, in general, the death rate is lower the more strenuous the exercise.

THE BODY AT REST
The physiological differences between active and inactive people are clear even when their bodies are at rest.

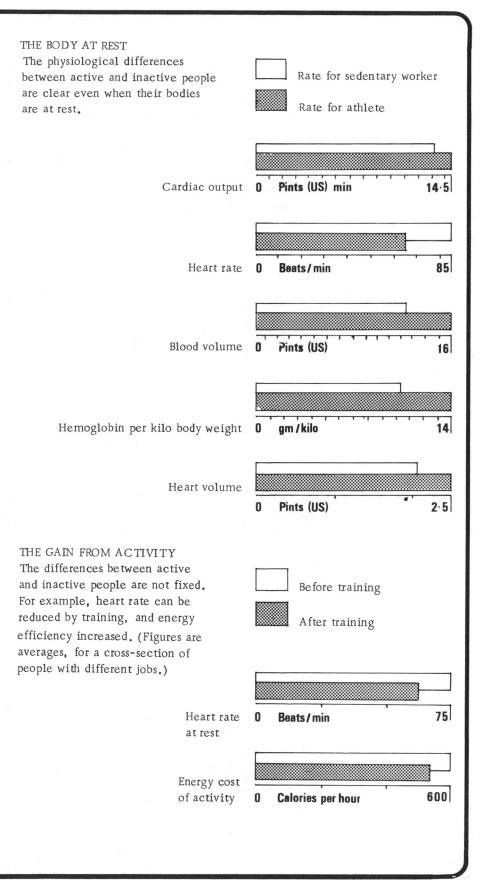

Rate for sedentary worker

Rate for athlete

Cardiac output   0   **Pints (US) min**   14·5

Heart rate   0   **Beats / min**   85

Blood volume   0   **Pints (US)**   16

Hemoglobin per kilo body weight   0   **gm /kilo**   14

Heart volume   0   **Pints (US)**   2·5

THE GAIN FROM ACTIVITY
The differences between active and inactive people are not fixed. For example, heart rate can be reduced by training, and energy efficiency increased. (Figures are averages, for a cross-section of people with different jobs.)

Before training

After training

Heart rate at rest   0   **Beats/ min**   75

Energy cost of activity   0   **Calories per hour**   600

©DIAGRAM

EXERCISE

NEED FOR EXERCISE

The natural conditions of our life no longer give us enough physical activity. Work and travel are increasingly sedentary - and leisure equally usually a matter of sitting watching something. (It has been estimated that on average 75 hours are spent by people in spectating - films, tv, sport, etc - for every 1 spent in physical participation.) So we have to plan and work, to achieve enough activity to keep us physically fit.

But if the need is there, so is the potential. A high level of physical fitness can often be reached in just a couple of months daily physical training - and maintained after that by exercising only on alternate days.

In theory, adequate fitness might perhaps be maintained every day by: walking briskly a mile or two to and from work; standing doing something for a couple of hours; using vigorous towelling movements after washing; stretching occasionally; and hurrying up a hill or stairs carrying a fairly heavy load. But in practice few have the time or taste or consistent discipline to organize even this; and most of us are well below desirable levels of fitness. So formal exercise, over and above the normal daily demands, becomes essential for the achievement of an efficient, strong, and durable body.

There are very many different approaches to physical fitness. The remaining pages of this chapter are concerned not with setting you a particular routine of exercise, but with telling you what kinds of exercise are available, what each can and cannot do, and how you should approach the general problem of keeping fit.

THE MAIN TYPES

Each component of fitness has its own corresponding forms of exercise. This brief survey gives examples for lone participants.

CIRCULO-RESPIRATORY EXERCISES (CR exercises) include brisk walking, jogging, running, cycling, swimming, and rope skipping and running on the spot; also cycling machines, rowing machines, etc. All are "aerobic", and therefore place demands on the circulation and respiration to supply oxygen to the muscles involved.

ISOTONIC EXERCISES consist almost entirely of weight-lifting. This gives the progressive resistance necessary to test, and increase, the maximum strength of a single joint movement.

ISOMETRIC EXERCISES are motionless. They use the force of one body muscle against another or against a fixed point such as a wall or bar. Towels or ropes provide simple accessories; more complex are machines (such as the Bullworker and Flexorciser) that provide variable resistance according to strength. Holding for 6 seconds or more produces the best strength gains, with 10 or 12 seconds recommended. Intensity of effort is more important than frequency.

MUSCULAR ENDURANCE EXERCISES

consist mainly of traditional "calisthenics", with or without equipment i.e. push ups, sit ups, squat leg thrusts, pull ups to a bar, bar dips. All these are usually isotonic - as is weight lifting done with a repetitive pumping motion (used for increasing muscle size rather than strength). They are only isometric if held locked in a tense position (i.e. against gravity).

FLEXIBILITY EXERCISES usually consist of stretching and rotating movements of parts of the body.

GENERAL FITNESS

The vital ingredients of a general fitness program are:
a) CR exercise;
b) endurance exercise for shoulders, arms, abdomen, back, and legs; and
c) flexibility and stretching exercises for neck and shoulders, back, and hips and hamstrings. Not especially useful are isometric exercises and single contraction isometrics (though isometrics may be needed to make weak muscles strong enough to begin endurance movements).

Endurance and flexibility exercises can be obtained from gym training, keep fit classes, some sports, or from personal routines. With personal routines, the desire to persist in a movement or position is often too weak for improvement.

Isometric exercises and single contraction isotonics (i.e. weight lifting) will not be considered in any further detail. There are many books on these subjects if required.

GENERAL ADVICE

a) Take some exercise daily, if possible - especially with general fitness routines. (Some CR exercises, though, may give as good results from 2 or 3 workouts a week as from 4 or 5. Isotonic exercises should alternate work days and rest days.)
b) Whatever frequency you can manage, make it regular: don't skip allocated days.
c) Help yourself by setting aside a regular time: whether (for example) before breakfast, or in the late morning or afternoon at work, or at home in the early evening, or just before going to bed.

CHOICE OF EXERCISE

The range of possibilities includes team sports, opponent sports, gym activities, keep-fit classes, companion activities (such as hiking or group jogging), lone indoor and outdoor routines, and extensions of normal activity such as walking to work or going up stairs carrying heavy shopping.
Factors in your choice of exercise will include:
a) your purpose;
b) the availability of time, money, facilities, equipment, and other people; and
c) your personal interests.
It is worth judging these things carefully before you make a decision.

PURPOSE

You may want to get generally fit, or to train for a sport, or perhaps to develop or strengthen a weak part of your body. The distinction is important because the function of different exercises is so specific.

Where training for sport is the intention, the best method is to practice the sport itself. Endurance running, for example, will not help too much with endurance swimming - though some exercises may provide a near substitute.
But some sports do not raise the level of general fitness much, and so additional fitness training may bring improved performance.
Similarly, strengthening a part of the body is best done in conjunction with a general fitness program.
(General fitness exercises also fulfil a useful warm-up function.)

REQUIREMENTS

Exercises vary greatly in their requirements. The same fitness may result from five minutes of daily exercise without equipment as from hours spent on (and traveling to) a specialized field with two teams and officials.
Choose an activity that you know you will be able to arrange regularly.

INTERESTS

What exercise you will keep to also depends on what you enjoy.
Different exercises offer varying elements of:
a) competition;
b) company;
c) being encouraged or goaded along by an instructor;
d) being pushed along by an abstract concept (such as time, or the desire to progress to the next level of fitness); and
e) associated factors, such as social life or the pleasure of being in the countryside.
You must judge which factors are most important to you. Remember that, to be enjoyable, some require success from you. Competition, or the eye of an instructor, or fighting against the clock, may not encourage you if you fail.

---

d) Also help yourself and avoid harmful effects by setting realistic targets. Start gently and progress slowly. Always underestimate your abilities. There is normally no long-term advantage in doing something "till it hurts" - and there are likely to be painfully obvious disadvantages. Something that seems all right for a day or two may result in sore muscles and the end of your resolutions. Your body is the best judge of what you can do. Take notice of it, and only adjust upwards when you are sure you are within your capabilities. Practical advice on determining your level of fitness and your safe levels of activity is given on pp. 190-1.

e) Read p. 185 on possible harmful effects, before you start. But never feel that your health entirely rules out the possibility of exercise. Only a minimum of organic health is needed for some kind of general fitness program to be possible and desirable. Take your physician's advice.

f) Be sure to give the body a general warm up, either before a routine or as part of it (eg beginning with the less difficult exercises, and not immediately using full speed or effort).

g) Do not be discouraged if you reach a stage at which no progress is made. Ease back to a slightly lower level of effort, and progress up again from there.

h) Once the desired level of fitness is achieved, less frequency is needed to maintain it, eg general fitness routines may only require 3 workouts a week. But keeping up daily exercise may still be a good idea for relaxation, digestion, sound sleep, etc.

i) Associated factors may be as important as exercise. Get adequate sleep. Try to find ways of forgetting work worries. Eat sensibly. If you smoke, stop if you can.

© DIAGRAM

HOW FIT ARE YOU ?

TESTS OF FITNESS

All components of fitness can be tested, eg isotonic strength by the weight of barbell that can be lifted, isometric strength using strain gauges and dynamometers. But only CR testing gives a good guide to general fitness. Muscular endurance tests may also be interesting for following progress.

CR FITNESS

The Tecumseh step test given here is safe for most age groups. It involves stepping up and down at a given rate between the ground and a single step, bench, or stool. Afterwards the pulse rate is taken. General rules for any step test are:
a) face the same way all the time and always step back to the ground on the same side;
b) "one step" is one complete ascent and descent;

c) keep the correct step rate;
d) take the pulse at wrist or throat (but if at the throat do not press too hard, as this can alter the rate);
e) do stop before the time limit if the test is too hard (count this as putting you in the lowest fitness category); and
f) to compare performances over time, try to repeat under similar conditions (eg time of day and time and size of last meal; recent physical activity, health, and sleep; and step rate).

Do not be discouraged by apparent lack of progress: temporary factors may be involved.

MUSCULAR ENDURANCE

Isometric endurance tests judge the duration for which a certain contraction can be held (eg how long a known weight can be held at arm's length). More relevant for general fitness are isotonic endurance tests, which judge the number of times a given movement can be repeated. Typical tests (with average performances of male college students in brackets) include pull ups (8 completed), push ups (25), sit ups (40), and bar dips (9).

HEART RATE AND EXERCISE

The Tecumseh step test gives you a guide to your general physical condition. You can use this to judge how high you should allow your heart rate to go during exercise.

Someone in the lowest ("very poor") category on the Tecumseh test (100 beats a minute or more) should begin exercising very gradually.
a) At the start of the first month he should exercise for 5 minutes a day at not more than 100 beats per minute. During that month he can gradually increase the time up to 10 minutes, but must keep the same heart rate limit.

b) During the second month the time allowed again increases from 5 to 10 minutes, but the heartrate limit is now 110 beats.
c) During the third month the same happens, but the limit is now 120.
d) Thereafter the person may allow his heart rate to rise to the desirable limit for his age.
Someone in the "below average" step test category should follow the same routine, but can begin at "b", and go on to "c" in the second month. Someone in the "above average" category can begin at "c".

AGE LIMIT

As the resting heart rate declines with age, the maximum possible and desirable rates also fall.

The estimated maximum possible heart rate for a healthy fit young adult is about 220 beats per minute. For a rough estimate of the maximum rate for anyone aged 30 years or older, the age should be subtracted from this figure. For example, the maximum heart rate of a man aged 40 is:
220 - 40 = 180 beats a minute.
To find the maximum heart rate that should be allowed in a fit person during exercise, the figure of 20 should be subtracted. For example, the maximum desirable rate for a man aged 40 is:
180 - 20 = 160 beats a minute.

Intensity	Heart rate	Walk/run (m.p.h.)	Cycling (m.p.h.)	Climbing (Grades)	Sports	Occupations
Maximum	200	13.0	20	12	running	digging
Very heavy	150	6.0	14	6	mountaineering	chopping wood
Heavy	140	5.5	12	5	tennis	pick and shovel
Fairly heavy	130	5.0	10	4	volleyball	gardening
Moderate	120	4.5	9	3	golf	house painting
Light	110	4.0	8	2	table tennis	auto repair
Very light	100	3.5	7	level	bowling	shopping

TECUMSEH SUBMAXIMAL TEST

For ages 10 to 69 (unless in poor health). Use 8in bench. Step at rate of 24 steps a minute for 3 minutes. Wait for exactly 1 minute after exercise. Then count heart beats for 15 seconds. Multiply that count by 4.

CATEGORIES
according to heart rate
(rates for women in brackets).

Excellent	under 68	(under 76)
Good	68-79	(76-85)
Above average	80-89	(86-94)
Below average	90-99	(95-109)
Very poor	100 plus	(110 plus)

Adapted from Montoye, Willis, and Cunningham, J of Gerontology, 1968.

USING HEARTRATE GUIDES

a) Count the number of pulse beats in 15 seconds and multiply by 4. Unlike the step test, count while still exercising if possible (eg if walking).

b) With a little practice it is easy to make a rough judgement of the speed of one's heart beat at any moment, and when it is getting a bit too high. This then becomes the most practical method.

The table gives some idea of levels of heartbeat in different common activities for fit young adults.

SPORT AND EXERCISE

Here we show the effect of different sports on fitness; eg, running has a high effect on CR capacity, but little on strength. The last column gives maximum recommended ages. (Note: badminton, handball, and tennis have less CR effect, and higher age limits, if played doubles.)

○ some effect
● considerable effect

	CR capacity	Muscular endurance	Strength	Power	Agility	Age
SOLO ACTIVITIES						
Archery		●	○			
Bicycling	●	○	○			
Calisthenics	●	●	○			
Canoeing	●	●	○			
Gymnastics		●	●	●		45
Hiking	●	●				
Jogging	●	●				
Skipping	●	●				40
Rowing	●	●	○			
Running	●	●				45
Skiing	●	●	○		●	45
Swimming	●	●	○	●	●	
TEAM SPORTS						
Basketball	●	●		●	●	30
Baseball	○	●		●		45
Football (US)	○	●	○	●	●	30
Hockey	●	●	○		●	30
Rowing	●	●	○			30
Soccer	●	●	○		●	45
Softball	○	○		●		50
Volleyball		○		●	●	
OPPONENT SPORTS						
Badminton	●	○			●	50
Bowling		○				
Canoeing	●	●	●			30
Golf	○			●		
Handball	●	○		●	●	45
Tennis	●	○		●	●	45
Skating	●	●	○		●	45
Skiing	●	●	○		●	45
Swimming	●	●	●			30

CR EXERCISE

For improvement in the cardio-vascular and respiratory (CR) systems, considerable demand must be placed on them by:
a) the duration of exercise; and
b) its intensity (as shown by the heart rate response).

The demand needed for improvement depends on age and existing fitness; but in a fit young adult, for example, exertion at 140 to 150 beats a minute would be needed for 8 to 10 minutes several times a week. With longer duration, though, a rate as low as 120 beats gives improvement.

Intensity depends on:
a) the proportion of body muscle involved (eg jogging is more intense than sit ups); and
b) the pace of exercise (eg running is more intense than jogging).

Very low intensity exercise (eg slow walking) may not give improvement whatever the duration. If the allowable heart rate is limited by poor fitness, this must be dealt with first; if by age, lengthy exercise of just enough intensity (eg steady walking) must be taken.

General fitness programs can give some CR fitness. But once some general fitness is achieved, the intensity of the program may not be very high and the duration insufficient. It is sensible to do additional CR exercises. An excellent source of CR programs is "The New Aerobics" by K. Cooper (Bantam).

If arranging your own CR program, start with 3 workouts a week for a fortnight, then 5 a week till desired fitness is reached. Then maintain with 3 or 4 a week.

© DIAGRAM

KEEPING FIT

A DAILY EXERCISE PROGRAM

A daily exercise program - even a gentle one - can work wonders for almost everyone. Automobiles and labor-saving equipment have made our lives easier in many respects, but by reducing general levels of physical exercise, they have helped produce a population that is chronically unfit. A body that is short of exercise is stale, sluggish, and generally inefficient. But the situation can be remedied - easily, and even enjoyably. There is no need to embark at once on an intensely vigorous exercise program - indeed, such a course of action would be positively unwise for most people today. Exercising regularly is the real key to improving fitness levels. A few simple exercises - such as those described here - can prove dramatically effective if they are carried out every day.

Stand feet apart - raise arms above head - bend and touch ground between feet

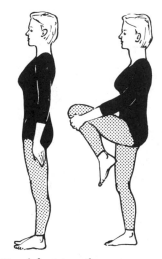

Stand feet together - grasp raised leg by knee and shin - stand - repeat with other leg

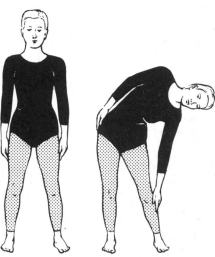

Stand feet apart, hands to sides - side bend to left - repeat with bend to right

Large circles with left arm, forward then back - repeat using right arm

Sit on floor grasping knees - bring knees toward chin and rock back - hold for 5 seconds

Lie face down, arms by sides,
legs together - raise upper
body and legs into position shown

Lie on back, knees bent, feet on
floor, arms back - swing arms
forward to sit and touch toes

Lie on side in position shown -
raise one leg - legs together - roll
onto other side and repeat

Lie face down with hands under
shoulders - push body off floor,
keeping knees on the floor

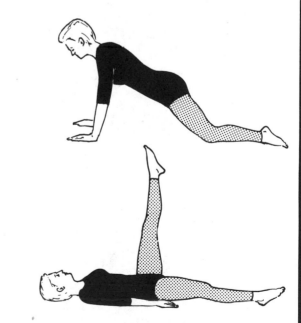

Lie on back, legs together, arms
by sides, palms down - raise leg -
lower leg - repeat with other leg

©DIAGRAM

CR ROUTINES

Here we give simple outlines, around which CR exercise routines can be constructed. Choose an exercise, and carry it out for the required amount of time each day (included repeated occasions if specified). Also for the required number of days per week.

Start at level A. If a range of choice is given, for the number of days per week at that level, start with the lowest number and build up to the highest.
Then very gradually, and without strain or exhaustion, build up toward level B (by increasing distance and/or speed and/or duration, whichever is specified).
In some cases, A1, A2, and A3 show how to start building up effort.
At level B, as indicated, fitness can usually be maintained with only 3 or 4 days' exercise per week. Exercises can be alternated for variety if desired.

	Distance	Rate	Time per day	Days per week
Walking 	A 1 mile	1 mile in 15 mins	15 mins	5
	B 3 miles	1 mile in 13 mins	39 mins	5
	or 4 miles	1 mile in 13 mins	52 mins	4
	or 5 miles	1 mile in 13 min	65 mins	3
Stationary running 	A	70 to 80 steps a min	1 min	5
	B	80 to 90 steps a min	20 mins	3 or 4
Running 	A 1) 1 mile	1 mile in 13 mins (walk)	13 mins	3 to 5
	2) 1 mile	1 mile in 11 mins (walk/run alternately)	11 mins	5
	3) 1 mile	1 mile in $9\frac{1}{2}$ mins	$9\frac{1}{2}$ mins	5
	B 2 miles	1 mile in $8\frac{1}{2}$ mins	17 mins	3
Combination 	A Walking			
	1) $\frac{1}{2}$ mile	120 steps per min		5
	2) 1 mile	120 steps per min		5
	B Alternate jogging and running			
	1) 1 mile	120 steps per min		5
	2) 3 miles	120 steps per min		5

	Distance	Rate	Time per day	Days per week
Swimming	A 1) 25 yds, 4 times*	25 yds in 35 secs	2½ mins	3 to 5
	2) 100 yds	100 yds in 2½ mins	2½ mins	5
	3) 100 yd increases	100 yds in 2½ mins	5 mins, 7½ mins etc	5
	B 1000 yds	100 yds in 2 mins	20 mins	3 or 4

* 4 separate occasions, with rests between

	Distance	Rate	Time per day	Days per week
Cycling	A 1 mile	12 mph	5 mins	5
	B 8 miles	20 mph	25 mins	3 or 4

	Distance	Rate	Time per day	Days per week
Bench stepping	A 8in bench	30 steps per min	3 mins	3 to 5
	B 15-18in bench	30 steps per min	5 mins	3 or 4

"One step" is one complete ascent and descent, involving four leg movement counts: first foot up, second foot up, first foot down, second foot down. At the rate specified above this gives 120 counts per minute i.e. 2 per second.

Tennis Also squash, badminton, basketball, handball (singles)

	Time per day	Days per week
A	10 mins	3 to 5
B	60 mins	3
or	45 mins	4

	Sequence	Times per day	Days per week
Rope skipping	A 1) ½ min skip, 1 min rest, ½ min skip	2	3 to 5
	2) ½ min skip, 1 min rest, ½ min skip, 1 min rest, ½ min rest	3	5
	etc. building gradually to:		
	3) 2 min skip, ½ min rest, 2 min skip	2	5
	4) 2 min skip, ½ min rest, 2 min skip, ½ min rest, 2 min skip	3	5
	5) 4-5 minute skip	1	3 or 4
	B 6 min skip	1	3 or 4

© DIAGRAM

KEEPING FIT 3

5BX PLAN

With most personal exercise routines, the incentive to persist in a movement or position is too weak for endurance improvement. The following plan overcomes this by adding the pressures of time and progressive grading.

The 5BX plan for men was developed by the Royal Canadian Air Force. It gives a balanced program, requiring little space or time and no special equipment, but progressing to the fitness levels of champion athletes.

The plan consists of 6 charts, each with 12 levels of fitness. An individual starts at the bottom of the first chart, works up to the top, and then progresses to the bottom of the next one. Within each chart, the required number of repetitions increases. From one chart to the next, the exercises become more complex and testing.

An individual's rate of progress is set:

a) by whether he can do the required exercises in the time allotted;

b) by his age (on Chart One, the older he is, the more days he must spend at each level - even if he can do the exercises comfortably in the allotted time); and

c) by his own desires and responses (he can progress more slowly if he wants - and should do if he gets stiff, sore, or unduly breathless, especially if in the older age groups).

The time allotted for the basic five exercises is, on all charts, 11 minutes. The time taken over individual exercises does not matter - providing all are done within 11 minutes. But there are alternatives - walking or running - that can be substituted for the fifth exercise. If so, the first four must be done in 5 minutes, and the walking and running in whatever time is specified at that level.

The program has a built-in warming-up routine, because a stretching and loosening exercise is always first. But an individual should also begin the first exercise slowly and easily, and gradually build up speed and vigor.

The maximum levels for average individuals are indicated on the charts. For example, on the first chart, a 6 year old boy would not be expected to get beyond level B, or a 7 year old beyond level A. Once a desired level of fitness has been reached, it can be kept up with only 3 sessions per week. After any period of inactivity, the individual should restart at any lower level that is comfortable.

The information on these pages about the 5BX has been reproduced by permission of the copyright holders: Canada Information, Penguin Books Ltd, and (for the USA) Simon and Schuster Inc. Each of these can advise about published sources from which this plan can be pursued.

SAMPLE CHART: Chart 1 (Beginners)

Exercise	1	2	3	4	5	5A	5B
	Repetitions					Minutes	
A+	20	18	22	13	400	5½	17
A	18	17	20	12	375	5½	17
A-	16	15	18	11	335	5½	17
B+	14	13	16	9	320	6	18
B	12	12	14	8	305	6	18
B-	10	11	12	7	280	6	18
C+	8	9	10	6	260	6½	19
C	7	8	9	5	235	6½	19
C-	6	7	8	4	205	6½	19
D+	4	5	6	3	175	7	20
D	3	4	5	3	145	7½	21
D-	2	3	4	2	100	8	21
Minutes for each exercise							
	2	1	1	6			

Alternatives to exercise 5: 5A, ½ mile run; 5B, 1 mile walk.

Minimum number of days at each level in Chart One.

Age	Days
under 20	1
20-29	2
30-39	4
40-49	7
50-59	8
60 and over	10

Exercise 2
Each chart uses the same 5 basic exercises; but they are slightly more complex and testing each time.

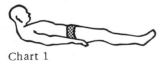
Chart 1

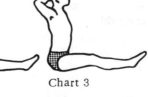
Chart 2   Chart 3

Chart 4   Chart 5

Chart 6

1a) feet astride, arms upward
 b) forward bend to touch floor
 c) return to upward stretch
 d) backward bend
Do not strain to keep knees straight.

2a) back lying, feet 6in apart, arms
at side
 b) sit up so can just see heels
 c) return to position a)
Keep legs straight. Head and
shoulders must clear floor.

3a) front lying, palms beneath
thighs
 b) raise head and one leg
 c) return to position
 d) raise head and other leg
 e) return to position
Above counts as one repetition.
Keep legs straight at knee. Thighs
must clear palms.

4a) front lying, hands under
shoulders, palms flat on floor
 b) fully straighten arms, lifting
body, keeping knees on floor
 c) bend arms and return so chest
touches floor
Above counts as one repetition.
Keep body straight from knees.

5 Stationary run. Lift feet about
4in from floor. Count one step
every time left foot touches floor.
Every 75 steps, do 10 scissor jumps.
Continue till required steps
completed.
Scissor jumps. Stand with right leg
and left arm stretched forward,
left leg and right arm back. Arm
height as shown. Jump up and
reverse arm and leg positions before
landing. Jump and return to first
position to complete one repetition.

©DIAGRAM

YOGA

Many thousands of women are now discovering the physical and mental benefits to be gained from the pursuit of yoga. This ancient discipline, developed in the East over thousands of years, has in recent decades won many enthusiastic followers in North American and European countries. Illustrated on these two pages are representative examples of popular yoga postures. Combined with breathing control, the attainment of such postures acts to produce deep levels of relaxation in both body and mind. Because of the precise nature of yoga postures, potential students are strongly recommended to attend classes to ensure that all postures are correctly learned. Hard work and dedication are essential for the serious yoga student, but the results will prove well worth the effort.

Examples of standing yoga postures.
1 Stand erect, feet together, with weight on both feet
2 Feet apart, legs straight, arms stretched, side bent to clasp ankle
3 Similar to 2 but with one arm above head and one knee bent
4 Feet apart, legs straight, hands behind back, head touching leg

5 Arms above head, one leg bent forward, other leg stretched behind
6 Similar to 5 but with body turned to side and arms outstretched
7 Feet together, legs straight, body bent, head touching legs, hands on floor behind feet
8 Stand on one leg, other leg bent to side, hands together

1

2

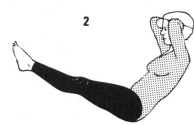

3

4

Examples of sitting and resting
yoga postures.
1 Basic sitting posture

2 Sitting posture, legs raised
straight and together, hands at
back of head

3 Sitting, legs to side, back
twisted, one hand behind back
4 Shoulder stand, back straight

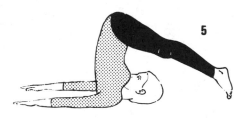

5

6

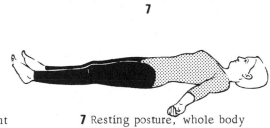

7

5 Lying posture, arms straight by
sides, hands palms down, feet
touching floor beyond head

6 Lying posture, arms straight
behind head, hands palms up, feet
touching floor beyond head

7 Resting posture, whole body
relaxed - used with breathing
exercises to end each session

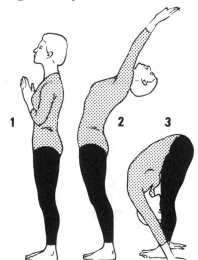

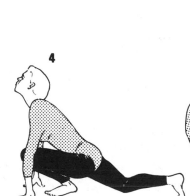

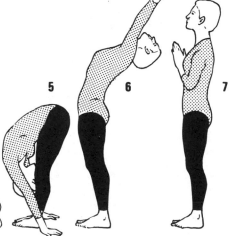

A combination of yoga postures
forming a short sequence.
1 Stand, hands and feet together

2 Inhale, adopt posture shown
3 Exhale, adopt posture shown
4 Inhale, adopt posture shown

5 Exhale, adopt posture shown
6 Inhale, adopt posture shown
7 Exhale, return to first posture

©DIAGRAM

5 Stress and drugs

Peer pressure and the
busy pace of our lives
lead us to indulge in
activities such as
drinking, smoking and
taking tranquillizers.
This chapter looks at the
risks behind alcohol,
tobacco and other drugs,
legal and illegal.

Left: The dawn of a new era:
women celebrating the repeal
of Prohibition in the USA, 1933
(UPI)

STRESS

STRESS

WHAT IS STRESS?

Stress is nervous tension. It may be conscious or unconscious - that is, the person may know he feels tense or he may not. It may be "environmental" or "psychological" - that is, the person may be reacting to a physical threat or a mental threat. And it may be "acute" or "chronic" - that is, the threat may be an event or a continuing situation. (These are just some of the variables.) But the result is that stress occurs in a person's reactions to certain situations which apparently threaten him or exert pressure on him. So whatever the cause, stress in the end depends on the person's reaction - not on the outside event.

STRESS AND HEALTH

The distinction between types of stress is important, though. It is commonplace to talk vaguely about the impact of stress on the body. But such talk usually confuses these different types, when their relationship with the body may differ considerably. It is especially important to distinguish between environmental and psychological stress, and acute and chronic. Combining these two criteria, examples of stressful situations might be:

a) acute environmental: being nearly run over by an automobile;
b) chronic environmental: working in extremely noisy conditions;
c) acute psychological: ranging from losing an argument to losing a wife; and
d) chronic psychological: unhappy at work, living in an unhappy marriage, etc.

ACUTE ENVIRONMENTAL STRESS

I am going slowly across the street, feeling tired. Suddenly I see an automobile bearing down on me: I leap to the sidewalk. Then my body begins to shake, and I start panting breathlessly. The "fight or flight" reaction syndrome has taken over.

FIGHT OR FLIGHT SYNDROME

When a threatening event occurs in the environment, the body has a specific reaction. Signals from the brain stimulate the autonomic nervous system (over which we have no conscious control). This causes the release of powerful hormones (including epinephrine), which key up the body for action. The resulting fight or flight syndrome includes:
increase in rate and strength of heartbeat;
constriction of blood vessels and rise in blood pressure;
increase in blood sugar and fatty acids;
dilation of nostrils and bronchi;
increased muscle tension;
and retraction of the eyeballs and dilation of the pupils.
All these speed the person's reactions and make him more capable of any supreme physical effort that the danger may demand - whether it is fighting or running away. Also blood coagulation time shortens, to reduce the effect of any wounds.

INAPPROPRIATE REACTION

In primitive man the fight or flight syndrome was appropriate to the dangers he faced.
But (apart from being run over!) few of our modern threats are ones we have physically to run from or fight. Imagine that, instead of being a harassed pedestrian, I am driving an automobile, and a lorry swerves in front of me. My danger may be just as great, and all the body processes of fight or flight instantly start working. But the only action I can take is to step on the brake. Even if that is successful, my body is left ready for a struggle that never happens.

DAMAGING EFFECTS

In fight or flight, the epinephrine has prompted the release into the blood stream of fatty acids - needed to fuel the muscles and make the blood clot more quickly if a wound occurs. When there is no struggle, these fatty acids are left circulating through the blood vessels, and may convert into cholesterol deposits. Moreover, repeated incidents may result in a constant high blood pressure.
In fact, research suggests that repeated environmental stress situations can be a major cause of atherosclerosis and related troubles - if no physical movement is possible to "use up" the fatty acids produced. For example, English bus drivers have been found to have a far higher incidence of heart disease than the men who go round the same buses selling the tickets - experiencing perhaps no less stress, but getting more exercise. (But of course there may have been different susceptibility to disease in the types of people who came to choose these different jobs.)

ACUTE PSYCHOLOGICAL STRESS

This refers to the psychological effect of particular incidents (though their impact may last over a long period of time). There are two main sources of such stress: conflict, and change.

CONFLICT

Acute conflict may be with a person, or with an abstract such as time. It produces emotions of either anger or anxiety, and both these are often accompanied by the physical changes of the fight or flight syndrome. The link is an automatic one, over which we can have little conscious control. Again, the reaction is usually physically inappropriate, and so the same physically damaging effects can occur.

CHANGE

Here no fight or flight syndrome is involved. But being faced with change and having to adapt to it does create psychological stress - and recent research has also shown that it consistently results in physical illness, even though the process is not understood. People who have experienced a high degree of stressful change in their lives in a limited period of time (whether one large source of stress or a number of small ones) typically develop some form of physical illness in the next few months. One researcher scored people on their recent experience of typical stressful events (ranging, for example, from changing one's eating habits to suffering the death of husband or wife). The top 10% of scorers suffered twice as much physical illness in the next four months as those in the bottom 10%.

CHRONIC STRESS

ENVIRONMENTAL

Every day a city bus driver may face several incidents of acute stress, when a traffic accident threatens. But he is also working under constant conditions of chronic environmental stress, even if no such incidents occur. This is due to:
a) the physical conditions of his work setting (unpleasantly cramped and noisy); and
b) the need for constant alertness and readiness, in case incidents of acute stress do occur.

In these circumstances the body reacts with what has been named the General Adaptation Syndrome: a long term adaptation to the presence of stress. One of the above causes alone is sufficient: for example, noise polluted work conditions tend to set up permanently high blood pressure (and perhaps also high blood cholesterol levels). Eventually the person feels exhausted, mentally and/or physically. Resistance to disease, or simply the desire to go on, break down, and physical, mental, or psychosomatic illness occurs.

PSYCHOLOGICAL

Chronic psychological stress depends much more on the person than on what happens to him. Someone promoted above his abilities will probably be under such constant stress - but some personalities may not be aware of the pressure, or of not being up to the job. On the other hand, someone in a position well within his abilities may carry it out with constant drive and tension. Perhaps it is because he does not realize he can do it easily; or because he wants to win promotion; or simply because that is his way of going about things.

Some heart researchers, in fact, have contrasted two personality types. Type A has intense ambition, competitive drive, and a sense of urgency. Type B, in contrast, has a content, unhurried personality. These researchers have also suggested that Type A is far more likely to get heart disease. But so far there has been no satisfactory proof of this. The high correlation between supposed "Type A" people and heart disease can equally be explained by higher cholesterol and blood pressure levels occurring for other reasons. Nevertheless, chronic psychological stress does often end in some form of illness. In cases where the constant tension is felt to be unpleasant, exhaustion may again lead to physical, mental, or psychosomatic symptoms. In other cases, the person is often trying to escape deep psychic conflicts by rushing through life. Here the process itself is less likely to end in illness. But what is likely is that the person's psychic balance eventually collapses, as outside events force the hidden conflicts to the surface. The conflicts (such as social or sexual fears) then expose themselves in neuroses and other mental illnesses.

PSYCHOSOMATIC ILLNESS

This is the name for illness that has its underlying cause in psychological stress. Illnesses sometimes brought about in this way include alcoholism, arthritis, asthma, constipation and indigestion, dermatitis, hair loss, migraine, ulcers, and stress symptoms in specific organs (such as heart pains and heartbeat irregularities).

©DIAGRAM

STRESS, DRUGS AND DRUG ABUSE

DEPENDENCIES

Stress threatens everyone to some extent; the conditions of our society make it hard to escape it. Despite this, many maintain an independent and intelligent approach to life - stress need not be something that defeats us. But others seek refuge in some form of dependence, a false center around which their lives can revolve.

Some of these patterns of dependence are encouraged by our society, such as overdependence on one's mate or children. Others are frowned upon (drugs or alcoholism); and yet others (excessive eating, procrastination) dismissed as symptoms of an underlying escapism.

DRUGS AND DRUG ABUSE

A drug is any chemical compound which can affect the body's functioning. Drug abuse is the use of any drug for a purpose other than a medically or socially accepted one. This includes the misuse of drugs obtained in medically or socially acceptable ways.

Especially relevant here are psychoactive drugs: those having effects on the body which bring about behavioral changes - relaxation, euphoria, hallucination, etc.

WHAT DRUGS DO

Some drugs have the following effects on nerve impulses:
1 They stimulate the neurons that try to stop messages.
2 They slow down or stop the production of the chemical transmitting agent.
3 They cause the agent to break down more quickly.
4 They reduce the effect of the chemical transmitter on the next neuron in the chain.
All or any of this has the effect of slowing down ("depressing") nervous activity. Fewer messages get through, and those that do get through more slowly and more weakly.
Other drugs have the opposite effect: they increase ("stimulate") nervous activity.

EFFECTS IN THE BRAIN

Different areas of the brain control different physiological and mental functions. It is because they stimulate or depress different specialized areas, that different drugs have different effects. For example, parts of the hypothalamus, when stimulated, give intense feelings of pleasure, while other areas control coordination, thought, sight and hearing, and so on.

VARIABLE FACTORS

A number of factors affect the impact of any drug. Some relate to the drug itself: the amount taken, its purity and concentration, and how it enters the body. Others concern the mental and physical state of the consumer. Mentally, drugs often heighten an existing psychic state or release a suppressed one; but the consumer's reaction to his immediate environment, and his expectations, are also important. Physically, the effects are likely to be increased by tiredness and (when the drug is swallowed) if there is little food in the stomach.

REASONS FOR TAKING DRUGS

SOCIAL CONFORMITY If the use of a drug is accepted in a group to which a person belongs, or which he identifies with, he will feel a need to use the drug to show that he belongs to the group. This is true of all drugs, from nicotine and alcohol to heroin.

PLEASURE One of the main reasons drugs are taken is to induce pleasant feelings - ranging from well-being and relaxation to mystic euphoria.

ESCAPE FROM PSYCHIC STRESS In a society which increasingly sees drugs as the answer to all physical problems, the use of drugs to escape one's psychological problems inevitably seems appropriate.

ALIENATION may underlie drug abuse. In social alienation, where the values of society are rejected, drug use may seem a valid symbol of opposition. In psychic alienation - when a person has rejected not only society but all alternatives, including himself, his hopes and his goals

the resulting feelings of meaninglessness, isolation, and inadequacy will predispose him to chronic drug abuse.

AVAILABILITY relates directly to drug use. Illegal use is highest where there is a ready supply (seaports, border towns), or where the market has attracted a ready supply (large cities, university towns). Legal drug use also increases with availability, eg alcoholism is common in the liquor trade.

CURIOSITY about drugs and why people take them can often start off drug taking.

AFFLUENCE AND LEISURE can produce boredom and loss of interest in meaningful activity. Drugs can then supply an easy answer to the desire for stimulation and escape.

THEOLOGICAL REASONS account for some drug use: both the practice of certain traditional religions and the personal search for self-identity and a reason for existence. Whether abuse occurs depends on the circumstances.

CLASSIFICATION

Psychoactive drugs fall into four major categories, according to their effects.

DEPRESSANTS ("downers") reduce nervous activity. They include alcohol, barbiturates, and opiates (opium, codeine, morphine, heroin). Taken in small doses, they have a sedative effect; in larger doses they bring on sleep. An overdose can kill: nervous activity is so reduced that vital functions such as respiration are impaired and may cease. Tranquilizers are a special category of depressants.

STIMULANTS include caffeine, nicotine, the amphetamines, and cocaine. They increase nervous activity, especially in the sympathetic nervous system, which mobilizes the body for action. So these drugs help prolong activity and take away the desire for sleep.

HALLUCINOGENS include mescaline, psilocybin, and LSD (lysergic acid diethylamide). They produce bizarre states of consciousness (which may resemble psychotic conditions). The interpretation of incoming sense stimuli is radically affected, and this produces hallucinations, delusions, and extraordinary reactions to normal situations and events.

MARIJUANA (cannabis) forms a separate, fourth category, although it is closely related to the hallucinogens.

Drugs can also be categorized as, eg legal or illegal, socially acceptable or unacceptable. But these really refer to drug use. A medically accepted drug may be illegal without a prescription; and social acceptability varies. One drug may be illegal but socially acceptable, another legal but unacceptable.

WHAT IS ADDICTION?

TOLERANCE to a drug occurs as the body gets used to it. As tolerance increases, so does the quantity of the drug needed to produce the original results.

PHYSICAL DEPENDENCE has occurred when the cells of the body have become used to functioning in the presence of the drug. When it is withdrawn, cellular activity is disrupted and the cellular need for the drug shows itself in withdrawal symptoms. Physical dependence can only be proved, therefore, when the supply of a drug is stopped.

PSYCHOLOGICAL DEPENDENCE is the need or compulsive desire to continue using a drug - whether or not there is physical dependence. It is not necessary for the person's cellular functioning, but it is necessary for his psychological functioning.

Any drug taking - and any other activity - can create psychological dependence. "ADDICTION" is now thought of as a general term, covering the above forms of dependence. It is usually a result of drug abuse, but it can occur through legal use (as in the case of nicotine). A drug's medical application may also give rise to addiction. Many of the first heroin addicts became addicted through the use of morphine as a pain killer in hospitals. The medical use of barbiturates has also produced many addicts.

HABITUATION refers to the repeated use of a drug when there is no form of dependence. Generalizations about habituation cannot be made for groups of drugs, because, depending on the individual, there may always be psychic dependence.

Drug		Dependence?	Tolerance?
Depressants:	Barbiturates	Yes	Yes
	Tranquilizers	Yes	Yes
	Opiates	Yes	Yes
Stimulants:	Amphetamines	Yes	No
	Cocaine	No	No
	Nicotine	Yes	?
	Caffeine	Yes	?
Hallucinogens		?	No
Marijuana		No	No
Alcohol		Yes	Yes
Wearing a hat		No	No

?: opinions differ

"Dependence" in the table refers to physical dependence.
All the above can create psychic dependence.
All those that do not create physical dependence allow habituation i.e. regular use without any dependence.
Alcohol can also be used regularly without any significant dependence.

This is because its use is normally kept in check by social norms: society has, over thousands of years, gained some experience in coping with its dangers. Yet, in the numbers of individuals it destroys, alcohol remains by far the most dangerous of all drugs.

©DIAGRAM

DEPRESSANTS

BARBITURATES

Barbiturates (nicknamed "barbs," "candy," or "goof balls") are made from barbituric acid. Like all depressants, they reduce the impulses reaching the brain. Because of this they have been medically prescribed for many years to relieve anxiety and tension and induce sleep; some have also been used as anesthetics. But with greater realization of their dangers, prescription is becoming less common.

TYPES Barbiturates vary in their immediacy and duration of effect, depending on the rate at which they are metabolized and eliminated; eg Seconal is a short-acting drug, Phenobarbital a long-lasting one.

SYMPTOMS Someone who has taken some barbiturates may well show signs of drowsiness, restlessness, irritability and belligerence, irrationality, mental confusion, and impairment of coordination and reflexes, with staggering and slurring of speech.

The pupils are constricted and sweating increases. The person experiences initial euphoria, followed by depression.

When an excessive amount is taken (an "overdose"), the depressive effect upon the nervous system is such that unconsciousness occurs, followed in extreme cases by death from respiratory failure.

WITHDRAWAL SYMPTOMS The barbiturates create tolerance and physical dependence. The effects of withdrawal in a chronic user can be worse than those of alcohol or heroin. They include irritability and restlessness, anxiety, insomnia, abdominal cramp, nausea and vomiting, tremors, hallucinations, severe convulsions, and sometimes death.

TYPES OF ABUSE Addicts are attracted by:

(a) the possibility of escaping from emotional stress, through sedation;

(b) the feelings of euphoria on initial ingestion, when large amounts of the drug are tolerated;

(c) the ability of barbiturates to counteract the effects of stimulants. This cyclical use of "uppers" and "downers" can lead to dependence on both.

The common prescription of barbiturates to induce relaxation and sleep has resulted in the largest group of dependent people being the middle aged, especially housewives. The same ready availability also makes the drug a common suicide weapon, while the combination of barbiturates' depressive effects with those of alcohol has brought many accidental deaths through taking barbiturates after heavy drinking.

OTHER TRANQUILIZERS Drugs such as Valium and Librium are increasingly prescribed in place of barbiturates to relieve anxiety and tension. They differ in derivation and mode of action, and an overdose is rarely fatal. But they do sometimes give rise to tolerance and physical dependence, and are subject to the same forms of abuse.

BARBITURATES

Drug	Description (but this can vary with dose and source)	Nickname
Amobarbital	Green blue	Blues, blue devils
Pentobarbital	Yellow	Yellows, nembies
Secobarbital	Red	Reds, red devils, red bird
Tuinal	Red blue	Rainbows, tooeys
Thorazine	Orange	
Miltown	White	
Librium	Green white	
Valium	Various	Goofers

OPIATES

The opiates are known in drug-taking circles as "the hard stuff." Opium itself and its derivative heroin are, in fact, the archetypal drugs of addiction. However, codeine and morphine, which are also derived from opium, are better known for their medical uses.

GENERAL ACTION

All depressants inhibit the activity of the central nervous system, impairing coordination and reflexes etc, but opiates especially affect the sensory centers, reducing pain and promoting sleep. As with alcohol, this nervous action may cause initial excitement, as inhibitions are removed.

In larger doses the opiates act on the pleasure centers of the hypothalamus, producing feelings of peace, contentment, safety, and euphoria.

General symptoms of opiate use include loss of appetite, constipation, and constriction of the pupils. An overdose of an opiate is likely to cause convulsions, unconsciousness, and death.

All opiates create tolerance and physical dependence. The symptoms of withdrawal from abusive use begin with stomach cramps, followed by diarrhoea, nausea and vomiting, running eyes and nose, sweating, and trembling. These are accompanied by irritability and restlessness, insomnia, anxiety and panic, depression, confusion, and an all-consuming desire for the drug.

OPIUM

Opium is the dried juice of the unripe seed capsules of the Indian poppy. The plant is cultivated in India, Persia, China, and Turkey, and opium is then prepared in either powder or liquid form. The poppy possesses its psychoactive powers only when grown in favorable conditions of climate and soil. Poppies produced in temperate climates have only a negligible effect.

Opium is traditionally smoked, using pipes, but it can also be injected or taken orally.

CODEINE

Codeine (methyl morphine) is the least effective of the opiates. It is white and crystalline in form, and is often used with aspirin for treating headaches. Because of the inhibiting effect on nervous reflexes it shares with all opiates, it is used in many cough medicines, and sometimes in the treatment of diarrhoea, since it reduces peristalsis (the automatic rhythmic contractions of the intestine). The risk of tolerance and abusive use are very small because of the large amounts necessary to produce pleasant effects.

MORPHINE

Morphine is the basis of all opiate action - it is opium's main active constituent. It was isolated from opium in 1805, and since then has been medically important as a pain killer. It is ten times as strong as opium, and must be administered with great care to avoid tolerance and physical dependence. However, instances of abuse are not too common, as drug users prefer heroin.

HEROIN

Heroin (diamorphine) was first isolated in 1898. It is three times as strong as morphine, and has a quicker and more intense effect, though a shorter duration. Among drug takers it is often known as "H," "horse," or "smack." In the USA it is not used medically. Its production, possession, and use are all connected with drug abuse. A grayish-brown powder in its pure form, for retail purposes it is mixed with milk or baking powder to add bulk. This results in a white coloring. The high cost of the drug, and its necessity to those who have become dependent on it, account for the high crime rate associated with its users.

The powder may be sniffed but is usually injected - normally into a muscle when use begins, but then into a major vein ("mainlining") as tolerance develops. Mainlining gives more immediate and powerful effects. Constant injection into the same vein causes hardening and scarring of the flesh tissue and eventual collapse of the vein. Unhygienic conditions and use of unsterilized needles can also cause infection, often resulting in sores, abscesses, hepatitis, jaundice, and thrombosis. Almost immediately upon injection intense feelings of euphoria and contentment envelop the user. The strength of these depends on the purity and strength of the heroin, and the psychological state of the user - the higher the previous tension and anxiety, the more powerful the subsequent feelings of pleasure and peace. It is the force of the initial pleasure that makes heroin more popular than morphine. In a chronic user, the ritual of injection is also important in the creation of pleasure.

Physical dependence on heroin is reached if one grain (60mg) of heroin is used in a period of up to two weeks. Withdrawal effects will then begin four to six hours after the effect of the last shot has worn off.

©DIAGRAM

STIMULANTS

AMPHETAMINES

The amphetamines ("pep pills" or "uppers") generally stimulate the sympathetic nervous system, which mobilizes the body for action with the "fight or flight" syndrome, including increase in epinephrine production, heart rate, blood sugar, and muscle tension.

EFFECTS The user experiences a sense of well-being and, with strong doses, euphoria. Alertness, wakefulness, and confidence are accompanied by feelings of mental and physical power. The user becomes talkative, excited, and hyperactive. Accompanying physical symptoms include sweating, trembling, dizziness, insomnia, and reduced appetite. Mood effects are probably due to stimulation of the hypothalamus, and sudden shifts to anxiety and panic can occur.

DEPENDENCE Amphetamines create tolerance, but are not considered physically addictive. However, psychic dependence is easily produced. The extra energy is "borrowed" from the body's reserves: when the drug's action has worn off, the body has to pay for it in fatigue and depression. This creates the desire for more of the drug to counteract these effects.

MEDICAL USAGE has become rarer since realization of the dangers. But amphetamines are still used for some purposes, eg: to prevent sleep in people who have to be alert for long periods; to treat minor depression; to counteract depressants; and to suppress the appetite in a few cases of obesity.

ABUSE of amphetamines is common because of the feelings of euphoria and alertness they give. The dangers include not only psychic dependence, but also: physical deterioration due to hyperactivity and lack of appetite; induced psychotic conditions of paranoia and schizophrenia, resulting from prolonged overdose; suicide due to mental depression following large doses; and death from overdose.

NICOTINE AND CAFFEINE

NICOTINE is a stimulant of the sympathetic nervous system. It is found in tobacco.

CAFFEINE is a stimulant of the central nervous system, found in coffee, tea, cocoa, and cola drinks. Its action combats fatigue, but it is a comparatively mild drug. It is also a "diuretic," i e. it increases the urine output of the kidneys. Medically, caffeine is often included in headache pills, to counteract the dulling effect of the painkilling ingredient. Abuse is unlikely because of the large quantities necessary, but those who drink considerable amounts of coffee probably have a mild psychic dependence, because of the feelings of tiredness experienced when the stimulation wears off.

AMPHETAMINES

Drug	Description (but this can vary with dose and source)		Nickname
Benzedrine		Red pink	Bennies
		Pink	Bennies
Dexadrine		Orange	Dexies
		Orange	Dexies
Methadrine		White	Speed, meth, crystal
Biphetamine		White	Whites
Edrial		White	
Dexamyl		Green	Christmas tree

COCAINE

COCAINE (often nicknamed "coke" or "snow") is a white powder obtained from the coca plant found in South America. Synthetic derivatives are also available.

EFFECTS Cocaine stimulates the central nervous system, dispelling fatigue, increasing alertness, mental activity, and reflex speed, and inducing euphoria. After an initial "rush," the effects become more steady. Accompanying physical symptoms include dilation of pupils, tremors, loss of appetite, and insomnia.

MEDICAL USE Although it is a stimulant, local application of cocaine has anesthetic effects. It is used for minor operations on the eye, ear, nose, and throat, and can also be used to anesthetize the lower limbs by injection into the spinal fluid.

DEPENDENCE Cocaine does not create physical dependence, but psychic dependence easily develops for the same reasons as with amphetamines.

ABUSE is the main use found for cocaine. As a powder it is inhaled, which eventually results in deterioration of the nasal linings and finally of the nasal septum separating the nostrils. Injection of a liquid form is an alternative, but using cocaine alone is unpopular, because of the violence of the sudden effects. So heroin and cocaine are often injected together. Cocaine is a short-acting drug and must be taken repeatedly to maintain the effects.

Dangers of prolonged use include insomnia, paranoia, hallucinations in the sense of touch known as "the cocaine bugs," and loss of weight and malnutrition through lack of interest in food. An overdose causes convulsions, and a dose of 12g or more at one time causes death by respiratory failure.

SUMMARY OF DRUG EFFECTS

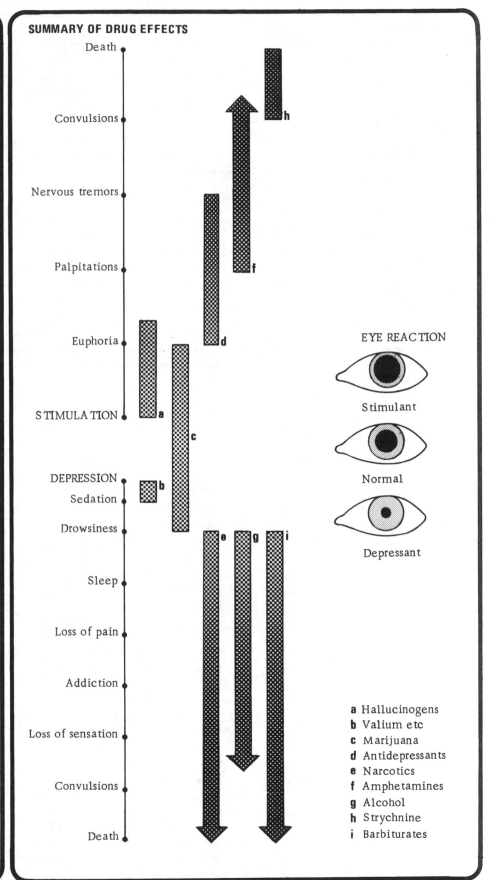

EYE REACTION

Stimulant

Normal

Depressant

a Hallucinogens
b Valium etc
c Marijuana
d Antidepressants
e Narcotics
f Amphetamines
g Alcohol
h Strychnine
i Barbiturates

©DIAGRAM

MARIJUANA AND HALLUCINOGENS

MARIJUANA

MARIJUANA (cannabis) comes from the "cannabis sativa" plant. It is prepared in two alternative ways. The flowers and top leaves of the plant can simply be dried and crushed; or the resin itself - which is the active ingredient - can be extracted. The first form is called simply marijuana (nicknamed "grass") and has the appearance of a dried crumbled herb. The second form is called hashish or cannabis resin (nicknamed "hash"), and appears as a hard brown lump. A liquid "oil" of the resin is also available.
CONSUMPTION Both main forms are normally smoked, usually in cigarettes called "joints" or "reefers". The dried flowers and leaves may be used alone, or mixed with tobacco. If the resin is used, it is heated and crumbled, and then the small pieces are mixed with the tobacco as the joint is made. Alternatively, both forms may be smoked in a pipe. A water pipe is popular, as it cools the smoke before it is inhaled.
The other mode of consumption is in drinks or food. This can give longer and smoother effects; also more extreme effects, as more of the drug can be taken into the body this way. However, control over dosage is diminished, and nausea may occur.

Marijuana is not now used medically, but it has been employed in the past as a mild anesthetic in many different parts of the world.
PHYSICAL EFFECTS The neurological effects of marijuana are not yet clearly understood. It is a general sedative of the central nervous system, but some areas of the sympathetic nervous system are apparently stimulated.
Particular symptoms of use include some increase in heart rate, inflammation of the conjunctiva* of the eyes, drying of the mouth, and increase in appetite.
PSYCHOLOGICAL EFFECTS The ability to experience the marijuana

HALLUCINOGENS

LSD

LSD (lysergic acid diethylamide) was originally obtained from the fungus "ergot", which is found in certain cereals. It is now produced synthetically.
Nicknamed "acid", it is the most widely used hallucinogenic drug. LSD can be injected, but is usually taken orally, either as a pill or as a drop of liquid on another substance (such as a sugar cube or a small square of blotting paper). Upon absorption it concentrates mainly in the liver, kidneys, and adrenal glands. Only about 1% is found in the brain.
ACTION The way in which LSD affects the nervous system is not completely understood. But it is known that certain chemical transmitters are interfered with and synesthesia occurs - crossconnecting of the sense responses, so one sees sounds and hears colors. Also LSD reduces the brain's ability to filter out unwanted data, so allowing more, unorganized stimuli to reach the consciousness.

PHYSICAL EFFECTS LSD causes an initial tingling in the extremities, goose pimples, and sometimes nausea, and muscle pain. The user feels cold and numb, and looks flushed. These effects soon wear off, leaving dilated pupils and increased heart rate, blood pressure, blood sugar level, and body temperature. Muscle coordination and pain perception are reduced.
THE LSD "TRIP" The psychological effects begin with the user becoming extremely emotional. This passes, but the senses are increasingly affected. Perception is enhanced and distorted: colors are more vivid, sounds more audible, inert objects seem to move. Synesthesia may occur, and orientation in time and space may be lost. Hallucinations varying in intensity and involvement are often experienced. At the same time, normal mental processes are impaired, and acquired modes of thought and behavior are disrupted. Emotional barriers are broken down, the past

may be seen in a new light, and repressed experiences may be released and relived.
The emotional impact of all this on the user varies greatly. Sometimes it is utterly terrifying (a "bad trip"), sometimes euphoric. This especially depends on: whether the user is with people that he trusts, in an environment that he enjoys, and in a happy mental state beforehand; and whether the mental disruption of the drug threatens the user's sense of self-definition.
Sometimes, in presenting a completely new sense of awareness, both of himself and of the external world, LSD can be of lasting benefit to the user.
A trip lasts from 8 to 12 hours.
DANGERS In the panic of a "bad trip", the user may cause himself and others physical harm. The mental impact can also be long-lasting: the release of repressed emotions and experiences may produce psychotic conditions in a previously unstable or neurotic person. Paranoia, schizophrenia,

"high" is an acquired technique. A new user must be taught what to look for, and how to enjoy it. The ritual and special language of marijuana use are part of this process. Marijuana is a uniquely social drug.

The effects vary with the dose, from those similar to moderate amounts of alcohol, to mild hallucinogenic effects. Also, like alcohol, marijuana tends to heighten the mood prevalent at the time of consumption.

Generally there is a pleasant feeling of mild euphoria. Inhibitions are lowered, and talking increases. Attention to the outside world is dulled, though thoughts seem rapid and involved. Sight and hearing are enhanced.

Coordination, flexibility of attention, and mental organization are impaired. Time and space are distorted and extended.

Experienced users can control their intake and maintain the pleasurable high without the depressant effects of the drug leading to sleep.

These effects last for three to five hours, reaching their peak after about 45 minutes. The user is left sleepy and hungry.

POSSIBLE DANGERS Marijuana does not create tolerance or physical dependence. However, if marijuana is mixed with tobacco, the user is also exposed to the dangers involved in tobacco smoking.

Psychologically, chronic marijuana use does seem in some cases to lead to loss of motivation and of social activity. But it does not bring the extreme mental (and physical) degeneration of alcohol. Because marijuana is illegal and therefore part of the traffic in drugs, the user may be brought into contact with other drug abuse. If so, he may be tempted to experiment with far more dangerous drugs.

and acute depression have all been caused through use of LSD.

In addition the power of the hallucinations can produce harmful actions; eg a belief that he can fly may cause the user to leap to his death.

"Flashbacks" are spontaneous recurrences of sensory disruption at a later date. They may occur at any time up to 18 months after the drug's use. They are especially dangerous where sudden loss of orientation may cause an accident eg driving a car, or crossing the road.

It is possible that LSD can cause chromosomal damage, but this is not yet certain.

DEPENDENCE LSD does not cause physical dependence, but the desire to repeat the experience may lead to psychic dependence in some users.

MEDICAL USES LSD has been used with varying success in the treatment of some psychotic disorders, such as schizophrenia and acute depression. It is used only in certain cases, and always in very carefully controlled conditions.

MESCALINE
Mescaline is produced from the peyote cactus of Mexico and the south-western USA. It is prepared as a powder, a capsule, or a liquid, and is usually taken orally though it may be sniffed or injected. It has effects very similar to LSD, though they are longer in taking effect. 600mg can give a 12 hour trip. There tend to be side effects, such as dizziness and nausea.

Peyote itself can be eaten, but about 50 times more is needed to produce the same effect. Also the presence of other substances makes the effects slightly different.

PSILOCYBIN
This is a hallucinogenic drug produced from the mushroom "psilocybe mexicana". It has been used for centuries for religious purposes, notably by the Aztecs. It causes hallucinations and distortion of perception for up to 8 hours, followed by physical and mental depression and continued disorientation.

OTHER HALLUCINOGENS
Fly agaric is another mushroom, which causes hallucinations and visual distortion followed by heavy sleep. As with all hallucinogenic mushrooms, vomiting and diarrhoea are possible side effects.

Morning glory seeds contain six derivates of lysergic acid, and have varying effects depending on their ripeness.

DMT (dimethyltryptamine) is an extremely short acting drug, produced as a powder and usually smoked with marijuana, tobacco, or tea.

STP or DOM is typical of a group of amphetamines that can also (or instead) produce hallucinogenic effects similar to LSD. Psychotic states often result, because of hallucinogenic misinterpretation of the usual body effects that amphetamines produce.

©DIAGRAM

ALCOHOL 1

WHAT'S YOUR POISON?
Bottles arranged in order of
increasing percentage alcohol

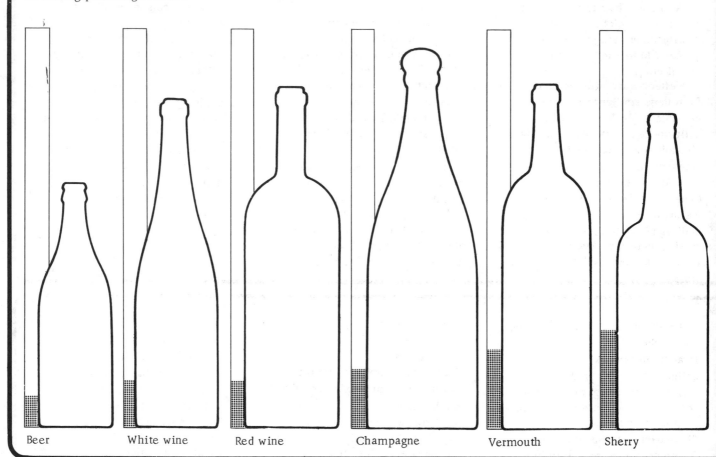

Beer   White wine   Red wine   Champagne   Vermouth   Sherry

THE CHEMICAL AND THE DRINK

Alcohols are volatile, colorless, pungent liquids, composed of three chemical elements: carbon, hydrogen, and oxygen. Ethyl alcohol (ethanol) is the type taken in alcoholic drinks. It may also be prescribed medically, to stimulate the appetite, or form a medicinal base in which other ingredients are dissolved.

Methyl alcohol (methanol, or "wood alcohol") is used commercially as a fuel and solvent. It is poisonous, and drinking it causes blindness and death.

ALCOHOLIC DRINKS
In the domestic and industrial production of alcoholic drinks, ethyl alcohol is produced by "fermentation": that is, the degeneration of a starch (such as maize, barley, rice, potatoes, grapes) by bacterial action. The drink that results depends on the starch used; eg malt and barley give beer, grapes give wine. Beers and wines are produced by fermentation alone. Only about a 15% level of alcohol is possible by this method. "Spirits", with their higher alcoholic levels (whiskey, gin, vodka, liqueurs, etc) also require "distillation". That is, the alcohol is evaporated off, leaving water behind, and resulting in a higher alcoholic concentration in the eventual liquid. Distilled alcohol may also be added to wines (sherry, port, etc) and beers, to strengthen them.

ALCOHOLIC STRENGTH
Commercially, the strength of an alcoholic beverage is expressed as so many "degrees proof". This refers to the liquid's specific gravity - not to the percentage of alcohol it contains. Proof measurement regulations vary between countries. With US proof measures, the percentage of alcohol is half the figure for "degrees proof". For example, a spirit that is 100 proof (written "100°") contains 50% alcohol.

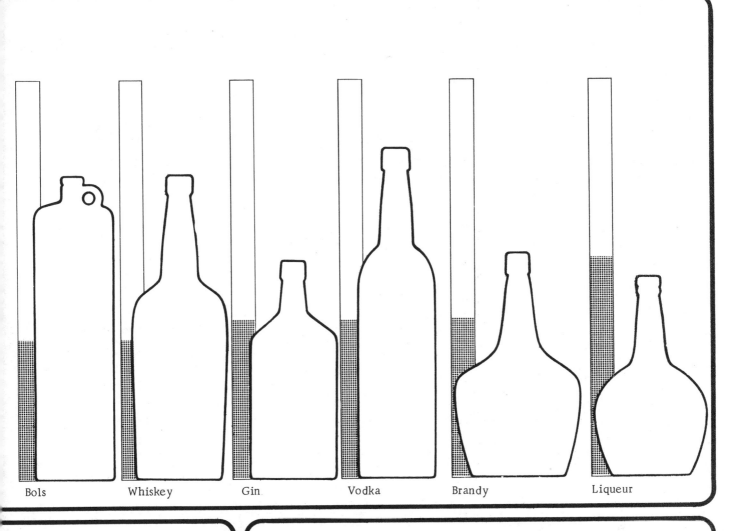

| Bols | Whiskey | Gin | Vodka | Brandy | Liqueur |

Typical alcohol contents

Lager	up to 8%
Beer	up to 8%
Cider	up to 8%
Wines	9 to 15%
Fortified wine	20%
Aperitif	25%
Spirits, liqueurs	40 to 50%

ALCOHOL IN THE BODY

About 20% of any alcohol drunk is absorbed in the stomach, and 80% in the intestines. It is then carried around the body by the bloodstream. The liver breaks down (oxidizes) the alcohol at an almost constant rate: usually about $2\frac{1}{2}$ bottles (1 pint) of beer or 1oz of whiskey per hour. This process eventually disposes of about 90% of the alcohol, forming carbon dioxide and water as end products. The remaining 10% is eliminated through the lungs and in the sweat. Alcohol in the body has four main effects:

a) It provides energy (alcohol has high calorific value, but contains no nutrients).
b) It acts as an anesthetic on the central nervous system, slowing it down and impairing its efficiency.
c) It stimulates urine production. With heavy alcohol intake, the body loses more water than is taken in, and the body cells become dehydrated.
d) It puts part of the liver temporarily out of action. After heavy drinking, as much as two thirds of the liver can be nonfunctioning - but it is usually fully recovered within a few days.

©DIAGRAM

ALCOHOL 2

BLOOD ALCOHOL LEVEL

The effect of alcohol on behavior depends on the amount reaching the brain via the bloodstream. This "blood alcohol level" is determined by several factors, apart from the quantity of alcohol drunk.

a) The size of the liver decides the rate of oxidation and elimination.

b) The size of the person decides the amount of blood in the system, because blood volume is proportionate to size. The larger the person, the greater the diluting effect of the blood on the alcohol consumed, and the more it takes to produce the same effect.

c) The speed and manner in which the alcohol is consumed is important. The longer one takes to drink a given quantity, the less effect it has.

d) Alcohol consumed on an empty stomach will have a greater and more immediate effect than that consumed during or after eating. Food acts as a buffer to absorption.

DRINK EQUIVALENTS

If we assume a person of average size (150lbs), drinking at an average rate on an empty stomach, then any one of the following would give a blood alcohol level of 0.03%:

$1\frac{1}{2}$oz (about one measure) of spirits;

$3\frac{1}{2}$oz (just over $\frac{1}{2}$ glass) of sherry or fortified wine;

$5\frac{1}{2}$oz (just under 1 glass) of ordinary (table) wine;

24oz (2 bottles) US beer; or $\frac{3}{4}$pt UK beer.

Twice these quantities will give twice the blood alcohol level (0.06%), and so on.

Alcohol equivalents

 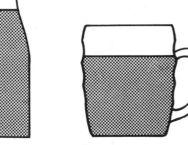   

2 bottles US beer $\frac{3}{4}$ pint UK beer   1 glass wine $\frac{1}{2}$ glass sherry   1 whiskey   1 vodka

BEHAVIORAL EFFECT

As the level of alcohol in the blood rises, the drinker's brain and nervous system are increasingly affected, and changes occur in his behavior.

0.02% Sense of warmth, friendliness. Visual reaction time slows.

0.04% Driving ability at speed impaired.

0.06% Feelings of mental relaxation and general well-being. Further slight decrease in skills.

0.09% Exaggerated emotions and behavior. Tendency to be loud and talkative. Loss of inhibitory control. Sensory and motor nerves increasingly dulled.

0.12% Staggering, and fumbling with words.

0.15% Intoxication.

0.20% Incapacitation, depression, nausea, loss of sphincter control.

0.30% Drunken stupor.

0.40% Coma.

0.60% + Lethal dose. Death through heart and respiratory failure. Fortunately, lethal doses seldom occur, as unconsciousness and vomiting force the drinker to stop.

DRINKING AND DRIVING

The behavioral effects of alcohol make drinking and driving very dangerous, both to the drinker and to others. Tests have shown that errors of judgement and control increase as soon as there is any alcohol in the blood stream. Therefore many countries prescribe a legal limit to the blood alcohol level of anyone in charge of a vehicle. In the USA this varies from 0.10 to 0.15%, except for Utah, where it is 0.08%. (Iowa, New Mexico, and Texas have no restrictions.) Limits elsewhere include 0.05% (Scandinavia), 0.08% (UK), and 0.15% (Australia).

HOW WE GET DRUNK

Alcohol is a physiological "depressant" i.e. as consumption occurs the transmission of impulses in the nervous system becomes slowed. First to be affected are the higher levels of the brain: inhibitions, worry, and anxiety are dissolved, resulting in a sense of well-being and euphoria. As the lower levels of the brain become affected, coordination, vision, and speech are impaired.

The small blood vessels of the skin become dilated (widen). Heat is radiated and the drinker feels warm. This means that blood has been diverted from the internal body organs, where the blood vessels are already constricted by the effect of alcohol on the nervous system. So, at the same time, the temperature of internal body organs falls.

Any increase in sexual desire is due to the depression of the usual inhibitions. Alcohol is not an aphrodisiac - physical sexuality is more and more impaired as blood alcohol level rises.

Eventually, the poisoning effect of excess alcohol causes nausea and possible vomiting, and may leave the drinker with the usual symptoms of a hangover.

ALCOHOL AND THE LAW

Chemically, alcohol is one of the most dangerous drugs known to man. But, over centuries of experience, society has managed to develop cultural attitudes which allow alcohol to be available without it causing too great disruption or harm. However, legal restrictions are needed to reinforce these attitudes. Most countries have a minimum age for its purchase, restrict the number of hours of the day during which it can be sold, control the number, ownership, and location of bars and liquor stores, and keep up strict observation of their orderliness.

HANGOVER

A hangover is the physical discomfort that follows the consumption of too much alcohol. Symptoms can include headache, upset stomach, thirst, dizziness, and irritability.

Three processes produce the hangover. First, the stomach lining is irritated by the excessive alcohol, and its functioning is disrupted. Second, cell dehydration occurs because the quantity of alcohol consumed exceeds the liver's ability to process it, leaving a prolonged level of alcohol in the blood. Third, the level of alcohol has a "shock" effect on the nervous system, from which it needs time to recover.

AVOIDANCE

The best way to avoid a hangover is not to drink too much. But there is less likelihood of a hangover if the alcohol is taken with meals: consumption and absorption are spread over a greater period of time, and the food acts as a barrier. Non-alcoholic drinks, taken at the same time or afterwards, dilute the alcohol; and there is usually less of an after-effect when alcohol is consumed in relaxed surroundings, and when cigarette smoking is cut to a minimum.

TREATMENT

The stomach is relieved by a fresh lining: milk, raw eggs, or simply a good breakfast! Only then should aspirin or other pain relievers be taken to help the headache. The danger of stomach irritation from pain-relieving drugs is always much worse when the stomach is empty. Fruit juice can be drunk for its vitamin C and refreshing taste, while fizzy drinks may have a soothing effect upon the stomach. Liquids of any sort help the dehydrated cells recover their fluid content.

Coffee or tea can be used to clear the head (the caffeine content stimulates the nervous system) and sugar can be taken to provide energy; but both these may leave the sufferer feeling worse when the immediate effects wear off. Similarly, for temporary relief, another alcoholic drink (in moderation) invigorates the sluggish nervous system and seems to dispel the unpleasant after effects. But this is only a postponement: the original hangover, and that of the extra alcohol, still await!

© DIAGRAM

ALCOHOLISM 1

THE PROBLEM

Alcoholism is the most serious drug problem in modern industrial society. In the United States in 1970 there were $5\frac{1}{2}$ million alcoholics and another 4 million pre-alcoholics. The total number of active narcotics addicts, in contrast, was 120,000 at most. In European countries there are usually not more than a few hundred narcotics addicts, but alcoholics number from 1% to 10% of the population. It is estimated that alcoholics in the USA have 7 times the normal accident rate, and $2\frac{1}{2}$ times the normal death and suicide rates, and that between 13% and 29% of the patients in general hospitals are alcoholics.

SOCIAL DRINKING

Alcohol's social uses are many. The relaxing effects of alcohol reduce inhibitions and relieve anxieties, so alcohol in small quantities acts as a social lubricant, decreasing self-consciousness and increasing congeniality, confidence, and "belongingness". As a result, it has been associated with every aspect of man's nature. Sometimes casual drunkenness is encouraged and viewed with amusement; and yet alcohol has a place in much of our social and sacred ritual, even up to the celebration of religious mystery and of birth and death. The varied uses of alcohol are an accepted part of the traditions integrating the social order. As individuals, we can worship or despise the release that alcohol can bring; but a normal attitude, for our society, is that, used sensibly, alcohol is one of the earth's gifts to man - and that, taken in moderation, alcoholic drinks give pleasure at little cost to body or soul.

DEFINITION

Alcoholism is hard to define, because its social manifestations vary. Some authorities define it, in effect, as any repeated drinking that is above the normal for a community. But if one looks at alcoholism as a process, what is significant is the factor of dependence. Like many other drugs, alcohol can produce psychological and, in extreme cases, physical dependence. In psychological "problem" drinking, the sufferer is constantly trying to escape his psychic difficulties into drink. In physical dependence, alcohol has become necessary for the body to function, and its removal produces extreme physical effects. The point at which drinking becomes alcoholism is not decided by the quantity drunk, or even by how far it dominates a person's social life. What distinguishes the alcoholic is that - whether he realizes it or not - his drinking is compulsive.

SYMPTOMATIC DRINKING

Today the conditions of our society seem to encourage people to turn more and more often to alcohol to escape stress - whether the pressure is of work or of their own psyche. This is called "symptomatic drinking". At first, the relief the drinker seeks is easily available. But gradually he achieves it only through greater and greater quantities of drink, as tolerance to the drug increases. Eventually, his psychological dependence is supported, and finally displaced, by physical dependence - with disastrous effects upon his body, his finances, and his family and social life.

WHO BECOMES AN ALCOHOLIC?

Alcoholism through symptomatic drinking is thought of by experts as an illness, relating to underlying personality disorder. It may even be linked with a metabolic defect of some sort. But also important are availability, social environment, and upbringing. Consequently, alcoholism predominates in certain social groups rather than others: senior executives and their wives, traveling salesmen, journalists, actors, and the children of alcoholics. Past cultural factors also modify society's influence: in the United States, alcoholism among Jews and Chinese is rare, that among Irish and Scandinavians much higher. And in all countries, many more men than women are alcoholics - usually in the ratio of 5 to 1. In the US, 7.3% of men are alcoholics, but only 1.3% of women. However, the proportion of women alcoholics is increasing.

HEALTH EFFECTS

One or two average-strength drinks a day are normally no hazard to health. But sometimes any drinking is inadvisable, eg: if the person is seriously overweight; or if he has recently taken sedatives, tranquilizers or anti-histamine (anti-allergic) tablets; or if he suffers from epilepsy, liver disorders, or stomach or duodenal ulcers.

INVETERATE DRINKING

There are also societies in which a very large intake of alcohol is considered normal among certain groups ("inveterate drinking"). Here intoxication and need may never be noticed: physical dependence is reached without any of the usual psychological symptoms or social and financial consequences. Yet in the inveterate drinker, alcohol may be present in the bloodstream every hour of the day and night. The physical consequences are present, too, and if the inveterate drinker is deprived of his normal intake (eg through having to go into hospital for some other reason), acute withdrawal symptoms will appear.

Inveterate drinking is most common in wine-producing countries, but it is also known in our society - among bartenders and dealers in alcohol, and in some business circles, where alcoholism may arise more through social custom and an expense account than through the pressure of psychic stress.

ALCOHOLISM IN SELECTED COUNTRIES

	Percent of population	Type of alcoholic	Drinking population	Main source of alcohol
Argentina	under 0.5%	Symptomatic & inveterate	Rural	Wine
Spain		Inveterate	Urban Rural	Wine
Brazil	0.5 to 0.9%	Symptomatic	Urban•	Spirits
Netherlands		Symptomatic	Urban•	Spirits
Czechoslovakia	1.0 to 1.4%	Symptomatic	Urban	Beer Spirits
England		Symptomatic	Urban	Beer
Ireland		Symptomatic	Urban Rural	Beer
Canada	1.5 to 1.9%	Symptomatic•	Urban•	Beer Spirits
Denmark		Symptomatic	Urban•	Beer
Norway		Symptomatic	Urban•	Spirits
Peru		Symptomatic•	Urban Rural	Beer Spirits
Scotland		Symptomatic	Urban•	Spirits Beer
Uruguay		Inveterate	Urban	Wine Spirits
Australia	2.0 to 3.0%	Symptomatic	Urban	Beer
Sweden		Symptomatic	Urban	Spirits
Switzerland		Symptomatic	Urban•	Wine Spirits
South Africa		Symptomatic	Urban	
Chile	4.0 to 5.0%	Inveterate•	Urban Rural	Wine
USA		Symptomatic	Urban	Beer Spirits
France	up to 10%	Inveterate•	Urban Rural	Wine Spirits

•Substantial minority category of other type

©DIAGRAM

ALCOHOLISM 2

SOCIAL DRINKING can lead to alcoholism: because the drinker starts to turn regularly to alcohol for relief from stress; or because his social drinking is so heavy that the beginnings of dependence are noticed.

EARLY ALCOHOLISM is marked by the beginning of memory blackouts. Increasing dependence is shown by surreptitious drinking and the urgency of first drinks. The drinker feels guilty, but cannot discuss the problem.

BASIC ALCOHOLISM The drinker can no longer stop unless forced to by intoxication. He bolsters himself with excuses and grandiose behavior, but his promises and resolutions fail. He starts avoiding family and friends, and neglects food, interests, work, and money. Physical deterioration sets in. Finally tolerance for alcohol decreases.

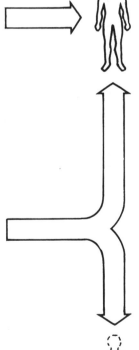

RECOVERY Procedures of treatment are detailed opposite. Psychologically, the patient regains the desire to be helped, thinks more rationally, and develops hope, moral commitment, outside interests, self-respect, and contentment in abstinence. Finally he recovers the respect of family and friends, and the confidence of employers.

CHRONIC ALCOHOLISM is marked by further moral deterioration, irrational thought, vague fears, fantasies, and psychotic behavior. Physical damage continues. The drinker has no alibis left, and can no longer take any step to recovery for himself.
Reaching this point may have taken from 5 to 25 years.

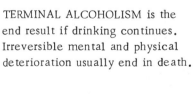

TERMINAL ALCOHOLISM is the end result if drinking continues. Irreversible mental and physical deterioration usually end in death.

PHYSIOLOGICAL EFFECTS

THE DIGESTIVE SYSTEM

The ordinary hangover shows how alcohol can irritate the stomach. In alcoholism, the stomach can be constantly inflamed (gastritis), and eventually the intestines too (gastro-enteritis). There are symptoms of nausea, abdominal pain, chilly sensations, and loss of appetite. The risk of ulceration is high.

VITAMIN DEFICIENCY

Through incapacitation and over-riding preoccupation with alcohol, the alcoholic neglects his food. Malnutrition and vitamin deficiencies result. Common manifestations include beriberi and pellagra - both due especially to B complex vitamin deficiencies. Beriberi involves inflammation of the nerves all over the body; pellagra affects the nerves, digestion, and skin.

CIRRHOSIS OF THE LIVER

The liver soon recovers from an occasional bout of alcohol. But alcoholism often produces the condition called cirrhosis: the organ shrivels, its cells are largely replaced by fibrous tissue, and its functioning deteriorates. Ten per cent of chronic alcoholics have cirrhosis of the liver, and 75% of people with cirrhosis have an alcoholic history. However, the main reason is not the alcoholic's self-poisoning with alcohol, but his neglect of nutrition: his cirrhosis is the product of protein and vitamin B deficiencies. There are few symptoms until cirrhosis is fairly advanced. Then the victim may complain of general ill health, loss of appetite, nausea, vomiting, and digestive trouble.

ALCOHOLIC MYOPATHY is a term for muscular decay resulting from alcoholism. Causes are lack of use of muscles, poor diet, and alcoholic damage to the nervous system. The heart, as a muscular organ, may be affected.

NEUROLOGICAL DEGENERATION

Alcoholism destroys brain cells and causes degeneration throughout the nervous system. Poor diet upsets the brain's metabolism, through lack of B complex vitamins particularly. The sufferer experiences short term and long term memory losses; inability to think clearly; muscular convulsions in the body and limbs; and trembling, emotional disturbance, hallucinations, and fits. Eventually, the decline of nervous functioning can result in pneumonia, kidney failure, or heart failure.

DELERIUM TREMENS (The "DTs") is a state in which the sufferer experiences extreme excitement, mental confusion, and anxiety, with trembling, fever, and rapid and irregular pulse. He may have hallucinations, especially ones involving animals approaching or touching him. The onset usually occurs when a period of heavy drinking has been followed by several days' abstention, and it is often preceded by restlessness, sleeplessness, and irritability.

TERMINAL ALCOHOLISM is the stage where the physical and mental damage done to the body is irreversible. Even if the person can be kept alive, his existence becomes that of a vegetable.

TREATMENT

The patient is deprived of his drug. Severe withdrawal symptoms follow: sweating, vomiting, body aches, diarrhoea, running nose and eyes, fits, convulsions, and hallucinations. Sedatives relieve these, but are terminated before they themselves become addictive. The patient's health is restored by good diet, and any physical problems due to the addiction are treated.

THERAPEUTIC TREATMENT

After detoxification, the underlying psychological causes are identified, if possible, and treated. The patient's motivation, self-confidence, and trust must be constantly strengthened.

Treatments, and their effectiveness, vary greatly. The following are the most widely accepted.

a) Aversion therapy tries to create conditioned reflexes of sickness and aversion at the presence of the drug. Techniques include electric shock therapy, and sensitizing drugs which produce severely unpleasant symptoms when the drug is taken.

b) Individual psychological therapy aims at removing the underlying psychological causes by bringing them to light and getting the patient to accept and face them for himself.

c) Group therapy aims at giving the patient objective outside views of himself, with which he must come to terms; and at the same time helps him to overcome his isolation, by giving him personal relationships, and contact with fellow sufferers.

In the case of alcoholics, Alcoholics Anonymous provides both group therapy and guidance by cured alcoholics. Their meetings also provide important support in later rehabilitation and continued abstention.

This depends above all on the patient's desire to be cured. Treatment is long-term, and its goal has to be lifetime abstention.

©DIAGRAM

TOBACCO 1

TOBACCO CONSUMPTION

CIGARETTES account for the bulk of tobacco consumption. The tobacco they use is usually flue-cured. This gives a neutral cigarette smoke which is easily inhaled i.e. taken down into the lungs.

CIGARS are generally made of air-cured tobacco. The smoke is more pungent and seldom inhaled i.e. it only enters the mouth and perhaps the throat - but much of the nicotine content is absorbed through the linings of the mouth.

PIPE tobacco is generally air, sun, or fire-cured. It too is seldom inhaled, and there is only a small amount of nicotine absorption through the mouth.

SNUFF is a powdered tobacco that is sniffed into the nostrils. Nicotine is absorbed through the linings of the nose, and some snuff probably passes down into the lungs.

CHEWING TOBACCO is a mixture of tobacco and molasses. Nicotine is absorbed through the mouth.

With the spread of cigarette smoking after World War I, other forms of consumption declined, especially the taking of snuff and chewing tobacco. But recently, as the dangers of cigarette smoking have been recognized, there has been some slight rise in the relative proportion of cigar and pipe smokers.

Average US consumption per person, per year

	1900	1970
All tobacco•	7.5	10.5
Cigarettes	50	4000
Cigars	110	59
Pipe tobacco•	1.6	0.5
Chewing•	4	0.5
Snuff•	0.3	0.2

•weight in lbs

THE PLANT AND PRODUCTION

Tobacco comes from the plant "nicotiana tabacum", and is produced in about 80 different countries, giving a world total of 8000 million lb a year. Half this is from the USA and China. After cultivation, tobacco is "cured" (dried) in one of four ways. Air curing takes place in a barn provided with a steady circulation of air, and takes about six weeks. Sun-cured leaves are first exposed to the sun, and then undergo a similar process.

Fire-cured leaves are hung over wood fires, and come into direct contact with the smoke.

Flue-cured leaves are also cured by the heat of a fire, but do not come into contact with the smoke. The method of curing affects the finished product. For example, flue-cured tobacco has a lower nicotine content than other kinds, and instead contains 15 to 20% more sugar. It also affects the leaf color. Air-cured leaves are reddish brown, sun-cured rather darker, fire-cured simply dark brown, and flue-cured light brown to yellow.

Up to 90% of tobacco is flue-cured. After curing the tobacco is left to mature, and then graded ready for manufacture. Grading is done by the size, color, and texture of the leaf. Different grades are used for different products.

CONSUMPTION BY COUNTRY

... showing consumption of cigarettes and all tobacco goods per adult per year.

1 Argentina
2 Australia
3 Austria
4 Brazil
5 Canada
6 Finland
7 France
8 W Germany
9 Greece
10 Iceland
11 India
12 Ireland
13 Italy
14 Japan
15 Mexico
16 Netherlands
17 Norway
18 Portugal
19 South Africa
20 Spain
21 Sweden
22 Switzerland
23 Turkey
24 UK
25 US

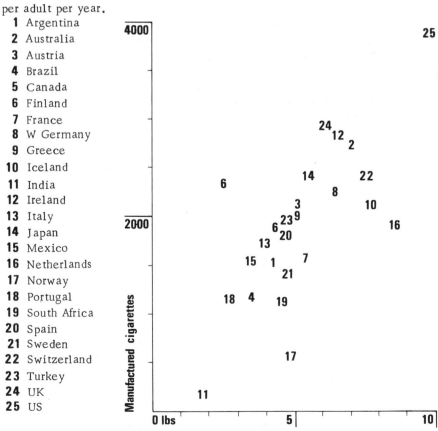

TOBACCO SMOKE

Tobacco smoke contains at least 300 different chemical compounds. These enter the lungs in the form of gases or solid particles. The solid particles condense to form a thick brown tar. (The inhaled smoke of 10,000 cigarettes i.e. under 2 years' smoking, at 15 cigarettes a day, yields about 3/8 lb of this tar.) It lines the passages down which the smoke travels, and collects in the lungs. The contents of tobacco smoke fall into five main categories.
CARCINOGENIC SUBSTANCES i.e. those that induce the growth of cancer. There are at least 15 carcinogens in tobacco smoke, including certain hydrocarbons, benzpyrene, and perhaps a radioactive isotope of polonium.
CO-CARCINOGENS, or cancer promoters, do not cause cancer themselves, but accelerate its production by the carcinogens. They include phenols and fatty acids.

IRRITANT SUBSTANCES disturb the bronchial passages, increasing mucus secretion but damaging the processes for expelling this mucus from the lungs. Many are also co-carcinogens.
GASES occur at dangerous levels in tobacco smoke. They include carbon monoxide at 400 times the level considered safe in industry, and hydrogen cyanide at 160 times the safe level.
NICOTINE is a powerful poison - a 70mg injection is enough to kill. One cigarette contains 0.5 to 2.0mg, depending on how the tobacco was cured. How much of this is absorbed depends on the method of smoking. Inhalation can absorb as much as 90%, non-inhalation as little as 10%. The nicotine that is absorbed affects the nervous system, including brain activity and the control of

epinephrine secretion and heart beat rate. Small amounts give the smoker a sense of stimulation, then leave him feeling more tired than before. Larger amounts sedate him. There is evidence that nicotine is the addictive constituent of tobacco i.e. nicotine gives tobacco its "kick" while the tar gives it its taste.
FILTER-TIP CIGARETTES
Seventy per cent of all cigarettes sold are now filter tipped. The filter is meant to remove the harmful substances in the smoke, but it is not necessarily successful. In fact, with any cigarette, the tobacco itself acts as a filter. But, as a result, smoking the last third of a cigarette releases as much dangerous material as the first two-thirds.

WHO SMOKES CIGARETTES?

Horizontal distances measure each age group in the population. Vertical distances show the group's smoking habits. For example, there are fewer men over 65 than women, but more of them smoke - so the column is narrower, but the shaded part taller.

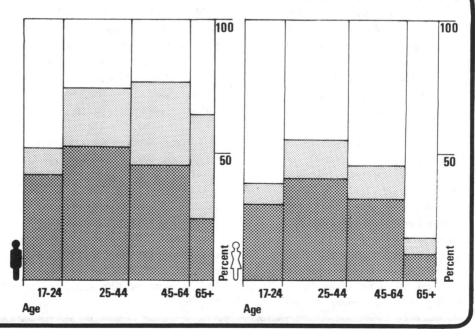

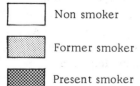

☐ Non smoker

▦ Former smoker

▨ Present smoker

©DIAGRAM

TOBACCO 2

REASONS FOR SMOKING

Despite the growing evidence of the dangers of smoking, old smokers go on smoking and new ones start. The reasons are many and often interdependent.

SOCIAL CONFORMITY Smoking is a socially accepted form of drug taking and for children it is an activity associated with a "grown up" way of life. Most smokers start in adolescence, copying schoolmates, friends, or workmates. Children from families in which both parents smoke are also more liable to do so themselves.

CURIOSITY Many first experiments in smoking follow from a child's natural curiosity to find out what so many adults and friends experience when they smoke.

PERSONALITY As with all drugs, the smoker's psychological make-up forms the basis of his habit and his dependence. Often those adolescents who start to smoke are those who feel unsuccessful or rebellious: smoking can give them a status symbol, a symbol of maturity. For others, of all ages, it may be used to combat or conceal nervousness, and give them confidence in company. Freudian psychology believes that smoking also provides sexual/oral gratification, acting as substitute for the loss of the mother's breast at weaning.

PLEASURE Once tolerance has developed, and smoking no longer gives unpleasant physical sensations such as dizziness, the smoker usually finds his habit comforting and pleasurable - if only because it takes away the desire experienced when he stops.

BOREDOM Smoking allows a certain amount of involvement, so it can give some release from boredom. Moreover, unlike many activities, it is socially acceptable for smoking to accompany or even interrupt work.

ESCAPE The sedative effects of smoking can provide some relief from anxiety and tension, while the physical activity of smoking can be an outlet for nervous energy, simply by giving the smoker something to do.

ADVERTISING The social acceptability of smoking means that the tobacco industry is allowed to advertise its products. Advertisements make smoking seem desirable in many ways. The constant sight of cigarettes may also result in a smoker being unable to give up.

GOVERNMENT INVOLVEMENT The high taxes imposed on tobacco products provide Governments everywhere with considerable revenue. This, and the large proportion of the population often supported by the industry, means that cigarettes will continue to be produced. As long as cigarettes are available, smoking will take place.

SOME EFFECTS OF SMOKING

DEPENDENCE Smoking gives many smokers a comforting habit that helps them relax and avoid stress. For others, it is just a meaningless activity that cannot be stopped. Both cases result from the dependence that cigarettes create. Withdrawal symptoms on stopping smoking may include intense craving, depression, anxiety, instability, restlessness, sleep disturbance, difficulty in concentrating, altered time perception, sweating, drop in blood pressure and heart rate, and gastro-intestinal changes. These symptoms seldom occur with any intensity, however: most people experience only very mild discomfort or none at all.

ACCIDENTS Smokers have four times more accidents than non-smokers. This may be due to a slowing of the reflex actions, lasting about 20 minutes, that follows smoking a cigarette. (It may also be linked with differences between the kind of people who become smokers and those who do not). Smokers also run a higher risk of death or injury by fire - the most common cause being smoking in bed.

ENDURANCE The physical endurance of smokers is lower the more the cigarettes they smoke and the more the time spent smoking.

EFFECTS ON NON-SMOKERS Smoking by others also affects non-smokers, causing eye irritation, headaches, and coughing in a smoke-laden atmosphere. It exposes them to the same health risks as smokers, though in a very much reduced way.

ECONOMICS Smoking can be very expensive on a personal level, but the greatest cost is to society. Illness resulting from smoking causes 20% of the annual loss of working days in the USA. The cost of smoking to a nation also includes the impaired abilities of its members, and the extra medical expenses incurred.

SMOKING AND HEALTH

Smoking is the largest single avoidable cause of ill-health and death. It can damage the cardio-vascular, respiratory, and digestive systems, and it encourages the growth of cancer in many parts of the body. Smokers run a much higher risk than non-smokers of illness and premature death.

For example:

-Cigarette smokers are twice as likely to die before middle age as non smokers. They run the same risk of death as someone 10 years older.

-Two out of five smokers die before 65. This happens to only one out of five non-smokers.

-The average smoker aged 35 has a life expectancy 5½ years shorter than a non-smoker.

How much damage smoking does depends on several things: the type of tobacco; the form it is smoked in; the temperature at which it is burned; the effectiveness of any filtration; whether inhalation occurs; the length of time the individual has been smoking; the amount he smokes; and the general state of his health. Smoking of all sorts is harmful, but usually cigarette smoking is the most deadly. The nicotine content of cigarette tobacco is often smaller, but the higher burning temperature and the greater tendency to inhale make up for this. Also the tendency to inhale favors lung damage and especially lung cancer, which is often not diagnosed till too late. Pipe and cigar smokers are more likely to develop the more noticeable - and so more curable - cancers of the mouth, pharynx, and larynx.

The convenience of cigarettes may also encourage more smoking than pipes or cigars.

SMOKING AND DEATH

All lines below the first show mortality rates for smokers.

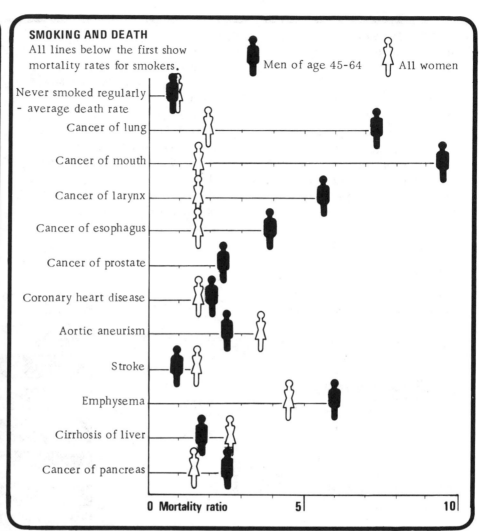

YOUR ODDS OF SURVIVING

into the next age group.
Smoker A: 25 or more cigarettes per day
Smoker B: 15-24 cigarettes per day
Smoker C: 1-14 cigarettes per day

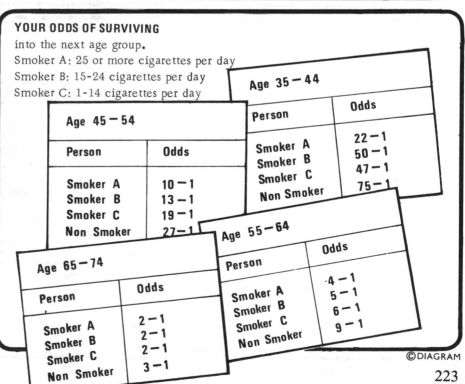

Age 35 — 44

Person	Odds
Smoker A	22 — 1
Smoker B	50 — 1
Smoker C	47 — 1
Non Smoker	75 — 1

Age 45 — 54

Person	Odds
Smoker A	10 — 1
Smoker B	13 — 1
Smoker C	19 — 1
Non Smoker	27 — 1

Age 55 — 64

Person	Odds
Smoker A	4 — 1
Smoker B	5 — 1
Smoker C	6 — 1
Non Smoker	9 — 1

Age 65 — 74

Person	Odds
Smoker A	2 — 1
Smoker B	2 — 1
Smoker C	2 — 1
Non Smoker	3 — 1

©DIAGRAM

223

RESPIRATORY SYSTEM

Smoking greatly reduces the efficiency of the lungs, especially in those who inhale.

In a normal lung, glands in the interior lining are constantly producing mucus. This captures dirt and bacteria, and the mucus and its contents are then forced out of the lungs by the action of cilia. These are small, hair-like projections that are constantly moving, pushing the mucus up into the throat, where it is swallowed. Inhaled smoke hinders the action of the cilia, whilst stimulating mucus production. As a result, mucus, tar, dirt, and bacteria collect in the lungs in festering pools, encouraging tissue degeneration and hindering gas exchange.

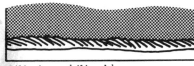

Bronchial cilia

Cilia immobilized by mucus

Inhaled tobacco smoke also tends to irritate the air passages, and to reduce air flow in the bronchi and bronchioles by making them contract.

SMOKER'S COUGH The constant cough that attends regular smoking is an attempt by the lungs to rid themselves of the tar and phlegm. Healthy lungs do not collect such phlegm, and only need the normal action of the cilia.

BRONCHITIS AND EMPHYSEMA
Bronchitis* is often triggered off by the irritation caused by cigarette smoke, and by the presence of bacteria in the lungs of smokers. Once established, it can progress rapidly from just a troublesome cough to a chronic condition which can kill. Emphysema* is also made more likely by the damage smoke and tar do to the lungs. Smokers, especially of cigarettes, run a much higher risk than non-smokers of contracting and dying from either of these.

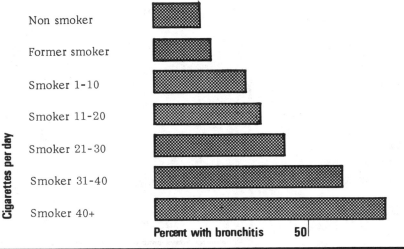

Cigarettes per day

- Non smoker
- Former smoker
- Smoker 1-10
- Smoker 11-20
- Smoker 21-30
- Smoker 31-40
- Smoker 40+

Percent with bronchitis   50   100

CANCER

Smoking is a direct cause of much cancer, because tobacco smoke contains carcinogens, and these can set off cancer wherever in the body the smoke reaches. As a result, cigarette smokers are 70 times more likely to have lung cancer than non-smokers. Smokers in general are also more likely to suffer from cancer of the mouth and pharynx (4 times more than non-smokers), larynx (5 times more) esophagus (2 times more), stomach (1½ times more), and bladder (1½ to 3 times more). Cancer of the mouth, pharynx, and larynx are the main forms in pipe and cigar smokers.

Cigarettes per day

- Non smoker
- Male smoker 1-10
- Male smoker 11-20
- Male smoker 21-40
- Male smoker 40+
- Female smoker 1-20
- Female smoker 20+

Mortality ratio   5   10   15

CARDIOVASCULAR SYSTEM
REDUCED OXYGEN INTAKE
Carbon monoxide is the most concentrated gas in tobacco smoke. Its affinity for blood hemoglobin is greater than that of oxygen i.e. it combines with it more readily than oxygen does. The greater concentration of carbon monoxide in a smoker's lungs means that hemoglobin which should be carrying oxygen to the tissues is now carrying useless carbon monoxide. The amount of oxygen in the bloodstream can be reduced by up to 8%. At the same time, the effects of smoking on the respiratory system also reduce the efficiency of oxygen intake. All this makes heart-strain a danger, as the heart works harder and harder to keep up the body's oxygen supply.

ATHEROSCLEROSIS AND THROMBOSIS
Atherosclerosis and thrombosis are more common in smokers than in non-smokers. Smoking raises the level of fatty acids and cholesterol in the blood, and encourages blood platelets (clotting bodies) to adhere to each other and to the blood vessel walls. Carbon monoxide in the bloodstream also seems to favour atherosclerosis.

OTHER EFFECTS
The action of nicotine on the endocrine system and sympathetic nervous system constricts blood vessels, raises blood pressure and blood sugar levels, and increases the heart rate. All this makes damage to the system likely.

CORONARY HEART DISEASE
All these factors greatly predispose the smoker to coronary heart disease of all kinds. Coronary heart disease occurs, on average, up to seven years earlier in smokers than in non-smokers.

THE RISK TO SMOKERS
Not only do smokers have a higher death rate than non-smokers, but the death rate is generally higher the more cigarettes smoked. These graphs show this for two age groups in two different surveys.

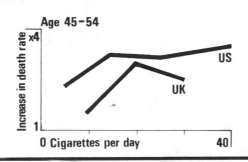

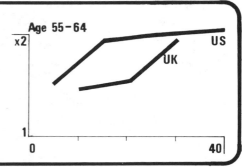

OTHER FACTORS
THE UNBORN CHILD Pregnant women who smoke may retard the growth of the fetus (the babies of smokers average 6oz lighter than those of non smokers). They are also more susceptible to miscarriages, still births, and the death of the child soon after birth.
ULCERS Those who smoke are five times more likely to suffer from gastric ulcers, and twice as likely to suffer from duodenal ulcers. This may be partly due to personality differences between non-smokers and smokers, but it has also been shown that smoking hinders the healing of these ulcers. Smoking also:
reduces the appetite (which often results in a weight gain on stopping); stains the teeth and fingers; increases the chances of periodontal disease; and impairs the senses of taste and smell.

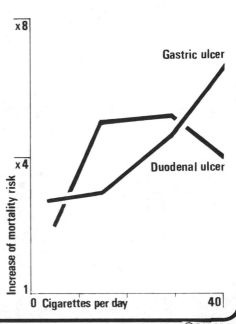

© DIAGRAM

6 Sex and sexuality

Few areas of human behavior have been masked in such mystery and taboo. Here is a straightforward, unbiased account of how our bodies function during the most intimate moments of our lives.

Left: Young couple (Spectrum Colour Library)

AVERAGE FEMALE EXPERIENCE 1

GROWTH OF SEXUAL EXPERIENCE

The information on this and the 3 subsequent pages is taken from the classic work by the American, Dr Alfred C Kinsey, and his associates ("Sexual Behavior in the Human Female"). Although their researches into the sexual behavior of American men and women are now some 30 years old, they are still the fullest and most reliable studies of their kind.

The first diagram shows what percentage of women have, by a given age, experienced either sexual arousal or orgasm; the second shows how these were first caused. Kinsey's findings showed just how rare so-called female "frigidity" is. Sexual arousal, and orgasm, begin in childhood. By the age of 35 only 2% of women have never experienced any sort of sexual arousal and only 9% have not experienced orgasm. For many women (40%) masturbation is the first source of orgasm, while petting is the main (34%) source of erotic arousal.

Percentage of women at each age who have experienced sexual arousal at some time in their lives. Also the percentage who have experienced orgasm.

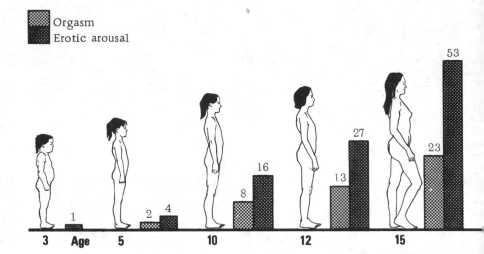

☐ Orgasm
■ Erotic arousal

Cause of first sexual response (percentages)

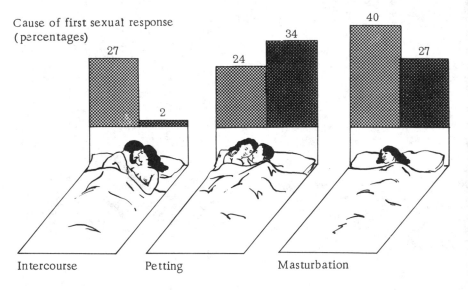

Intercourse   Petting   Masturbation

These graphs compare female and male sexual activity. Their growth of heterosexual experience corresponds closely, but the graph showing female orgasm rises more slowly and does not reach a peak till later.

Female orgasm

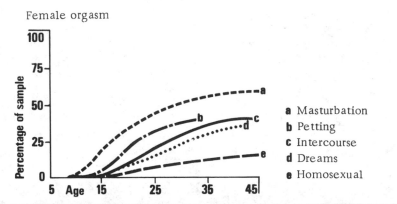

a Masturbation
b Petting
c Intercourse
d Dreams
e Homosexual

228

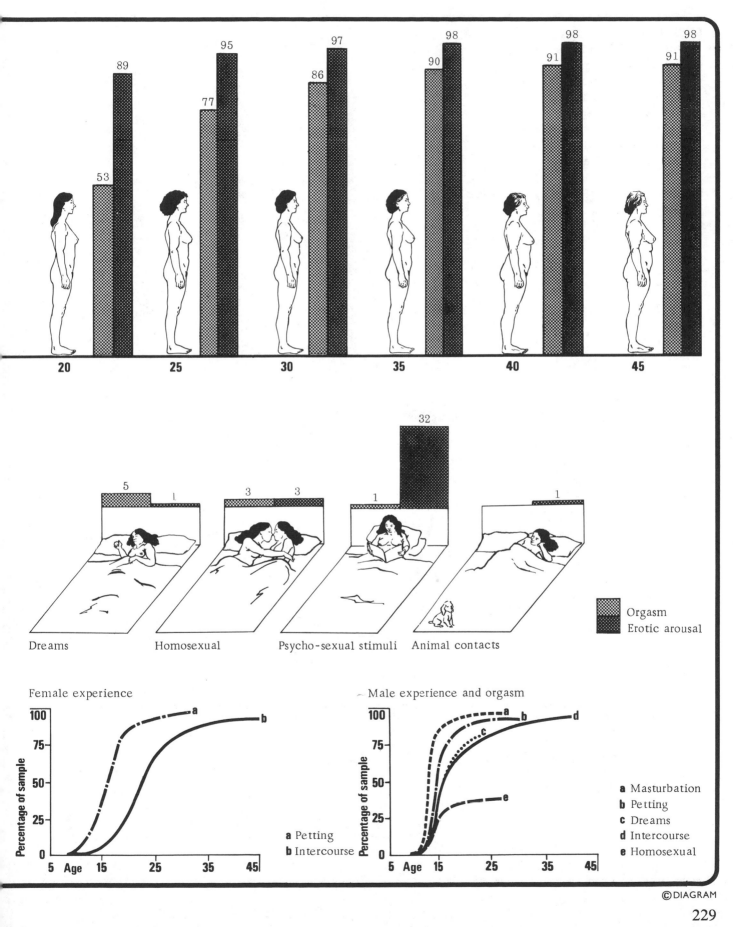

20 25 30 35 40 45

89 95 97 98 98 98
53 77 86 90 91 91

Orgasm
Erotic arousal

Dreams   Homosexual   Psycho-sexual stimuli   Animal contacts

5 1   3 3   1 32   1

Female experience

Percentage of sample

100
75
50
25
0

5 Age 15 25 35 45

a Petting
b Intercourse

Male experience and orgasm

Percentage of sample

100
75
50
25
0

5 Age 15 25 35 45

a Masturbation
b Petting
c Dreams
d Intercourse
e Homosexual

©DIAGRAM

229

AVERAGE FEMALE EXPERIENCE 2

ORGASM AND AGE

The diagram shows what percentage of sexually active women achieve orgasm at different ages. Response changes with age; most women reach their sexual peak during their their late 20s and 30s. For many women there is then little change until the late 40s or early 50s.

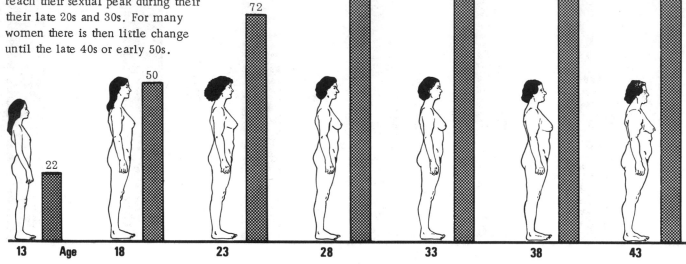

SEXUAL OUTLETS

Types of sexual outlet also vary with age. Taking 5 of the most likely activities, the graphs show what percentage they are of total sexual outlet. For example between the ages of 13 and 20 masturbation accounts for over 50% of a woman's sexual outlet; by her mid-30s this has declined to about 14%, and nearly 80% of her sexual activity is in heterosexual intercourse.

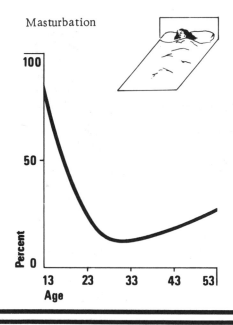

Masturbation

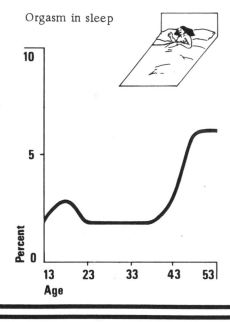

Orgasm in sleep

FEMALE AND MALE PATTERNS

On average men have orgasms more frequently than women. Also their maximum sexual responsiveness occurs earlier, by their late teens. It then gradually declines, in contrast to the later sexual development and decline in women.

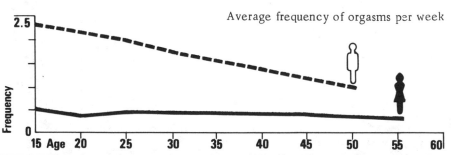

Average frequency of orgasms per week

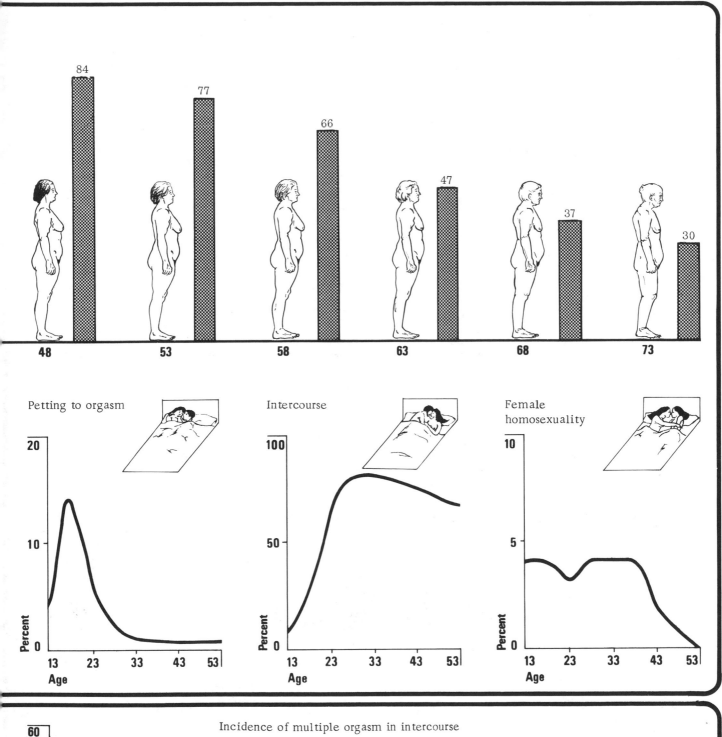

Petting to orgasm

Intercourse

Female homosexuality

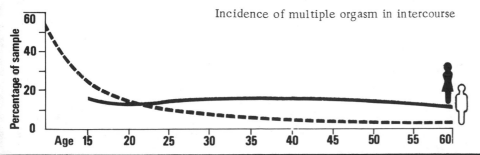

Incidence of multiple orgasm in intercourse

The graph compares the male and female experience of multiple orgasm. Kinsey found that 14% of the women with some sexual experience in his sample were capable of multiple orgasm.

©DIAGRAM

AVERAGE MALE EXPERIENCE 1

GROWTH OF EXPERIENCE

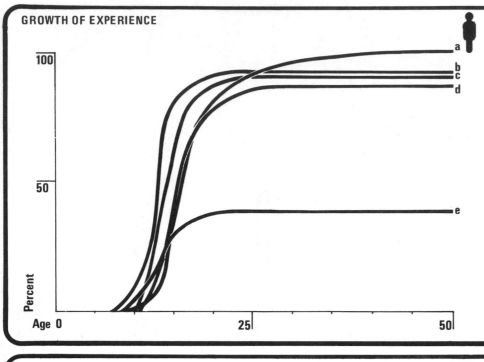

100

50

Percent

Age 0   25   50

a
b
c
d

e

We have numerous scientific studies on the sexual life of remote tribes - and almost none on our own society. The data on these four pages is taken from the classic work by Kinsey and his associates "Sexual Behavior in the Human Male" These first two tables look at different kinds of sexual activity, and show what percentage of the population have experienced them at least once, by a given age. All material K33-41 by permission Institute for Sex Research Inc. See

CHILDHOOD SEX PLAY

40

20

Percent

Age 5 6 7 8 9 10 11 12 13 14 15

Percentage of children experiencing sex play at each age.

Heterosexual only

Heterosexual and homosexual

Homosexual only

FIRST EJACULATION

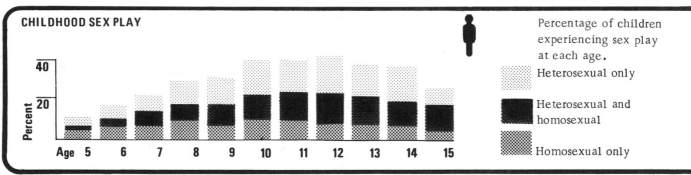

Percent   68   13   11 5 3

Source   M   N   I   P O

Sources of first male experience of ejaculation.

M   Masturbation
N   Nocturnal emission
I   Intercourse
P   Petting
O   Others

FREQUENCY OF ORGASM

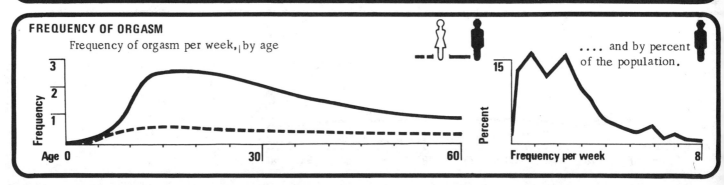

Frequency of orgasm per week, by age

3
2
1

Frequency

Age 0   30   60

.... and by percent of the population.

15

Percent

Frequency per week   8

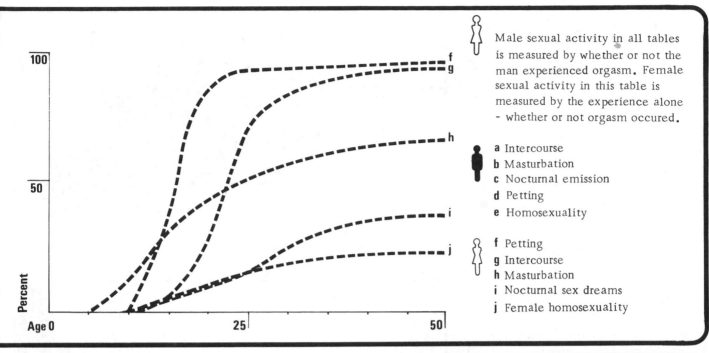

Male sexual activity in all tables is measured by whether or not the man experienced orgasm. Female sexual activity in this table is measured by the experience alone - whether or not orgasm occured.

a Intercourse
b Masturbation
c Nocturnal emission
d Petting
e Homosexuality

f Petting
g Intercourse
h Masturbation
i Nocturnal sex dreams
j Female homosexuality

SOURCES OF ORGASM

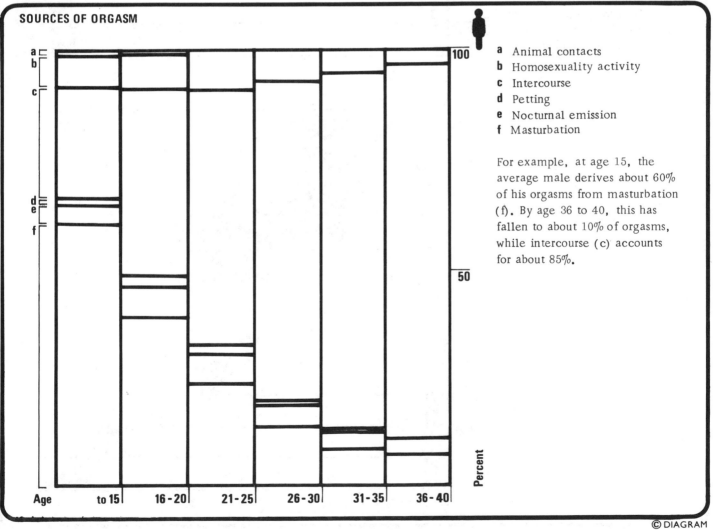

a Animal contacts
b Homosexuality activity
c Intercourse
d Petting
e Nocturnal emission
f Masturbation

For example, at age 15, the average male derives about 60% of his orgasms from masturbation (f). By age 36 to 40, this has fallen to about 10% of orgasms, while intercourse (c) accounts for about 85%.

© DIAGRAM

233

AVERAGE MALE EXPERIENCE 2

ORGASM AND AGE

SOURCES

Here we show how the sources of orgasmic outlet vary with age, for single and for married men. For example, in single men, masturbation (a) averages about 60% of total outlet in young adolescents (15 and under), but only 25% by age 31 to 35, when non-marital intercourse (d) has become the main outlet. Homosexuality (e) also increases with age - because more heterosexual men get married. In married men, marital intercourse (g) is by far the main outlet.

FREQUENCY

Frequency of different outlets varies in the same way. For example marital intercourse (c) declines in married men from almost 4 times a week (age 16 to 20) to only about twice a week (age 36 to 40).

a Masturbation
b Nocturnal orgasm
c Petting to orgasm
d Non-marital intercourse
e Intercourse with prostitutes
f Homosexual activity
g Marital intercourse

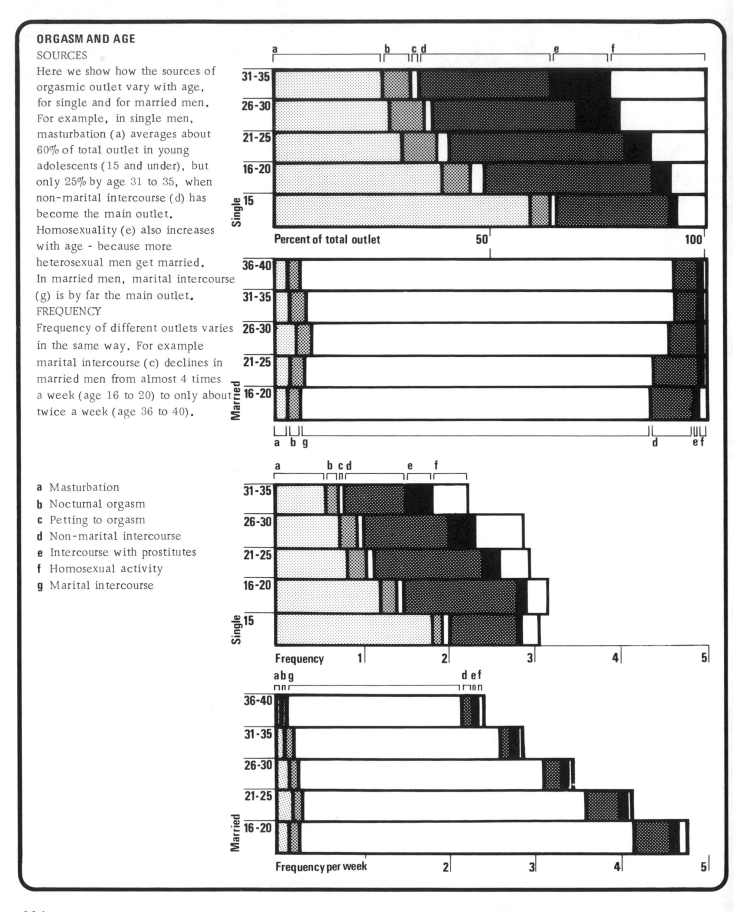

234

SEXUAL RESPONSE

Here we show how sexual response changes with age. For example, the percentage lacking all sexual response declines from almost 100% at age 5 to almost 0% at 35, and then begins to rise again.

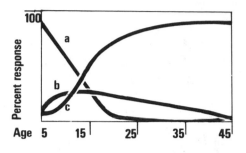

Sexual response in men
a No sexual response
b Homosexual response
c Heterosexual response

CAPABILITY AND AGE

Morning erection

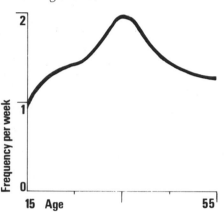

Duration of erection

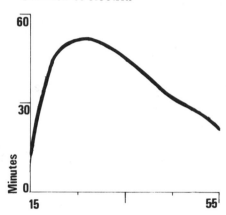

Angle of erection

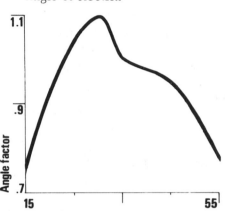

With age:
a) the experience of waking with an erection becomes less frequent;

b) the length of time for which an erection can be maintained shortens; and

c) the angle of the erect penis becomes less steep.

MULTIPLE ORGASM

This compares, for different ages, the percentage of men and women who can reach orgasm more than once during a single session of intercourse. The numbers are calculated as percentages of all those who have intercourse at each age. (For example, at age 10, of the very few males who have intercourse, more than half can have multiple orgasm.) In men, the percentage declines with age - falling rapidly from puberty to 25. In women, it rises between 25 and 30, because for them multiple orgasm is partly a learned response.

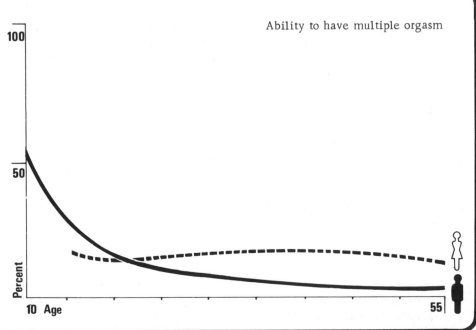

Ability to have multiple orgasm

©DIAGRAM

SEXUAL AROUSAL

EROGENOUS ZONES

These are the most erotically-sensitive areas of the body, although, of course, they vary considerably from individual to individual. Stimulation by hand or mouth (or other light object) of any of these sensitive areas is not only an important part of intercourse, but can also be sexually satisfying in itself. It can lead to mutual masturbation or oral-genital sex, even though traditionally, western-style lovemaking follows a pattern of foreplay, intercourse, and orgasm.

Many people - women and men alike - prefer sexual activity to end with the particular closeness of intercourse, and indeed many believe that lovemaking must always finish with vaginal intercourse; but to consider that stimulation of the erogenous zones of the body is only a prelude to, or substitute for, orgasm through vaginal intercourse, or even oral-genital sex-play, is to limit the pleasures that it can bring.

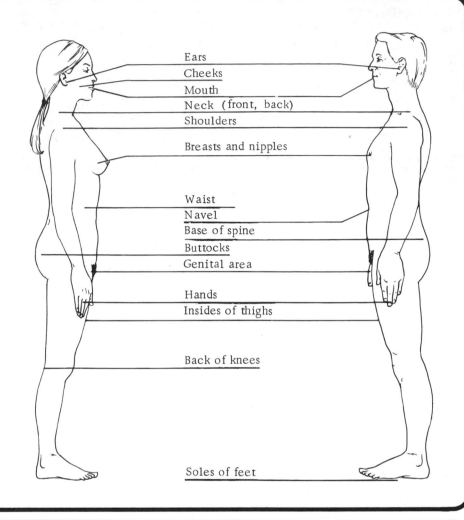

Ears
Cheeks
Mouth
Neck (front, back)
Shoulders
Breasts and nipples
Waist
Navel
Base of spine
Buttocks
Genital area
Hands
Insides of thighs
Back of knees
Soles of feet

ORAL SEX

The mouth and genitals are potentially the two most erotic areas of the body. This role of the mouth is clear from the practice of of deep kissing. But though oral sex includes lip and deep kissing, and any oral contact with the body, it most often refers to oral-genital contact, i.e. cunnilingus and fellatio.

CUNNILINGUS

This is the stimulation by the mouth of a woman's genitals. The whole area from the tip and shaft of the clitoris to the anus, including the vaginal entrance, is highly sensitive and in most women responds even more intensely to oral contact than to stimulation by the fingers. The types of oral techniques used include kissing, licking, and sucking, and if these are used, most women attain orgasm very easily.

FELLATIO

This refers to oral stimulation of the male genitals. Techniques again include licking, kissing, and sucking, but also friction of the shaft and tip while the penis is inside the mouth.

ATTITUDES

Although oral sex is a source of considerable pleasure for both sexes, many people object to it strongly. Various objections have been raised against oral-genital sex, mainly that it is unnatural, sinful, or unhygienic. In fact, it is a common aspect of sexuality, widely used since ancient times. Its use today is as widespread, although often unadmitted. From a hygienic point of view, provided the genitals are kept clean, there are fewer and less harmful bacteria there than in the mouth. Likewise, both vaginal fluid and semen are just body fluids and usually tasteless. In female homosexual practice, in particular, use of oral sex is widespread - not only because vaginal penetration rarely occurs, but also because it is the most effective way for a woman to attain orgasm with a partner.

THE FEMALE ORGASM

The female orgasm has possibly been the subject of more debate and literature than any other area of human sexuality. But, fortunately, since the work of Masters and Johnson in the 1960s, many of the myths and mysteries surrounding it have gone. It is now known that, although the experience and intensity of orgasm may vary considerably, the actual physical process of orgasm is always the same. The distinction between "vaginal" and "clitoral" orgasm is a myth.

REACHING ORGASM

Women vary greatly in what they respond to sexually, but the mons pubis, labia minora, clitoris, and vaginal entrance are almost always important. The clitoris is the most sexually responsive part of a woman's body, and in most cases fairly continuous clitoral stimulation is needed for orgasm. However, as the tip of the clitoris is extremely sensitive, constant direct touch can become painful. So for the majority of women, manipulation of the whole genital area of the mons pubis - by hand, tongue, or vibrator - is more pleasurable During intercourse, movement of the penis in and out of the vagina provides continual clitoral stimulation by moving the labia minora backward and forward over the clitoral tip. The anus is another potentially erotic area, but it is not as easily penetrated as the vagina. For anal intercourse, lubrication - KY jelly, for example - is generally needed.

THE FEELING OF ORGASM

The time needed to reach orgasm varies from woman to woman and from occasion to occasion. Just before orgasm there is a feeling of tension lasting possibly 2-4 seconds when all the small muscles in the pelvis surrounding the vagina and uterus contract. This is followed by the orgasm itself, which may last 10-15 seconds. It is felt as a series of rhythmic muscular contractions, occurring every 0.8 seconds, firstly around the outer third of the vagina and spreading upward to the uterus. Both uterus and rectum also contract. In a mild orgasm there may be 3-5 contractions; in an intense one 8-12. Also during orgasm the muscles of the abdomen, buttocks, arms, face, legs and neck may contract. Breathing is more rapid, and blood pressure climbs. All these return to normal and orgasm is usually followed by feelings of relaxation and peace.

MULTIPLE ORGASM

Women, unlike men, are capable of multiple orgasm. That is, immediately or shortly after a first orgasm, if a woman maintains her sexual excitement at the plateau level, she can move directly into a second orgasm. Some women can experience 3-5 orgasms within a few minutes, and up to 12 in one hour have been recorded

EXPERIENCE OF ORGASM

Virtually all women are physically able to attain orgasm. But a recent US survey, the Hite Report, suggests that as few as 30% regularly achieve orgasm through intercourse. Kinsey's data, shown below, indicates that such orgasm does gradually become more frequent in long-term relationships - but that even after 20 years of a relationship, 11% of women never reach orgasm.

Nevertheless, it seems that the vast majority of women prefer intercourse to any other sexual activity, because of the closeness and affection associated with it. Yet, despite the "sexual revolution" of the 1960s, women still seem to feel guilt about their own sexuality, and are reluctant either to initiate sexual activity or to communicate their sexual needs to their partners.

The diagram shows what % of women experience orgasm in marital intercourse, and how often they do so.

Legend:
- ☐ Never
- Rarely
- Often
- ■ Always or nearly always

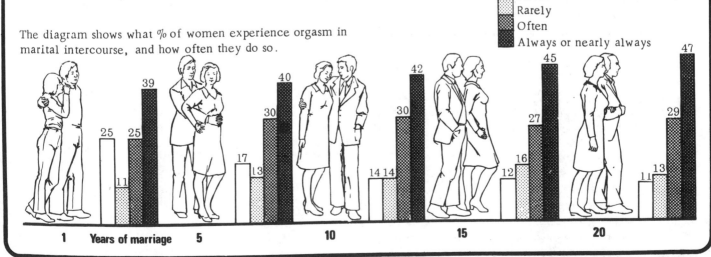

Years of marriage: 1 · 5 · 10 · 15 · 20

©DIAGRAM

THE SEXUAL PROCESS (FEMALE)

PHYSICAL CHANGES DURING LOVEMAKING

Important changes take place in the body during lovemaking, as a result of muscular tension and the swelling of certain tissues with blood. These processes were first described in detail by William Masters and Virginia Johnson, in their book, "Human Sexual Response," (1966).

The act of lovemaking can be divided into four stages: the excitement phase, the plateau phase, the orgasm, and the resolution phase.

EXCITEMENT PHASE

The duration of this phase varies according to the amount and effectiveness of the stimulation. General muscular tension begins, and heart rate and blood pressure start to increase.

CHANGES IN THE BREAST

Early in the excitement phase the nipples become erect. Later, increased definition of the vein pattern in the breasts becomes obvious, and there may be an increase in the size of the breasts themselves.

In the plateau phase, the breasts continue to enlarge, and the areolae become prominent, so engulfing the nipples. Also, a pink mottling known as the sex flush may appear on the breasts. In the resolution phase, the areolae subside leaving the nipples prominent, and the breasts gradually resume their normal size.

a Areola
b Nipple

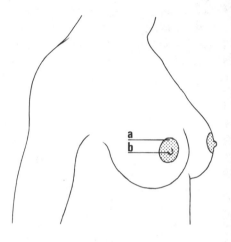

EXTERNAL GENITAL CHANGES

In the excitement phase, the clitoris increases in length and diameter. The labia majora open and spread flat while the labia minora swell and extend outward. In the plateau phase the clitoris shortens and may disappear under its hood. The labia majora swell further, and the labia minora change color from pink to red, or from red to deep red in a woman who has had children.

At orgasm, no particular changes are discernible, but during the resolution phase the labia resume their usual color and size, and the clitoris returns to normal.

a Clitoris
b Urethra
c Vagina
d Labia minora
e Labia majora

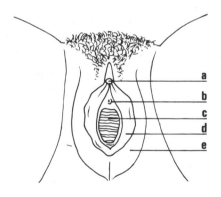

INTERNAL GENITAL CHANGES

During the excitement phase, the vagina becomes moistened with a lubricant "sweated" through its walls. The uterus and cervix pull away from the vagina, and the inner $\frac{2}{3}$ of the vagina expands. During the plateau phase, the expansion of the inner vagina continues while the outer $\frac{1}{3}$ contracts, gripping the penis. The uterus continues to move away from the vagina.

At orgasm the uterus and the lower $\frac{1}{3}$ of the vagina experience a wave of contractions.

During the resolution phase, the uterus, cervix, and vagina return to normal.

a Clitoris
b Uterus
c Cervix
d Vagina

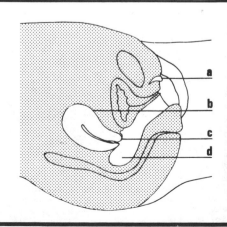

PLATEAU PHASE
This phase is an extension of the excitement phase. If effective stimulation continues, sexual tensions increase and the desire for release of the tensions in orgasm is intensified.

ORGASM
The orgasmic phase usually lasts only a matter of seconds. Orgasms do, however, vary in intensity and duration from woman to woman and occasion to occasion.

RESOLUTION PHASE
When orgasm is over, the resolution phase begins. There is gradual muscular and physiological relaxation, and within 30 minutes the body returns to its unstimulated state.

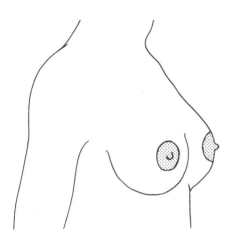
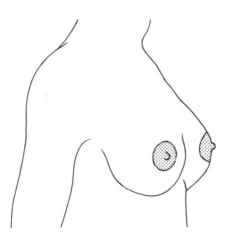
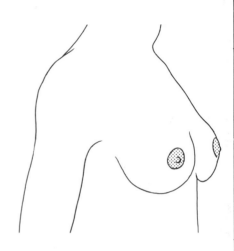

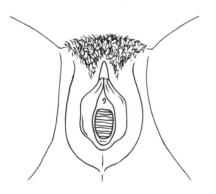
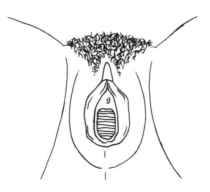
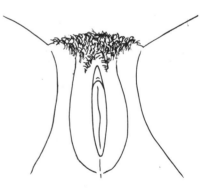

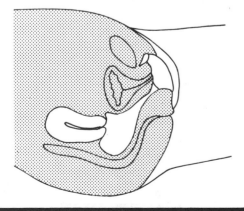
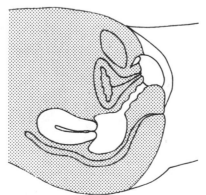
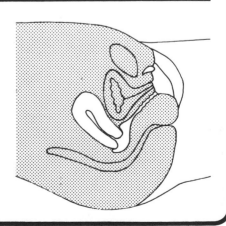

©DIAGRAM

THE SEXUAL PROCESS (MALE)

SEXUAL AROUSAL

The sexual process takes the male sexual system from its normal, inactive state, to orgasm. On orgasm, semen is discharged from the penis.

The stimulus that first arouses the system can be purely psychological - the thought of sex. But the system normally needs physical pressure on the skin surface of the penis to reach orgasm. In copulation this is provided by the contact of the penis with the female genitals. For both arousal and for orgasm, conditions need to be right. The system can be inhibited, or the process reversed, by:

a) adverse physical conditions in the surroundings, eg cold;

b) psychological distractions (eg worry, or sudden disturbance);

c) adverse body states (eg tiredness, or too recent orgasm).

BODILY RESPONSES

The following responses accompany the male sexual process:

a) blood pressure, and rates of breathing and heart beat, all rise often, on orgasm, to about $2\frac{1}{2}$ times the normal level;

b) muscular spasms may affect groups of muscles in the face, chest, and abdomen;

c) contractions of the rectum occur.

In some men there is also:

d) swelling and erection of the nipples;

e) flushing of skin color around chest, neck, and forehead;

f) perspiration from soles, palms, and body, and sometimes from head, face, and neck.

ORGASM

For both men and women, the experience of orgasm can be one of very intense sexual excitement and emotional release. Yet the physical process, of irritation and spasm, can be compared with that of sneezing.

In intercourse, orgasm is usually accompanied by convulsions of the body, involuntary movements, and sounds such as sighs and groans. But orgasm can occur at various levels of sexual excitement. Evidence, in men, of a high level of excitement, are very high rates of breathing, and flushing of the skin.

1 Heart rate quickens rapidly
2 Heart rate levels off
3 Sharp rise as orgasm approaches
4 Gradual return to normal

a Penis erects
b Scrotum thickens
c Testes rise
d "Sex flush" appears
e Penis tip and testes swell
f Ejaculation, heavy breathing, and muscular spasms
g "Sex flush" disappears
h Loss of erection
i Penis returns to normal state

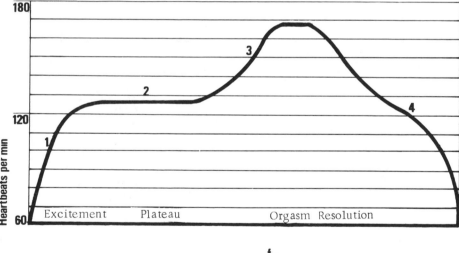

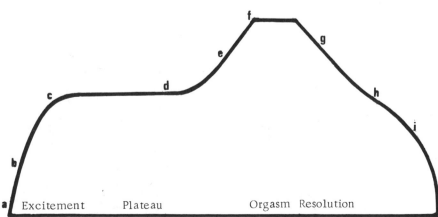

BEHAVIOR OF THE PENIS

1) EXCITEMENT

The sexual stimulus triggers off an automatic reflex, which sends blood flowing into the spongy tissue of the penis.

The spongy mass swells and presses against the sheath of skin. As a result, the penis becomes stiff and sticks out at an angle from the body, usually pointing slightly upward.

Muscular contraction pulls the testes closer in to the body. This stage can be maintained for long periods, and can be lost and regained, without orgasm, many times.

2) PLATEAU LEVEL

The testes are drawn still closer to the body. The penis increases slightly in diameter, near the tip, and the opening in the tip becomes more slit-like. The tip itself may change color, to a deeper red-purple.

3) ORGASM

The muscles around the urethra give a number of rapid involuntary contractions. This forces semen out of the penis at high pressure (ejaculation).

There are usually three or four major bursts of semen, one every 0.8 seconds, followed by weaker, more irregular, muscular contractions.

4) RESOLUTION

Often there is:

first, a very rapid reduction in penis size, to about 50% larger than its normal state;

followed by a slower reduction back to normal.

But each of these stages may be prolonged, eg if the penis remains inserted in the female genitals.

The response patterns shown here and in J05 were first described by WH Masters and VE Johnson in "Human Sexual Response" (1966).

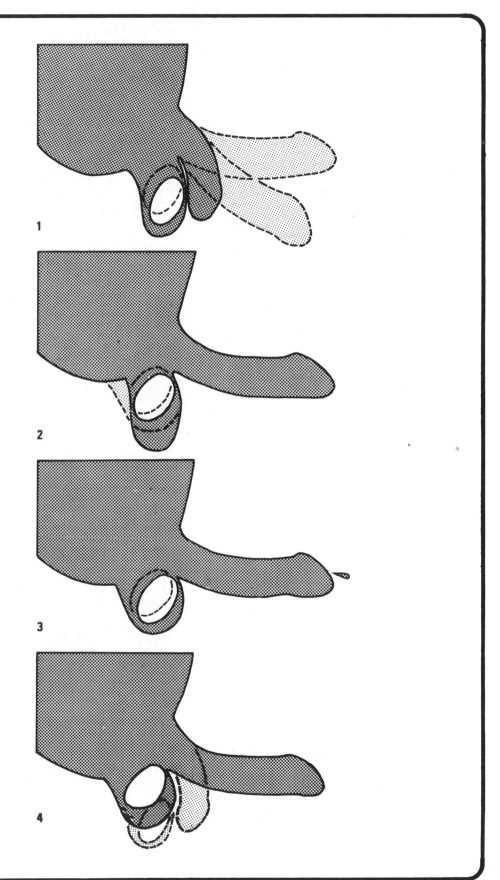

© DIAGRAM

SEXUAL INTERCOURSE 1

ENTRANCE

The couple take up one of the positions that allow sexual intercourse to take place. The penis tip points at the entrance to the vagina.

Then the man may only need to push forward from his hips: his penis slides immediately into the woman's vagina. Alternatively, in other positions, the woman lowers her vagina onto the man's penis.

But very often the following techniques are used:

a) either partner holds the penis, to help direct it into the vaginal entrance;

b) either partner holds the woman's labia apart to help the penis slide between them;

c) the penis is inserted only very gradually, beginning with the tip and progressing at first with several small forward movements and half retreats. This spreads the vaginal lubrication over the penis surface, and helps the vagina to accommodate itself gradually to the penis' size. (Also it is used as a tantalizing technique, to heighten the sensation of entrance.)

DELAYING ENTRANCE

Once physiological readiness has been reached, the penis can be inserted into the vagina. But a couple may choose to delay this for many minutes. During this time they continue lovemaking without intercourse. In fact, during lovemaking, the highest levels of sexual excitement before intercourse are usually not reached until the body has been physiologically ready for some time. This is especially true of women, but also of men.

FACE TO FACE POSITIONS

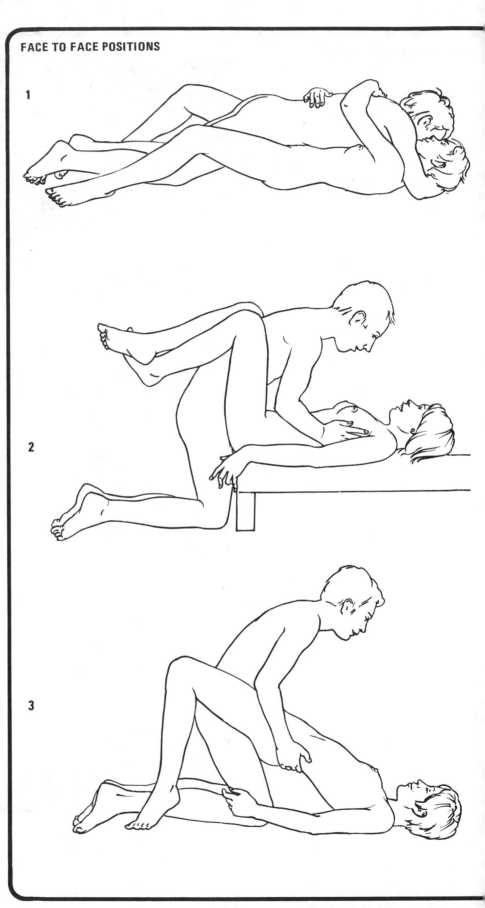

1

2

3

The number of positions for intercourse is almost endless. The choice of position depends on the mutual tastes and preferences of the couple. Among the most common, however, are front-entry positions with either the male or the female on top. Some typical examples are shown here.

1 The so-called "missionary position" involves the couple lying face to face with the man on top. Penetration is easy and the close proximity of the bodies allows the exchange of caresses.

2 A variation of this position can be achieved when the woman's body is supported on a bed while the man kneels on the floor.

3 A further variation of the "missionary position" involves the man supporting the woman's body with his hands.

4 The female can take a more active part in lovemaking when she is above the man. This can be a satisfactory position for couples where the woman cannot bear the weight of her partner's body on top of her. The woman has considerable freedom of movement and her partner can caress her clitoris during intercourse.

5 The position in which the couple kneel facing each other with the woman on top again allows the woman to take an active part in lovemaking. It has the disadvantage that the man needs to support his weight on his hands so that he cannot easily caress his partner.

6 The partners face each other both in a half-lying position with the weight supported on their hands. This has the disadvantage that neither partner can easily caress the other.

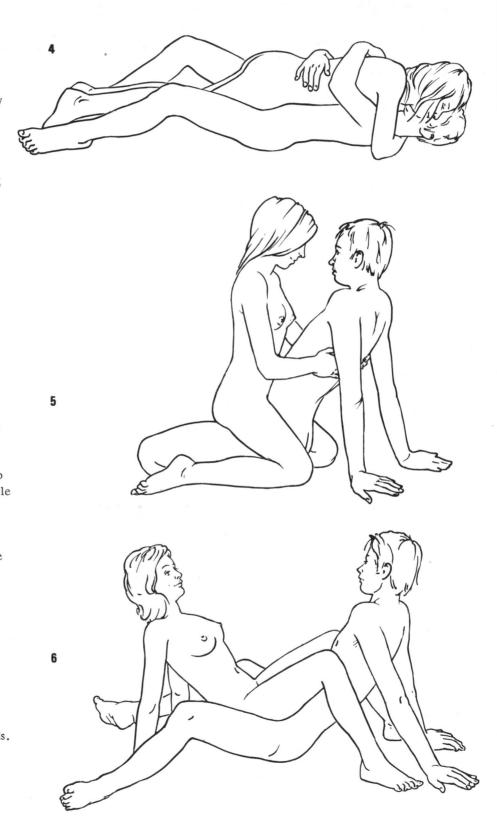

©DIAGRAM

SEXUAL INTERCOURSE 2

STANDING POSITIONS

For successful intercourse in a
standing position, both partners
must be about the same height.
If the woman is rather short, she
can stand on a low stool, a pile
of books, or even one step up some
stairs, until she is at the correct
height for easy penetration.
Alternatively, her partner can pick
her up and she can clasp her hands
around his neck, and her legs
around his waist.

SIDE POSITIONS

If the couple lie side to side, they
can caress freely as neither has
to support the weight of the other.

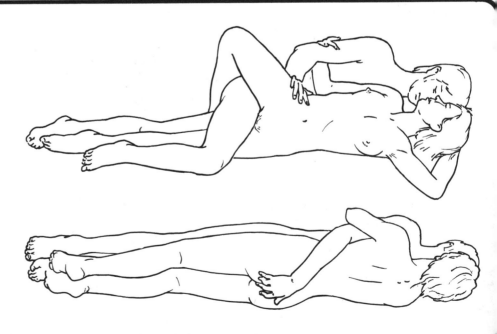

REAR-ENTRY POSITIONS

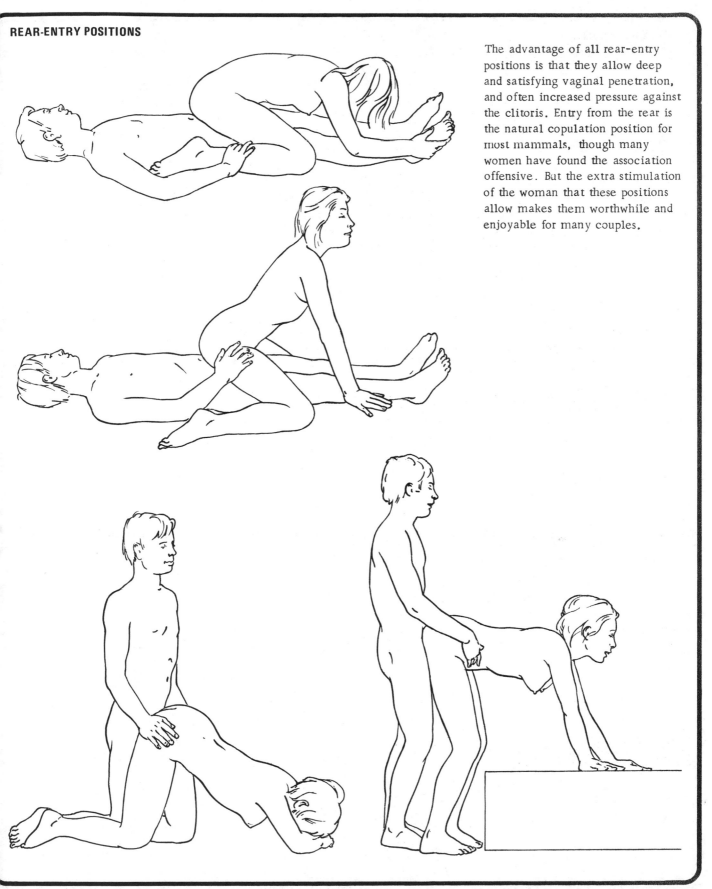

The advantage of all rear-entry positions is that they allow deep and satisfying vaginal penetration, and often increased pressure against the clitoris. Entry from the rear is the natural copulation position for most mammals, though many women have found the association offensive. But the extra stimulation of the woman that these positions allow makes them worthwhile and enjoyable for many couples.

© DIAGRAM

SEXUAL INTERCOURSE 3

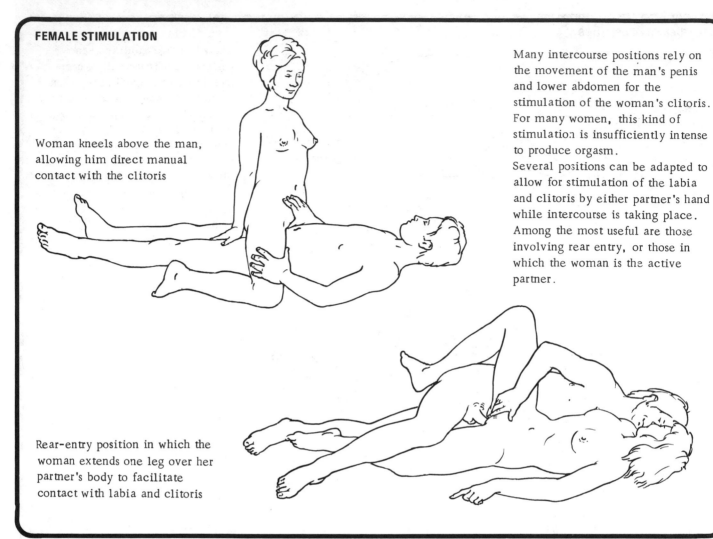

FEMALE STIMULATION

Woman kneels above the man, allowing him direct manual contact with the clitoris

Many intercourse positions rely on the movement of the man's penis and lower abdomen for the stimulation of the woman's clitoris. For many women, this kind of stimulation is insufficiently intense to produce orgasm.

Several positions can be adapted to allow for stimulation of the labia and clitoris by either partner's hand while intercourse is taking place. Among the most useful are those involving rear entry, or those in which the woman is the active partner.

Rear-entry position in which the woman extends one leg over her partner's body to facilitate contact with labia and clitoris

INTERCOURSE FOR CONCEPTION

To increase the chance of fertilizing an egg, the woman's body should be tilted so that the vagina is in a vertical position.

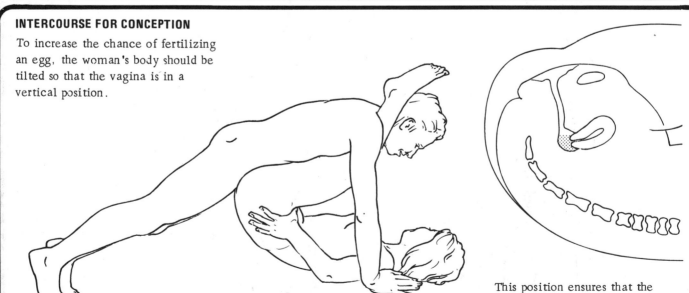

This position ensures that the ejaculated semen lies in a pool at the upper end of the vagina, near the cervix.

INTERCOURSE IN PREGNANCY

For women who have had a previous miscarriage, intercourse in the first 3 months of a pregnancy can sometimes be unwise. The doctor will advise on this when the pregnancy is first confirmed. Later in pregnancy, the woman's thickening waistline may make intercourse in more conventional positions uncomfortable if not impossible.

A variety of positions can be adapted for use during pregnancy. Among the most suitable are rear-entry positions and those in which the woman can control the depth of penetration.

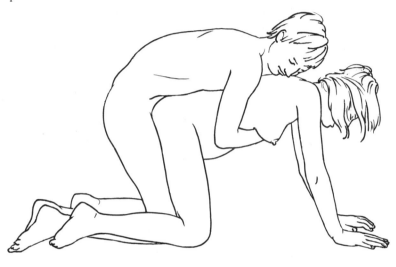

1 Rear-entry position with both partners kneeling

2 Face-to-face position on a chair, the woman seated above the man, allowing her to control the degree of penetration

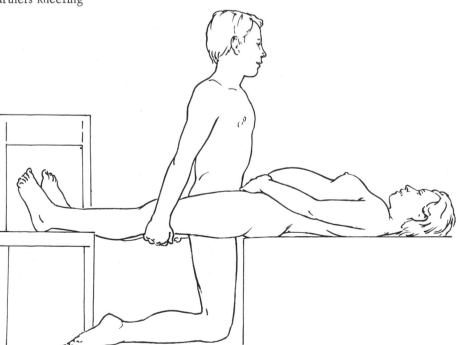

3 A suitable position for the final months of pregnancy. The woman's body and feet are supported and there is no pressure on her abdomen

© DIAGRAM

MASTURBATION

"Masturbation" means stimulation of the genitals other than by intercourse - usually with the hands, but sometimes with other parts of the body or with objects, and usually with the intention of achieving orgasm. It usually means self-stimulation - but not always. It also includes stimulation of a partner's genitals, whether or not the couple also experience sexual intercourse, and whether in heterosexual or homosexual circumstances. So its uses range from the first experiments of schoolchildren to the mature relations of lovers.

Such masturbation occurs throughout society, and masturbation to orgasm is experienced at some time in their lives by about 93% of all men and by over 60% of all women.

TRADITIONAL ATTITUDES

In our society, masturbation has been a subject more taboo than interpersonal sex. There were, perhaps, three reasons for this.

First, masturbation is the way in which the sexuality of their offspring forces itself on parents' attention. Second, for most people masturbation seems an admission of failure, compared with finding sexual expression in a relationship with another person. (This is especially true whenever sexuality is dominated by concepts of male "conquest" of the female.) Third, acceptance of masturbation means recognizing the power of the sex urge - beyond (if need be) both social pretense and communication.

So the traditional attitude was one of disapproval, reaching psychotic levels in the later 19th century. Children were terrified with stories that masturbation would bring physical and mental illness. Some were even forced to wear physical contraptions designed to prevent wearers from "abusing" themselves.

MODERN ATTITUDES

First, it is now generally recognized that masturbation causes no special physical or mental deterioration. Physically excessive masturbation in men can ca prostate gland trouble - but so can excessive intercourse, or total abstinence. Mentally, excessive masturbation may well form a part of some failure to face up to reality - but so can excessive eating. Masturbation is here a sign of failure, not a cause; at most, it reinforces the sense of failure. Second, past masturbation is not a cause of difficulty when interpersonal sexual relations begin. There may be fears about sex, which express themselves in a preference for masturbation. But the cause of these lies in personality disorders, due to childhood experience, bad sexual education, a traumatic incident, or or a repressive family atmosphere. The masturbation is not the cause. Nowadays masturbation - though still little talked about - is more generally accepted for what it is:

TECHNIQUES — FEMALE

Women have various ways of masturbating, although the most commonly used rely on stimulation of the clitoris. During her life, a woman may use one or more of the methods described below.

CLITORAL AND LABIAL

Well over 80% of women who masturbate regularly, concentrate on direct stimulation of clitoris or labia. This is done in various ways. One or more fingers can either be rubbed over or around the clitoris, or the whole hand may be used to apply steady and rhythmical pressure. (Few women masturbate by actually rubbing the clitoral glans; more commonly they massage the shaft or general clitoral area.) A pillow, vibrator , or continuous stream of water from a faucet, may be used instead of the hands. Alternatively, some women masturbate by crossing their legs, applying steady pressure from their thighs onto the genital area. In addition, while using any of these techniques, a finger or similar object may be inserted into the vagina, although few women rely on vaginal insertion alone. Slightly different from these methods is a technique used by about 5% of women. In this a woman assumes a prone position similar to that of a woman-on-top position in intercourse. Although there may be some direct stimulation of the genitals, possibly with a pillow, it is usually very slight, and a climax is achieved by rhythmical pelvic thrusting combined with a build-up of muscular tension similar to that during intercourse.

OTHER METHODS

In most women the breasts and nipples are highly sensitive and in a few cases their stimulation alone is sufficient for orgasm. Likewise about 2% of women achieve orgasm through fantasy.

part of our normal sexual experience. First, it has functioned (especially before the development of the pill) as a way of diverting some of the sexual drive that otherwise would result in immature partnership commitments. Second, it releases acute tension due to an unsatisfactory sexual life or the temporary absence of a partner. Third, it alleviates the sexual loneliness of old age. Certainly, it is not something that parents should try to create guilt feelings about, in their offspring.

WHEN MASTURBATION BEGINS

Some kind of stimulation of the genitals occurs in many infants only a few months old. One survey found that genital play, or rocking to and fro, on the genitals, occurred in more than half of infants under one year old. Some such cases have been reported in which stimulation was even carried to orgasm (though not, of course,

ejaculation): the youngest a boy of 5 months and a girl of 3 months. More conscious auto-eroticism can begin at any age from 5 years on, but for most men it begins sometime between the ages of 9 and 18, with the great majority in the time just after puberty (i.e. for most men, 13 or 14). For in men, with the onset of puberty, masturbation becomes a necessary release of sexual tension and sperm production, rather than just a pleasant sensation. Masturbation, in fact, provides the first experience of ejaculation for $\frac{2}{3}$ of all boys. The average age of this is 13.9 years. Thereafter, masturbation is normally the main sexual outlet of early adolescence. Those women who masturbate before adulthood tend to start a little earlier than men. But by the age of 20, only 40% have tried masturbation, compared with 92% of all men. Masturbation provides the first orgasm for only just over $\frac{1}{3}$ of women.

MASTURBATION IN MALES

Male masturbation frequencies range from once a month or less to two or three times a day or more. (Women, on average, masturbate much less frequently.) The average male frequency, among those who do masturbate, is about twice a week at the age of 15. Thereafter, it declines throughout adulthood. (With women, it increases up to middle age.) Masturbation is responsible for between 5 and 10% of the outlet of married men, depending on age. It is mainly accounted for by temporary absence of the partner, conflicts, or psychological difficulties. Among single men, it varies from as low as 20% to as high as 80%. It is highest in the young and the highly educated. Techniques of masturbation vary considerably between different men: especially in the part of the penis stimulated, the speed of hand movement, the hardness of touch, and whether stimulation continues during ejaculation.

OTHER SOURCES OF ORGASM

Other forms of deliberate self-induced orgasm are rare. Only 2 or 3 men in a 1000 are able to suck their own penises. Very few men can reach orgasm by stimulating other parts of the body. Only 3 or 4 men in every 5000 can achieve orgasm by fantasy alone. But involuntary orgasm, during sleep, accompanied by a sexual dream, is fairly common in men from adolescence on. Almost every man experiences it at some time (and about $\frac{2}{3}$ of women). As a source of orgasm in men it typically varies between 2 and 8% of total outlet (2 to 3% in women). It tends to happen more frequently when masturbation is the main form of sexual outlet.

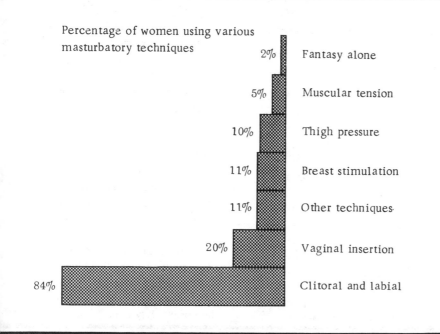

Percentage of women using various masturbatory techniques

2%	Fantasy alone
5%	Muscular tension
10%	Thigh pressure
11%	Breast stimulation
11%	Other techniques
20%	Vaginal insertion
84%	Clitoral and labial

HOMOSEXUALITY

Homosexuals are men and women who are emotionally and physically attracted to their own sex.

In our society, conventional images of male homosexuals are that they are in some way "feminine", or perhaps that their sex interests make them a danger to children. The vast majority of homosexuals fit neither of these ideas. They are just ordinary people - not identifiable by body type, mannerisms, dress, or occupation. Most non-homosexual people will, without realizing it, have friends and relatives who are homosexual. However, today many homosexuals do declare themselves openly, and refer to themselves as "gay".

INCIDENCE

It is very difficult to state the number of homosexuals in a population. Because of guilt and secrecy, surveys need not reveal the truth. Also it is important to distinguish between occasional incidents and a real preference. "Homosexual" activities can occur in preadolescent sex play, while homosexual feeling, perhaps accompanied by mutual masturbation and group masturbation, is typical of one stage of adolescence.

Roughly 33% of men (and 20% women) have at least one homosexual experience to orgasm between adolescence and old age. But only 2-5% of adults in western society are exclusively homosexual.

Between these numbers are those who have considerable adult experience of both heterosexuality and homosexuality - whether in the same or different periods of their lives. Many such individuals may be married and have children.

MALE HOMOSEXUAL ACTIVITY

There is still considerable ignorance and fear among non-homosexuals about homosexual procedures. However, apart from the impossibility of genital intercourse and the changes of emphasis this causes, homosexual activity is basically parallel to heterosexual. Many techniques are are common to both, and in both there are mutual kissing and caressing, and the general processes of getting used to one another and giving affection.

The main sexual techniques used include mutual masturbation, fellatio, anal intercourse, and interfemoral intercourse (where the penis is stimulated by being moved between the partner's thighs). The first two of these are quite common in heterosexual activity, while anal and even interfemoral intercourse do also occur.

Most non-homosexuals also think that homosexual partners take on exclusively passive (female) or active (male) roles. In reality, the roles often alternate.

ANAL INTERCOURSE

This involves inserting the penis into the partner's anus. Normally some kind of artificial lubrication is used.

Two rings of muscles, called sphincters, are sited at the opening to and inside the anus. They are normally closed, to prevent waste products coming out involuntarily. The outer sphincter can be relaxed or tensed at will, but the inner is less under conscious control. However, homosexual men have to learn to relax both these muscles, to avoid pain during anal intercourse. A dilator can be used, which will gradually stretch them, but most men find that both these muscles do relax if they are at ease with their partners.

When the penis has passed through the two sphincters it reaches the rectum, which is wide and does not register pain.

The passive partner in anal intercourse often experiences intense pleasure. This is partly because of the erotic sensitivity of the anus (which is familiar to heterosexuals too), but also because his partner's penis tends to rub against his prostate gland, which is sited near the rectum. Stimulation of the prostate seems to produce an especially prolonged and intense orgasm.

During orgasm the anus automatically tightens and puts pressure on the penis inside it, so intensifying the pleasure of the active partner.

One objection to anal intercourse is a hygienic one. But the rectum is normally free of waste products, and contains only about as many bacteria as the mouth. These are of a different kind, though, and the genitals should be washed carefully after anal intercourse (and especially before any oral-genital contact or heterosexual intercourse). Objections to being the passive partner in homosexual intercourse include the fear of pain and of "subjection" and "femininity". Anal intercourse can in fact be very painful if the sphincters are not properly relaxed. But the usual exchange of roles between partners prevents the development of superior/inferior roles or psychological typecasting.

INCIDENCE OF TECHNIQUES

Temporary adolescent homosexuality does not usually go beyond mutual masturbation. When fellatio or anal intercourse occur, it is a likely sign that a homosexual preference will continue in adulthood. Among adults, mutual masturbation and fellatio are common to both incidental and preferred homosexuality, but anal intercourse is usual only where homosexuality is preferred.

©DIAGRA

PROBLEMS OF HOMOSEXUALITY — MALE AND FEMALE

Homosexuals themselves may articulate reasons for their preferences; but society in general tends to see homosexuality as an abnormal state needing explanation in terms of some physical or psychological defect. Physical theories put forward include:
a) neurophysiological - in which homosexual behavior is due to malfunction or stimulation of certain parts of the brain;
b) hormonal - in which it is due to hormone imbalance (eg, in a male homosexual, a lack of male hormones or an excess of female ones); and
c) genetic - in which some inherited defect is blamed.
Bits of evidence for each of these include:
a) a recent very experimental technique of transforming homosexuals into heterosexuals by using tiny electric shocks to knock out certain cells in the brain.
b) evidence of hormone imbalance in a few homosexuals; and

c) evidence that homosexuals tend to have been born to elderly mothers, a situation that is known to be one in which chromosome abnormality often occurs.
But research findings for all these theories are mainly ambiguous or contradicted by other studies, and there is no scientific agreement about any of them.

PSYCHOLOGICAL FACTORS
Even if physical factors are involved, psychological ones seem more important. It is significant that the potential for homosexuality seems to exist in all men; and that homosexual feeling is typical of one stage of early adolescence, and only unusual if prolonged. This prolongation seems to happen, in our society, when certain family emotional patterns have occurred in childhood. In particular, combination of an unaffectionate, even hostile, father, with an emotional, over-intimate mother, may mean that a growing boy lacks or reacts against identification with a male model, and is drawn into

his mother's femininity.
The situation may be made worse if the mother accompanies her mother-son intimacy with subconsciously ambiguous physical caresses, out of frustration or resentment against the father.
The child's sexuality may be awakened; but as he cannot imagine himself taking a sexual initiative in relation to his mother, women may come to seem unobtainable or alarming.
Even where a firm break with the mother is made, a fear of her power may remain as a fear of the power of women in general. This is especially true where the mother's force is not one of intimacy compared with a distant father, but of outright dominance compared with a weak one. This is another situation in which homosexuality can arise, for the growing boy sees no advantage in becoming "male".
All this depends, though, on a pervasive influence of the traditional roles of men and women.

THEORIES ABOUT MALE HOMOSEXUALITY

These mainly arise because society treats homosexuality as abnormal.
First, homosexuals have difficulty admitting their sexual preference even to themselves. On average, it takes 6 years from the first feelings of homosexual attraction to outright self-admission of "being homosexual" - and some may avoid it for a lifetime, suffering instead a lingering dissatisfaction with heterosexuality.
Second, self-admission is often followed by acute depression and shame. However militant some "gays" may now be, the typical features of "psychological coming out" in the past have been suicide

attempts and psychiatric treatment.
Third, self-admission often leads to fear of the opinion of others, if they find out. In particular, fear of parental knowledge may produce acute guilt feelings.
Fourth, the homosexual will often need to suppress ordinary expression of his sexuality (eg, in most jobs heterosexual flirting is acceptable, but homosexual is not.)
Fifth, the homosexual may find it hard to meet others like himself. Homosexual organization is more common now, in big cities, but many individuals remain isolated.
Sixth, homosexual alliances tend to be fragile and transient, because:

a) the inability to reproduce removes the binding purpose of raising another generation, and leaves an empty sexuality;
b) the partners are often guilty and insecure; and perhaps
c) any alliance suffers the less clearly defined the partner's separate roles.
Finally, alliance instability, lack of reproduction, and a resulting inevitable emphasis on sexuality and youth, leave many older homosexuals to face years of appalling loneliness - which may lead to the desperation of a sexual offence.

HOMOSEXUALITY 2

LESBIANISM

The word "homosexual" is used, as with a man, to refer to women who are attracted, emotionally and physically, to members of their own sex. But, more frequently, female homosexuals are known as lesbians, a name derived from the Greek island of Lesbos, which, some 2600 years ago, was the home of the poet Sappho. Many of her poems were beautifully and movingly addressed to women, and from them comes the term "sapphic love" to describe love between women. It has been suggested that the term "homoemotional" more accurately describes lesbianism, as such love between women is more frequently characterized by its intense emotionalism than by its sexual aspect. But, in our society, no matter which term is used, the conventional interpretation of lesbianism has been incorrectly one of perversion.

INCIDENCE

Statistics of actual numbers of lesbians in a population are almost impossible to obtain. Because of secrecy, and even guilt, surveys need not reveal the truth. Also, it is important to distinguish between occasional incidents and a real preference. Some degree of "homosexual" activity is common, either in pre-adolescent sex play or in the schoolgirl crushes of adolescence. It has long been thought that there are fewer lesbians than male homosexuals in the population, although this is unlikely. Perhaps 20% of women have at least one homosexual experience to orgasm during their lives. But only perhaps 2-5% of adult females in western society are exclusively homosexual. Although today more women are displaying their lesbianism more openly and declaring themselves "gay," the vast majority of lesbians are not distinguishable from other women. Popular images of the masculine-type "dyke" are, in reality, incorrect; and most lesbians are not identifiable by body type, mannerisms, dress, or occupation.

LESBIAN ACTIVITIES

Lesbian lovemaking has always been seen in two completely different ways. On the one hand there is considerable ignorance and fear among non-homosexuals about lesbian procedure; while on the other hand, distorted presentations of lesbian activity have long been accepted as a source of erotic arousal in classical art, and today in hard-core pornography. Both views incorrectly assume that lovemaking between women is a preliminary to, or a substitute for, heterosexual intercourse. In fact, the reverse is true. What is notable is that lesbians can achieve a far greater sexual satisfaction with each other than many other women achieve in their heterosexual relations. Nearly all lesbian lovemaking results in orgasm for both women, a fact which may explain why men have tended to feel threatened by lesbianism. Many sexual activities are common to both lesbians and heterosexuals. In both there is mutual kissing and caressing, particularly of the breasts and genitals, and the general procedures of getting used to one another and of giving affection. The main techniques used are mutual masturbation, and cunnilingus (kissing, sucking, and licking, of the clitoris). Vibrators are commonly used for clitoral stimulation; the dildo, or artificial penis, is not used: in fact its use has long been a particularly pernicious and long-lasting myth. By definition, lesbians are attracted to other women and their love-making does not center on penis-substitutes. One additional practice which is sometimes used is that of tribadism, where one woman lies on top of another and moves in such a way as to stimulate the clitoris of each.

A further myth is that each partner takes an exclusively "butch" (active male) or "femme" (active female) role. In sexual practice if such roles are assumed, they usually alternate. This stereotyped view has confused non-homosexuals and lesbians themselves, not only sexually but also in a social context. And, this confusion has, to some degree, resulted in the ideological rejection of such roles as advocated by radical lesbians. A final and overriding myth is, perhaps, that all lesbians are first and foremost sexual beings, whereas they are no more sexually active or sexually obsessed than are most people. By comparison with male homosexuals, sexual promiscuity is rare among lesbians, as is prostitution. At the same time, particularly with the recent emergence of "gayness," and of the more open expression of female sexuality, promiscuity is neither more nor less frequent among lesbians than among heterosexual women.

THEORIES ABOUT LESBIANISM

Until recently there have been few studies exclusively of lesbianism. Physical explanations such as hormone imbalance or congenital defects have been put forward, but findings are ambiguous and not generally accepted. In the past, lesbianism has also been defined in terms of neurosis or of immature development, but these are generalizations based on lesbians who have sought psychiatric help. One of the most sympathetic and recent theories is that of the psychiatrist, Dr Charlotte Wolff. She emphasizes the bisexual nature of lesbianism, seeing an essentially bisexual element in the female anatomy (the clitoris).

She explains lesbianism as a recognition, and rejection, from early childhood, of the "second sex" emotional attitudes and social position of women. Because of this she also says that lesbians are ideally suited to lead a move for the equality of women and the rejection of female/male stereotypes - which to some extent is the basis of Radical Lesbian ideology.

PROBLEMS

Society has so far tended to see lesbianism as abnormal. As a result, a woman who feels attracted to other women may fear the opinion of others, both within the family and society, and may even also have difficulty in admitting her sexual preference to herself. There has been some liberalizing of attitudes in society recently; also, with the development of the women's movement, some women have felt a commitment to lesbianism as part of a general rejection of any need for the male. However, other lesbians still feel the need to repress ordinary expression of their sexuality, and perhaps to play a token part in heterosexual relations. (In some cases such activity may be part of a genuine bisexuality.) Although lesbianism is now more organized, many lesbians remain isolated, unable to meet others.

ATTITUDES TOWARD LESBIANISM

Human societies have shown ambivalent and contradictory attitudes towards lesbians. On the one hand female homosexuality has been regarded as a deviation, while on the other it has been ignored or ridiculed. These last attitudes, which reflect the approach of male-orientated societies toward women, have meant that female homosexuals have significantly escaped the same degree of legal persecution suffered by males. There are also far fewer historical references to lesbianism. It was accepted in ancient Greece and Rome. And it is also known that the Mohave Indians of North America recognized and accepted a class of homosexual women. But references to lesbians are scanty even during the Middle Ages when persecution of homosexuals reached fanatical proportions. In most European countries homosexuality was a capital offense. But significantly more sentences were carried out on males. Likewise, during the 1930s, in New York, more than 700 males were convicted on homosexuality charges, but only one female.

MODERN ATTITUDES

During the 20th century lesbians have therefore had to fight two battles, one for basic recognition and one for acceptance. This began in the late 1920s with the publication in England of Radclyffe Hall's "The Well of Loneliness," which aroused public disgust, and has continued more recently through the women's movement and the rise of Radical Lesbianism. Today lesbianism is out in the open for the first time. Despite its relative legal freedom (in Europe it is a crime only in Austria and Spain), lack of understanding about lesbians and discrimination against them - socially and economically - are still widespread.

BISEXUALITY

Most of the confusion and disgust surrounding homosexuality exists because it is too frequently assumed that homosexuals deviate markedly from normality. There is very little evidence to support this. Instead it is believed by many, including Freud, Kinsey, and more recently, psychiatrist Dr Charlotte Wolff, that each person is inherently bisexual, and depending on conditioning, may choose a partner from the opposite sex (heterosexuality), the same sex (homosexuality), or both (bisexuality).

©DIAGRAM

FEMALE SEXUAL PROBLEMS

SEXUAL PROBLEMS

A wide range of problems can affect a woman's enjoyment of her sexuality. They may be physical or psychological in origin, but the result is the same - distress for both partners.

A very few women are truly frigid - that is, incapable of responding to any kind of sexual stimulation. But there are many who are temporarily unable to achieve orgasm. Some women find penile penetration or intercourse painful or even impossible; while others have a partner with some kind of sexual difficulty. The purely physical problems are usually easy to identify; psychological problems can also be treated but the origin of the problem must be traced first. Common underlying causes of psychological problems are feelings of fear, shame, or guilt about sexual response, or anxieties connected with pregnancy.

Important in the treatment of such problems are the revolutionary sexual therapy techniques developed by Masters and Johnson.

FRIGIDITY

Frigidity, or general sexual dysfunction, is a complex female complaint in which a woman derives little or no erotic pleasure from sexual stimulation. Treatment often takes the form of sensate focus therapy - a technique developed by Masters and Johnson. In sensate focus therapy the couple refrain from intercourse and orgasm for a period. During this time they learn to caress each other's bodies until the woman is sufficiently aroused to initiate intercourse.

PROBLEMS WITH ORGASM

For women, the most common sexual complaints are those to do with the difficulty of reaching or inability to achieve orgasm.

REASONS

There are many complex reasons why women may experience difficulty in achieving orgasm, not only with a partner but also by themselves in masturbation. Shame, about their sexuality, fear of actually examining their genitals, and therefore ignorance about their function, are among the main reasons for some women's avoidance of masturbation. Ignorance about the physiology of sex can equally prevent women from achieving orgasm with their partner. But in a female/male situation, other problems arise. Some of the most important are probably the refusal or inability of women to express clearly what they want from sex, and a deference to, or overprotectiveness towards, the male orgasm, at the loss of their own. This often leads to the practice by some women of faking orgasm within their relationship - a practice which ultimately can cause great strain on both people.

MASTERS-JOHNSON THERAPY

Masters-Johnson sex therapy divides the female problems with orgasm into two main categories:

a) primary orgasmic dysfunction - women who have never experienced orgasm; and

b) secondary orgasmic dysfunction - women who have previously experienced orgasm but whose ability to achieve it has since stopped.

Such women who have problems with orgasm are often very sexually responsive, enjoy intercourse, but are still unable to continue their response beyond the plateau phase. Therapy consists essentially of encouraging an inorgasmic woman to bring herself to orgasm by masturbation either by hand or with an electric vibrator - a phallic-shaped sexual aid which when held against the clitoral area can bring a woman to orgasm. Once a woman can achieve orgasm easily by herself, her partner is brought into the therapy. Normal intercourse takes place during which the woman makes no attempt to achieve orgasm. Instead her partner brings her to climax either manually or with the vibrator. In general after only a few sessions, the woman will be achieving orgasm during intercourse.

FEMINIST THERAPY

Feminists in the USA have started pre-orgasmic groups to tackle the same problem of female orgasmic dysfunction. They also concentrate on encouraging women to discover their own physical sexuality, but at the same time are examining some of the male-dependence roles mentioned above.

PAINFUL INTERCOURSE

Discomfort during intercourse can make lovemaking an unpleasant experience for some women. There are several possible explanations for the discomfort.

VIRGINITY

In a woman who has not had intercourse, the hymen may be intact and unstretched (a). The first few times intercourse takes place there may be a little pain or even slight bleeding as the hymen is stretched or torn to accommodate the penis (b). A very few women have tough hymens which need minor surgery.

VAGINISMUS

This is a comparatively rare disorder in which the muscles surrounding the vaginal entrance go into spasm when penetration is attempted. Treatment involves the woman learning, over a period of time, to insert first one, then two, fingers into her vagina without experiencing muscular spasm. Her partner takes part in this, and when the woman's confidence is established, intercourse can take place.

DISPAREUNIA

This is a general term for painful intercourse. The pain may be caused by several factors.

Vaginal infections and irritations can be exacerbated by the friction of the penis moving in the vagina. Insufficient vaginal lubrication can cause pain. During normal stimulation the vaginal walls secrete a lubricating fluid which facilitates entry by the penis. Pain results if the man attempts entry before the woman is sufficiently aroused or if he is wearing a condom. In the first case, entry must be delayed until the woman is fully aroused, and in the second, a lubricating jelly, or even saliva, will help solve the problem.

A common cause of pain in the pelvis is the penis hitting the cervix during particularly deep thrusts (c). Pelvic pain can also be caused by infections of the uterus, cervix, or Fallopian tubes, cysts or tumors on the ovaries, or tears in the ligaments supporting the uterus (following childbirth).

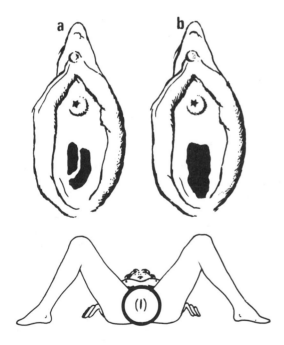

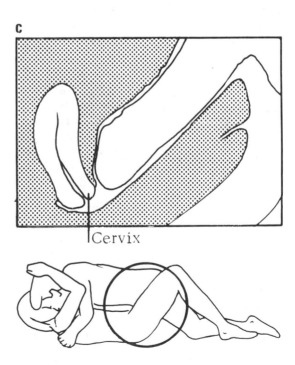

Cervix

©DIAGRAM

MALE SEXUAL PROBLEMS

DISORDERS

PAIN DURING INTERCOURSE

There are several possible causes:
a) inflammation of the foreskin;
b) chronic prostate infection;
c) scars in the urethra, due usually to untreated gonorrhea;
d) over-tight foreskin, that makes erection painful; or
e) allergy of the penis to vaginal fluids or contraceptive chemicals.
The first, second, and third can be treated by dealing with the infection; the fourth (and again the first), by circumcision; the fifth by use of a condom.

PSYCHOLOGICAL IMPOTENCE

An occasional incident of impotence is quite normal. It should only be thought of as a problem if it is a regular occurrence. The general nature of impotence is dealt with on pp. 94-5 together with possible physical causes causes. But 9 cases in 10 are psychological - as is often shown by an ability to have erection (and orgasm) from self-masturbation or during sleep, but not in a sexual relationship. Such impotence typically involves men who have experienced intercourse to orgasm as the norm at a past stage in their lives, and who may eventually return to this. However, in a few cases of deep psychological disturbance, a man may never have had an erection in any circumstances.

The psychological causes of impotence can arise from various levels of motivations. Conscious or nearly conscious causes can be:
a) fear of the consequences of intercourse - pregnancy or venereal disease;
b) feelings of resentment, disgust, or dislike towards the partner; and especially
c) fear of sexual failure - the feeling that one is on trial. For example, fear that one will fail to "perform impressively", or fail to please the partner; especially, fear of premature ejaculation, and of impotence itself. Such fear may be set off by a single incident or particular situation, or by the impact of one's own or one's partner's sexual difficulties.

More deep-rooted causes are usually due to early experience, including especially the relationship with one's parents, or to traumatic experiences when first attempting intercourse. They can result in:
a) distaste for sexual activity;
b) feelings of resentment towards women in general;
c) inability to reconcile sexuality with an idealistic image of women;
d) fear of unacceptable incest fantasies, through failure to progress beyond a childhood "Oedipus complex" (a desire to kill one's father and have intercourse with one's mother);
e) fear of the vagina as a castration instrument (i.e. belief that it could chop off one's penis); and
f) general neurotic personality disorders.

But by far the most common cause is simply the fear of failure. Premature ejaculation, for example, can set up a self-consciousness that eventually ends in impotence. Thereafter, a vicious circle of impotence and fear of impotence may be established.

PREMATURE EJACULATION

This is when the man reaches ejaculation too quickly, for the woman to be sexually satisfied. Occasional premature ejaculation is normal - it can happen simply because of prolonged lack of sexual outlet. Besides, few women are likely to reach orgasm at every intercourse. Consistent premature ejaculation is a serious problem that can lead to partner dissatisfaction, self-consciousness, and impotence.

Sometimes early experiences can set up conditioning in over-rapid ejaculation; for example, intercourse with prostitutes, need for speed in semi-private places such as automobiles, and lack of concern for the partner during one's sexual learning. Even in maturity, possible causes of premature ejaculation can include real lack of consideration for the woman. However, in our culture, and especially in stable relationships, a man is likely to feel that the ability to satisfy his partner is as much a symbol of his sexuality as his ability to reach orgasm.

As a result, the major cause of premature ejaculation is simply anxiety that premature ejaculation will occur. The situation is the same as with impotence. Once there is anxiety about one's sexual performance, a vicious circle is set up; for anxiety immediately inhibits a true sexual response. In addition, with premature ejaculation, the situation is likely to be complicated by bitterness between the sexual partners. If the man is impotent, neither partner can find sexual release. But if he ejaculates prematurely, he seems to have found it at the expense of his partner. So lack of consideration may be the accusation, even when it is not the cause.

EJACULATORY INCOMPETENCE

This is more rare. The man has no difficulty with erection, but cannot reach orgasm inside the vagina. As with other difficulties, it can lead to anxiety, self-consciousness, and finally impotence. The cause usually lies in the past, often in particular traumatic incidents, and often against a background of a sexually restricted upbringing. The result is a psychological attitude that sees the woman as repulsive or contaminating or threatening.

TREATMENT

Difficulties that have psychological causes are nevertheless amenable to physical therapy - if this is built on a basis of communication and love. The techniques described here are those developed by Masters and Johnson, the researchers into human sexual biology, and described in their book "Human Sexual Inadequacy". No precis is really sufficient, though, and they are not responsible for the descriptions here.

The techniques for treating these disorders may seem emotionally degrading - to those who do not have the far greater burden of the disorders themselves.

BASIC PRINCIPLES

Masters' and Johnson's basic principles are that sex is a form of communication, and that treatment can only occur in the context of a sexual partnership (typically, the married couple). Successful therapy depends on the goodwill of both partners, their ability to learn to relax, and their realization that sex is not primarily a matter of successful performance.

SENSATE FOCUS EXERCISES

These are simple but vital. The couple lie naked, stroking and feeling each other's bodies - but not the breasts or genitals. No intercourse is allowed. The couple learn to relax and experience sensual pleasure free from any demand (whether of end-point release, self-explanation, reassurance, or immediate return of pleasure received). At first, clumsiness, self-consciousness, and embarrassed humor are likely; but usually genuine enjoyment soon begins. This undermines the crippling tendency to sexual self-evaluation. Later, sexual expression is allowed. Breasts and genitals may be touched, and the couple guide each other and explain what is most pleasurable.

During these exercises a lotion is used to prevent roughness and also help some get more used to genital fluids. Rejection of the lotion (eg, as immature) was found to be a very good guide to who would fail to benefit from therapy. Later, specific treatment for the disorders begins. Where impotence exists as a result of another underlying disorder, the impotence is treated first.

PSYCHOLOGICAL IMPOTENCE

Here the principles of communication and demand-free activity are especially important. After sensate focus, the couple progress to more specific manual manipulation, with the man guiding the woman's hand. There is still no attempt at intercourse; if erection occurs, it is allowed to go again, then to re-establish itself. Once erection is easily obtainable, the woman uses a position astride the man. First she only tries to keep the penis within her vagina; later she begins gentle, non-demanding thrusting. Finally the man joins in slow thrusting, but with no goal of ejaculating or satisfying his partner. Orgasm is just accepted if it happens.

PREMATURE EJACULATION

Common techniques for this include:
a) distracting the mind with mental tasks or physical pain;
b) avoiding any touching of the male genitals before intercourse begins, while manually stimulating the woman almost to orgasm; and
c) use of anesthetic creams and jellies, tranquilizing drugs, and even excessive doses of alcohol. None of these necessarily works, and they are anyway hardly symbols of a happy sexuality. Masters and Johnson found this problem fairly easy to deal with. In their therapy, the couple first use a sitting position - the woman

leaning back against the bedhead, the man, between her legs, leaning with his back against her. The woman masturbates the man till he is too near to orgasm to stop it for himself (2 to 4 seconds before ejaculation). At his warning, she presses the penis tip firmly between her thumb and first two fingers for 3 or 4 seconds, and the man loses the urge to ejaculate. This is repeated 4 or 5 times a session. After thorough practice, the couple use a position of intercourse with the woman above. They remain motionless, with the man's penis inside the woman's vagina, and the woman intervenes with squeeze control if necessary. Also the husband thrusts if necessary to maintain erection. Later a lateral position is used, so the woman can still intervene if the man's orgasm begins too soon. Squeeze control is used regularly, for 6 months to a year, before at least one intercourse a week. After that it is used as needed.

EJACULATORY INCOMPETENCE

The Masters and Johnson therapy begins with masturbation of the man's penis by the woman, discovering what the man finds stimulating and bringing him to orgasm. There should be no attempt to hurry this. Once the man can identify the woman with sexual pleasure, the couple use one of the positions with the woman above.

The man is again brought to imminent orgasm manually, and then the woman thrusts to take the penis into her vagina. The technique is repeated until the man gets used to ejaculating into the vagina. Later he also gets used to inserting his penis into the vagina when his sexual excitement is low. Moisturizing lotion prevents irritation during the stages of manual stimulation.

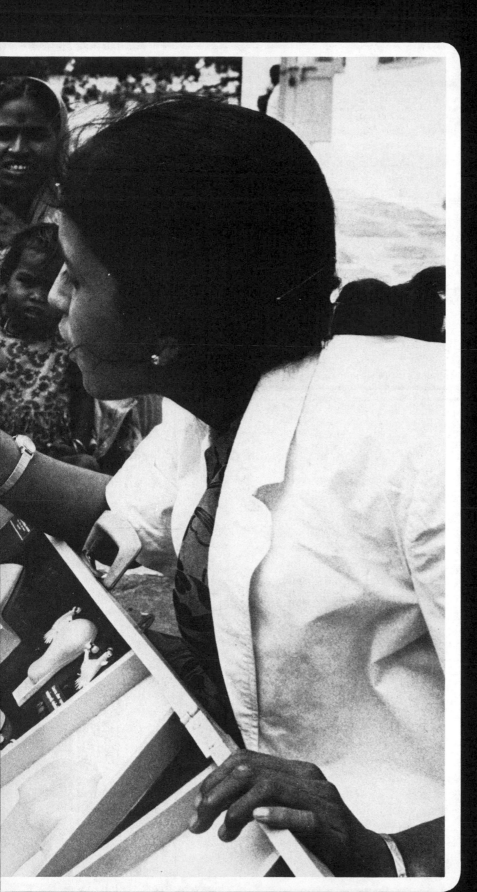

7 Contraception, infertility and abortion

As in the West, methods of birth control are spreading all over the Third World in an attempt to control the population explosion.

Left: A doctor explaining the use of the contraceptive loop to her family-planning clinic in Aurangabad, India (International Labor Office)

CONCEPTION AND CONTRACEPTION

CONCEPTION AND CONTRACEPTION

For pregnancy to occur, several conditions must be fulfilled:
semen from the man must enter the woman's vagina;
the semen must contain healthy male sperm;
the sperm must find conditions in the vagina in which they can live;
the living sperm must make their way into the woman's uterus and (possibly) the Fallopian tubes;
they must find an egg there ready for fertilization;
and the egg, once fertilized, must be able to implant itself in the uterus.

By preventing any one of these, contraception is achieved. But it is important to note three things.

First, that sperm may reach the vagina even if the penis does not enter it. Sperm ejaculated onto the vulva or surrounding skin can still swim into the vagina.
Second, that, although conditions in the vagina are hostile to sperm, they can live there for 6 hours or more. So any barrier to prevent sperm moving up into the uterus must last at least this long after intercourse.
Third, that once sperm have reached the uterus they can live 4 to 5 days or more. So, to avoid conception, there must be at least this time gap between the arrival of sperm in the uterus and the arrival of the egg.

A normally fertile woman experiencing regular intercourse with a normally fertile man stands about a 60% chance of becoming pregnant in any one month. Therefore, for the woman who intends to have heterosexual intercourse, but not to have babies, some form of safe, effective contraception is essential.

THE MAIN TYPES OF CONTRACEPTIVE

There is a wide variety of contraceptive techniques in use today - none of which is ideal. Many concentrate on keeping sperm out of the uterus. Caps and condoms aim to provide a physical barrier; spermicides a chemical barrier. Withdrawal modifies the sex act, to try to keep sperm out of the female tract completely. Other techniques - generally more effective - concentrate on interfering with the ovum. The oral contraceptive pills usually affect the ovum's development and release. Intrauterine devices (IUDs) are thought to prevent it implanting in the uterus.
Finally, there are two other types of technique. "Rhythm" methods simply aim (not necessarily successfully) to avoid intercourse at those times of the month when sperms might find an ovum ready for fertilization.
Sterilization methods are surgical operations to make one partner incapable of having children.
Techniques can be combined to give more effective contraception.

CONTRACEPTIVE TECHNIQUES
The first part of this chapter will consider in detail the forms of contraception introduced here and located on the diagram opposite.
CONTRACEPTIVE PILL Oral contraceptives consist of small pills, one of which a woman takes every day for most or all of each month. There are various kinds and they have the effect of preventing ovulation or of creating a barrier to sperm.
WITHDRAWAL This is a simple but not particularly effective form of birth control. The man withdraws his penis from the woman's vagina just before his orgasm, and so prevents the semen from getting into the vagina.
RHYTHM METHODS There are 2 rhythm methods of birth control - calendar and temperature. In both the woman abstains from intercourse from between 10 and 14 days of each month so as to avoid intercourse on the days when she is most likely to conceive.

1 CONDOM is a rubber sheath placed on a man's erect penis. Ejaculated semen is trapped in it and so prevented from entering the vagina.
2 DIAPHRAGM is a rubber cap inserted by the woman into her vagina. It covers the entrance to the cervix, acting as a barrier to sperm.
3 SPERMICIDES are chemicals. Placed inside a woman's vagina, they both kill and act as a barrier to sperm.
4 INTRAUTERINE DEVICE (IUD) is a small device inserted into the woman's womb on a long-term basis. While in place, its effect on the uterus prevents implantation.

5a FEMALE STERILIZATION This consists of an operation on the Fallopian tubes, to prevent eggs passing from the ovaries to the womb.
5b MALE STERILIZATION (or "vasectomy") is a minor operation on the man's vas deferens which prevents sperm being ejaculated in the semen.

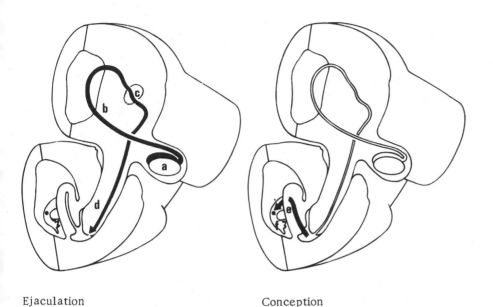

Ejaculation   Conception

The diagrams show the genital coupling and route of the semen. Ejaculation occurs in 2 stages. Rhythmic muscular contractions begin in testes and epididymides (a) and continue along the vas deferens (b) also involving the seminal vesicles and prostate (c). Sperm and seminal fluid collect in the urethra inside the prostate. Then a sphincter relaxes, letting this semen pass down toward the penis. Contractions along the length of the urethra (d) cause ejaculation. Inside the woman's body, semen passes from vagina into uterus (e) and Fallopian tubes (f).

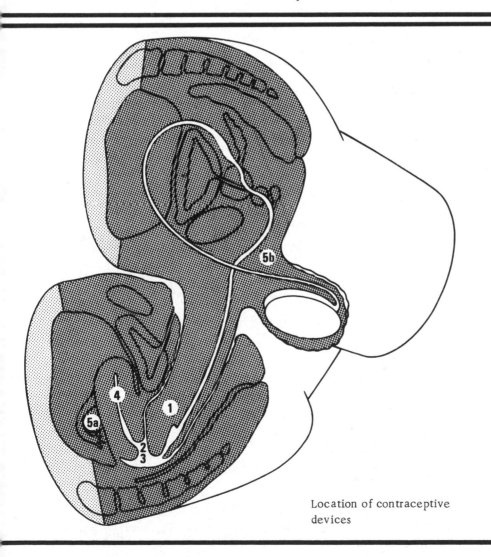

Location of contraceptive devices

Most women change the type of contraceptive they use at least once or twice during their life.
BEFORE CHILDREN When a woman starts heterosexual activity she will probably rely on the man to practice withdrawal, or to use a condom. Once regular relations are established, most young women prefer to use more effective and continuous methods like the Pill or one of the new IUDs.
FAMILY PLANNING Many women take the Pill between having children. Others use an IUD, or (less effectively) diaphragm or condoms with spermicides. (An IUD can be removed by a doctor when conception is desired.) Some rely on the rhythm method, despite its failure rate.
AFTER HAVING CHILDREN Once a woman's family is complete, she may either return to a contraceptive that she has tried and liked, or she may at this point decide to be sterilized.

©DIAGRAM

THE PILL

CONTRACEPTIVE PILL

No other form of contraception has been as revolutionary as the "Pill." It is easy to use, reversible, and nearly 100% effective.

The Pill uses synthetic forms of the hormones estrogen and progesterone. These are produced naturally in the body for a few days in each menstrual cycle - and continuously during pregnancy. In each case, they have the effect of inhibiting output of FSH and LH hormones. FSH and LH are needed if follicles are to ripen for ovulation and this is why no ovulation occurs during pregnancy. The contraceptive pill has a similar effect; and as no ovulation occurs, no ovum is available for fertilization by sperms.

There are three main types of Pill:

a) the combination pill - so called because each active pill in the package contains both hormones;

b) the sequential pill, in which $\frac{3}{4}$ of the active pills contain just estrogen, and only the rest contain both hormones; and

c) the continuous pill, in which all pills contain progesterone alone, and which works rather differently.

The sequential pill has recently been withdrawn from use in some countries.

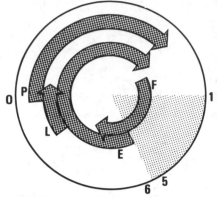

NORMAL MENSTRUAL CYCLE

- ▦ Menstruation
- ➡ Hormone action
- **F** FSH
- **E** Estrogen
- **L** LH
- **P** Progesterone
- **O** Ovulation-about day 15

TYPES OF PILL

COMBINATION PILL This is the most widely-used and effective type. The woman takes one standard pill each day for 21 days, starting on the 5th day after menstruation begins, and ending on the 25th day. There is a gap of 7 days during which no hormone is taken, and menstruation occurs; then a new package is started.

As well as preventing ovulation, the combination pill:

a) affects the uterus lining, so implantation could not occur;

b) causes the cervical mucus to thicken, forming a chemical barrier to sperm.

SEQUENTIAL PILL This is closer to a woman's natural cycle, but less effective. Again, 21 pills are taken, starting on the 5th day after menstruation. But the first 14 pills contain estrogen alone; only the remainder contain both hormones. Ovulation is prevented, but the uterus lining and cervical mucus are unaffected.

CONTINUOUS PILL There are 28 pills in each pack, all active and all containing synthetic progesterone only. One is taken every day, even during menstruation. They work mainly by their effect on the uterus lining and cervical mucus, rather than on ovulation.

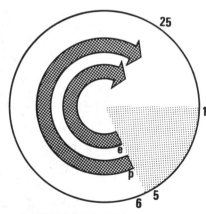

COMBINED PILL CYCLE

- ▦ Menstruation
- ➡ Hormone action
- **5** First pill
- **25** Last pill
- **e** Estrogen
- **p** Progesterone

SIDE EFFECTS

Most women experience some side effects on the pill. There may be headaches, nausea, swollen or tender breasts, heavier periods, and vaginal discharge. But not all women experience these, and most symptoms disappear within the first few months. If they do not, a change of brand may remove any unpleasant side effects.

No woman should take the Pill without consulting a doctor. All pills, and especially high-estrogen ones, carry a risk of blood clotting. The resulting thrombosis may be fatal. This is more likely in women over 35, and pregnancy itself carries higher risks. Other disorders a doctor must consider before prescribing the Pill include hepatitis, diabetes, migraine, and epilepsy. There is still no proof that the Pill causes cancer. But estrogen can aggravate some types of existing

TAKING THE PILL

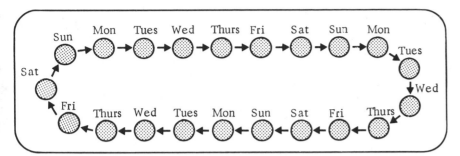

Taking the pill is quite easy, the problem is to remember to do so. Most pills come in packets of 21 which are designed to aid memory. To start oral contraception the first day of a period counts as day 1. Pill-taking begins on day 5, whether bleeding has stopped or not. It continues until day 25 when the last pill is taken. A gap of 7 pill-free days follows before the next course, during which menstruation occurs. For women who have difficulty in remembering this sequence, combined and sequential pills are available in packs of 28. But the extra pills are dummies. The first packet of pills may not give complete protection and for the first 2 weeks a second contraceptive should be used. If a combined pill is forgotten, it should be taken within 12 hours of the usual time even if it means taking 2 in 1 day. If more than 2 are missed and the gap between pills is more than 36 hours, then the pack should be finished but a second contraceptive also used. It is important not to forget to take the continuous pill, and it must be taken at the same time every day.

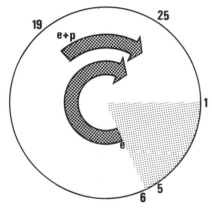

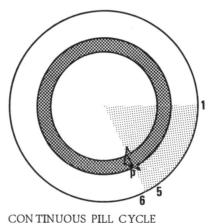

SEQUENTIAL PILL CYCLE
- **5** First estrogen-only pill
- **19** First estrogen/progesterone pill
- **25** Last pill

CONTINUOUS PILL CYCLE
- **p** Progesterone only

cancer. Cervical smears are an important part of the medical examination that accompany the Pill, and a cancerous condition would be found in good time. Doctors still disagree on how long a woman should stay on the Pill. On average, women tend to use it for 3 to 4 years. To regain fertility, a woman only needs to stop taking the Pill. But it may be some months before her ovaries are functioning normally and conception can occur.

DEPO-PROVERA

This is based on similar hormonal principles to the oral pill, but is given by injection: 150mg is injected every 3 months, or larger doses every 6 months. It is given to an estimated 1 million women in about 70 countries - mostly developing ones. But in the USA and UK, use of the drug is only very rarely allowed. In its favor is its effectiveness, when other methods fail through lack of motivation or care. Against it are its possible side effects and links with disease. Symptoms in some women may include:
a) disruption of menstrual bleeding, which may be prolonged, heavy, unpredictable, or absent altogether;
b) vomiting, dizziness, moodiness, headaches, and weight-gain; and
c) rectal bleeding.
Links with disease include:
a) an established link with subsequent infertility in some women;
b) established links with blood-clotting disorders; and
c) possible links with breast and cervical cancer.

©DIAGRAM

IUDS AND CONDOMS

INTRAUTERINE DEVICE (IUD)

The IUD, also known as the "coil" or "loop," is a small plastic device for insertion into the uterus. It may be left there for several years, and while in place works as a contraceptive almost as effective as the Pill without requiring much attention. Intercourse can occur without restriction. Once the IUD is removed, fertility returns in 1 to 12 months.

Comparable practices date back to biblical times, when camel drivers inserted pebbles into the uteri of female camels, to keep them from becoming pregnant.

Yet how an IUD works is uncertain. Theories include:
a) that an IUD makes the ovum pass down the Fallopian tube too rapidly for either fertilization or implantation;
b) that it interferes with the lining of the uterus, so that implantation cannot take place; and
c) that it interferes directly with the implantation process.

The IUD is very effective while in place - though some doctors advise use of a spermicide as well around ovulation. But it may fall out, especially during the first few months or at menstruation. Early IUDs were too large for women who had not had children. Recently smaller designs, usable by all women, have appeared. However, the failure rate may be a little higher.

Many women experience side effects with IUDs - usually heavy periods and/or pain. For this reason, about 25% of women fitted have their IUD removed. IUDs may also aggravate infection, or (rarely) cause it. There is also a slight risk of perforation of the uterine wall, but significant damage is rare.

If pregnancy does occur, the IUD can be removed, as it increases the risk of miscarriage. Otherwise, though, most types have no effect on a fetus.

Many doctors advise renewal of the IUD every 2-3 years.

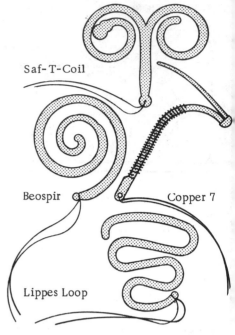

Saf-T-Coil

Beospir   Copper 7

Lippes Loop

Lippes Loop and Saf-T-Coil are the most commonly used IUDs for women who have had children. For those who have not, the Copper 7 (and the similar Copper T) are the most common. These both have copper wound round the stem, and this aids their contraceptive effect.

CONDOM

The condom ("sheath," "rubber," "French letter") is still probably the most widely used contraceptive. It has a long history, dating back some hundreds of years, and was popularized as a protection against venereal disease. When used carefully, preferably with a spermicide, it is an effective means of birth control.

The condom consists of a thin rubber sheath, about 7in (18cm) long, open at one end and closed at the other. It fits tightly over the man's erect penis. When he ejaculates, his semen is trapped in the sealed end. This prevents sperms from entering the vagina.

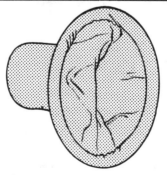

USING A CONDOM
The condom is taken out of the packet rolled up, and is unrolled onto the erect penis just before intercourse. (The actions of this can be incorporated into lovemaking.)
At least 1in (2.5cm) at the tip

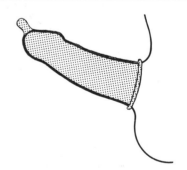

should be left empty of air to help prevent bursting or leakage. After ejaculation, the man withdraws his penis before his erection subsides. While withdrawing, he should hold the condom so that it does not come off.

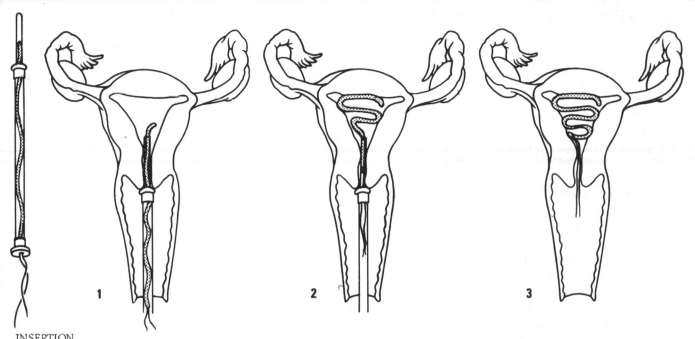

INSERTION

1 An IUD must only be fitted by a trained person. Some require anesthetic, but most are packed in a thin plastic inserter, which may be passed without difficulty through the cervical canal into the uterus. This is easiest during or just after menstruation.

2 The IUD is then pushed through the inserter and takes up its normal shape inside the uterus. The insertion takes only a few minutes. Some women experience discomfort similar to a heavy period pain. This can last for 24 hours with some slight bleeding.

3 The IUD has nylon threads (or a stem projection) left hanging through the cervix into the vagina. As a result, a woman can - and should - make regular checks to ensure her IUD is still in place.

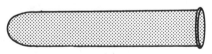

Plain ended

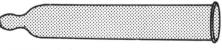

Teat ended

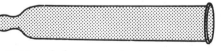

Teat ended

TYPES

There are various types of condoms - plain ended or teat ended - and they can be of different colors. Some people complain that condoms reduce sensitivity: lubricated brands claim to be an improvement.

For easier insertion, it is better for a woman to use spermicidal cream or jelly, which gives the advantage of extra contraceptive effectiveness. Condoms have kept their popularity largely because they do not need medical supervision and can be obtained and carried around easily. They are sold in sealed packets and have a maximum shelf life of 2 years, away from heat. Their chief disadvantage is that due to the annoyance of interrupting lovemaking, some couples may decide to "take the risk" of intercourse without contraception.

©DIAGRAM

THE RHYTHM METHOD

RHYTHM METHOD

With this method, a couple do not have intercourse during the part of the woman's menstrual cycle during which she can conceive i.e. when a fertilizable egg is available.

The menstrual cycle lasts (in principle) 28 days. During this, the egg is available for fertilization for only about 1 day - the 24 hours that follow ovulation. However, there is no direct sign of ovulation, only of menstruation. Ovulation typically occurs halfway between the menstruations - on about the 15th day. So a woman can count forward 14 days from the start of her last menstruation, to guess when ovulation will occur. But the menstrual cycle is seldom perfectly regular. In most women, menstruation is erratic when periods return after the birth of a baby, and in a quarter of women it is always fairly erratic. (Other women may have a record of regular menstruation for years, followed by sudden unexpected irregularity.) Also, even where menstruation is regular, ovulation need not occur at the mid-point, the 15th day. It can occur anywhere from 16 to 12 days before the start of the next menstruation. In fact, ovulation is sometimes induced by the stimulus of sexual intercourse. Finally, sperms can live in the woman's cervix for up to 72 hours and sometimes longer - so even if intercourse is four days before ovulation it may on rare occasions cause conception.

Therefore the "calendar rhythm method" - just based on dates - is not very effective, even if several days are kept free from intercourse around the likely date of ovulation. The "temperature rhythm method" is better. A woman normally has a sudden rise of about 1°F (0.6°C) in body temperature during the day of ovulation, due to increased progesterone production. Use of a thermometer and a record chart should show this rise.

By the time the temperature rise is actually recorded, the possible fertilization period is usually over. (Also the action of the progesterone makes the cervical mucus unfavorable to sperm penetration.) This gives a "safe period" after ovulation, from the temperature rise up to (and including, if desired) the next menstruation. But it gives no safe period after menstruation, since there is no way of telling when the next ovulation will occur. Between menstruation and ovulation, only the calendar method gives any indication of safety.

TEMPERATURE RHYTHM METHOD

The temperature should be taken first thing on waking, before any activity (even getting out of bed). A rectal thermometer is preferable to an oral one.

The circle shows the pattern of temperature during the menstrual cycle. Low temperatures are at the outside, high at the center. There is an abrupt rise at ovulation (about day 14.)

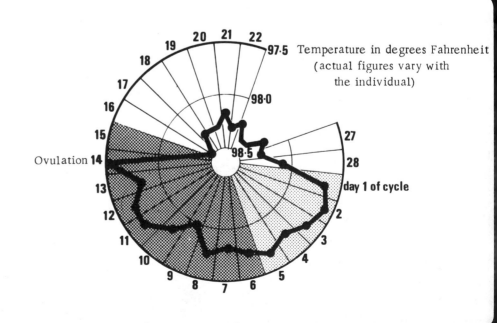

Temperature in degrees Fahrenheit (actual figures vary with the individual)

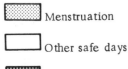

Menstruation

Other safe days

Unsafe days

266

CALENDAR RHYTHM METHOD

REGULAR MENSTRUATION

Suppose a woman had menstruation regularly every 28 days. Ovulation would be most likely on the 15th day of the cycle, but could happen anytime from the 13th to the 17th - a period of 5 days. Since sperm can live 72 hours and even longer, 4 days before this are also unsafe. And since the egg may still be fertilizable 24 hours after ovulation, the day after the 5-day period is unsafe too. This gives a total of 10 unsafe days, from the 9th to the 18th days of the cycle inclusive. Some women have cycles as short as 21 days, others as long as 38 days - but this does not matter if the cycles are still regular. The woman still abstains for a period of 10 days, starting 20 days before the next menstruation is expected.

IRREGULAR MENSTRUATION

A woman with irregular menstruation should keep an accurate record of her menstrual cycle for a year beforehand, and note the shortest and longest cycles. She must then calculate as follows:

a) she subtracts 19 from the number of days in her shortest cycle; and

b) she subtracts 10 from the number of days in her longest cycle.

Figure (a) gives the earliest day on which pregnancy can occur, counting from the start of the last menstruation; figure (b) gives the latest.

For example, if the shortest cycle is 25 days, and the longest 29:
a) 25-19=6; and b) 29-10=19.
So the unsafe days are from the 6th to the 19th days inclusive, after the start of the last menstruation. Perhaps 15% of women have menstrual cycles so irregular that the calendar rhythm method cannot be applied.

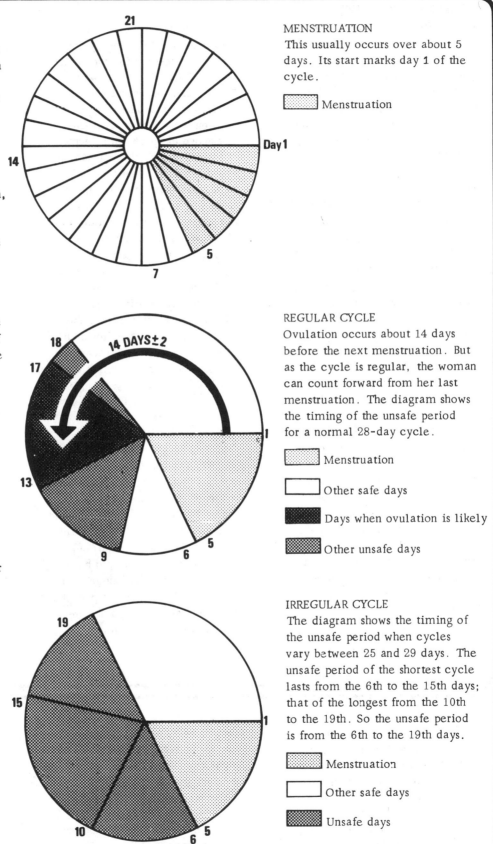

MENSTRUATION

This usually occurs over about 5 days. Its start marks day 1 of the cycle.

░ Menstruation

REGULAR CYCLE

Ovulation occurs about 14 days before the next menstruation. But as the cycle is regular, the woman can count forward from her last menstruation. The diagram shows the timing of the unsafe period for a normal 28-day cycle.

░ Menstruation

☐ Other safe days

■ Days when ovulation is likely

▓ Other unsafe days

IRREGULAR CYCLE

The diagram shows the timing of the unsafe period when cycles vary between 25 and 29 days. The unsafe period of the shortest cycle lasts from the 6th to the 15th days; that of the longest from the 10th to the 19th. So the unsafe period is from the 6th to the 19th days.

░ Menstruation

☐ Other safe days

▓ Unsafe days

©DIAGRAM

STERILIZATION AND SPERMICIDES

FEMALE STERILIZATION

Sterilization is the most effective form of birth control. But it is also the most final - a last solution. As yet, no reversible method has been perfected, which means that a person considering the operation must be absolutely certain of her or his decision before undergoing sterilization.

FEMALE STERILIZATION

For women who are quite certain that they do not want any more children, sterilization is becoming more popular. The operation consists basically of cutting, tying, or removing all or part of the Fallopian tubes. As a result, eggs can no longer pass from the ovaries to the uterus and the sperm is unable to reach the eggs. Provided that the operation is done correctly - there have been rare instances of the Fallopian tubes rejoining - sterilization is 100% effective.

Afterward there are no obvious changes. Sexual interest should remain unchanged and the menstrual cycle continues as normal. Some women, in fact, gain increased enjoyment from sex once the fear of pregnancy has been so completely removed. There are a number of different ways in which a woman can be sterilized. All require hospitalization but the time needed for recovery varies.

TUBAL LIGATION

This is the most commonly used method of sterilization for women. It can be done in various ways. Traditionally a general anesthetic is given and a 2-3in (5-7.5cm) incision made in the abdomen, just above the pubic hair. A piece is cut out of the Fallopian tubes, and the ends are then tied and folded back into the surrounding tissue. This type of operation is often performed immediately after childbirth. After some days in hospital, it is then necessary for a woman to rest for some weeks (which for a working woman with children may be difficult).

The same operation can be done by making a much smaller incision in the upper vagina. No scar is visible and the recovery period is much shorter. But it is a much more skilled and difficult operation, and therefore rare.

ENDOSCOPIC TECHNIQUE

This is a fairly recent development in female sterilization. It involves the use of an instrument known as a laparoscope, which consists of a fine tube which conducts light and is connected to a telescope. This is inserted through a small cut in the abdomen - it can also be inserted through the vagina - and is used to light up and inspect the Fallopian tubes. Very fine forceps are inserted, either through the

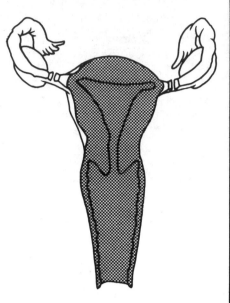

Fallopian tubes cut and tied

same cut or through a second smaller one, and an electro-current is passed along them which cauterizes the Fallopian tubes. After, there are only 2 tiny scars and the recovery time is very short.

HYSTERECTOMY

Until quite recently hysterectomy, which involves the complete removal of the uterus.
was a fairly widespread means of sterilization. However, it is not now generally recommended for birth control purposes, though it is an operation that many women unfortunately have to undergo for other reasons.

WITHDRAWAL

Withdrawal (or coitus interruptus) is the oldest and simplest method of birth control. The man takes his penis out of the woman's vagina just before his orgasm. His semen is ejaculated outside her body. Used with great care, withdrawal may be effective, but only if every drop of semen is not only kept out of the vagina but also right away from the vaginal lips. It is impossible to be sure of this, because:
a) some fluid containing live sperm may "weep" from the penis before orgasm; and
b) in the pleasure of orgasm, the man may not withdraw properly. Continued use of withdrawal can also be frustrating. The woman in particular may not be able to relax through fear that the man may not withdraw.

MALE STERILIZATION

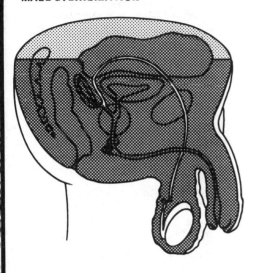

Vas deferens cut and tied

VASECTOMY

A vasectomy is a safe, simple, surgical operation in which each vas deferens - the duct leading from each testis to the penis - is cut and tied off. As a result the semen a man ejects no longer contains sperm.

Apart from instances where the cut tubes have rejoined, the operation is always completely effective. It does not alter a man's ability to have an orgasm or to ejaculate. But the operation is rarely reversible, which again means that a man must be absolutely sure before undergoing a vasectomy. The operation is quite short and generally lasts under half an hour. For the majority of vasectomy operations, a local anesthetic only is needed. Either

1 or 2 very small cuts are made on or near the scrotum. A piece about 1cm long is removed from each duct, the cut ends then being folded back and tied. Once the operation is over the man can generally return straight home and can be back at work within 2 or 3 days. The most common aftereffects are likely to be some soreness and bruising. A vasectomy is not immediately effective as there are usually some sperms stored in the seminal vesicles, above the cut. For this reason, a second method of contraception must be used, until two successive follow-up tests of the semen show negative sperm counts (perhaps 2 or 3 months after the operation).

SPERMICIDES

These are chemical products which are inserted into the woman's vagina before sexual intercourse. They act in 2 ways: by killing the sperm, and by creating a barrier of foam or fluid through which sperm cannot pass into the uterus.

Spermicides come in various forms: creams, jellies, aerosol foams, foaming tablets, suppositories, and C-film, a a fairly recent spermicide-impregnated plastic. But used by themselves, none of these is at all reliable as a contraceptive. If used, they should be combined with another method, such as cap or condom.

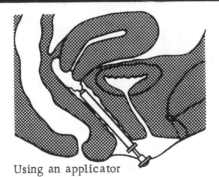

Using an applicator

Creams, jellies, and aerosols are sold with a special applicator. Using this, the woman squirts the chemical high up into her vagina. This should be done as near to intercourse as possible, and certainly no more than one hour before, as effectiveness is only temporary.

Suppositories and tablets come in solid form and must be inserted by hand deep into the vagina. Suppositories are cone-shaped and melt at body temperature; tablets dissolve and foam in the vagina's moisture. Both should be inserted 15 minutes before intercourse. The new C-film consists of a small square of soluble plastic which can either be inserted into the vagina or placed on the tip of the man's penis before it enters the vagina. It dissolves, releasing spermicide; but is no more reliable than other spermicides (and less so than some).

Some women find that spermicides irritate their genitals.

© DIAGRAM

269

THE CAP (DIAPHRAGM)

CONTRACEPTIVE CAPS

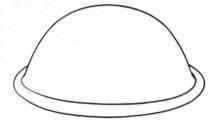

Diaphragm

Cervical cap

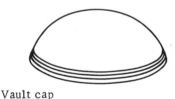

Vault cap

DIAPHRAGM (DUTCH CAP)

The diaphragm is the best known example of caps that fit across a woman's cervix to act as a barrier to sperm. The diaphragm is a dome-like rubber device. Its rim contains a coiled spring. By itself the cap is not particularly safe but used carefully in conjunction with spermicides it is quite effective. For most women before the Pill was introduced, the diaphragm was the safest method of contraception available to them.

Putting a diaphragm in place is not very difficult, though at first it needs practice. The woman holds its edges together and pushes it by hand into the vagina so that the bottom edge rests against the rear of the vagina and the top edge rests against the vaginal wall behind the bladder. The spring causes the diaphragm to regain its circular shape so that it is held in place. Before insertion 2-4in (5-10cm) of spermicidal cream or jelly should be squeezed onto the inside (closest to the cervix) or both sides of the cap. The cap should be put into the vagina not more than 2-3 hours before intercourse. After intercourse it should be left in place for at least 6-8 hours while the sperm die. If intercourse occurs again in that time, more spermicide should first be introduced into the vagina without disturbing the cap.

Diaphragms vary in size. An initial fitting by a doctor or nurse is essential, and the cap should be checked for fit every 6 months, after a pregnancy, or if more than 10lb (4.5kg) is gained or lost in weight. At home, the cap must be washed after use, according to instructions, and checked carefully for holes.

CERVICAL CAP

This is much smaller, and fits onto the cervix. It is no longer used except in special cases.

VAULT CAP

A vault cap is much more rigid than other caps. It fits across the top end of the vagina and is held in place by suction.

UNSATISFACTORY METHODS

MORNING-AFTER PILL

This should only be seen as an emergency measure. It consists of giving a woman large doses of estrogen about 3 days after unprotected intercourse. The estrogen affects the uterus lining and implantation is prevented. The side effects are undesirable and may be harmful.

LACTATION

Breast-feeding mothers were once thought to be unable to conceive in the first 6 weeks after childbirth, if their periods had not returned. But in fact, though the likelihood of conception is reduced, contraception is still necessary.

DOUCHE

Douching - washing out the vagina after intercourse - is a completely ineffective method of birth control. Not only can sperm reach the cervix within 90 seconds, but the effect of squirting liquid into the vagina could be to help the sperm on their way.

NO ORGASM

It was once believed that if a woman did not have an orgasm she would not conceive. This is obviously untrue; many women do not have orgasms but still become pregnant.

AMERICAN (GRECIAN) TIPS

Claimed to increase sensitivity, these are short rubber condoms that fit over the tip of the penis only. Not only do they fail to increase sensitivity, they are also likely to come off in the vagina.

GAMIC APPLIANCE

This consists of a small rubber bag attached to the end of a thin rubber tube. The tube is pushed down the urethra of a man's penis, so that when he ejaculates the semen is held in the bag. Very rarely used, this method is likely to cause damage to the urethra or infection.

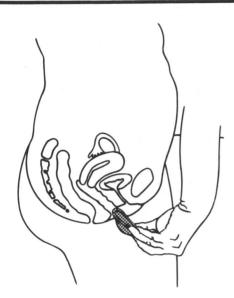

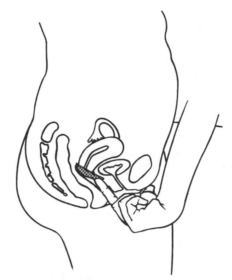

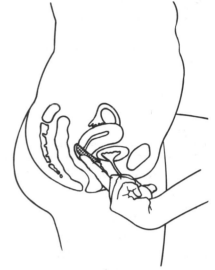

Diaphragm being inserted into the vagina

Placing the diaphragm over the cervix

Checking the diaphragm

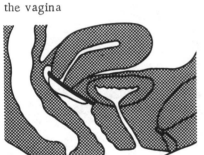

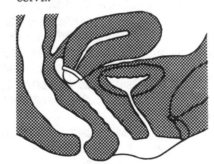

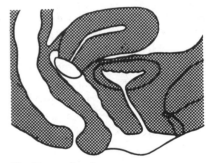

Diaphragm in place

Cervical cap in place

Vault cap in place

EFFECTIVENESS OF TECHNIQUES

It is misleading to make a hard and fast statement about the effectiveness of contraceptives. As the table shows, there is a difference between effectiveness in theory and success in practice. For example, the combined pill has a theoretical effectiveness of 99.9%. In practice, this drops to between 95% and 98%, as out of 100 women using it for a year, between 2 and 5 become pregnant. Failures happen for various reasons. The method itself can fail, or a couple can fail either to use it correctly or even to use it at all.

THE ODDS OF BECOMING PREGNANT

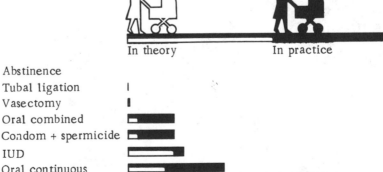

In theory   In practice

Abstinence
Tubal ligation
Vasectomy
Oral combined
Condom + spermicide
IUD
Oral continuous
Condom
Cap + spermicide
Withdrawal
Spermicidal foam
Rhythm calendar

0   5   10   15   20   25   30   35
Pregnancies per 100 women

©DIAGRAM

WORLD VIEW

CONTRACEPTION TODAY

Contraception today, even with its faults, has reached a fairly high level of sophistication. Yet of the world's fertile women, probably less than a third actually use contraception regularly. There are various reasons for this discrepancy. Outside Europe and North America, women in most parts of the world know little about modern contraception. Anti-contraceptive laws still exist in some countries, while in others social or religious attitudes create further barriers. Despite the enormous increase in family planning programmes, facilities are often inadequate. Statistics on this page are based on surveys done by the International Planned Parenthood Federation and give some idea of the state of contraception today.

WORLD USE OF CONTRACEPTIVES

The diagram indicates what percentage of fertile couples in the world are regularly using contraception. In the world as a whole the proportion is just under one-third but there are vast regional differences. As may be expected, the greatest use is in North America where 80% of fertile couples regularly use contraception. In East Africa, however, the percentage is well under 2%. Figures like these can only be approximate, as information from some parts of the world is both scanty and unreliable.

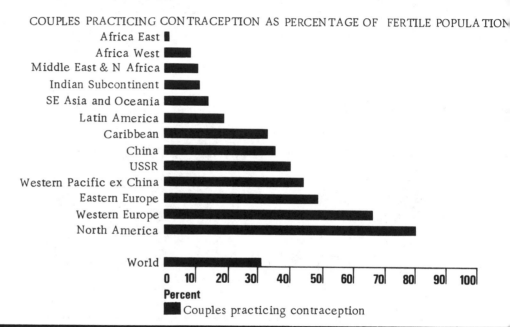

COUPLES PRACTICING CONTRACEPTION AS PERCENTAGE OF FERTILE POPULATION

- Africa East
- Africa West
- Middle East & N Africa
- Indian Subcontinent
- SE Asia and Oceania
- Latin America
- Caribbean
- China
- USSR
- Western Pacific ex China
- Eastern Europe
- Western Europe
- North America

- World

Percent

■ Couples practicing contraception

AVAILABILITY OF CONTRACEPTIVES

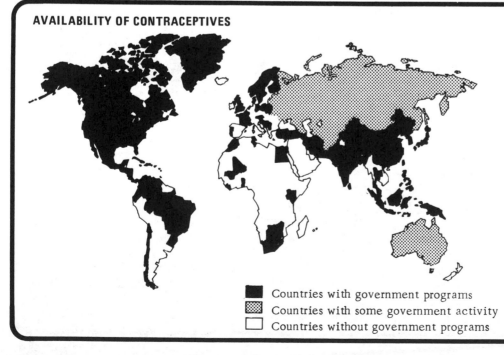

■ Countries with government programs
▦ Countries with some government activity
☐ Countries without government programs

Before 1960 most organized contraceptive services were provided by voluntary associations concerned with the effects of continuous childbearing in women. In the last 16 years voluntary work has increased and some 120 countries now have family planning associations. State interest has also grown, largely due to concern about population growth. The map shows countries which, by the end of 1975, had government programs. Nearly all have been set up since 1960. There are still over 30 countries, mostly in Africa, with no organized family planning.

CONTRACEPTIVE EDUCATION

Unwanted pregnancies are common among teenage girls, yet most are reluctant to use birth control. To encourage them, UK family planning workers produced a conventional comic strip - part of which is shown here - giving accurate information about contraception.

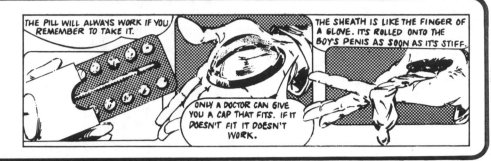

DISTRIBUTION OF METHODS USED BY PRACTICING COUPLES

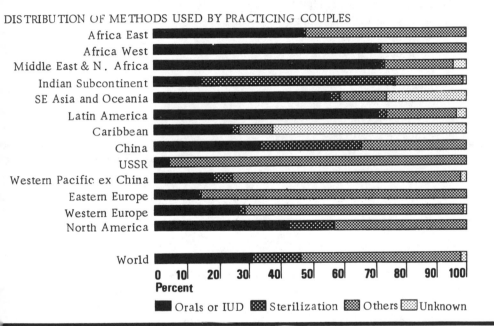

Africa East
Africa West
Middle East & N. Africa
Indian Subcontinent
SE Asia and Oceania
Latin America
Caribbean
China
USSR
Western Pacific ex China
Eastern Europe
Western Europe
North America

World

0 10 20 30 40 50 60 70 80 90 100
Percent

■ Orals or IUD ▓ Sterilization ▓ Others ▓ Unknown

Taking the world as a whole, traditional methods of contraception - cap, condom, rhythm, withdrawal - are used by just over half the couples who regularly use contraception. But this varies from region to region. Sterilization is used by 61% of those using contraceptives in India; but in other parts of the world it is fairly uncommon. In Latin America, West and North Africa, and the Middle East, if contraception is used at all it is mainly the pill or IUD. Perhaps surprisingly, the traditional methods are still the most widely used in Europe.

Contraception and family planning services are available in most European countries. There are only four - Greece, Malta, Eire, and Spain - where anti-contraceptive laws still restrict either sale or publicity. Government involvement is fairly recent, however. In Britain, where voluntary services had been available since the 1920s, the government did not assume responsibility until 1974. And it was not until 1975 that contraceptive services became part of social security in France and Portugal.

☐ Countries with family planning associations
■ Countries without family planning associations

© DIAGRAM

INFERTILITY 1

INFERTILITY

A practical indicator of infertility is a couple's failure to achieve conception in a year or more of intercourse, if no contraceptive measure is used. Out of every 100 couples, 10 cannot have children, and 15 have fewer than they wish. So a quarter of couples are below normal fertility. The causes can involve either or both partners: the woman in 50 to 55% of cases, the man in 30 to 35%, and both in about 15%.

CAUSES IN WOMEN

The most common is failure to ovulate, due to hormonal failure. This may result from actual disorders of the hormone mechanism, or from emotional stress and other psychological factors. Hormonal imbalance can also prevent a fertilized egg attaching itself to the wall of the uterus, while emotional stress may operate directly by setting up spasms in the Fallopian tubes to prevent them transporting the egg. A second group of causes concerns the vaginal and cervical fluids. These may be inadequate for sperm transport, or even actively hostile to sperm movement or survival. (Again, hormonal imbalance may be involved.) A third group are congenital, including possession of a hymen or vagina too tight for penetration, and common malformations such as: fusion of the small vulval lips; vagina divided in two or totally absent; uterus divided in two; or uterus and cervix absent.

Another possible cause in this group is tilting (retroversion) of the uterus, so that the sperm do not normally find their way in; but most authorities dismiss this.

Finally, there is infertility that is linked with other disorders in the sex organs, including: infection with venereal disease, cystitis, etc; growth of fibroids, polyps, cysts, or cancer; and effects of exposure to high doses of radiation. These may affect the ovaries, block the Fallopian tubes, etc.

INVESTIGATION AND TREATMENT FOR WOMEN

Diagnosis may prove simple. But if necessary a woman's doctor may refer her to a team of specialists using advanced techniques of investigation and treatment.

POST-COITAL TEST This is usually made 6-18 hours after intercourse, and as near as possible to the day of ovulation. A mucus specimen is taken from the cervix. Microscopic study of this shows the quantity and quality of the sperm present - which depend on both the material originally ejaculated and the condition of the cervical mucus. The mucus should be most receptive to sperm at ovulation, but hormonal imbalance may distort this. Also infertility sometimes results from incompatability between mucus and sperm.

SCRAPING The health of and hormonal influence upon the uterus lining can be tested by dilation and curettage. The resulting stretching of the cervix helps sperm penetration, and so may itself aid fertility.

SALPINGOGRAPHY is X raying of the uterus and Fallopian tubes, to reveal their internal condition, by introducing into them an X ray-opaque oily or water-soluble dye. It is done before ovulation to avoid possible ovum damage. Again, sometimes the procedure itself helps: the dye unsticks adhering tube walls, and pregnancy follows.

GAS TEST (INSUFFLATION) Carbon dioxide is blown through the Fallopian tubes, revealing and sometimes even clearing blockage. But it gives no detailed information, and some consider it outmoded.

UTEROSCOPY is use of a periscope instrument to give an internal view of the uterus via the vagina.

LAPAROSCOPY gives a good external view of uterus, tubes, and ovaries, without a large abdominal incision. Carbon dioxide gas is blown through a hollow needle into the abdominal cavity. This distends the abdominal wall, and allows a clear view of the reproductive organs through a laparoscope introduced through a tiny abdominal slit.

Various specialists may be involved.

a SURGICAL GYNECOLOGIST specializing in investigating and operating on the female reproductive organs.

b MEDICAL GYNECOLOGIST advising on non-surgical treatments.

c HISTOLOGIST analyzing tissue samples taken by surgical gynecologist from ovaries, uterus lining, etc.

d RADIOLOGIST interpreting X rays of (for example) the Fallopian tubes.

e ENDOCRINOLOGIST looking for disturbances in the hormones of the endocrine system.

f BIOCHEMIST providing the endocrinologist with precise measurements of hormone levels.

g PSYCHIATRIST to help overcome psychological barriers to pregnancy.

h GENETICIST assessing risks of inherited abnormality, and advising abortion if necessary.

i UROLOGIST specializing in disorders involving the urinary tract.

j ANDROLOGIST specializing in investigating the male reproductive system.

Medical investigations can often reveal the cause of infertility, and lead to successful treatment. Just the knowledge that something is being done may help build an atmosphere in which fertility can occur. Some clinics now claim 65% success within 4 years of attendance.

The diagram shows possible faults in the female genitals:

a small labia fused;
b hymen too strong;
c vagina narrowed or divided;
d cervical mucus unreceptive;
e uterus tilted or divided;
f Fallopian tubes blocked by infection;
g ovaries failing to produce eggs.

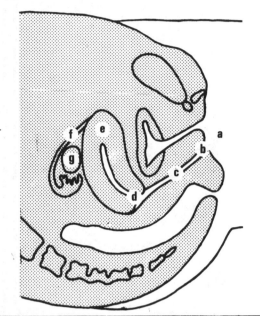

a Labia
b Hymen
c Vagina
d Cervix
e Uterus
f Fallopian tubes
g Ovary

The diagram shows working relationships within the medical team.

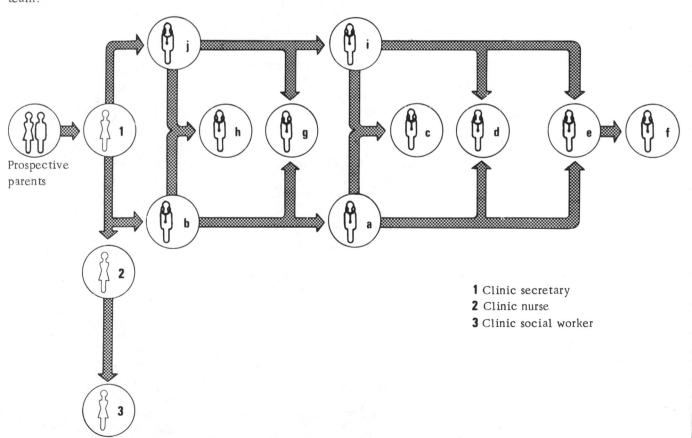

Prospective parents

1 Clinic secretary
2 Clinic nurse
3 Clinic social worker

© DIAGRAM

INFERTILITY 2

TIMING OF INTERCOURSE

Chances of conception may be improved by concentrating intercourse on the woman's fertile phase of each month. If the charts that she keeps show that she has regular 28-day periods, her fertile phase will usually lie between days 11 and 16. With irregular cycles of between 27 and 35 days, chances of pregnancy improve if intercourse occurs on five alternate days, starting with the 13th day of the cycle. (Intercourse on all the fertile days would exhaust the man's sperm output.)

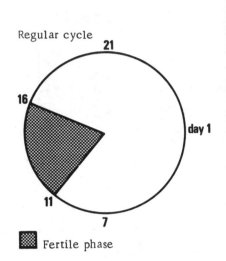

Regular cycle

21

16

day 1

11

7

▨ Fertile phase

TREATING INFECTION

The cervix is often affected by infections involving vaginal discharge. The cause of some of these is unknown, but some are due to venereal disease or other diagnosable infection, or to foreign bodies such as forgotten tampons or contraceptive devices. Treatment usually involves a course of pessaries, or sometimes antibiotics. (Occasionally the cervix is cauterized under anesthetic, to burn away chronically infected tissue.) Infection may impair fertility. Risk of spreading infection may also prevent investigatory techniques.

TECHNIQUES OF INTERCOURSE

FULL PENETRATION Sometimes infertility is due to poor coital connection. When obesity is the cause, if the woman hooks her knees over the man's shoulders during intercourse, this flexes the hips and allows deeper penetration. (Care must be taken in case pain occurs.) When the cause is vaginal tightness, then (perhaps after treatment by dilation) the woman should squat over the man, as he lies on his back, and slowly lower herself onto his penis.

RETROVERTED UTERUS In about 10% of women, the uterus is tilted back, and the cervix forward (retroverted uterus). The typical position of intercourse may then not bring the semen into contact with the cervix. Most have no difficulty in becoming pregnant despite this. Fertility is helped in such cases, though, if the woman uses a face-down positon (lying or kneeling), with the man entering her from behind. The lying position should make the pool of semen bathe the cervix - but may cause cystitis if the penis bruises the woman's bladder. Instead, the woman can change from a kneeling position to a lying one after intercourse ends.

AFTER INTERCOURSE Fertility in any woman is usually improved if the woman remains fairly still for at least half an hour after intercourse ends.

Retroverted uterus

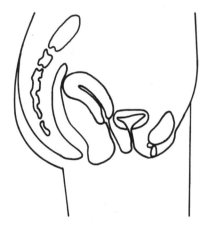

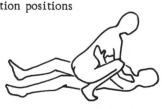

Full penetration positions

Recommended for retroverted uterus

276

IMPROVING FEMALE RECEPTIVITY

Mechanical devices and ointments may be useful for improving the sperm receptivity of various parts of the woman's genitals.

DILATION of the vagina with glass dilators or with the fingers helps in many cases where vaginal tightness prevents successful penetration.

DOUCHING Making the cervical mucus more alkaline may improve the chances of the cervix taking up viable sperm. The woman sits in a bath and douches the upper part of the vagina with warm water containing bicarbonate of soda. For this purpose she uses a douche can equipped with tube and nozzle. This method should only be used when a post-coital test has shown that it is appropriate.

VAGINAL ACIDITY Although alkaline cervical mucus can be desirable, it also sometimes helps if the mucus of the vagina itself is made more acid: vaginal acidity seems to make sperm move up towards the cervix. Most vaginas have natural acidity, but occasionally an ointment is prescribed to enhance this. The ointment is inserted on the day before intercourse; use on the actual day of intercourse may produce acid conditions strong enough to kill the sperm.

SYNTHETIC HORMONES The receptivity of the cervix to sperm depends largely on the hormonal balance. Lack of estrogen may make it unreceptive. Synthetic estrogen given for 4 or 5 days around ovulation may improve receptivity.

FERTILITY DRUGS FOR WOMEN

These have given remarkable results in recent years.

CLOMIPHENE is usually the first drug tried when failure to ovulate is suspected. Just how it works is unclear (in fact, it was originally tested as a contraceptive, and found to have the reverse effect!). The woman may need no more than a single 5-tablet course taken during one menstrual cycle: with luck, ovulation follows less than a fortnight after. If no pregnancy occurs after a month, a second course may be tried, and so on - with intervals - up to 6 courses. About 30% of those given the drug conceive. Only one ovum is released at a time, so multiple pregnancies are unlikely (twins occur in about 7% of cases). Ovarian cysts are a possible side effect; otherwise the drug seems harmless.

PERGONAL is more controversial. It is an extract of FSH and LH hormones obtained from menopausal women. (FSH and LH stimulate the ovaries to produce estrogen and progesterone, the hormones that prepare the uterus for pregnancy. Patients receive the drug by injection under closely controlled hospital conditions. Good supervision should yield no more than one or two ova; miscalculation may produce either none or a large number. Most unwanted multiple births due to fertility drugs stem from Pergonal, and such births carry increased risks to mother and babies alike.

HYPOTHALAMIC RELEASING FACTORS These are substances produced by the hypothalamus at the base of the brain. They stimulate the pituitary to produce FSH and LH. Such hypothalamic releasing factors are now made synthetically, and have cured infertility which was due to poor pituitary action.

ANTI-SPASMODICS These do not promote ovulation: they are used simply to get existing ova to their destination by preventing the Fallopian tubes going into spasm. (In some women, such temporary spasms block ova transport.)

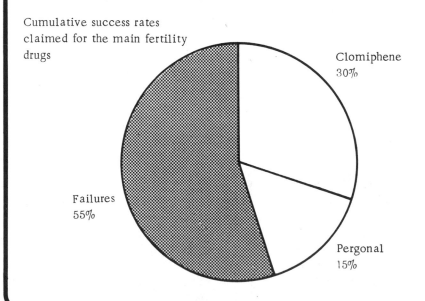

Cumulative success rates claimed for the main fertility drugs

Clomiphene 30%

Pergonal 15%

Failures 55%

© DIAGRAM

INFERTILITY 3

OPERATIONS FOR FEMALE FERTILITY

A number of operations may be performed to improve fertility in appropriate cases.

VULVA Operations on the vulva include opening cysts and abscesses, and dividing the small lips of the vulva if these have fused.

HYMEN Where this is too thick and rigid to be stretched, it may be cut and stitched back.

VAGINA A narrow vagina may be widened and lined with a skin graft; and the longitudinal division that sometimes separates the two sides of the vagina can be removed.

CERVIX A badly torn cervix can be treated by plastic surgery. Repeated abortions between the 14th and 28th weeks are sometimes prevented by stitching the cervix. Dilation may aid sperm penetration.

UTERUS A backward-tilting (retroverted) uterus can be brought forward by inserting a special plastic pessary into the vagina, or by shortening the ligaments between uterus and groin. Fibroids can be cut away from the uterus wall. The wall that sometimes divides a uterus can be removed.

FALLOPIAN TUBES cannot be cleared when totally blocked, as from some old infection. But the "petals" at the ovarian ends of the tubes may need teasing out; and if the uterus end of a tube is blocked, this can be cut away and the shortened tube rejoined to the uterus.

Sometimes tubes are intact but hampered by bands of tissue that prevent them moving to collect the the ovum from the ovary. These bands can be cut away.

OVARIES Cutting a wedge from the ovaries allows egg release, when this is blocked by certain ovarian cysts. Removal of other cysts also aids pregnancy. Probably a third of gynecological operations are for ovarian cysts.

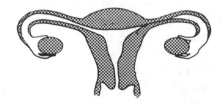

Fallopian tube blocked

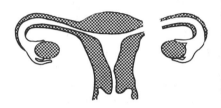

Blocked section removed

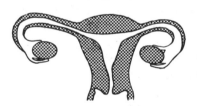

Shortened tube rejoined

PSYCHIATRIC HELP

This may be relevant. First, there are women in whom long-term failure to ovulate derives ultimately from psychological stress. These cases include those showing clear mental symptoms (eg severe depression), and those with no surface symptoms but nevertheless some underlying unhappiness or insecurity. Second, there are those men and women in whom long-term trauma interferes with the actual sexual act. Third, there are couples in whom the desire and struggle for fertility has itself given rise to mental tension, with emotional or physical consequences (eg male impotence).

ARTIFICIAL INSEMINATION

This involves the artificial transference of semen onto the cervix - usually by syringing through a tube on 3 or 4 successive days around ovulation. After insemination the woman remains lying down for about half an hour. Artificial insemination (AI) with the partner's sperm (AIH) may be used: when physical or psychological causes prevent normal intercourse or ejaculation; or when the man's sperm count is low, or the woman's cervical mucus is hostile to his sperm. AIH can be carried out in a clinic or surgery, or (using special equipment) by the couple at home. AI using the sperm of an anonymous donor (AID) is much more common. It is used when the partner is totally infertile. or when his chromosomes are known to carry hereditary disorders. The donor is carefully chosen to match the partner in appearance and to be free from disease.

The success rate from AID is about 66% within the first three months. Babies conceived by AI develop no differently, of course, from those conceived in the ordinary way.

INFERTILITY IN MEN

Male infertility is of two kinds:
a) cases where there is no ejaculation (i.e. impotence: see N 10);
b) cases where the quality of the ejaculate is poor - as shown by sperm concentration, shape, and mobility.

SPERM CONCENTRATION This depends not only on sperm production, but also on the amount of fluid - as shown by the total amount of semen. Both very small and very large amounts are unfavorable to fertility. A small amount suggests that sperm production is also low. It also fails to buffer the sperm against the acidity of the vaginal fluids. A large amount dilutes the semen too much, and makes it more likely to spill out of the vagina.

SPERM SHAPE The higher the number of abnormal forms, the less the likelihood of fertility. For example, fertility is probably impossible if the tapering shapes rise above 8 or 10%.

SPERM MOVEMENT The length of life of the sperm (as shown by their movement) is significant not only because they may need time before encountering an egg to fertilize. For some reason not yet understood, sperm also need to survive for some time in the female reproductive tract before they are capable of fertilizing an egg.

CAUSES OF MALE INFERTILITY
Causes of poor sperm production can include:
a) heat around the testicles, due, for example, to tight underclothing, obesity, or working conditions;
b) factors of general vitality, such as poor health, inadequate nutrition. lack of exercise, excessive smoking and drinking, etc;
c) emotional stress; and
d) too prolonged abstinence (this can increase the number of abnormal sperm).

More specialized factors (some of which can cause sterility) include:
a) some birth defects;
b) failure of the testes to descend before puberty;
c) some childhood diseases, and some other illnesses (eg mumps if it occurs in adulthood rather than childhood);
d) some hazards such as exposure to X rays, radioactivity, some chemicals and metals, gasoline fumes, and carbon monoxide; and
e) some genital disorders, such as varicocele and blocked ducts, and tuberculous infection of the prostate.

AIDS TO FERTILITY
Many of these causes of infertility are treatable. But, more generally, we do not yet know of any substances that will improve male fertility. Severe lack of vitamins will impair fertility; but no special vitamin intake seems to raise the fertility level of a well-fed person. As for hormones, the pituitary hormones have only limited effect, while testosterone actually hinders sperm production - it is only useful where infertility is due to genital underdevelopment or impotence. However, there are techniques to aid fertility. One is the medical technique of artificial insemination. Another is a practical sexual technique. It seems that the second half of a man's ejaculate - the fluid from the seminal vesicles - is actually likely to harm the sperm, while that of the prostate, in the first half, protects it. So a couple can increase their chances of parenthood if the man withdraws from the vagina halfway through his ejaculation. (They should also abstain from further intercourse for 48 hours afterwards.)

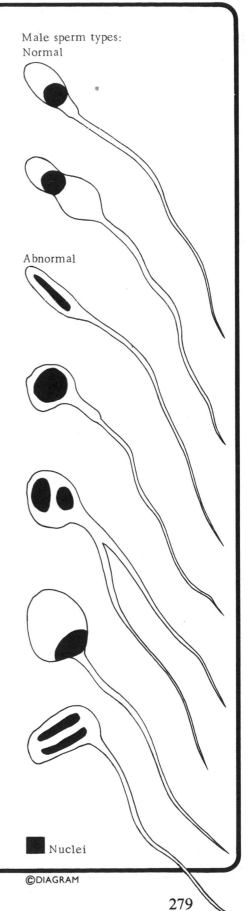

Male sperm types:
Normal

Abnormal

■ Nuclei

©DIAGRAM

ABORTION

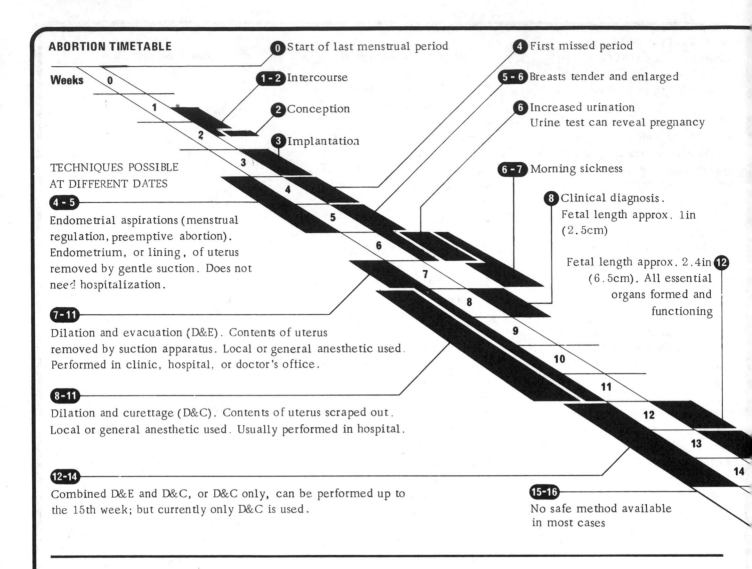

ABORTION TIMETABLE

Weeks 0

0 Start of last menstrual period

1 - 2 Intercourse

2 Conception

3 Implantation

4 First missed period

5 - 6 Breasts tender and enlarged

6 Increased urination
Urine test can reveal pregnancy

6 - 7 Morning sickness

8 Clinical diagnosis.
Fetal length approx. 1in
(2.5cm)

Fetal length approx. 2.4in **12**
(6.5cm). All essential
organs formed and
functioning

TECHNIQUES POSSIBLE
AT DIFFERENT DATES

4 - 5
Endometrial aspirations (menstrual
regulation, preemptive abortion).
Endometrium, or lining, of uterus
removed by gentle suction. Does not
need hospitalization.

7 - 11
Dilation and evacuation (D&E). Contents of uterus
removed by suction apparatus. Local or general anesthetic used.
Performed in clinic, hospital, or doctor's office.

8 - 11
Dilation and curettage (D&C). Contents of uterus scraped out.
Local or general anesthetic used. Usually performed in hospital.

12 - 14
Combined D&E and D&C, or D&C only, can be performed up to
the 15th week; but currently only D&C is used.

15 - 16
No safe method available
in most cases

Abortion - the deliberate termination
of pregnancy - is a controversial
and emotive issue. The decision to
end a pregnancy is rarely easy. For
most women, however, whether the
pregnancy has resulted from lack,
misuse, or failure of contraception,
the problems of continuing with
an unwanted pregnancy are generally
greater than those of terminating it.
CONFIRMING PREGNANCY
A missed period is usually the first
sign of pregnancy. Others may be
feelings of sickness, revulsion
against some foods, and frequent
urination. Fourteen days after the
first missed period, a urine test can
confirm pregnancy. Although there
are home-testing kits, these are not

reliable, and a sample of early-
morning urine in a clean container
can be taken to a doctor, clinic,
hospital, or pregnancy-testing
association. Results are often known
within a few hours.
CHOOSING AN ABORTION
Once a woman has decided to have
an abortion, time becomes
important. Although abortions can
be carried out until the 28th week
(the fetus is then legally considered
to have a separate existence) they
are not often performed after the
12th week, and only very rarely
after the 20th.
ABORTION LAWS
Abortion is the most widely used
method of birth control in the world.

It has been estimated that nearly 1
in 4 pregnancies are terminated
either legally or illegally. There has
been a gradual liberalization of
abortion laws and today well over
75% of the world's population live in
countries where abortion is, to a
greater or lesser extent, legal.
Laws and facilities vary. In the US
abortion up to 12 weeks has been
available "on request" since 1973;
in Britain abortion has been legal
since 1967 but various socio-medical
grounds must be presented, and
ultimately the decision is not the
woman's alone.
In only a few countries, such as
Belgium and Indonesia, is abortion
completely outlawed.

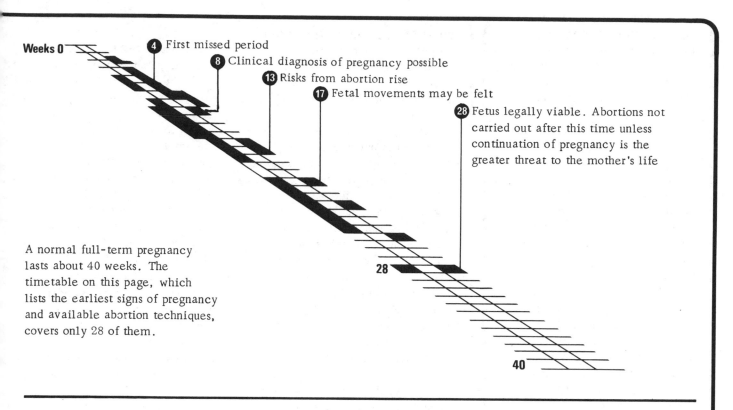

Weeks 0

4 First missed period

8 Clinical diagnosis of pregnancy possible

13 Risks from abortion rise

17 Fetal movements may be felt

28 Fetus legally viable. Abortions not carried out after this time unless continuation of pregnancy is the greater threat to the mother's life

A normal full-term pregnancy lasts about 40 weeks. The timetable on this page, which lists the earliest signs of pregnancy and available abortion techniques, covers only 28 of them.

28

40

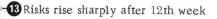

13 Risks rise sharply after 12th week

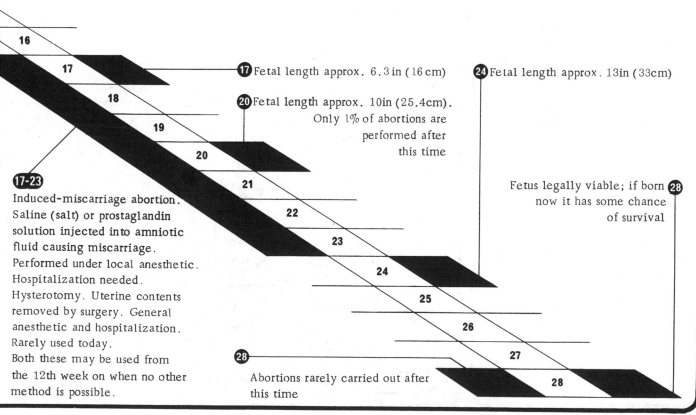

16

17

18

19

20

21

22

23

24

25

26

27

28

17 Fetal length approx. 6.3 in (16 cm)

24 Fetal length approx. 13in (33cm)

20 Fetal length approx. 10in (25.4cm). Only 1% of abortions are performed after this time

Fetus legally viable; if born now it has some chance of survival **28**

17-23
Induced-miscarriage abortion. Saline (salt) or prostaglandin solution injected into amniotic fluid causing miscarriage. Performed under local anesthetic. Hospitalization needed. Hysterotomy. Uterine contents removed by surgery. General anesthetic and hospitalization. Rarely used today. Both these may be used from the 12th week on when no other method is possible.

28
Abortions rarely carried out after this time

©DIAGRAM

METHODS OF ABORTION 1

ENDOMETRIAL ASPIRATION

Known also as interception, menstrual regulation, and menstrual suction, this is a preemptive abortion technique - meaning that it can be performed for up to 2 weeks after a period was due, i.e. before a pregnancy can be positively confirmed.

EQUIPMENT

This consists of a small, flexible plastic cannula (tube) about 4-5mm long attached to a suction source - usually an electrical or mechanical pump. A syringe may be used in very early pregnancy.

PROCEDURE

The cannula is passed through the cervix into the uterus, its small size making much dilation unnecessary. The endometrium, or lining of the uterus, is gently sucked out, and with it the fetal tissue. The process takes only a few minutes and local anesthetic is rarely needed. Interception, although not yet widely available, is generally performed in a clinic or doctor's office.

ADVANTAGES AND DISADVANTAGES

Interception is fast and there are apparently few risks - the flexible cannula reduces risk of damage to the uterus. Also, being carried out so early in pregnancy, emotional strain on a woman is lessened. However, for the same reason, the method may be used unnecessarily, on a woman who is not pregnant; or, as the fetus is so tiny at this stage, there may be uncertainty about its complete removal.

MENSTRUAL EXTRACTION

In some women's self-help groups, the above process is used, not as an abortion technique, but as a means of avoiding menstrual discomfort by extracting the uterus lining in one quick operation. It is then given the different name of "menstrual extraction."

DILATION AND EVACUATION (D&E)

Also called vacuum curettage, suction abortion, or STOP (suction termination of pregnancy), D&E is the most common method of abortion today. It is quick, easy to perform (it is increasingly carried out in outpatient clinics), and there is low risk of complication. Essentially the fetus is removed, by suction, from the uterus through a narrow tube inserted through the cervix. D&E is normally carried out between 7-12 weeks from the last menstrual period, after which time the fetus is too large for D&E to be performed safely.

PREPARATION

Very little preparation is needed. The woman's blood type is checked, and she should not eat for about 6 hours before the operation. Pubic hair need not be shaved. After an internal examination, the speculum is inserted and the patient given an anesthetic - local or general.

RECOVERY

The abortion takes about 10 minutes; the rest period afterwards about 2-3 hours. When D&E is performed in hospital, patients often stay in overnight. Recovery is fast, though strenuous activity should be avoided for a couple of days. There is usually some bleeding, possibly with mild cramps, for up to 7 days. The normal period starts 4-6 weeks after the abortion. Most doctors advise that tampons and sexual intercourse should be avoided for 2-4 weeks to prevent possible infection.

DILATION AND CURETTAGE (D&C)

Before the development of suction abortion, D&C was the standard method used for pre-12th-week abortions. It is still a standard gynecological procedure, and is the only abortion technique used from the 12th to 15th weeks. After dilating the cervix, the uterus contents are scraped away with a curette. D&C is more complicated to perform than D&E, is more painful, requires general anesthetic, and carries more risks of perforation and infection.

DILATION

The diagram shows the cervix being dilated. A series of polished metal dilators (a) are used, the largest being about the width of a finger. A speculum (b) holds the vaginal walls open.

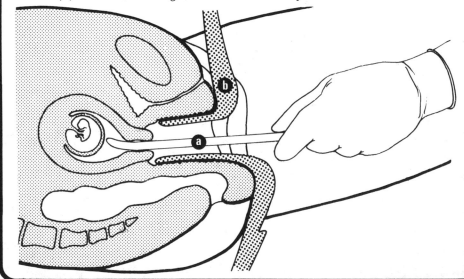

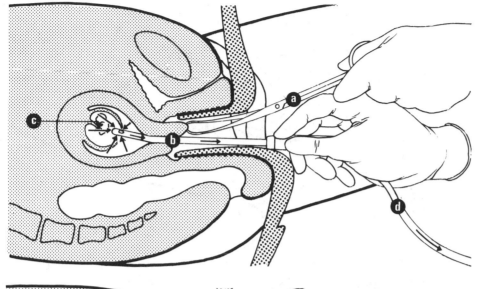

EVACUATION ONE
Once dilation is complete the cervix is held steady with a tentaculum (a). A vacurette or suction curette (b) is inserted into the uterus until it touches the fetus (c). The vacurette, which is about 1/3 in (8 mm) wide, has 2 side openings and is attached (d) by transparent plastic tubing to a suction machine or aspirator.

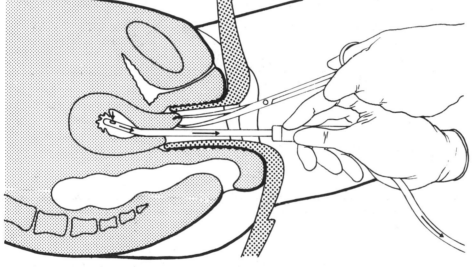

EVACUATION TWO
With the vacurette inside the uterus the suction machine is turned on. The fetal material breaks up and is gently suctioned through the tip of the vacurette into a vacuum bottle. The suction tube is moved around until the uterus is empty, and then removed. Aspiration takes about 2-5 minutes and afterwards the doctor usually scrapes the inside of the uterus (see below) to ensure that none of the fetus or placenta has been left behind.

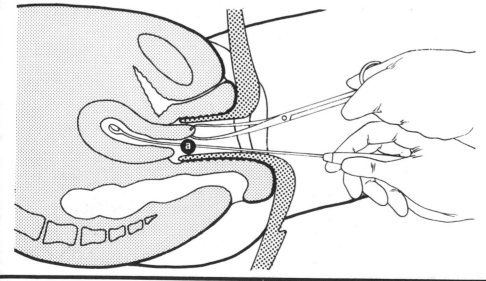

CURETTAGE
The diagram shows the uterus being scraped with a curette (a). When curettage only is used as an abortion technique, the cervix is first dilated as described above. A curette - a thin metal instrument with a spoon-like tip - is then inserted into the uterus. The fetus and placenta are scraped loose and are removed with forceps.

©DIAGRAM

METHODS OF ABORTION 2

INDUCED-LABOR ABORTIONS

Currently the usual technique for late abortion is to induce miscarriage. Being so similar to normal childbirth, this can be a much more distressing experience than earlier abortion methods. There is also more potential risk, eg hemorrhage, shock, infection, incomplete abortion. Therefore, late abortions are always carried out in hospital, and are rare.

PROCEDURE

Under local anesthetic an amniocentesis is performed. Amniotic fluid is withdrawn (a) and replaced (b) by a miscarriage-inducing agent, normally a concentrated saline (salt) solution. The fetus dies and in 6-48 hours the cervix dilates. Contractions occur and the fetus and placenta are expelled. Recently use has also been made of prostaglandins (naturally occurring hormones). These stimulate contractions of the womb, causing miscarriage more rapidly than saline solution. Also, despite side effects, prostaglandins are safer.

a Amniotic fluid is withdrawn

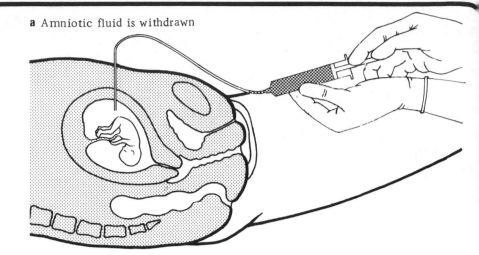

b Saline solution is injected into the uterus

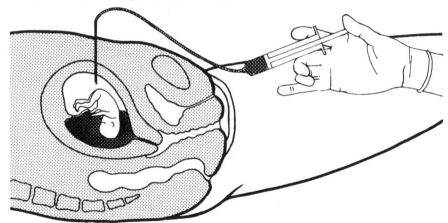

HYSTEROTOMY

Hysterotomy (not to be confused with hysterectomy) is a method of late abortion that is rarely used today. It is similar to a mini-Cesarian section, and involves major surgery and hospitalization.
Hysterotomy is the most complicated of all abortion techniques, and carries the highest risks. It is generally used only when a saline-solution abortion has failed. The resulting scar may rupture in a subsequent pregnancy.

a Under general anesthetic, incisions are made in the abdominal wall, usually below the pubic hairline.

b The contents of the uterus - fetus and placenta - are removed through the incisions, which are then sewn up.

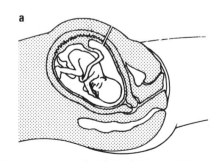

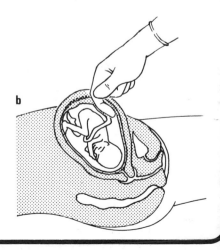

ILLEGAL ABORTIONS

The diagram shows some of the physical damage that can result from illegal abortions. "Back-street" and self-induced abortions are still common - worldwide they cause 30-50% of all maternal deaths from pregnancy and childbirth. Various techniques are used, most of them either unsuccessful or highly dangerous. Inserting objects or pumping fluid and air into the uterus are among the most common methods. They are often fatal. But there has always been a demand for abortion, legal or otherwise. One argument for complete legalization is to prevent catastrophes that result from crude, unhygienic abortions. Ironically it was, in some part, for this reason that abortion was outlawed in the 19th century.

Possible physical consequences of illegal abortion

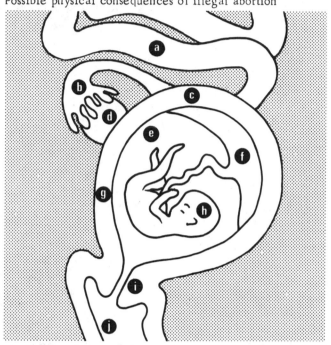

a Punctured intestine, possibly leading to peritonitis, septicemia, and intra-peritoneal hemorrhage
b Infection of Fallopian tubes
c Perforation into the peritoneal cavity
d Infection of the ovaries
e Intrauterine infection
f Perforation through placental site causing internal hemorrhage
g Blood clot (possibly infected) or air embolism (possibly lethal)
h Fetal malformation
i Laceration of cervix
j Laceration of vagina

RISKS

The diagram compares mortality rates from legal abortions with those resulting from pregnancy and childbirth. It also compares the mortality rates of different types of abortion.

As the diagram shows, with techniques up to 12 weeks, it is actually much safer to terminate a pregnancy medically than to continue with it. After this time, when the fetus is becoming too large to be removed by either suction or D&C, the risks do rise sharply. Even so, with the exception of hysterotomy and saline induction, legal abortion is, on average, safer than full-term pregnancy.

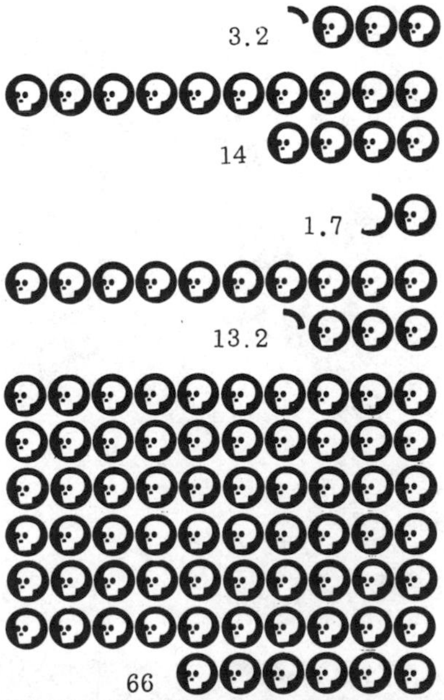

ABORTION AGAINST LIVE BIRTH
Legal abortion (all techniques)

3.2

Live births

14

DIFFERENT ABORTION TECHNIQUES
Suction abortions

1.7

Amniotic fluid exchange

13.2

Hysterotomy

66

Maternal deaths per 100,000 pregnancies ended by each method shown

©DIAGRAM

8 Pregnancy

The miracle of life before birth—a time of sharing and expectation for every member of the family. This chapter deals with the facts behind the complex changes undergone by mother and fetus during these special nine months.

Left: Pregnant mother and child
(Rex Features Ltd.)

PREGNANCY

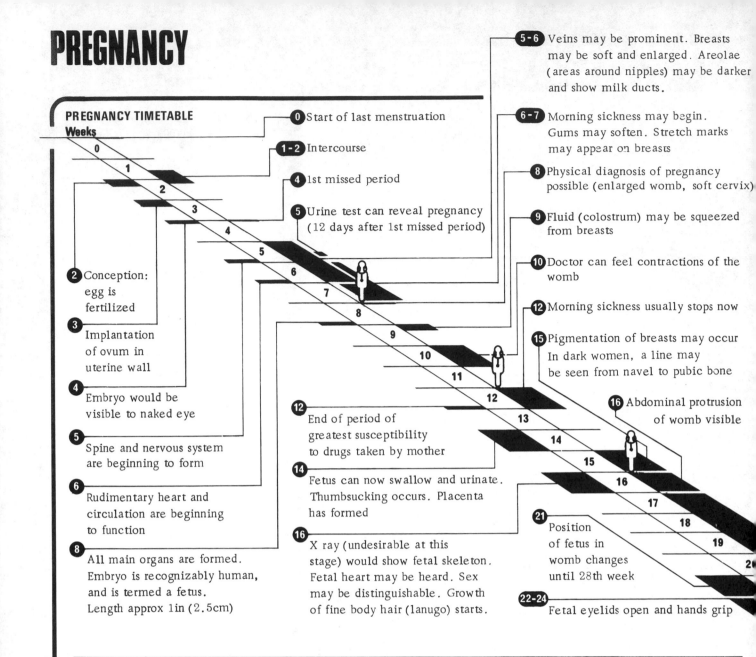

PREGNANCY TIMETABLE
Weeks

0 Start of last menstruation

1-2 Intercourse

4 1st missed period

5 Urine test can reveal pregnancy (12 days after 1st missed period)

5-6 Veins may be prominent. Breasts may be soft and enlarged. Areolae (areas around nipples) may be darker and show milk ducts.

6-7 Morning sickness may begin. Gums may soften. Stretch marks may appear on breasts

8 Physical diagnosis of pregnancy possible (enlarged womb, soft cervix)

9 Fluid (colostrum) may be squeezed from breasts

10 Doctor can feel contractions of the womb

12 Morning sickness usually stops now

15 Pigmentation of breasts may occur In dark women, a line may be seen from navel to pubic bone

16 Abdominal protrusion of womb visible

2 Conception: egg is fertilized

3 Implantation of ovum in uterine wall

4 Embryo would be visible to naked eye

5 Spine and nervous system are beginning to form

6 Rudimentary heart and circulation are beginning to function

8 All main organs are formed. Embryo is recognizably human, and is termed a fetus. Length approx 1in (2.5cm)

12 End of period of greatest susceptibility to drugs taken by mother

14 Fetus can now swallow and urinate. Thumbsucking occurs. Placenta has formed

16 X ray (undesirable at this stage) would show fetal skeleton. Fetal heart may be heard. Sex may be distinguishable. Growth of fine body hair (lanugo) starts.

21 Position of fetus in womb changes until 28th week

22-24 Fetal eyelids open and hands grip

This timetable gives a rough guide to certain milestones and occasional problems in pregnancy. Some events relate to the mother-to-be, while others involve the development of the fetus. Most people think of pregnancy lasting 9 months. But calendar months vary in length. Our chart irons out these variations by showing a timespan of 40 weeks. These can be divided into 9 31-day months (279 days) or 10 28-day (lunar) months. Traditionally most doctors use lunar months for pregnancy calculations, dating the onset of pregnancy from the first day of the last menstrual period, despite the fact that conception occurs about 14 days later.

Some dates shown are only approximate. There are large variations in the timing and even appearance or non-appearance of certain signs or symptoms. One factor that especially affects some variations is whether the mother-to-be is expecting her first baby or a second or subsequent child.

 Visits to doctor or antenatal clinic

● Week of pregnancy

At some stage in their lives most women experience the desire to have children. But it need no longer be the unplanned, haphazard process that it once was. Today a woman can decide firstly whether to have children, and if so, when.

But for women embarking on their first pregnancy, fear of the unknown is still a common emotion. Probably no other event causes as much apprehension and anxiety. For a woman to avoid this it is important for her to understand what is happening to her body and to her unborn child during the 9 months of pregnancy and the birth that follows. This chapter describes the various stages from fertilization to labor including some of the potential problems.

FERTILIZATION AND IMPLANTATION
After sexual intercourse one male sperm fertilizes a female egg. A week later this attaches itself to the lining of the uterus and the embryo starts shaping into a miniscule human being.

EMBRYO TO FETUS
By the 8th week the embryo is recognizably human. Now termed a fetus, during the next 7 months its organs will increase 120 times in weight.

PREGNANCY
The absence of menstruation is the first sign for most women. Other changes follow such as morning sickness, breast enlargement, and swelling of the abdomen. Good antenatal care aims to prevent any complications and to ensure the good health of mother and child.

BIRTH
For many women this is the most alarming aspect of pregnancy. Relaxation and an understanding of the stages of labor and delivery help to make childbirth easier and allow a woman greater participation.

COMPLICATIONS AND RISKS
Complications may be unavoidable, eg miscarriage. But some risk factors that may affect the unborn child, such as smoking, can be controlled by the mother.

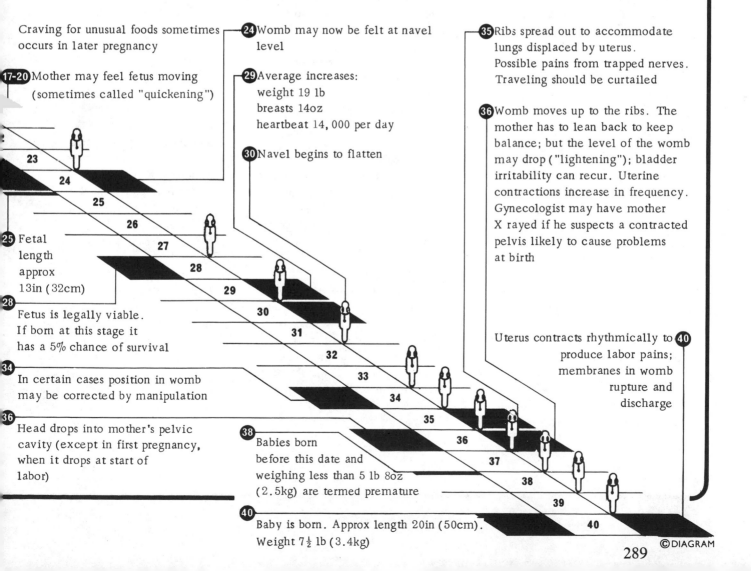

Craving for unusual foods sometimes occurs in later pregnancy

17-20 Mother may feel fetus moving (sometimes called "quickening")

23
24
25
26
27
28
29
30
31
32
33
34
35
36
37
38
39
40

25 Fetal length approx 13in (32cm)

28 Fetus is legally viable. If born at this stage it has a 5% chance of survival

34 In certain cases position in womb may be corrected by manipulation

36 Head drops into mother's pelvic cavity (except in first pregnancy, when it drops at start of labor)

24 Womb may now be felt at navel level

29 Average increases:
weight 19 lb
breasts 14oz
heartbeat 14,000 per day

30 Navel begins to flatten

38 Babies born before this date and weighing less than 5 lb 8oz (2.5kg) are termed premature

40 Baby is born. Approx length 20in (50cm). Weight 7½ lb (3.4kg)

35 Ribs spread out to accommodate lungs displaced by uterus. Possible pains from trapped nerves. Traveling should be curtailed

36 Womb moves up to the ribs. The mother has to lean back to keep balance; but the level of the womb may drop ("lightening"); bladder irritability can recur. Uterine contractions increase in frequency. Gynecologist may have mother X rayed if he suspects a contracted pelvis likely to cause problems at birth

Uterus contracts rhythmically to **40** produce labor pains; membranes in womb rupture and discharge

©DIAGRAM

PREGNANCY

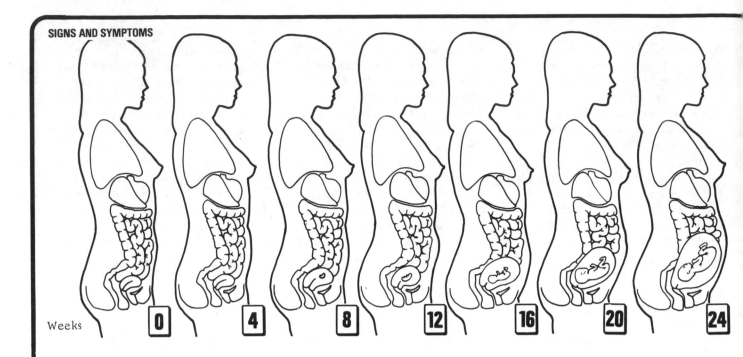

Weeks 0 4 8 12 16 20 24

Usually the first sign of pregnancy is amenorrhea (absence of menstruation). But if periods are normally irregular, the time of ovulation is uncertain and so amenorrhea is not a definite diagnosis of pregnancy.

The most noticeable physical manifestation of pregnancy after the 3rd month is the swelling of the abdomen as the uterus expands beyond the pelvis. (The swelling causes stretchmarks which often remain after birth.)

Between the 4th and 5th months the mother feels the fetal movements ("quickening") for the first time. The sensations are faint at first but get stronger. From the 5th month the fetal heart can be heard with a stethoscope and fetal movements seen from the outside.

The mother's weight gradually increases (on average by between 25 and 30 lb - about $11\frac{1}{2}$ to $13\frac{1}{2}$kg). She will begin to feel tired because of her shape and size, and so become lethargic, increasingly so towards the end of pregnancy. Her posture changes, as she has to lean back to balance the baby's weight. Because of this, backache is often experienced. Eventually she may have to walk with a waddling movement, with her legs slightly apart.

Most symptoms of pregnancy, some of which cause discomfort (in turn causing insomnia), result from the changed hormone levels and the increased pressure of the growing fetus.

HORMONAL EFFECTS

a) Emotional changes. The altered hormone levels of pregnancy cause changes in emotional states. There seems to be a general pattern common to most women (though not all). In the first 3 months, there are often extreme changes in mood, with an ambivalent response to pregnancy. During the 2nd trimester, the woman has accepted the fetus and prepared for it: she has adjusted to the hormonal changes.

b) Morning sickness.

About $\frac{2}{3}$ of women experience this, usually from the date of the first missed period until the 2nd or 3rd month, when it often ceases abruptly. It varies in severity, from nausea in the morning only, to vomiting during the day. The exact cause of morning sickness is not known, though it is thought that the increase in estrogen is responsible. The body eventually adjusts, but while nausea continues, it is a good idea to eat small frequent meals rather than large ones. Dry toast in the morning may help, and greasy spiced foods should be avoided.

c) The breasts start to enlarge in preparation for lactation. They may itch, tingle, or feel heavy, and are sometimes painful. Their veins

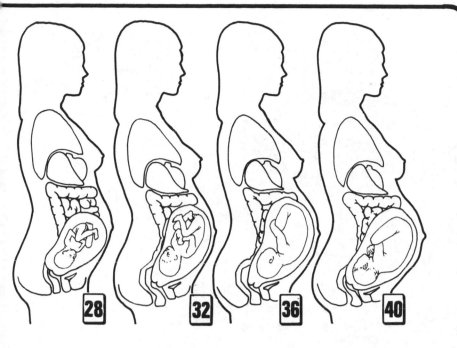

28 **32** **36** **40**

become prominent. By the 16th
week they start to secrete a thin
fluid from the nipples (colostrum).
The areolae become mottled due
to increased pigmentation.
(Increased pigmentation may also
appear on the face and external
genitals, and a dark line may run
from the navel to the genitals.)
d) Appetite. As the fetus grows,
so does the mother's appetite.
But pressure and reduced
motility of the stomach induced by
hormones reduces the capacity for
large meals. There may be cravings
for certain foods. With some women,
this craving extends to the truly
unusual - coal for example - and is
then known as pica. By contrast,
certain foods and substances may
become repulsive for some women.
Coffee, meat, alcohol, wine, and
greasy foods are examples.
e) Constipation. Reduced motility
of the large intestine increases the

possibility of constipation, and
therefore hemorrhoids. Dried fruit
or bran will ease constipation.
Laxatives should be avoided.
f) "Heartburn." Relaxation of the
esophagus sphincter can cause
regurgitation and heartburn. A
good diet should ease this situation.
FETAL PRESSURE
a) Frequent urination. The pressure
of the uterus on the bladder makes
urination more frequent. This
happens in the 2nd and 3rd months
and also near term, when the fetus
settles down into the pelvis
("lightening," or engagement).
b) Varicose veins. Fetal
pressure on the main leg veins in
the groin may cause varicose veins.
The veins in the legs dilate as a
result of the pressure of blood trying
to return to the heart.

PREGNANCY TESTING
HCG TEST
The test for pregnancy is made on
the woman's urine. It takes about
2 minutes to carry out and is 95%
accurate after the 40th day of
pregnancy. The test is based on the
fact that the placenta secretes
large quantities of Human Chorionic
Gonadotrophin (HCG) within 40
days of the last menstruation.
For the test, a drop of a substance
which neutralizes HCG (anti-HCG)
is combined on a glass slide with a
drop of the woman's urine. A
minute later another substance is
added (latex rubber particles with
HCG). If there is no HCG in the
urine the anti-HCG will fix onto
the HCG in the latex rubber particles
forming milky lumps or "curds."
This is a negative result. But if the
woman is pregnant, the HCG in her
urine will be fixed by the anti-HCG
in the first mixture and when the
rubber and HCG is added there will
be no anti-HCG left to combine
with the added HCG. The particles
will not form lumps but remain
smooth.
The test should not replace clinical
diagnosis, as mistakes can and do
occur.
CLINICAL DIAGNOSIS (6-10th week)
The two stages of the examination
are quite painless; they cause mild
discomfort only if the woman is
not relaxed. A speculum is inserted
into the vagina in order to look at
the cervix, which is a bluish color
in pregnancy. Then, after removal
of the speculum, the doctor gently
inserts two fingers into the vagina,
while pressing on the abdomen with
the other hand, in order to feel
whether the uterus has enlarged.
The test will also show up any
abnormal swellings in the uterus,
while a Pap smear is often
taken at the same time to check
for cancer of the cervix.

©DIAGRAM

EJACULATION AND CONCEPTION

EJACULATION

The diagram shows the expulsion of semen from the male sex organs at ejaculation.

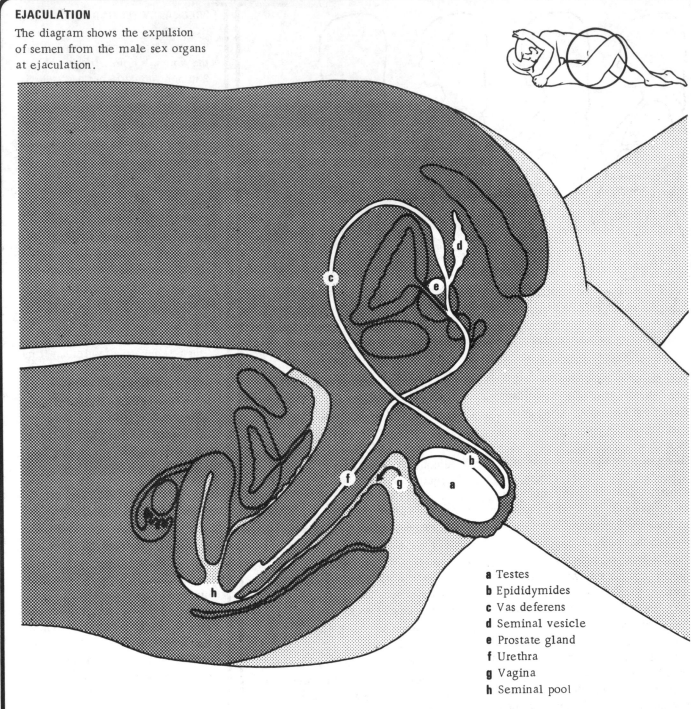

a Testes
b Epididymides
c Vas deferens
d Seminal vesicle
e Prostate gland
f Urethra
g Vagina
h Seminal pool

EJACULATION

Sperm have more than 1ft (0.3m) to travel before reaching the female vagina. At ejaculation muscular contractions in the testes (a), epididymides (b), and along the vas deferens (c) propel the sperm towards the penis. On their way they mix with the seminal fluid secreted from the seminal vesicles (d), and prostate (e). The resulting semen is then propelled through the urethra (f) into the woman's vagina (g, h).

SPERM PRODUCTION

Sperm cells are formed inside the testes at a rate of about 200 million a day. While developing, they are stored in the epididymides. Each sperm is about 1/500th of an inch long, and takes 60 to 72 days to mature. By contrast, in a woman just one mature egg is produced a month.

CONCEPTION

The diagram shows the route of the sperm from ejaculation to conception.

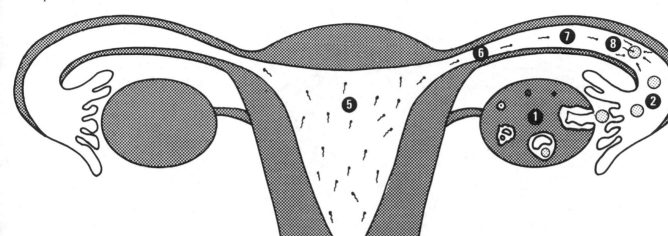

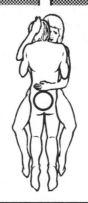

1 Development of the female egg. In the 2 weeks before ovulation a number of egg follicles have been maturing in the ovary. A week before ovulation one of these suddenly accelerates its growth.

2 Ovulation. The mature egg bursts from its follicle. Muscular contractions propel it along the Fallopian tube. If not fertilized within 24-48 hours the egg will degenerate.

3 Intercourse takes place. About 400 million sperm are ejaculated by the man into the female vagina. Of these, only one sperm will fertilize the ovum. The sperm travel fast, possibly covering 1in (2.5cm) in 8 minutes; also muscular spasms may aid them.

4 The sperm arrive at the cervix. The seminal fluid has liquified, and about half the original sperm have died in the acidic conditions of the vagina. The remainder pass through the cervical mucus. Normally a barrier to sperm, at ovulation the mucus can be easily penetrated.

5 Sperm reach the top of the uterus. There are possibly only 6000 of the original number left, but it has taken them well under an hour to arrive. About half the sperms now turn into the wrong Fallopian tube.

6 Remaining sperm swim into the top of the Fallopian tube that contains the matured female ovum. Conditions are favorable and sperm may survive here for up to 72 hours. Should ovulation not have taken place, sperm can therefore wait for a newly-developed ovum to arrive.

7 A few hundred sperm complete their journey along the Fallopian tube to the female ovum. Enzymes released by the sperm heads now break down the ovum's outer wall.

8 Fertilization. One male sperm penetrates the ovum. The cell wall immediately hardens, preventing other sperm from entering, and the nucleii of the two cells fuse together: a new human life is conceived.

©DIAGRAM

293

FERTILIZATION AND IMPLANTATION

THE FIRST FOUR WEEKS

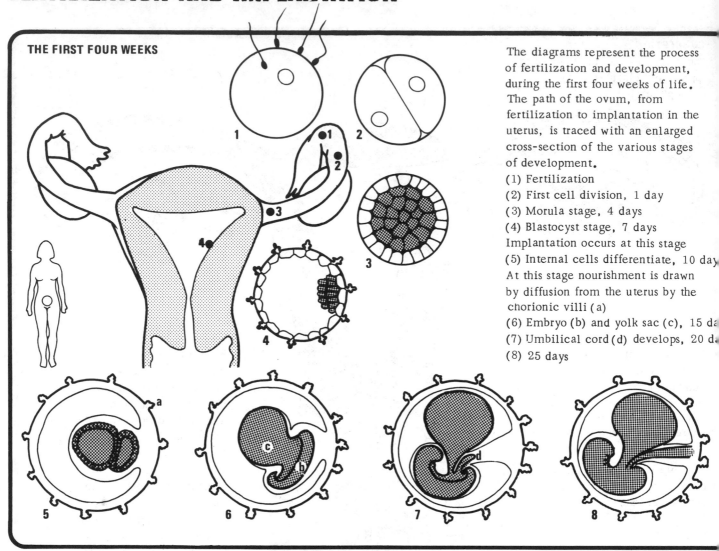

The diagrams represent the process of fertilization and development, during the first four weeks of life. The path of the ovum, from fertilization to implantation in the uterus, is traced with an enlarged cross-section of the various stages of development.

(1) Fertilization
(2) First cell division, 1 day
(3) Morula stage, 4 days
(4) Blastocyst stage, 7 days
Implantation occurs at this stage
(5) Internal cells differentiate, 10 day
At this stage nourishment is drawn by diffusion from the uterus by the chorionic villi (a)
(6) Embryo (b) and yolk sac (c), 15 da
(7) Umbilical cord (d) develops, 20 d
(8) 25 days

MULTIPLE FERTILIZATION

Since the advent of fertility drugs, multiple births have become increasingly common. The drug, which stimulates the growth of follicles, may cause the release of more than one ovum from the ovary.

A single ovum may also split 3, 4, or 5 times - still utilizing a single placenta, as in the case of identical twins.

Triplets, quads, and quins may develop from 3, 4, and 5 ova, with 3, 4, and 5 placentae, and 3, 4, and 5 amniotic sacs respectively. But other combinations can, and do, occur; eg triplets may be the product of one ovum plus one that has split, as in identical twins.

Quads may be the result of two split ova, or of two single ova plus one split one. The sharing of the placenta is dependent upon whether or not the ovum has split. On rare occasions when the uterus is stretched to its ultimate capacity, the same amniotic sac may be shared. IDENTICAL TWINS are the result of one ovum splitting soon after fertilization. (Siamese twins are the product of a splitting which for some reason has been arrested before completion.) The fetuses lie within separate sacs of amniotic fluid, though they share the same placenta. The latter means that

they will be of the same blood group. The splitting of the ovum means that they will share the same genetic structure, i.e. be of the same sex with very similar features, hair, etc.
NON-IDENTICAL TWINS are the result of two ova being fertilized by two sperm. Thus they may or may not be of the same sex, blood group, etc. They will only share the general resemblances of any two children born of the same parents. Multiple births tend to be a trait inherited and carried by women rather than men, i.e. the tendency to release more than one ovum from the ovary is an exclusively female distinction.

Fertilization occurs in the Fallopian tube within a day of ovulation. There may be as many as 100,000 sperm in the Fallopian tube, or as few as 100. But more than one sperm is needed to produce enzymes to break down the ovum's wall. The nuclei fuse (1) and the ovum wall hardens, preventing the entry of other sperm. Soon after fertilization the ovum begins to divide, first into two (2), then into four, and so on. The first division takes about 24 hours. Subsequent divisions take less time. The small bundle of cells, now called a morula, looks like a mulberry (3). The ovum at this point will normally be about to enter the uterine cavity.

Helped by a little uterine fluid, the cells of the morula are separated by a small space. The outer cells flatten into a cellular wall, the trophoblast, and the remaining cluster of cells, the blastocyst, moves to one side (4). The amniotic sac, placenta and fetus develop from these cells.

By about the 7th day small projections, the chorionic villi, will have formed on the trophoblast. These burrow into the uterus wall. The embryo undergoes continual cell differentiation (5-8). Cell differentiation takes place with each cell division. Thus this is a vital stage of development. Seemingly disproportionate repercussions, i.e. the stunted growth of an organ of the body, can occur from the damage or loss of one cell alone.
Implantation in the uterus establishes a basis of embryonic nutrition. After about 18 days the nervous system begins to form and it continues to develop until a few weeks after birth.
By the end of the first month the embryo is about 4mm long, about the size of a tapioca grain, now with millions of cells intricately organized to carry out specific functions. A primitive heart is now formed. The embryo is already 10,000 times bigger than the original ovum.

Implantation

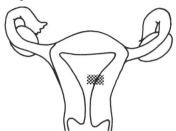

The enlarged sections of the blastocyst (below) show how it burrows into the uterus wall (endometrium).

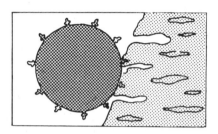

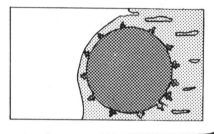

Non-identical twins, with separate placentae

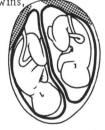

Identical twins, sharing the same placenta

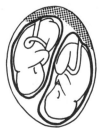

▨ Placenta

ECTOPIC PREGNANCY

If the fertilized egg does not reach the uterus within seven days, the tiny arm-like protrusions (chorionic villi) which will have formed by then will burrow into the wall of the Fallopian tube. The latter will become sorely distended as it can only stretch to a limited extent. The chorionic villi will continue to burrow into the wall in search of nourishment - which is obviously restricted. Eventually they will break through the muscular wall or into an artery causing bleeding, pain, and the loss of the embryo. (Surgery is always necessary.) Occasionally, however, the embryo will escape into the cavity of the abdomen and the chorionic villi will burrow into the wall where eventually a placenta will develop. Healthy babies have occasionally been delivered (by Cesarian section) which have developed within the abdomen.
Ectopic pregnancies are not uncommon, and since the same hormones are secreted as in a normal pregnancy causing the naturally anticipated reactions, they are not always detected until discomfort is felt. One in every ten women who has had an ectopic pregnancy is liable to have another. They are often due to prior inflammation of the Fallopian tube.

©DIAGRAM

THE EMBRYO

THE UTERUS

The tiny embryo has embedded itself in the wall of the uterus. Up to the 8th week the uterus contains the growth without enlarging.

Between the 4th and 8th weeks of life the embryo develops from a small limbless object resembling a white kidney bean 4mm long into a miniscule but complete human being all of 40mm from head to toe.

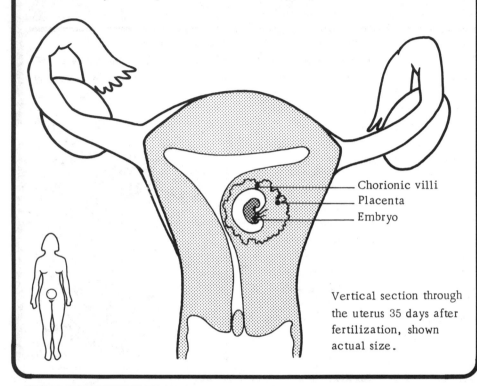

Chorionic villi
Placenta
Embryo

Vertical section through the uterus 35 days after fertilization, shown actual size.

DEVELOPMENT OF THE EMBRYO

By the end of the first 2 months the initial formation of the organs is complete. The sex of the embryo is apparent by the 50th day.

The embryo starts off as soft tissue. But by the 40th day the skeleton of cartilage is growing and by the 45th day the first bone cells appear.

We can follow the development, for example, of the arms, as an external feature, and the heart, as an internal organ.

The arms appear as buds at 30 days. By the 40th day they differentiate into hands, and lower and upper arms. The fingers are in outline only. By the 50th day the arms are growing and the fingers have separated.

(The legs and feet develop in the same way as the arms but correspondingly later.)

The heart continues to form for about 2 months but at 30-35 days it takes over circulation of the blood, which had hitherto been circulated via the umbilical cord and the placenta.

THE SUPPORT SYSTEM

THE PLACENTA develops during the first 10 weeks of life from the spot where chorionic villi first burrowed into the uterine wall. The remaining chorionic villi surrounding the embryo die.

The placenta, which looks like a bath cap roughly 8in (20cm) in diameter, and about 1 lb (0.45kg) in weight when fully developed, has a maternal (outer) and embryonic (inner) surface.

The outer surface is divided into roughly 20 lobes of chorionic villi and tiny blood vessels. The flow of blood to and from these vessels is supplied from the mother's uterine artery and vein. The inner surface, covered by a layer of amnion, has a series of tiny vessels radiating out from the umbilical cord at the center.

The umbilical cord, which links the embryo to the placenta, supports and protects two arteries and a vein which carry blood to and from the embryo.

The placenta acts as both a pool and a filter. The cells of the maternal surface fill with blood from which the blood vessels on the embryonic surface draw not only oxygen but also, by diffusion, proteins and vitamins which are essential to growth. Waste products will be drawn from the embryo's blood vessels in the same way. But though there is this free exchange, the blood systems of the mother and embryo are quite separate.

Although the placenta is largely protective in function, there are some drugs and viruses against which it has no defense.

The placenta also produces the hormone, progesterone, upon which the pregnancy depends. The placenta, or afterbirth, is expelled after the birth of the baby.

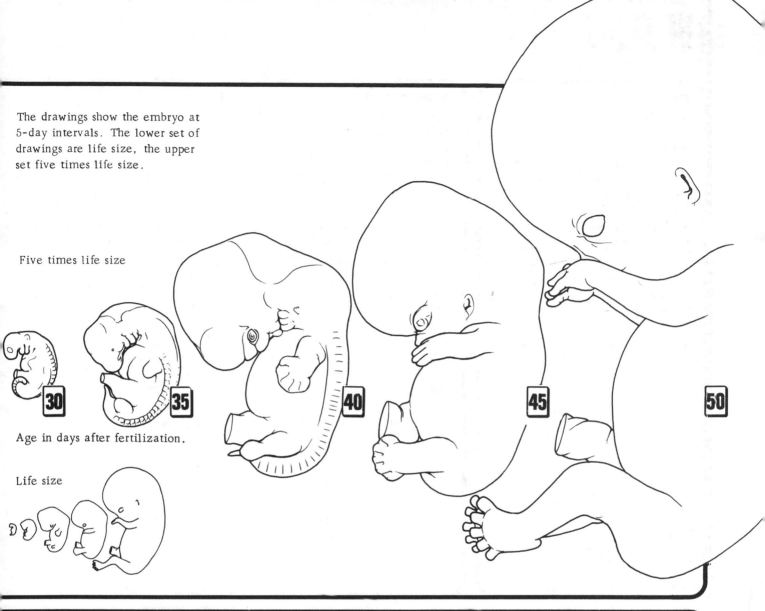

The drawings show the embryo at 5-day intervals. The lower set of drawings are life size, the upper set five times life size.

Five times life size

30 **35** **40** **45** **50**

Age in days after fertilization.

Life size

THE AMNIOTIC SAC (or bag of waters) is a sac of tough membrane, the amnion, within which a fluid (largely water with some protein) is contained. The sac forms around the embryo soon after it has become attached to the uterus wall. In it, the embryo has complete freedom of movement until about the 30th week. It is the growth and gentle pressure of the amniotic sac which slowly enlarges the uterus, and so the abdomen, giving the overt sign of pregnancy.

The fluid cushions the embryo from knocks, etc. It maintains a constant temperature and thus insulates the embryo, providing a level of water-conditioned central heating. It absorbs the waste excreted by the embryo, and is also the medium with which the fetus first learns to swallow.

In cases of multiple birth each fetus normally develops in its own sac. On average, at the 36th week of pregnancy, the sac contains about 2.4pt (1.1 liters). By the 40th week, however, roughly $\frac{1}{3}$ of this will have been lost.

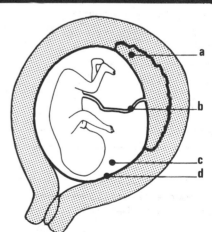

Cross-section through the uterus
a Placenta
b Umbilical cord
c Amniotic fluid
d Amnion

©DIAGRAM

THE FETUS

HORMONES

At ovulation, the follicle which releases the ovum changes its function to become the corpus luteum, which produces estrogen and progesterone.

Estrogen prevents more ova developing, while progesterone prepares the lining of the uterus for a possible implantation.

When the fertilized ovum implants it secretes human chorionic gonadotrophin. This maintains the corpus luteum until the placenta - which takes over the function of producing hormones after 3 months - has developed sufficiently to maintain pregnancy.

FETAL GROWTH

After the sex of the embryo has been determined, it is referred to as a fetus.

During the next 7 months the organs of the fetus grow. By the time the baby is born, all the organs have increased their weight 120 times. Weight increases from about 1oz (28.3gm) at 8 weeks to $7\frac{1}{2}$ lb (3.4kg), on average, at birth. It has increased 5000 million times since fertilization. Over the next 20 years weight increases only 20 times.

Length increases from 40mm at 8 weeks to 20in (50.8cm), from crown to heel, at birth. It has increased $12\frac{1}{2}$ times.

Growth gradually slows down just before birth, but rapidly speeds up after the first few days of birth.

Besides general fetal growth, head and body hair, and nails, are growing by the 18th week. By the 30th week, fat is deposited under the skin, making it smoother and more rounded, and less red and wrinkled.

Eyelids, which have been growing, close over the eyes in the 9th week, to open again in the 22nd.

On the internal front, at 14 weeks the heart pumps 6pt (2.83 liters) a day. By the 18th week, it can be heard externally by placing an ear on the mother's abdomen. At around 38 weeks the heart pumps 720pt (340 liters) per day. The total blood content is 0.6pt (0.28 liters).

Muscular reflexes develop on eyelids, palms, and feet, and the swallowing reflex starts at the 14th week. Thumbsucking also takes place around this time.

Fetal movements, or "quickening," can be felt at the 18th week. Urination into the amniotic fluid begins around the 14th week.

Premature live birth is possible at 22 weeks, but survival prospects are poor. Births at 24 weeks, though, can often be kept alive in intensive care units, by open-heart surgery and use of special respirators.

RHESUS INCOMPATIBILITY

There are various systems of grouping blood types. Within a system the groups are usually incompatible. If an incompatible group is introduced into the body (i.e. in blood transfusion), antibodies will be produced to inactivate the influence of the "foreign" blood type. In the case of blood incompatibility the consequences can be far-reaching.

One system of grouping is called the rhesus system, being discovered in experiments on rhesus monkeys in 1940.

Blood is denoted rhesus positive if it contains the rhesus factor. If it does not contain the factor it is rhesus negative. Most people (85%) are rhesus positive. Normally this is of no consequence; however, problems can arise in pregnancy when a rhesus negative woman is carrying a rhesus positive child. (If the father is rhesus positive, this is likely in three out of four pregnancies). In the course of labor or birth it is quite likely that the child's blood will get into the mother's blood system. If this happens, antibodies will be produced in the mother to protect her against the rhesus positive blood of the baby. But these antibodies are small enough, in a subsequent pregnancy, to pass through the placenta, and so inactivate the blood of the fetus, if it is again rhesus positive.

The dangers increase with each subsequent pregnancy. In a severe case the fetus may suffer from anemia, jaundice, or a weak heart. One in 200 pregnancies is complicated in this way.

If the child is likely to be moderately affected, it is usually given a complete transfusion of rhesus negative blood shortly after birth. As there is no rhesus factor in negative blood, no antibodies are formed, and within 40 days the baby's own rhesus positive blood will have replaced the transfused blood, which is broken down as normal in the liver. In serious cases the fetus can be transfused while still in the uterus.

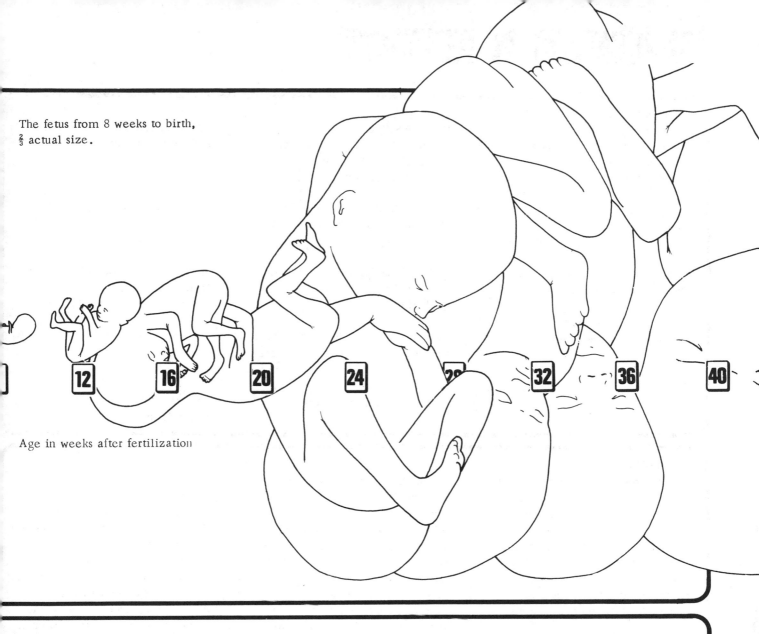

The fetus from 8 weeks to birth,
⅔ actual size.

Age in weeks after fertilization

12 16 20 24 28 32 36 40

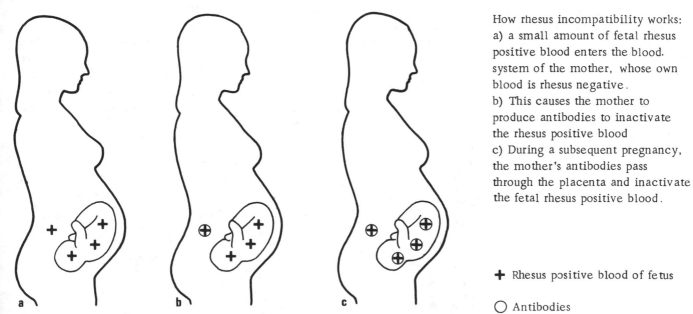

How rhesus incompatibility works:
a) a small amount of fetal rhesus positive blood enters the blood system of the mother, whose own blood is rhesus negative.
b) This causes the mother to produce antibodies to inactivate the rhesus positive blood
c) During a subsequent pregnancy, the mother's antibodies pass through the placenta and inactivate the fetal rhesus positive blood.

✚ Rhesus positive blood of fetus

◯ Antibodies

a   b   c

©DIAGRAM

COMPLICATIONS IN PREGNANCY

COMPLICATIONS IN PREGNANCY

The majority of pregnancies are completely normal, but there are some (about 30%) in which complications may develop. If left unattended these conditions can become serious, but the main purpose of antenatal care is to detect potential dangers and, where possible, to prevent them from happening.

The chart shows which unusual symptoms should be reported immediately to a doctor and what the possible causes of such symptoms may be.

DANGER SIGNS	POSSIBLE CAUSES
Severe abdominal pain, possibly with slight bleeding, in first few weeks of pregnancy.	Ectopic pregnancy
Vaginal bleeding with or without abdominal pain in the first 28 weeks of pregnancy.	Threatened miscarriage
Vaginal bleeding with or without abdominal pain after the 28th week of pregnancy	Premature separation of placenta (abrupto placenta, if pain; placenta praevia, if painless)
Severe swelling of fingers and face, with blurred vision and headaches, after the 20th week of pregnancy,	Toxemia of pregnancy
Gush of water from the vagina at 28-36th week	Rupture of membranes (bursting of amniotic sac)

MISCARRIAGE

About 1 in 6 women miscarry, and a threatened miscarriage is the usual cause of bleeding in the first half of pregnancy. Known more correctly as a spontaneous abortion, it occurs most often at the 6th or 10th week. The fetus detaches itself from the uterus and is expelled. Most common reasons are:
a) major abnormality in the fetus (about 50% of aborted fetuses are found to be abnormal);
b) death of fetus;
c) faulty hormone production;
d) anatomical defect or functional abnormality;
e) illness or infection;

The diagram shows three main types of miscarriage:
a Threatened
b Inevitable
c Incomplete

f) defective sperm or ovum;
g) psychological conditions.
There are different kinds of miscarriage at different stages of pregnancy: threatened, inevitable, complete, incomplete, and missed are the most usual. The symptoms of a threatened miscarriage will generally appear during the first few weeks. Bleeding, red or brown, without pain, occurs (a). At this stage it is uncertain whether the miscarriage will occur, and in 80% of cases the threat passes and pregnancy continues. But if the cervix opens, then the miscarriage is considered inevitable (b). A complete abortion means the uterus empties itself of the entire pregnancy. Incomplete abortion (c) leaves varying amounts of tissue in the uterus and a D and C is required. In a missed pregnancy, the fetus has died but remains in the uterus. It is eventually aborted and a D and C is usually given. A miscarriage can be a tense and despairing time, and many women still feel expectant of the birth even though they are no longer pregnant. But miscarriage is often a sign that the fetus was defective, and so rejected; and, after a first miscarriage, the chance that the next pregnancy will be successful is high.

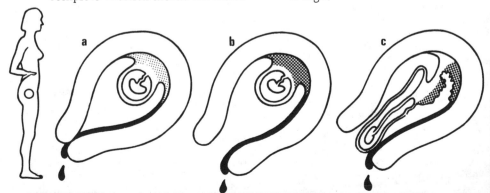

ECTOPIC PREGNANCY

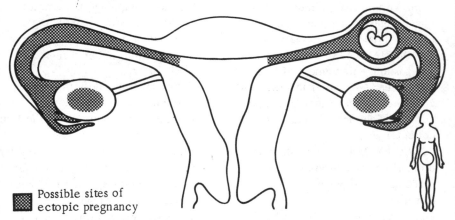

Possible sites of ectopic pregnancy

The diagram shows an ectopic pregnancy and possible areas where one might develop. Bleeding and pain in early pregnancy (6th-12th week) may be caused by an ectopic pregnancy. In this condition the fertilized egg has failed to reach the uterus and has implanted within a Fallopian tube.

DISPLACED PLACENTA

Hemorrhage after the 28th week may be caused by 2 fairly rare conditions:
a placenta praevia, in which the placenta lies in the lower part of the uterus; and
b abruptio placenta, in which the placenta separates prematurely from the uterus.

A woman with placenta praevia will be hospitalized after the 28th week. A Cesarian may be necessary, but in 20% of cases delivery is normal. Most cases of abruptio placenta continue normally. and in only 25% is separation too great for the infant to survive.

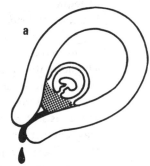

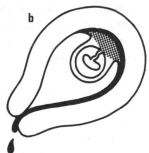

RUPTURE OF AMNIOTIC SAC

A sudden gush of water from the vagina after the 28th week generally means that the amniotic sac in which the baby grows has burst and amniotic fluid is escaping. If this occurs before the 28th-36th week it may often precede premature labor.

The woman is hospitalized and drugs or sedatives may be given to discourage labor. After the 36th week labor will be allowed to continue or will be induced as the baby is sufficiently mature to survive.

TOXEMIA OF PREGNANCY

Toxemia of pregnancy is a serious condition that can occur in late pregnancy. It affects 7-12% of women having their first baby and 3-6% of those having subsequent children.

SYMPTOMS AND SIGNS
There are 3 main warnings:
a) edema (swelling due to water retention) of fingers, face, or legs;
b) raised blood pressure; and
c) protein in the urine.
Excessive weight gain is also associated with toxemia.

PROGRESS
Toxemia goes through 2 stages.
a) Pre-eclampsia. This rarely occurs before the 20th week and the conditions are those mentioned above. (Some degree of edema, particularly of the legs, is, however, common in pregnancy. It becomes dangerous when associated with other symptoms.) As pre-eclampsia progresses, vision becomes blurred and the woman will suffer severe headaches.
b) Eclampsia is the final and most severe stage, and may be fatal. Fits, followed by unconsciousness or coma, are characteristic. It is particularly dangerous for the fetus.

TREATMENT
Today toxemia rarely develops to a final stage. This is almost entirely due to antenatal care, where the symptoms can be detected early on. For this reason alone regular attendance at the antenatal clinic is vitally important.
Bed rest and a restricted diet is generally sufficient to prevent toxemia from developing. Diuretic tablets may be given to get rid of excess water and salt. For more severe cases hospitalization is necessary so that the condition can be checked.

©DIAGRAM

301

ANTENATAL CARE

ANTENATAL CARE

Antenatal care is to ensure that every pregnant woman maintains good health, learns about child care, has a normal delivery, and bears healthy children. She (and hopefully her partner) will learn what is happening to her and what to expect.

Antenatal care has reduced maternal and infant mortality.

CLINICAL VISITS

The woman goes to her doctor or antenatal clinic to have her pregnancy confirmed.

The approximate date of the birth will be determined and the doctor will carry out:

a) a full consultation entailing discussion of past illnesses and operations, of present health, of any complaints now that she is pregnant, and consideration of her anxieties and questions; and

b) a general and then obstetrical examination, which will reveal any conditions which may affect the pregnancy, for which treatment will be given, and which enables the doctor to anticipate possible complications. Any previous pregnancies, miscarriages, or abortions will be considered.

The physical examination entails a general medical check up, a urine test, blood test, and blood pressure test, and obstetrical abdominal and pelvic examinations.

URINE TEST is taken to see if albumin, signifying a kidney disorder, or sugar, which may suggest diabetes, are present.

BLOOD TEST determines blood type, rhesus factor and iron content, and any presence of syphilis.

BLOOD PRESSURE TEST (which is taken at every antenatal visit) will show whether toxemia of pregnancy may occur. Its cause is unknown, but its effects can be severe. The arteries supplying the uterus go into spasm, reducing the blood supply to the placenta, with possibly fatal results to the fetus.

ABDOMINAL EXAM for muscle tone and possible enlargement of liver and spleen. (Also after the 12th and 28th week to check that growth and position of the fetus in the uterus is correct.)

PELVIC EXAM for pelvic structure and dimensions, and an internal test made as part of pregnancy confirmation.

Breasts and nipples will also be examined, and legs for signs of varicose veins.

The mother-to-be continues her antenatal visits every month until she is 7 months pregnant, then every 2 weeks until she is 9 months, with a weekly check up till the birth. Visits will be more frequent if any previous illnesses (eg diabetes, heart disease, hypertension) are likely to cause complications. At each visit the baby's position in the uterus will be checked.

When the fetal head settles down into the pelvic cavity ("lightening"), this suggests that the mother's pelvic shape and size are normal. An examination will be made to ascertain the position of the fetal head, and a cervical check made at the same time.

ILLNESSES

Any fever, chill, heavy cold, or other illness during pregnancy should be reported immediately to the doctor.

German measles contracted up to the 12th week of pregnancy may interrupt the development of the fetus and lead to deafness and heart defect in the child.

WEIGHT GAIN

A gain of 25 to 30 lb (11.3 to 13.6kg) from conception to birth is normal. Any more than this is unnecessary and even undesirable. At term the fetus weighs 7 to 8 lb (3.2 to 3.6kg), and the amniotic sac and placenta $2\frac{1}{2}$ lb (1.1kg). The mother carries the balance as fats and fluids in her tissues.

DIET

Eating for two is definitely out! A woman's average Calorie requirement is about 2300; the fetus requires only an extra 300. So intake should only increase slightly. It is important that extra emphasis is put on proteins, vitamins, and minerals. Whole foods are preferable to refined. But vitamin and mineral supplements should only be taken on the doctor's advice.

GENERAL CARE

Bathing is safe and relaxing. Water in the vagina is best avoided, but is only dangerous if forced in under pressure.

The genitals and breasts should be kept clean, as secretions become heavier during pregnancy. Dental care is also important, as the gums become softer and so more easily injured by food and toothbrushes. Injured gums are susceptible to infection which can cause loss of teeth.

Drinking and smoking should be kept to a minimum, and it is best not to smoke at all. Some women who normally smoke find that they do not want to during pregnancy. Douching is unnecessary, as the vagina is self-cleansing, and in pregnancy it can be dangerous. Unless there is a history of miscarriage, or possibility of complications, intercourse during pregnancy will not harm the fetus. Near term, intercourse in some positions may be uncomfortable for the woman, disturbing for the fetus, and difficult to achieve; but see Intercourse in pregnancy, p. 247.

ANTENATAL EXERCISES

RELAXATION helps to relieve tension during pregnancy and labor. A comfortable position is important for practicing it. The floor, a bed or chair are all suitable. In the later stages of pregnancy lying on your side may be more comfortable than on the back. Concentration on each part of the body is necessary to learn relaxation. After practice a sensation of "floating" can be felt.

BREATHING exercises should be practiced during pregnancy to gain control over the different muscles involved in breathing, which will be used in various ways in labor. In early labor, slow, deep breathing is used, relaxing during contractions. Later on, rapid, shallow panting is used, speeding up as each contraction intensifies.

THE POSTURE of a pregnant woman is altered as the abdomen enlarges. Strain on the back and abdomen can be avoided by learning a good posture, which can be obtained by pressing the whole spine length against a wall, tucking the buttocks and abdomen in, keeping the head up and shoulders back, and maintaining this posture. Humping and hollowing the back mobilizes it and prevents it aching.

LABOR POSITIONS can be practiced to strengthen the inner thigh muscles and control of breathing.

1 Relaxation practice
2 Breathing exercises
3 Posture exercises
4 Labor exercises

1a   1b

2a   2b

3a   3b   3c

4a   4b

©DIAGRAM

PREPARING FOR CHILDBIRTH

PREPARING FOR CHILDBIRTH

Labor and delivery are, for many women, the most alarming aspects of pregnancy. As with the physical changes of pregnancy itself, an understanding of the processes involved helps to relieve anxiety. A woman has 9 months to prepare for birth, and in order to participate fully in the experience she should become acquainted with all the available possibilities. Each woman's experience is personal and individual: whether she delivers at home or in a hospital, with or without drugs, should in the final analysis be for her to decide.

NATURAL CHILDBIRTH

Natural childbirth is the process of giving birth without the automatic use of drugs or obstetrical techniques. The idea was popularized in the 1930s by the English doctor, Grantly Dick-Read. It is based on the assumption that much of the pain of childbirth is caused by tension, itself due to the woman's anxieties and fears about labor. If these are eliminated, tension is relieved and pain will be lessened. The keynote to relaxation in labor is an understanding by the woman of all aspects of pregnancy and birth. Armed with this knowledge, Dick-Read maintained, the woman can approach labor with confidence.

PSYCHOPROPHYLAXIS

The psychoprophylactic method of childbirth was introduced by a French doctor, Fernand Lamaze. He felt that relaxation was not enough, and introduced pre-learned muscular and breathing exercises to be used by the woman during labor. With these a woman is no longer helplessly passive but can actively participate in the process of birth.

Although the theory and exercises are the basis of childbirth preparation today, few women actually give birth without drugs - nor should a woman feel she has failed if she is not one of those few.

BREATHING EXERCISES

The diagram shows four types of breathing used during labor:

a deep chest breathing for early first stage;

b shallow chest breathing used in the middle stage;

c shallow rapid breathing (panting) for transition; and

d expulsion, in which the breath is held while the woman bears down to push against the baby.

At the beginning of labor, deep abdominal breathing helps to relieve pressure. Once contractions increase so that they harden the abdominal wall, the woman switches to deep chest breathing. She continues to change as labor progresses, each time alternating her normal breathing (between contractions) with the learned form used during the contractions. Once delivery begins, expulsion breathing is used.

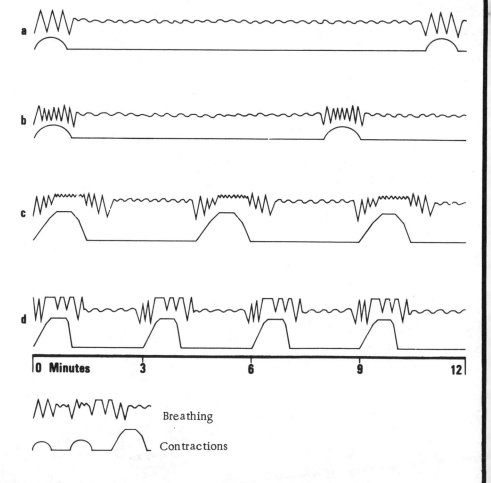

0 Minutes   3   6   9   12

Breathing

Contractions

HOME OR HOSPITAL

Forty years ago it was normal for a woman to have her child at home. In Holland over half of all deliveries are still carried out in the home, with absolute safety. In the US and Britain, however, home births are now extremely rare. The dying practice of midwifery and the increasing accent on the safety and well-being of both mother and child has meant that most doctors prefer to deliver in hospital where the facilities for any possible complications are immediately available. However, given an absence of complications, there is no reason why a woman who wishes to do so should not deliver her child at home. The advantages are obvious - your own room, a familiar atmosphere and possibly the presence of friends, can greatly ease the doubts and tensions of labor.

Most women feel safer delivering in a hospital, though. It is always recommended that a woman have her first child there. After the first birth it is often possible to predict whether the next will be normal. The risks of pregnancy and labor rise for the third and subsequent births, and for mothers under 17 and over 35 - so these are usually delivered in a hospital. A pregnant woman will also be admitted for a hospital birth if there is any evidence or suspicion of possible difficulty or danger - toxemia, diabetes, rhesus incompatibility, prematurity, multiple birth, difficult fetal position, small pelvis. Unfortunately the choice of a home or a hospital delivery today largely depends on the facilities available. The main aim of doctors has been to lower the risks, and even though hospital birth is not essential for many women, it has become part of a routine for greater safety.

DRUGS IN LABOR

A woman who delivers in a hospital can use a variety of pain-relieving or pain-killing drugs. There are 2 kinds: analgesics and anesthetics.
ANALGESICS These relieve pain and may be taken orally, inhaled, or injected. During early labor, painkillers are rarely necessary, but narcotics such as Demeral (Pethidine) may be given. Barbiturates, also used in the past, are now not generally given, as they can cause breathing to stop in some cases.

INHALANT ANALGESICS can be self-administered and the intake controlled as needed. There are 2 types: trilene, and nitrous oxide ("laughing gas"), and they are used for dilation and delivery.

REGIONAL ANESTHETICS are widely used. Injected into the woman's body at a specific point, the anesthetic completely blocks off pain. The epidural block is probably the most efficient and is becoming more available. Anesthetic is injected into the epidural cavity which lies between the spinal cord and its covering, the dura. It numbs the entire region from the lower abdomen to the feet, and can be readministered during labor. Although an effective pain-reliever it can lessen the mother's ability to push and may result in a forceps delivery.
Other regional anesthetics include the pudendal nerve block, given immediately before a forceps delivery, and the caudal block. Both anesthetize the pelvis. A paracervical block is injected into the plexus, making the cervix and upper vagina completely insensitive. A regional anesthetic is also given before an episiotomy.
Drugs used in labor are carefully supervised but most do cross the placenta and can make the newborn infant drowsy and slow to suck. These effects, although undesirable, are temporary and should be weighed against the possible long-term effects on a woman of a painful and distressing labor.

The diagram shows the nerve supply for the uterus, cervix, and vagina. Most of the nerves for this area collect together at the paracervical ganglion or plexus (a). When local anesthetic is injected here, the cervix and upper vagina are desensitized, without impairing the ability of the uterus to proceed with labor.

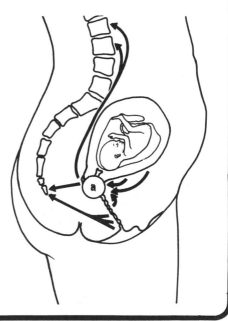

©DIAGRAM

EXPERIENCE OF CHILDBIRTH

LABOR

Women often find that their experiences during labor differ. However, most labors follow a similar general pattern.

Hours

Frequency and intensity of contractions

Stage	FIRST STAGE Effacement of cervix	
Frequency of contractions	15-30 minutes	2-5 minutes
Duration of contractions	40 seconds	40-90 seconds
Drugs	Tranquilizers, narcotics, epidurals	
Sensations	Low backache with contractions	Pain becomes abdominal

Position of child

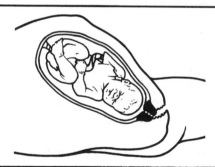 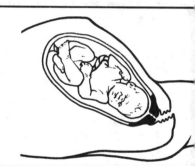

ALTERNATIVE BIRTH STYLES

The possibly damaging effects on a woman of a prolonged, painful, and distressing childbirth are now well recognized. But recently the accent has moved from the effects of birth on the mother to the effects of birth on the child. It is argued that the methods of delivery commonly used can have a lasting and detrimental effect on a child. There are 3 main advocates of a new approach to birth: R D Laing, the Scottish psychologist; Frederick Leboyer, the French obstetrician; and the American psychoanalyst, Elizabeth Fehr. All concern themselves primarily with the well-being of the child and consider that not only is it aware of its time in the womb but also of its arrival into the world. They feel that its initial impressions are critical to its future development, and that being dangled upside down and slapped on the bottom are both unnecessary and damaging.

R D LAING

Laing argues that cutting the umbilical cord too soon is a major cause of birth traumas. The cord linking mother and newborn child still carries on its functions of providing blood, oxygen, and nutrients, even after the actual birth. Immediate severance causes unnecessary shock to the baby, which can be avoided by leaving the cord uncut until it has naturally ceased to function. It used to be common practice not to cut the cord until it stopped pulsating. Doctors today tend to cut it while it is still performing transitional duties. Laing believes that if the cord is left for 4 or 5 minutes until the baby's own circulation has taken over the process is more natural and non-traumatic.

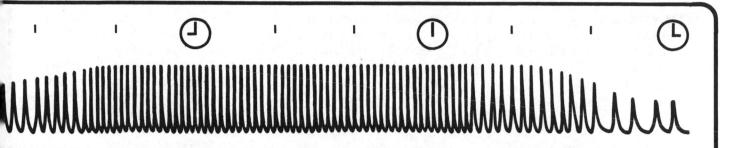

	SECOND STAGE	THIRD STAGE
Dilation of cervix	Delivery	After birth
2-3 minutes	2-5 minutes	5-10 minutes
40-90 seconds	60-90 seconds	
Epidural	Trilene, nitrous oxide, paracervical block	Ergometrine
The "show" Amniotic sac ruptures (if not earlier)	Possible nausea "Bearing down"	

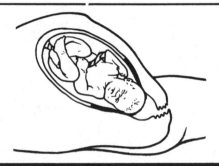

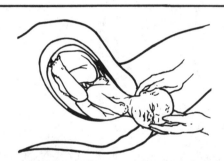

FREDERICK LEBOYER
In his book "Birth Without Violence" Leboyer also advocates not cutting the umbilicus until it has ceased to function. In addition he believes that the newborn child's eyes, ears, and skin are hypersensitive, and should be treated gently and with respect. Struggling out of the womb into bright lights and noise, being put on hard scales and hung upside down (which immediately forces the spine into an unaccustomed angle) are all alien to a being who has been 9 months in the womb. Leboyer suggests that lights and noise in the delivery room should be at a minimum, that the child should be placed on the mother's stomach before cutting the cord, and that after the cord is cut, the child should be placed in a bath of water at body temperature and **allowed** to move and "open up" in a calm, unhurried way. Certainly Leboyer has noted that a baby born in this way opens its eyes and begins smiling immediately - one cry, as opposed to a torrent of tears and red-faced rage, having satisfied everyone of its ability to breathe.

ELIZABETH FEHR
The late Elizabeth Fehr believed that a traumatic birth left a permanent impression. She introduced into psychoanalysis the process of "rebirthing," by which a person retraces their life backward toward birth. As a result of her investigations, she concluded that auditory hallucinations suffered by schizophrenics might well be related to sounds heard by the child as it struggled from the womb. Her work has given added impetus to the movement for gentler birthstyles.

©DIAGRAM

LABOR

FETAL POSITIONS

These are the positions in which the baby can lie in the mother's pelvis just before labor begins. The face down position may cause remolding of the skull. A transverse lie can usually be manipulated into another position; uncorrected, it requires a Cesarian section.

Frequency of birth positions

Vertex: head down   95%

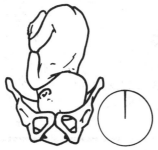

Vertex: face down   0.5%

Breech   3.5%

Transverse   1.0%

DURATION OF PREGNANCY

Pregnancy normally lasts from between 36 and 42 weeks from fertilization.

LABOR

Throughout pregnancy the uterus undergoes slight contractions known as Braxton Hicks contractions. As the pregnancy nears term these contractions become more frequent and intense.

Labor itself has three distinct stages. In the first stage, the cervix is "effaced" and "dilated," to allow the fetus to pass without damaging it. The second stage is the actual delivery of the baby; the third, delivery of the placenta.

FIRST STAGE

Before delivery can begin the uterus must undergo a change in shape to permit the fetus to pass through the cervix. The upper uterus pulls the lower uterus and cervix up around the head of the fetus.

This process takes about 8 hours for women having their first child (primigravidae), and may take 4-5 hours for those having their subsequent children (multigravidae). By the time effacement is completed, contractions are occurring about every 3 to 5 minutes, and lasting 40-90 seconds. The mucous plug lodged in the cervix throughout pregnancy is displaced, as the cervix begins to dilate to allow the baby a free passage through. The process, a continuation of effacement, reveals the amnion surrounding the baby's head. If the amnion has not been ruptured already, it is usually ruptured during dilation, either by the baby's head or by the doctor delivering. This releases a quantity of amniotic fluid. To allow the baby to pass, the cervix must dilate to accommodate its head, which is about 4in (10cm) in diameter.

As dilation proceeds, contractions become more frequent and intense; by full dilation they will be occurring every 2-3 minutes and lasting 60-90 seconds.

Dilation takes from 3-5 hours for a woman having her first child, and less for subsequent children.

FIRST STAGE OF LABOR

The cervix and uterus as labor begins.
Contractions every 15-30 minutes.

Partial effacement
Contraction and retraction of the uterus shorten the neck of the cervix.
Contractions every 10 minutes.

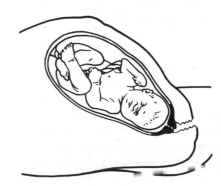

INDUCED LABOR

Labor may be induced artificially if the health of the mother or fetus is in danger. There has also been a tendency for births to be induced for the convenience of hospitals. Labor will normally begin within 24 hours of induction, and tends to be shorter than a spontaneous labor. But the contractions follow the same pattern as spontaneous labor, even though each stage takes much less time.

The frequency of induction varies greatly from place to place. The following are the main conditions in which induction is justified.

PRE-ECLAMPSIA is the major medical cause of induction, accounting for 50% of such cases. It is characterized by high blood pressure, edema, and proteinuria in the woman. Although the danger to her is slight, danger to the fetus increases with severity. Should eclampsia (a type of epilepsy) develop, the mortality rate is very high for mother and fetus.

POST-MATURITY accounts for 35% of inductions. If the fetus remains in the uterus after term, it will continue to grow, making for a difficult birth. Placental function begins to fall off after the 40th week, and there is a tendency toward mental damage in post-mature babies.

HEMORRHAGE (10% of inductions) is caused by the placenta separating from the uterus before birth. In difficult cases Cesarian section is required.

RHESUS INCOMPATIBILITY necessitates induction in severe cases.

DIABETES in the mother, if untreated, results in a high mortality rate for both mother and fetus. But antenatal checks will reveal diabetes if it has not been detected previously, so mortality is not a problem. Treatment must be carefully controlled. Even so, fetuses tend to be puffy and fat, and thus difficult to deliver. They will usually require induction by the 38th week.

METHODS OF INDUCTION
There are two methods, surgical and medical.

SURGICAL INDUCTION involves the artificial rupture of the amniotic sac, below the fetus. About a pint of fluid is drained off, and labor usually begins within 24 hours. The rupturing of the sac does not in itself make for any difficulty in delivery.

MEDICAL INDUCTION involves intravenous infusion of oxytocin, to stimulate uterine contractions. The oxytocin is given throughout labor - though not for more than 10 hours.

Many procedures in obstetrics, including induced birth, have become controversial issues in some countries: their justification has been in question. Lack of information as to the reasons for their use exacerbates the situation, causing resentment between women and hospital staff.

Full effacement
Contractions every 5 minutes

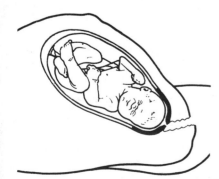

Partial dilation
Continued contraction and retraction dilate the cervix.
Contractions every 2-5 minutes

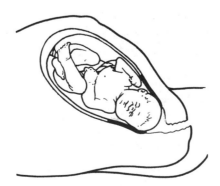

Full dilation
The fetus is able to pass through the cervix without damaging it.
Contractions every 2-3 minutes

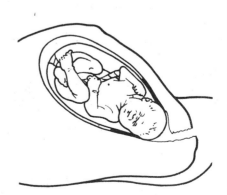

© DIAGRAM

BIRTH

DELIVERY

TRANSITION TO SECOND STAGE

The transition to the 2nd stage is characterized by feelings of pressure in the low pelvis, backache, and often nausea and leg cramps.

At this point, when tension rises, it is helpful and comforting for the woman's partner or friend to prompt her to do her ante-natal breathing exercises.

Contractions continue, once every few minutes, and there is an increasingly uncontrollable desire to push or bear down - like trying to relieve severe constipation.

However, it is not safe to bear down until full dilation has been achieved, as the cervix may tear.

SECOND STAGE

When the cervix is fully dilated, bearing down can begin. The baby is now being pushed out of the uterus and down the vagina, and will be delivered in anything from 5 to 40 minutes.

If there is danger of the perineum (B01) being torn by the baby's head, an episiotomy may be performed: the doctor makes an incision from the vagina obliquely down towards the anus. This cut is sewn up after the delivery is completed.

The fetus began its journey on its side, head first (but see p. 308).

Contractions of the uterus force the fetus down into the pelvis. The head is rotated downward beneath the pubic arch and, as the head is born, it rotates back to its original position. The shoulders and then the breech follow the same pattern of rotation as they are delivered, and the baby is born.

The 2nd stage is now completed - in primigravidae it takes up to 1 hour, in multigravidae less.

Mucus is extracted from the mouth and nose of the baby, who may be suspended upside down to drain mucus from its lungs. The umbilical cord is clamped and cut, sometimes immediately and sometimes after some minutes; in a week the stump of the cord will dry out and fall off.

The baby on delivery is wet and covered in a fatty substance, vernix. As oxygen begins to circulate in the lungs, the baby's color will change gradually from bluish to pink.

THIRD STAGE

The placenta is delivered within 30 minutes of the baby. As birth occurs, the uterus retracts quite markedly. The placenta is not capable of contraction or retraction, and shears away from the uterus. Light traction on the cord aids its delivery.

Once delivered, the placenta is checked to ensure none is left inside the uterus, since this could lead to infection and hemorrhage.

CHECKS

The baby is checked just after birth for congenital malformations. It will be given a vitamin K injection to help blood clotting in case of hemorrhage, and silver nitrate eye drops to help protect the eyes. Heart and lung functions are checked. A hand or foot print is taken to check for mongolism, and a mouth smear to check for PKU.

SECOND STAGE OF LABOR

SECOND STAGE OF LABOR

Full dilation signifies the beginning of delivery. The woman "bears down" to help expel the baby. Contractions every 2-5 minutes.

The baby's head passes through the cervix and rotates to squeeze beneath the pubic arch.

The head is born, and rotates back to its previous position. The baby's shoulders rotate to pass through the pelvis.

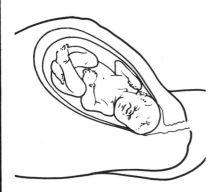
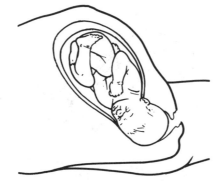
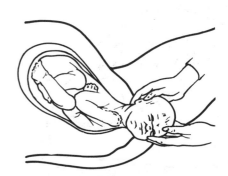

POSITIONS FOR DELIVERY

Dorsal

Left lateral

Lithotomy

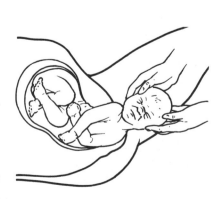

Semi-upright

The diagrams show the positions currently used for delivery. The dorsal, in which the woman lies flat on her back with her knees up and separated, and the left lateral in which she lies on her left side, knees towards her chest, are the most typical in many countries. But the lithotomy position is the one usually used in the USA.

FIRST STAGE

During the early stages there is little active work a woman can do, and generally she remains up and about waiting for the contractions to increase in frequency. Once labor is fully established and she is in bed, she should choose the most comfortable position. During the first stage it is not advisable for her to lie flat on her back. Sitting propped up on pillows or lying on one side improves the flow of blood through the uterus, providing more oxygen for the baby. During transition itself a woman can squat, sit upright, use any position which helps her the most.

SECOND STAGE

This is the time when the mother can most actively participate in labor, and she should choose the position in which she can work the best - changing at any time if she wishes. Many women find they can push best by clasping their legs, drawing them up to their abdomen.

THIRD STAGE

Once the baby is born, the dorsal position is generally used for the delivery of the placenta.

The right shoulder, then the left, is born.

The baby breathes spontaneously. Mucus is cleaned from its face and air passages. The umbilical cord is clamped.

THIRD STAGE

The placenta is delivered within 30 minutes of the baby.

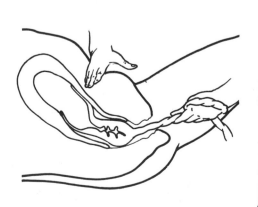

©DIAGRAM

AFTER THE BIRTH

PROCEDURES AFTER BIRTH

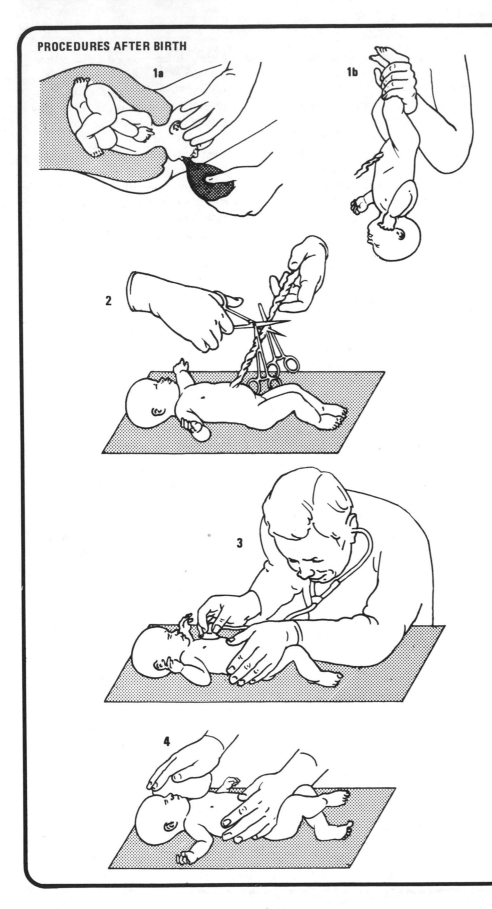

1 BREATHING
To aid breathing, air passages are cleared with a mucus extractor as the head emerges (**a**).
The baby may be held upside down to prevent inhalation of mucus (**b**). Oxygen may be given through a mask or, if necessary, through a tube passed into the trachea.

2 CUTTING THE CORD
The umbilical cord may be severed as soon as the baby's air passages are quite clear and regular breathing is established. One clamp is placed on the cord about 6in (15cm) from the baby and a second 3in (7.5cm) nearer the mother. The cord is then cut between these two points.

3 PRELIMINARY CHECKS
Immediately after birth, the baby's condition is briefly checked. This initial inspection includes a chest examination with a stethoscope to listen to the heartbeat and breathing rate.

4 WASHING
It is no longer usual to bathe a newborn baby immediately as it is thought that the vernix may give some protection against infection. Generally, only the hands and face are washed, either with warm water or with a very mild antiseptic.

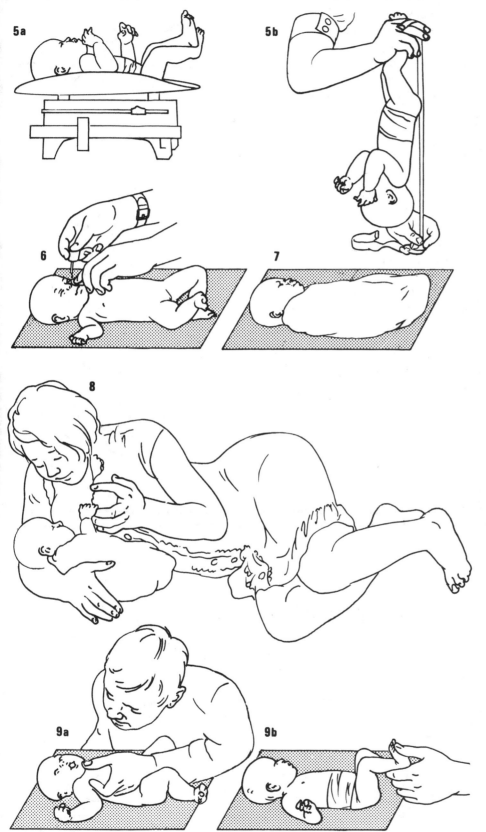

5a **5b** **6** **7** **8** **9a** **9b**

5 LABELING, WEIGHING, AND MEASURING

The baby is labeled with his family name, the time and date of birth, and any relevant details of the delivery. He is then weighed (**a**) and his body length (**b**) and head circumference are measured and recorded.

6 PREVENTING INFECTION

A drop of weak silver-nitrate solution may be put into each eye to eliminate any chance of serious infection.

7 KEEPING HIM WARM

A newborn baby's temperature can drop dramatically after birth, so it is essential that he is wrapped in a blanket or placed in a warm crib or incubator.

8 FEEDING

If the mother has decided to breast-feed and the condition of the baby is satisfactory, he may be given to his mother after delivery for a feeding. This is desirable as the baby's suckling stimulates milk production in the breasts, and also helps the uterus to contract.

9 MEDICAL EXAMINATION

Within 24 hours of delivery a baby is usually given a full medical examination. This includes a check for cleft palate (**a**) and club foot (**b**). Also checked are the head, spine, hips, and genitals. A blood test is performed to detect a rare but easily treated cause of mental handicap - phenylketonuria (PKU).

©DIAGRAM

PROBLEM BIRTHS

EPISIOTOMY

An episiotomy - cutting of the mother's perineum - is sometimes carried out during labor to ease the passage of the baby's head.

It is a simple operation carried out under local anesthetic; the perineum is cut along the mid-line or slightly obliquely. The incision is stitched after the delivery.

An episiotomy is needed during a breech delivery, forceps delivery, and if the baby is being delivered very quickly, as in a premature or multiple birth. It is also given if the tissues of the perineum are so rigid that they are delaying the second stage of labor, or if the entrance to the vagina is so tight that tearing is inevitable.

Whether or not an episiotomy is desirable in all deliveries is a matter of dispute among doctors.

BREECH BIRTH

Normally the fetus moves from breech to vertex position between the 24th and 28th week. However, some fail to do so, and 3.5% of fetuses remain in breech position till birth.

A normal-sized baby in breech position will usually be delivered with no problems for mother or child. But a small pelvis or a large fetal head may lead to difficulty.

The duration of delivery can be critical - a long delivery may result in oxygen starvation if the head squeezes the umbilical cord. A short delivery may cause damage to the fetus and mother.

Breech delivery is in three stages: the breech and legs are born first, then the shoulders, and finally the head. Forceps are usually used to help ease the head out gently and avoid injury.

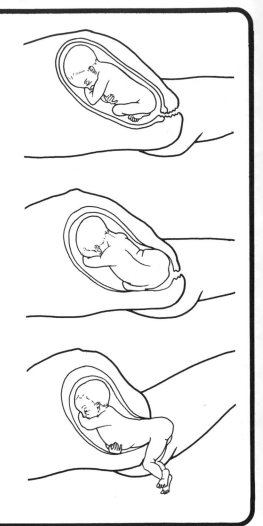

FORCEPS DELIVERY

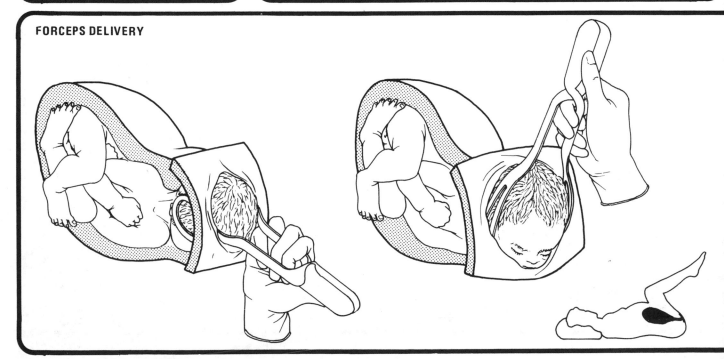

MULTIPLE BIRTHS

Twins are born on average once every 85 births, triplets once every 7,500 births, quadruplets once every 650,000, and quintuplets once every 57,000,000 births Difficulties may arise in multiple births. They tend to be premature, and so must be delivered in a hospital. Labor is usually straightforward as each baby is small. The birth canal is dilated after the birth of the first baby so that subsequent ones are born easily.

Toxemia of pregnancy and anemia occur more frequently in a multiple pregnancy.

The maternal death rate in twin pregnancies is 2-3 times greater than a single pregnancy. In one in 14 twin births, one twin dies, and with larger multiple births the likelihood of fetal death rises steeply.

CESARIAN BIRTH

Cesarian section is an operation carried out on a pregnant woman to deliver her baby, if this is not possible through the vagina.
Reasons for it include:
a) fetal distress;
b) low-lying placenta;
c) very small pelvis;
d) obstructive fibroids;
e) transverse fetal position; and
f) previous uterine injury.
General anesthetic is given before the operation. A cut is made below the navel into the abdomen and uterus, and the baby is delivered through this.
It is possible for a woman to have several Cesarian sections, but 4 is thought to be enough.
It is only in the last 25 years that Cesarian section has become a safe operation. Now it accounts for 10% of all deliveries in the USA. It is often used in preference to a difficult forceps delivery.

VACUUM EXTRACTION

Vacuum extraction is used as an alternative to simple forceps delivery, and is very popular in Scandinavia and the UK.
Vacuum extraction can be started before the cervix is fully dilated. A metal cup is inserted into the vagina and placed against the fetal head. It is connected to suction equipment, the vacuum formed being strong enough to allow the fetus to be gently pulled out of the uterus.
Scalp tissues are sucked into the cup, but within a few hours of delivery any swelling subsides.

EMERGENCY BIRTH

In 80% of cases a woman can deliver without any problems; but if a birth begins unexpectedly, the help of a doctor or hospital should always be sought. If no help is available, then it is best simply to give encouragement to the mother, and let nature take its course without interference. Let the mother "bear down" (push against the baby) as soon as she wants to; do not worry whether full dilation has occurred, for any damage to the cervix can be repaired later in hospital. Show her how to push during contractions, by holding her breath, raising head and shoulders, and pulling the knees up. Pain is not normally a problem, if the atmosphere is kept calm and quiet.
After the birth, clean the mucus from the baby's mouth and nose. Breathing should begin within 30 seconds. Do not cut the cord. Keep mother and baby warm.
More than $\frac{1}{2}$pt of blood from the uterus (not the placenta) signifies hemorrhage. In this case only, the abdomen should be massaged to try to ease the bleeding.

Frequency of forceps delivery varies greatly from country to country. In the USA an average of 50% of births involve forceps. About 6% of births in the UK are forceps deliveries.
Forceps are used in the 2nd stage of labor to aid the progress of the fetus, and are used in the following circumstances:
a) slow or no fetal progress;
b) maternal distress, eg pre-eclampsia.
exacerbated by the effort required during labor;
c) fetal distress.
APPLICATION
Forceps consist of two curved blades that interlock and fit closely around the fetal head. One blade is inserted into the uterus and located in position around the head. The other blade is then inserted, and when positioned, locked into the first blade. Gentle traction draws the fetus down through the vagina. Local anesthetic and episiotomy may be needed in forceps delivery.
CONDITIONS
a) The cervix must be fully dilated to allow insertion of the forceps blades. Damage will be caused to the cervix and vagina if the fetus is pulled through before dilation.
b) The amniotic membranes must be ruptured, if they are not already, and bladder and rectum empty.
c) The forceps can only be applied to the fetal head.

©DIAGRAM

THE PREMATURE BABY

PREMATURITY

Any baby who weighs less than 5½lb (2kg) at birth is generally considered premature. Strictly, however, premature babies fall into two groups - those born before week 35 of pregnancy (pre-term) and those born at full term but who weigh less than 5½lb (low birth weight).

REASONS FOR PREMATURITY

Factors causing premature labor can include: lack of adequate antenatal care; poor maternal health; maternal disease such as diabetes; fetal congenital abnormality; a multiple pregnancy; and an accident or severe shock. Low birth weight babies may simply be the product of small parents (genetically small), or may have received insufficient nourishment while developing in the uterus.

BIRTH

Because of his smallness, a premature baby is delivered very quickly. As there is a slight risk of brain damage if the soft bones of the baby's head suddenly strike the mother's hard perineum, an episiotomy and forceps are often used to ensure the baby's safe delivery.

APPEARANCE

Apart from being small and light, a premature baby's body seems ill proportioned. His eyes are widely spaced and usually shut, and his nails very short. His skin has no fat and is red and wrinkled. His cry is feeble and his movements weak.

DEVELOPMENT

Premature babies gradually catch up both physically and mentally with full-term babies. But a very small premature baby - and some with a birth weight of less than 2lb (0.9kg) now survive - may take some time to do so.

CARE OF THE PREMATURE

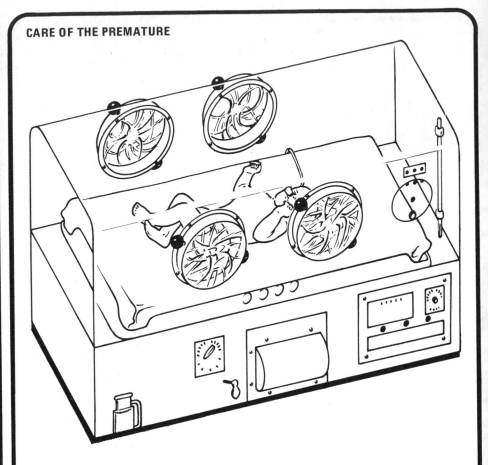

The body systems of a premature baby are generally immature. Breathing, feeding, keeping warm, and resisting infection all present problems to both pre-term and low birth weight babies. The use of an incubator improves the nursing of a premature baby and greatly increases his chances of growing healthily and developing normally. Temperature control is especially difficult for a premature baby who lacks the insulating layer of fat beneath the skin. An incubator provides extra warmth and allows the careful control of both humidity and air temperature. The isolation of an incubator also prevents the transfer of serious infection.

An immature respiratory system and poorly developed cough reflex make breathing and clearing air passages difficult and the baby may need help from special apparatus to maintain steady breathing. Feeding can also be problematic for a premature baby because his ability to suck is poor. He is fed on a strict three-hourly schedule through a dropper or tube or, if he is able to suck, from breast or bottle.

The inevitable limitation of physical contact between the premature baby and his parents can cause some distress in the early days. The parent-child bond, however, begins to be established just as soon as the baby no longer needs special care - usually when he has reached 5½lb (2kg) - and can be freely handled by his parents.

THE NEW BABY

THE BIRTH EXPERIENCE

In recent years, obstetricians and psychologists have taken an increasing interest in the effect of the birth experience on a baby. Two in particular - R.D. Laing and Frederick Leboyer - have emphasized the importance of not severing the umbilical cord until the baby's own circulatory system is working fully. This, they argue, makes the birth process more natural and less traumatic.

In his book "Birth Without Violence," Leboyer suggests that lights and noise in the delivery room should be at a minimum, that the baby should be placed on the mother's abdomen before the cord is cut, and that after this the baby should be placed in a bath of warm water and allowed to "open up" in an unhurried way.

CIRCUMCISION

Circumcision is the surgical removal of the foreskin covering the tip of the penis. It can be carried out for medical reasons, and is also commonly performed for religious reasons. Medical thinking in some countries used to favor circumcision for reasons of hygiene, but most doctors now argue that a normal foreskin is quite easy to keep clean and free from infection.

Circumcision involves a simple operation, usually carried out straight after the birth or about one week later. (The period between is avoided as the establishment of feeding may be interrupted.) For a few days after circumcision the penis requires special care, about which the doctor will give the necessary advice.

THE NEWBORN BABY

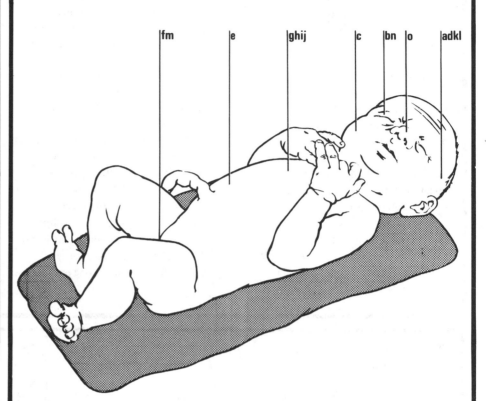

COMMON CHARACTERISTICS

Some aspects of a newborn baby's appearance may surprise or even alarm a new parent. All are quite normal and usually disappear within a day or so. They include:

a misshapen or swollen head due to pressure in labor;

b swollen, puffy eyes;

c tiny yellowish-white spots covering the face;

d throbbing fontanel;

e distended abdomen;

f swollen genitals;

g bright pink, purple, or even blue skin color due to temporary lack of oxygen during delivery;

h vernix - a greasy, white protective substance - on some parts of the body;

i lanugo - fine, downy hair - on some areas of the body;

j maternal blood smeared on the body due to tearing or cutting of vaginal tissues.

LESS COMMON CHARACTERISTICS

A newborn baby may display other characteristics. These features are less common but need not worry a parent as they are of no significance. They include:

k instrument marks on the head after a forceps delivery;

l a cephalhematoma - swelling caused by bruising during birth - on the head (sometimes the swelling hardens but it always remains harmless and subsides in a few months);

m vaginal discharge or bleeding in newborn girls, caused by the transfer of maternal hormones via the placenta before birth;

n crossed eyes; red spots or broken veins in the whites of the eyes caused by pressure on the neck during delivery;

o snuffly breathing and frequent sneezing caused by a shallow nose bridge.

©DIAGRAM

POSTNATAL CARE

AFTER THE BIRTH

For about 10 days after birth, there is a steady loss of a bloody substance, called lochia, from the vagina, as the placental site and uterine lining break down.

The breasts produce colostrum for the first few days. This is then replaced by milk. Sometimes the breasts are overfull and painful.

For a day or two after the birth, the mother may experience some constipation, and difficulty in urinating; or she may urinate involuntarily, especially when coughing or laughing. This is caused by muscle slackness in the pelvic area and is best treated by early mobilization and reassurance.

Changes in hormone balance often cause the mother to be depressed and weepy for a short while after giving birth. This is called the "3rd or 4th day blues," after the time it usually occurs.

Menstruation normally returns after about 24 weeks if the mother is breastfeeding, or 6-10 weeks if not. Ovulation starts in the first case after about the 20th week. Women who do not breastfeed can therefore become pregnant much sooner; while of those who do, the longer they breastfeed, usually the lower the likelihood of pregnancy, although this is not always the case.

POSTNATAL EXAMINATION

This takes place after 6 weeks. The position of the uterus is checked, and the mother is asked if she has any pain or discomfort in the abdominal area, or any vaginal discharge. Often a blood test is made to check for anemia. Blood pressure is always measured.

A vaginal inspection is made to see that any stitches from an episiotomy have healed, and if there is any inflammation, or erosion, of the cervix (25% of mothers have it to some degree after giving birth). Mostly it is self-healing but treatment is required if it persists.

POST-NATAL EXERCISES

These are important for most women, as they retone muscles (especially those of the pelvic floor), stimulate blood circulation, and promote good posture.

They should be done as many times as possible a day as soon as the mother is up and about. Some abdominal exercises can be performed while feeding the baby (see exercise 1) or around the home (see exercise 2).

1

2

3 a

3 b

Exercise 1
Tighten abdominal muscles while sitting in correct posture position.
Exercise 2
Stand straight, pull in abdomen and buttocks. Tighten up inside.

Exercise 3
Lie relaxed on floor, knees bent, feet flat (a). Draw in abdomen tightly, then raise head (b). Hold few seconds, lower head slowly. Repeat 10 times.

CHANGES IN THE UTERUS

The puerperium is the time when the uterus and other genital organs gradually return to their normal size.

The uterus, cervix, and vagina undergo immense stretching during pregnancy and labor; but within 6 weeks of the birth evidence of the pregnancy is difficult to find. The uterus weighs, after birth. about $2\frac{1}{2}$ lb (1kg) and after 2 weeks, about 11oz (350g). In rare cases the uterus retroverts following pregnancy.

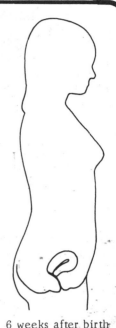

After birth   1 week after birth   6 weeks after birth

Exercise 4

Alternately hollow (a) and hump (b) back, abdominal muscles held tightly.

Hollowing back, move head and hip first to right, then to left (c).

Exercise 5

Lie flat on floor, back straight (a). Feet must be held by another person or a heavy item of furniture. Cross arms on chest, raise body to forward position (b), then lie back (c). Arms can be stretched forward above head before lying back.

Exercise 6

Lie on back legs straight. Move feet up and down, and round in circles (a). Tighten kneecaps, tense leg muscles (b). Ankles crossed, press thighs together, tighten up inside (c).

4 a

5 a

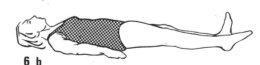

6 a

4 b

5 b

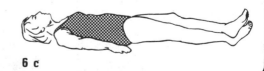

6 b

4 c

5 c

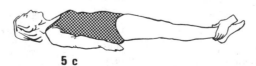

6 c

©DIAGRAM

THE MOTHERS AFTER BIRTH

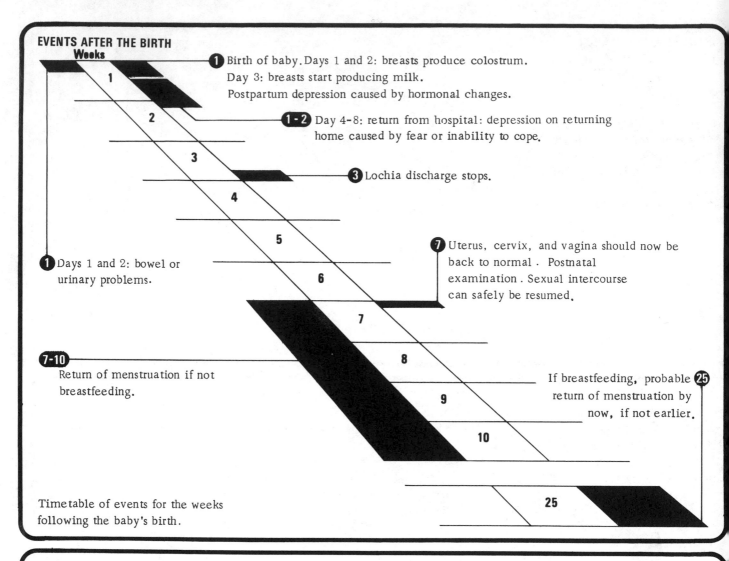

EVENTS AFTER THE BIRTH
Weeks

1 Birth of baby. Days 1 and 2: breasts produce colostrum.
Day 3: breasts start producing milk.
Postpartum depression caused by hormonal changes.

1-2 Day 4-8: return from hospital: depression on returning
home caused by fear or inability to cope.

3 Lochia discharge stops.

1 Days 1 and 2: bowel or
urinary problems.

7 Uterus, cervix, and vagina should now be
back to normal. Postnatal
examination. Sexual intercourse
can safely be resumed.

7-10 Return of menstruation if not
breastfeeding.

If breastfeeding, probable **25**
return of menstruation by
now, if not earlier.

Timetable of events for the weeks
following the baby's birth.

POSTPARTUM DEPRESSION

Postpartum depression, or "baby blues," is rapidly being acknowledged as a significant aftereffect of childbirth. Though its intensity ranges from mere anxiety to severe psychosis, most women experience depression of some kind during the postnatal period.

It is not confined to first-time mothers: some women experience depression after the birth of each of their children.

Almost every mother goes through a "low" period about 3 days after the birth, roughly coinciding with the time the breasts begin to produce milk rather than colostrum. Many more women, however, experience severe depression on return from hospital. These feelings may last only a matter of days, but, particularly in a woman who is physically run down, may persist for a few months.

Among the feelings most commonly experienced during depression are confusion, shock, insecurity, inadequacy, fear of inability to cope with the baby, and even disappointment about its sex or appearance. Many women are frightened because they cannot rationalize their anxieties, and many fear a deterioration of their relationship with their partner. Postpartum depression is often attributed to hormonal imbalance following childbirth, but evidence is as yet inconclusive since depression has been noted in adoptive as well as natural mothers. Probably the single most important cause of postpartum depression is society's glorification of motherhood which sets up uncertainty and guilt in women who doubt their ability to be loving, caring mothers.

Treatment for the depression can involve drugs, but it is usually preferable to treat the cause rather than the symptoms. If the mother receives help, support, and constant reassurances from her family, friends, and other mothers, the chances of a quick recovery are high.

INTERCOURSE

Medical opinion generally favors delaying the resumption of intercourse until after the postnatal examination. Some people, however, argue that problems are unlikely provided that there is no vaginal discomfort and the discharge of lochia has ceased.

In any case, it is wise to wait until you feel ready. Reduced sexual interest after childbirth may be due to emotional upheaval or a lowered estrogen level. Other problems include muscular cramps during intercourse or pain from stitches after an episiotomy. Pregnancy is possible before menstruation resumes or during lactation. A diaphragm used before the birth will no longer fit, and the pill should not be used if breastfeeding. Condoms with spermicides are recommended.

BREASTFEEDING

More and more women in Western countries are choosing to breastfeed their babies. Most doctors welcome this, for they regard breastfeeding as the safest and most natural method of infant feeding, and many mothers agree that it is an enjoyable and rewarding experience. Some women, however, are uncertain about breastfeeding. Perhaps they have commitments that would make it impossible; or they may find the whole idea distasteful. And for some women who had planned to breastfeed, problems arising after the birth force them to turn to bottle feeding. Current breastfeeding propaganda may make mothers who are bottle feeding their babies feel inadequate and uncaring. This should be ignored; although breastfeeding is preferable for most babies, the vital physical contact between mother and child can be as intimate, warm, and loving whether the baby is fed by breast or bottle.

Mothers who do decide to breast-feed should ensure that they take sufficient rest - tension and the inability to relax can reduce the milk supply.

Diet is another important factor. The lactating mother should ensure that she eats a high-Calorie diet with particular emphasis on foods rich in protein, vitamins, and calcium.

BREAST OR BOTTLE?

BREAST FEEDING	BOTTLE FEEDING
Milk instantly available, at correct temperature, and sterile	Milk needs mixing and (usually) heating. Equipment must be sterilized
Antibodies protect the baby against some infections for first 6 months	No equivalent
Breast milk is cheaper than formula milk	Some expense - bottles, milk, teats must be purchased
Mother cannot tell how much milk the baby has taken without test weighings	Mother can see at a glance how much milk the baby has taken
Mother's health and well-being affect the milk supply	Milk supply independent of the mother
Milk supply usually adjusts itself to the baby's needs but cannot always meet the occasional need for extra milk	Extra feeds present no problem but can overfeed
Some drugs can be passed to the baby via the milk	Mother's medications do not affect the baby

© DIAGRAM

RISKS IN PREGNANCY 1

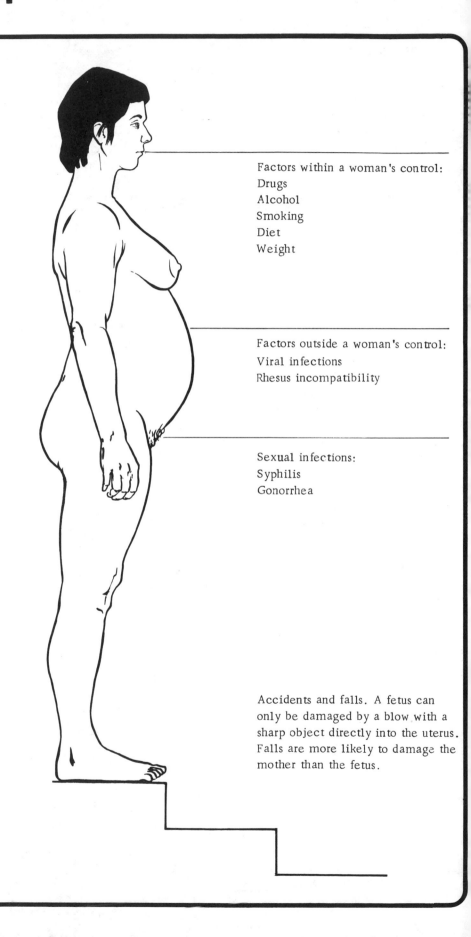

RISKS IN PREGNANCY

Abnormalities at birth can be caused by various factors - some hereditary, some environmental - and every expectant mother worries about them. On this page we list some of the chief dangers in pregnancy. Although the fetus is well protected its development can be seriously affected by many of the risks mentioned.

Hereditary factors. Some genetic disorders can be inherited by an infant. Examples are hemophilia, sickle-cell anemia.

Age. Congenital abnormalities are more likely to occur in babies born to women under 16 and over 40.

Vaccination. Smallpox and German measles vaccinations should not be given during pregnancy. Others should be avoided, but some can be given after the first 14 weeks.

Environment. There are many factors in the environment that can put an unborn infant at risk. Air pollution, water pollution and a degree of radiation are unavoidable today. A woman can lessen this risk by avoiding X rays, although today these are generally only given in order to detect and prevent a greater risk.

Factors within a woman's control:
Drugs
Alcohol
Smoking
Diet
Weight

Factors outside a woman's control:
Viral infections
Rhesus incompatibility

Sexual infections:
Syphilis
Gonorrhea

Accidents and falls. A fetus can only be damaged by a blow with a sharp object directly into the uterus. Falls are more likely to damage the mother than the fetus.

AMNIOCENTESIS

This is the method by which amniotic fluid is extracted from the uterus, and analyzed for possible fetal abnormalities. Such abnormalities can now be detected early in pregnancy.

Amniocentesis is best carried out between the 12th and 16th weeks of pregnancy. A local anesthetic is given and a needle inserted into the uterine cavity. About 10-20ml of amniotic fluid is withdrawn and the cells in it studied for defects. There is probably a less than 1-in-2000 risk to the fetus, though this may rise if it is carried out later.

A growing number of fetal abnormalities can now be detected, including Down's syndrome, and spina bifida. Those who might want the test are:
a) women who have already given birth to a defective child;
b) women carrying a serious disorder;
c) women aged over 40, since they have approximately a 1-in-50 chance of delivering a child with congenital abnormalities.

The diagram shows amniotic fluid being extracted from the uterus.

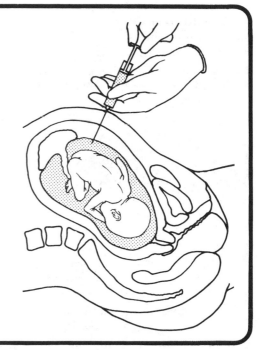

THALIDOMIDE

Widely prescribed as a non-addictive tranquilizer, thalidomide was taken by thousands of pregnant women in the late 1950s. As a result, some 8000 seriously malformed children were born and thalidomide turned into a tragedy.

Thalidomide was synthesized in West Germany in 1956. It had been tested, insufficiently as it turned out, on animals and humans, and was manufactured under the name of Contergan. As a sleeping pill and tranquilizer, especially recommended for pregnant women (it eased morning sickness), it became popular and was sold in Britain (as Distaval), various European countries, Canada, Australia, New Zealand, and Japan. In the USA it was judged unsafe and was not sold. By 1959 seriously deformed babies were delivered in Germany. They were suffering from phocomelia - a rare condition in which the hands, feet, or both start immediately from the main joint like seal flippers. By 1961 it was obvious that thalidomide was responsible and it was withdrawn in every country.

The thalidomide disaster tragically demonstrated the dangers of taking drugs during pregnancy. Most of them cross the placenta and, as was shown, the effects can be disastrous. Thalidomide, which causes malformations only between the 5th and 8th weeks, also demonstrated the particular dangers during the early stages when the fetal organs are developing.

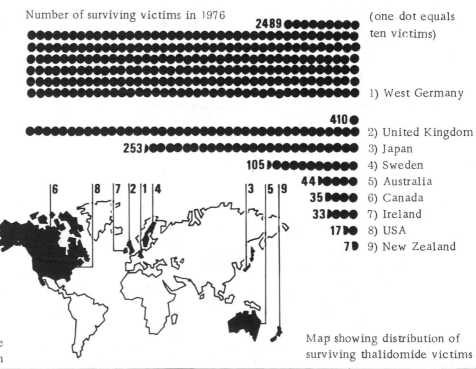

Number of surviving victims in 1976

(one dot equals ten victims)

2489 — 1) West Germany
410 — 2) United Kingdom
253 — 3) Japan
105 — 4) Sweden
44 — 5) Australia
35 — 6) Canada
33 — 7) Ireland
17 — 8) USA
7 — 9) New Zealand

Map showing distribution of surviving thalidomide victims

©DIAGRAM

RISKS IN PREGNANCY 2

INHERITED DISORDERS

The chromosomes that the fetus inherits from its parents carry many thousands of "genes," or units of genetic information. These decide the characteristics and activities of every cell in the body: some cells respond to some genes, some to others. Inherited defects arise in two ways. First, if any of the parents' genes are faulty, the relevant body cells may not respond in a desirable way ("gene abnormality"). Second, even if all the genes are healthy, the chromosomes may have been muddled or broken in the original pairing of ovum and sperm ("chromosome abnormality").

GENE ABNORMALITIES

There are several examples.

a) Sickle cell anemia. The red blood cells are abnormally shaped and so are destroyed by the body.

b) Phenylketonuria (PKU). Failure to produce one enzyme causes inability to process a vital amino acid contained in milk. Severe mental subnormality results unless a special diet is followed from birth.

c) "Wilson's disease." Defective metabolism of copper results in deposits of excess copper in the brain, liver, and eye. Unless treated it leads to mental derangement and cirrhosis of the liver.

Some gene abnormalities show up if just one parent has the faulty gene, others only if both parents have it.

SEX-LINKED GENE ABNORMALITIES

This special category includes hemophilia, red-green color blindness, and two forms of muscular dystrophy. The faulty gene responsible is carried on the X sex chromosome - for this has other functions apart from determining sex. So in a woman the disorder does not usually show up, as her other X chromosome usually supplies a healthy gene: the the healthy gene is "dominant." But in a man, there is no other X chromosome, only a Y chromosome, with no gene responsible for the defective function; and so the disorder appears.

The diagram shows the effects of this process of inheritance: a woman who does not herself show signs of the disorder can pass it on, so her sons may suffer and daughters carry it.

If a woman is found to be such a "carrier," there is risk not only to her own subsequent children, but also to those of her female relatives on the maternal side - because they may also have inherited the defective gene.

If no previous family history is found after careful check, it is likely to be an isolated mutation - in either mother or child. If in the mother, she can still pass it on to subsequent children.

CHROMOSOME ABNORMALITIES

There are two types. First, those involving the X and Y sex chromosomes - so that the fetus, instead of having XX or XY chromosomes, has XXX, XXY, XYY, or X alone. All are linked with disorders - mostly involving abnormal genitals. Second, those involving other chromosomes; eg Down's syndrome (mongolism), where an extra chromosome results in physical and mental abnormalities.

■ Sufferer ○ Normal gene
▨ Carrier ● Defective gene

Patterns of inheritance of a sex-linked gene abnormality

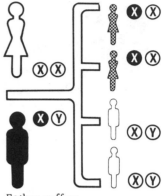

Father suffers
All girls are carriers
No boys suffer or carry

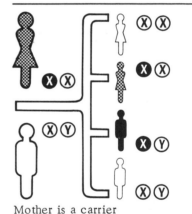

Mother is a carrier
Half girls likely to carry
Half boys likely to suffer; none carry

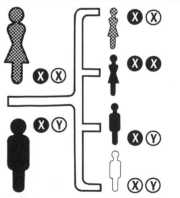

Mother is a carrier
Father suffers
Half girls likely to suffer, half to carry
Half boys likely to suffer; none carry

MALNUTRITION

Malnutrition in the mother during pregnancy can seriously affect the fetus. It may even die from undernourishment - revealed in autopsy not only by low weight, but also by stunting of each individual organ and low cytoplasm content of the body cells. More usually, the baby is born alive but underweight - and low birthweight carries with it increased risk of cerebral palsy, epilepsy, autism, blindness, deafness, mental subnormality, and neonatal death. Some sources estimate that a third of all long-term childhood handicaps are associated with low birthweight. In England, in 1973, babies weighing under $5\frac{1}{2}$ lb (2.5kg) at birth accounted for 7.1% of births, but 60% of stillbirths and deaths under 1 month.

Low birthweight is not totally due to malnutrition, but it is a major factor. In Guatemala, a nutritional program reduced low birthweight from 20% to 5.1% of births. This is below that of many industrialized countries, so it is not just in "developing" countries that there is room for nutritional improvement. But careful nutrition must begin before the 20th week of pregnancy. Any later improvement in diet has only limited effect.

MINIMUM DIET

Ideally a pregnant woman should follow the normal daily food guide, increasing Calorie intake and weight. But many women find that their appetite in pregnancy is small or unpredictable; and poverty may be a factor. At the very least, wholewheat bread and milk are preferable to tea or coffee and biscuits. But if at all possible the minimum diet shown below should be followed each day. (Cold food may be more easily faced than heated.)

MINIMUM RECOMMENDED DIET

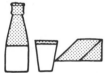

1pt milk, or $\frac{1}{2}$pt and 2oz (57g) of cheese

A portion of meat or fish, or an egg

A portion of root or raw green vegetable

Fruit, orange juice, and/or potato

Wholewheat bread

1pt of water

Liver or oily fish once a week

White fish once a week

Bran to avoid constipation

INFECTION

The amniotic sac protects the fetus against most bacteria. But virus infections in the mother's blood stream may cross the placenta and sometimes cause damage.

RUBELLA (German measles) in a woman in early pregnancy can cause fetal death or fetal deformities such as deafness, blindness, brain damage, and retarded growth. Rubella in the first 3 months of pregnancy causes defects in up to 50% of babies, and is often considered grounds for abortion. There is also slight risk in the 4th and 5th months. But if the mother has previously had the illness, or a vaccine against it, she is unlikely to contract an infection strong enough to spread to the fetus. (Vaccination cannot be given during pregnancy, as live vaccine is used.)

CMV (cytomegalo virus) is spread between adults by close personal or sexual contact. It is found on the cervixes of up to 25% of pregnant women, usually without noticeable symptoms. It can cause low birthweight, prematurity, deafness, and mental retardation if it infects the fetus. No vaccine is available. But only when it is contracted for the first time during pregnancy is it usually strong enough to infect the fetus; and only 1 infected baby in 10 suffers permanent damage.

SEXUAL INFECTIONS Syphilis in the mother can cause death, deformity, or disease in the fetus, unless the mother is treated before the 16th week of pregnancy.

OTHER INFECTIONS Mumps, chickenpox, and fevers (as in influenza) are sometimes suspected of causing deformities.

©DIAGRAM

SMOKING IN PREGNANCY

SMOKING IN PREGNANCY

As well as affecting a woman's own health, smoking may affect the health of an unborn baby. It is widely accepted that there is a connection between smoking and complications in pregnancy, and studies have shown that smoking while pregnant results in a lighter baby and an increase in perinatal mortality. Effects of smoking in pregnancy have also been found to persist into childhood. Probably the dangers are most severe if smoking continues after the first 3 months, and not all effects have been totally proven. Unlike many factors affecting her child's health, it is the mother's own decision whether or not she will smoke.

INCREASE IN SMOKING

Despite the possible ill effects of smoking while pregnant, a study of 18,631 pregnant women in Cardiff, UK, shows that smoking in pregnancy has become more common. In 1964 only 4% of Cardiff pregnant women smoked 20 or more cigarettes a day, compared with 15% in 1970.

SMOKING AND FETAL HEALTH

Fetal breathing movements are a direct measure of fetal health and well-being. Smoking has been shown to reduce the incidence of fetal breathing movements and must therefore be considered damaging to fetal health. A study of 18 normal pregnant women has shown that two cigarettes smoked consecutively produce a dramatic reduction in fetal breathing movements. The diagram shows the percentage of time that fetal breathing movements could be observed during a test period.

Percentage of pregnant women smoking 20+ cigarettes a day (Cardiff UK survey)

1964

1970

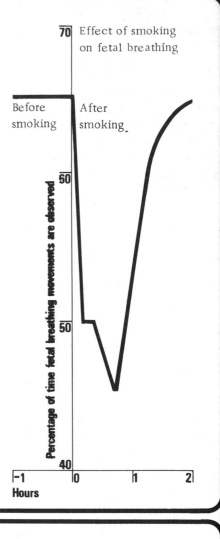

Effect of smoking on fetal breathing

Before smoking   After smoking

Percentage of time fetal breathing movements are observed

70

60

50

50

40

-1   0   1   2
Hours

BIRTHWEIGHT

Statistics suggest that women who smoke when pregnant tend to produce lighter babies than non-smoking mothers.
A study in Cardiff, UK, showed that the average birthweights of babies born to mothers who smoked in pregnancy was 5 lb 14½oz (2.70kg), compared with 6 lb 4½oz (2.83kg) for non-smoking mothers.
Only 4% of the live babies born to non-smoking mothers in Cardiff weighed less than 5 lb 8oz (2.5kg) compared with over twice that percentage born to mothers smoking at least 20 cigarettes a day when pregnant.
A study made in California included the smoking habits of both parents. It showed that babies weighing less than 5 lb 8 oz (2.5kg) at birth were most common when both parents were smokers.

The diagram shows the difference in average birth weights of babies born (a) to non-smokers (2870g) and (b) to smokers (2700g).

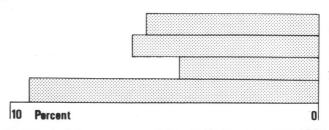

Percentage of babies who weigh less than 2.5kg at birth

Neither smokes
Father smokes
Mother smokes
Both smoke

10   Percent   0

PERINATAL MORTALITY

The relationship between smoking and unsuccessful pregnancy (miscarriage, stillbirth, neonatal death) is not yet completely clear. In findings at Cardiff, UK (a), the rate of stillbirth and neonatal death with mothers smoking over 20 cigarettes a day was 3.8%, with non-smoking mothers only 2.5%. But in a Californian study (b), smokers' babies had a lower mortality rate than non-smokers'. (This study, though, only considered babies born live and with birthweight under 5½ lb - 2.5kg.) However, a study at Sheffield, UK, clearly showed that, among women of similar blood pressure, smokers' pregnancies were much more likely to be unsuccessful (c). The confusing factor pinpointed was that smoking is also associated with low blood pressure (which favors successful pregnancy). But why there is this association is unknown. Does low blood pressure favor smoking, or smoking favor low blood pressure? Both seem unlikely.

Smoking and unsuccessful pregnancy

a Cardiff study

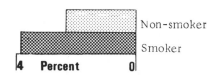

Non-smoker
Smoker

4 **Percent** 0

b California study

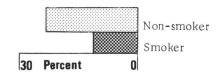

Non-smoker
Smoker

30 **Percent** 0

c Sheffield study

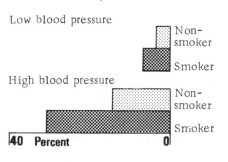

Low blood pressure

Non-smoker
Smoker

High blood pressure

Non-smoker
Smoker

40 **Percent** 0

FETAL MALFORMATION

Although there is no firm evidence that smoking is associated with fetal malformation, findings at Cardiff suggest that there may in fact be such a link.

The incidence of congenital heart disease among babies born to women who smoked during pregnancy was 0.73% compared with 0.47% among babies born to non-smokers. Hare lip and cleft palate were also found slightly more commonly among babies born to women who smoked.

Percentage of babies with congenital heart disease

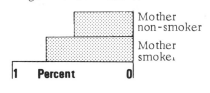

Mother non-smoker
Mother smoker

1 **Percent** 0

POSTNATAL DEVELOPMENT

Babies born to women who smoked while pregnant have been shown to grow more rapidly than babies born to non-smokers. The difference in average weekly growth rates is most marked from birth to six weeks. But over a year the difference in growth between smokers' and non-smokers' babies is negligible. So, in the end, non-smokers' babies tend to preserve any size advantage they had at birth.

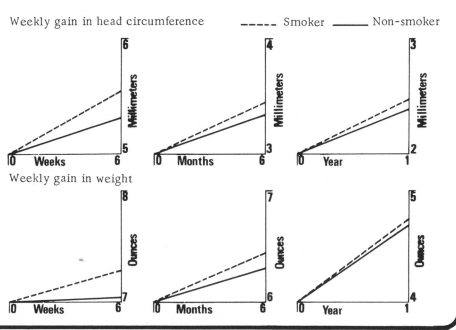

Weekly gain in head circumference   ----- Smoker ——— Non-smoker

Weekly gain in weight

©DIAGRAM

SMOKING, ALCOHOL, DRUGS IN PREGNANCY

LONG-TERM EFFECTS

The diagrams below are based on a survey taken in the UK of 17,000 children whose progress had been followed from birth. They compare the mental abilities of 11-year-old children of mothers who smoked during pregnancy with those of non-smoking mothers.

Children of mothers who smoked up to 9 cigarettes daily were some 5-5½ months behind in school progress, and those of mothers who smoked 10+ cigarettes daily were 5½-7 months behind, compared with children of the same age of non-smoking mothers.

In addition, the survey noticed that children of smoking mothers tend to be 0.5-1cm shorter than the children of non-smokers.

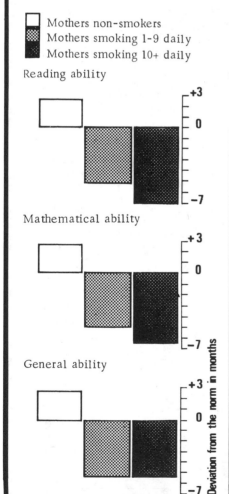

☐ Mothers non-smokers
▨ Mothers smoking 1-9 daily
■ Mothers smoking 10+ daily

Reading ability

Mathematical ability

General ability

Deviation from the norm in months

ALCOHOL IN PREGNANCY

An occasional drink does no harm during pregnancy. But recent studies from the USA and Germany suggest that an increasing number of babies are being born seriously deformed or mentally retarded because their mothers have been drinking heavily during pregnancy. It has even been claimed that 1 in 3 women who drink heavily must expect their child to be mentally or physically defective. But so far evidence is mainly from actual alcoholics. The diagrams below are based on a US study of 23 children born to chronically alcoholic mothers. They were compared with children of non-alcoholic mothers and matched for socioeconomic group, race, maternal age, etc.

a The diagram shows mortality rates among the newborn babies. 17% of those born to alcoholic mothers died within 1 week as opposed to 2% of those born to non-alcoholic mothers.

b Growth deficiencies. The diagram compares birthweight, length, and head circumference of both sets of infants. Comparisons were based on a measurement of normality that generally only 3% of the overall baby population fails to achieve. In the survey, 13% to 32% of the children of alcoholic mothers failed to reach the measurement.

c IQ performance. At the age of 7, 44% of the children born to alcoholic mothers had an IQ of 70 or under.

a Percentage of babies who died in the first week of age

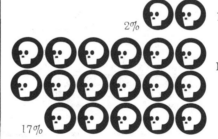

2% Mother was a non-alcoholic

Mother was an alcoholic during pregnancy

17%

b Percentage of babies below a given measurement of size

■ Mother was an alcoholic during pregnancy
▧ Mother was a non-alcoholic

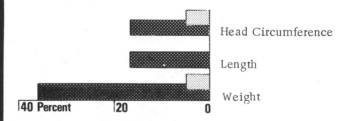

Head Circumference

Length

Weight

40 Percent   20   0

c Percentage of 7-year-old children with an I.Q. below 79

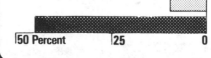

Mother was a non-alcoholic

Mother was an alcoholic during pregnancy

50 Percent   25   0

DRUGS AND PLACENTA CROSSING

Most doctors today recommend that women avoid all drugs during pregnancy unless they are absolutely essential for the mother's well-being. The fetus maintains its hold on life through the umbilical cord and the work of the placenta.

Oxygen and nutrients pass from the mother's circulation into that of the fetus via the placenta and umbilicus, and carbon dioxide and other waste products pass back the same way.

As a result, most substances in the mother's bloodstream will reach the fetus. In the case of drugs, recent studies show that, as with thalidomide, effects on the fetus can be disastrous. This is particularly true during the first 3 months when the fetal organs are forming.

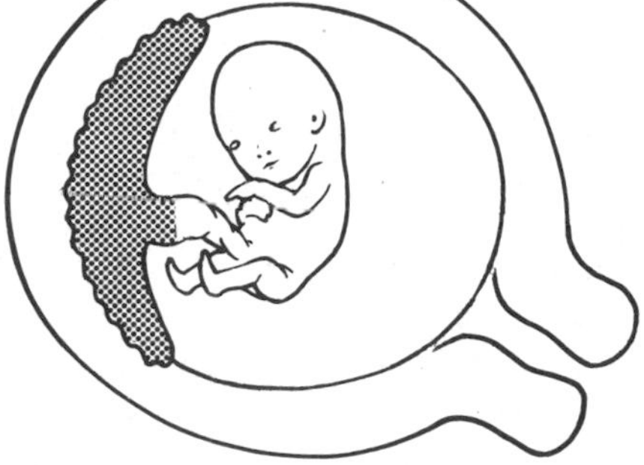

PLACENTA

DRUGS	EFFECTS ON FETUS
Caffeine (in coffee)	Stimulates fetal nervous system
Tannic acid (in tea)	Depressant
Sleeping pills	Possible malformations; but some now designed specifically for safe use by pregnant women
Tranquilizers	
LSD and other psychedelics	Increased risk of miscarriage; possible chromosome damage
Cocaine; Amphetamines	Acts as stimulant on fetus
Heroin; Morphine	Possible fetal addiction; can mean blood transfusion needed at birth
Aspirin	Large amounts can cause miscarriage or hemorrhage in newborn
Phenacetin	
Antibiotics	Possible damage to fetal kidneys
a) Streptomycin, gentamycin	
b) Sulphonamides (long-term)	a) Associated with deafness in infants
c) Tetracycline	b) Can cause jaundice
	c) Possible deformities; stains teeth
Antihistamines	
Cortisone	Possible malformations; some now designed specifically for safe use by pregnant women
Progesterone (for hormone deficiency and possible miscarriage)	Fetal and placental abnormalities – possibly stillbirth; cleft lip
Antithyroid	
Marijuana	Genital abnormalities in female infants
	Possible goiter
	As yet no proof of ill effects

The diagram shows some of the drugs that cross the placenta and what the effects on the fetus might be.

©DIAGRAM

329

DEFECTS AT BIRTH

DEFECTS AT BIRTH

These are anatomical defects present at birth. Thirty live births in every 1000 have some kind of congenital malformation. They may be so severe that life is not possible, eg anencephaly, or they may be so trivial that life is not interfered with, eg an extra finger. Many, but not all, defects, are obvious at birth. Some, such as defects in the heart or kidneys, may be discovered within a few days while others are only detected after many years or by chance during surgery or autopsy. Malformations are either of genetic origin or due to external factors which affect the pregnant woman, eg infection, drugs, or high-energy radiation. Most ova are never fertilized, and many of those that are fail to be implanted in the uterus; but 20% of those that result in pregnancy are aborted within the first 12 weeks. Some of these spontaneous abortions are of empty sacs with no embryo, others are of defective embryos. Congenital defects are one of the most important causes of death in the first and later weeks after birth. They mainly affect the central nervous system (brain and spinal cord).

Embryo development is a continuous process following a strict sequence. The initiation of each step in the process depends on the successful completion of the one before. Any interruption or disorganization of these processes at any time may result in a malformation. Usually the earlier the interruption occurs, the more severe the defect.

The most common anatomical defects in the new-born

DEFECT	RATE*	DESCRIPTION
Double ureter	300	Two ureters from one kidney. Usually without symptoms or significance. Very occasionally, obstructed urine flow, causing infection. Genetic. Surgery if necessary.
Male inguinal hernia	80	Hernia in the groin, between the muscles of abdomen and thigh. Developmental. Surgery needed.
Mental subnormality (except mongolism)	17	Varying degrees of defect from a variety of different causes.
Spina bifida, often with hydrocephalus	10	Spina bifida - defect leaving spinal cord exposed; hydrocephalus - obstruction in skull causing collection of cerebro-spinal fluid under pressure. Genetic, or result of antenatal injury or infection. Surgery to prevent paralysis or death.
Anencephaly	6	Absence of brain and top part of skull. Replaced by fibrous tissue. Invariably fatal. More common with very young or old mothers.
Cleft lip and palate	5	Lip and palate not fused. Difficult breathing, feeding, and speaking. Partly genetic. Associated with thalidomide, rubella. Plastic surgery required.
Down's syndrome (mongolism)	3.5	Rate rises to 200 if mother is over 40. Caused by extra chromosome. Characterized by mental retardation, heart defects, Mongoloid features, protruding tongue.
Coeliac disease	2.5	Disorder of unknown cause, producing inability to assimilate some foods. Chronic diarrhoea and malnutrition. Treatment dietary. Recovery usual, but slow.

* Rate per 10,000 live births

DEATH AT BIRTH

MATERNAL MORTALITY

Medical advances have made childbirth safer today than ever before. A hundred years ago abnormal presentation, protracted deliveries, hemorrhage, and puerperal fever resulted in a mortality rate of up to 250 per 1000 live births in hospital; the rate for home deliveries was, however, much lower. Developments in antiseptic and operative techniques have controlled puerperal fever and helped overcome complications of pregnancy and birth.

Today the main causes of maternal death are: illegal abortion, and miscarriage (35%); ectopic pregnancy (15%); hemorrhage (usually postpartum) (15%); toxemia (10%); and traumatic labor (10%). Other deaths mainly result from maternal disease (eg heart and lung complications).

FETAL MORTALITY

Though much higher than the maternal death rate, the fetal death rate has also fallen dramatically over the last 100 years. This is due to improved techniques, and better care and diet during pregnancy. Most stillbirths are due to prematurity, placental insufficiency, and congenital defects.

INFANT MORTALITY

The infant period covers the first year of life. A hundred years ago 200 infants in every 1000 died. Since then the rate has fallen to 23 per 1000 live births. This is mainly due to improved antenatal and postnatal care, better delivery technique, and advances in medical understanding. Congenital defects are the largest cause of death, and rhesus incompatability (D12), diabetes, eclampsia, and heart disease can also be fatal.

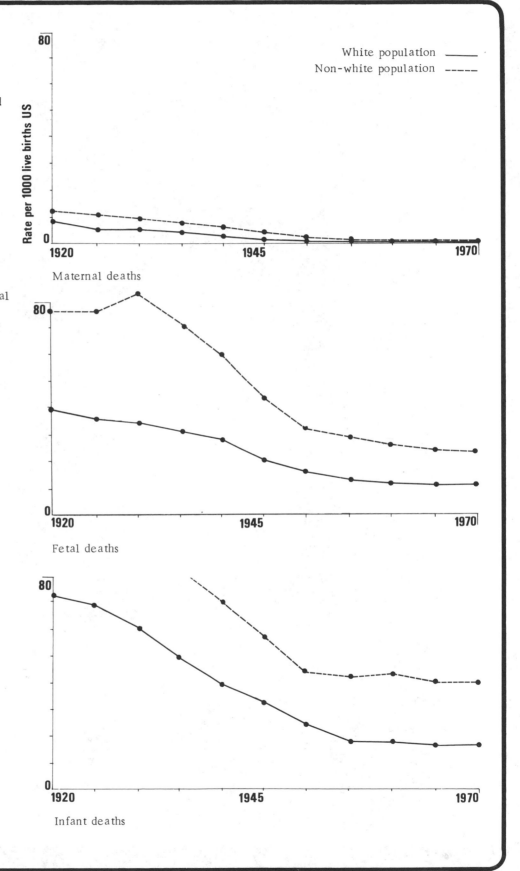

White population ——
Non-white population -----

Maternal deaths

Fetal deaths

Infant deaths

©DIAGRAM

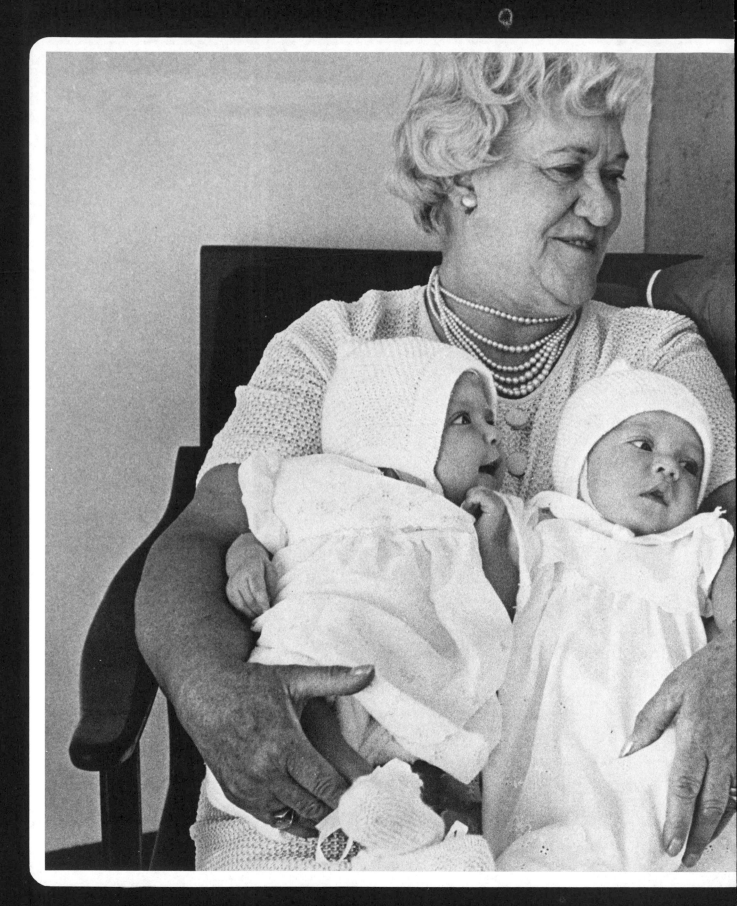

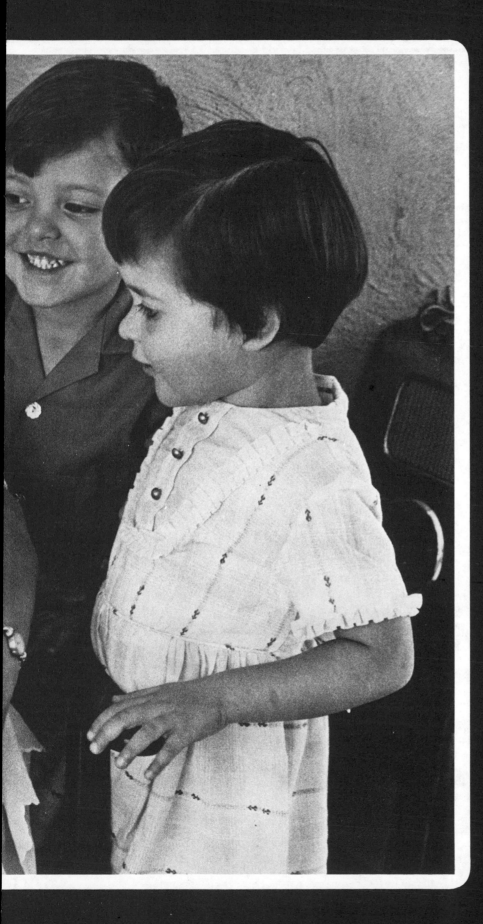

9 Growing old

Understanding the inevitable changes our bodies undergo as we grow older helps to conquer our fears and allows us to continue leading our lives to the fullest.

Left: A happy grandmother showing off the twins (Camera Press Ltd.)

WOMAN AND THE MENOPAUSE 1

THE MENOPAUSE

The menopause (also called the climacteric, or change of life) marks the end of the reproductive part of a woman's life. Its chief outward sign is the cessation of the monthly flow of menstrual blood. Some women also experience a variety of symptoms due to change in hormonal balance. Parts of the body may begin to age noticeably, but most women should be able to remain physically, mentally, and sexually as active after the menopause as before it. Some see the menopause as a time of regret. For others, it represents a welcome release from unwanted biological demands on the body.

ONSET OF THE MENOPAUSE

The menopause begins at different ages in different individuals. The usual age is around 50, though a few women start the menopause in their 30s, while onset in others is delayed into the 50s. But about half of all women lose their capacity to bear children by the age of 50, and only 5% remain capable beyond 55.

The diagram shows the percentage of American women who have experienced a natural menopause by different ages.

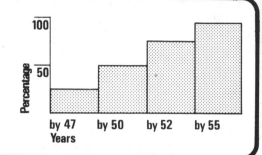

VARIATIONS

There are many reasons for variation in the time of onset. Race is one factor. White women of northern European stock tend to reach the climacteric late if they entered puberty early, and early if they entered puberty late. With women of Mediterranean or colored origin, both puberty and menopause come relatively early. Then, too, women of some families cease menstruating early, others late, for their racial group. Thus heredity plays a part. At one time, climate also seemed to be implicated, though, in fact, it makes no difference. But a high living standard does tend to prolong a woman's reproductive life, while poor conditions shorten it. Childbirth is also relevant; if a woman has a child after she has passed 40, her menopause may be delayed. The menopause may be hastened in women who have never given birth.

ABNORMAL VARIATIONS
Some women experience exceptionally early menopause for no apparent reason. In the majority of cases, however, exceptionally early menopause occurs as a result of disease or medical treatment. Disease of the pituitary gland can trigger the event. So can medical irradiation of the ovaries, or their surgical removal.

Factors influencing the relative time of onset

	Early menopause		Late menopause
Race	Black; southern European		White; northern European
Living standard	Poor		Rich
Motherhood	Non-childbearing		Childbearing
Weight	Fat		Thin
Puberty	Late puberty		Early puberty
Surgery	Ovaries removed		Ovaries intact
Heredity	Heredity may favor either early or late menopause		

HOW THE CHANGE OCCURS

BEFORE THE CHANGE

The top diagram reminds us how ovary and uterus function in a normal 28-day menstrual cycle. FSH and LH hormones (produced by the pituitary gland at the base of the brain) stimulate one of the ovaries to release a ripened egg into the nearest Fallopian tube, so making that egg available for fertilization by male sperm. Meanwhile the follicle inside which the egg has developed is producing hormones too, first estrogen and then also progesterone. These cause the lining of the uterus to thicken in preparation for implantation if the egg is fertilized. If the egg is not fertilized, the thickened lining breaks down, and the woman experiences her menstrual period.

THE CYCLE BREAKS DOWN

In middle age, the aging ovaries cease to respond to FSH and LH, though secretions of these increase. As a result:

a) fewer follicles are formed, and fewer release eggs;

b) estrogen and progesterone output from the ovaries falls off;

c) the uterus lining ceases to thicken, and menstrual bleeding changes pattern and eventually stops; and

d) uterus and ovaries start to shrink.

Once egg production has ceased entirely, the woman is infertile.

AFTER THE CHANGE

Small quantities of FSH and LH are still produced by the adrenal glands. Some androgenic (male) hormones are produced by the ovaries.

FSH Follicle stimulating hormone
LH Luteinizing hormone

E Estrogen
P Progesterone
▨ Uterine lining

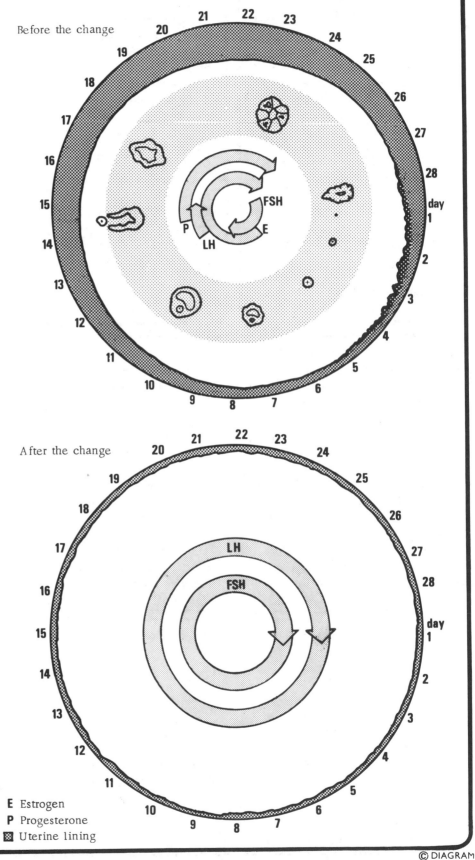

Before the change

After the change

© DIAGRAM

WOMEN AND THE MENOPAUSE 2

SURGICAL MENOPAUSE

If a woman's ovaries have to be removed, she will undergo a "surgical menopause." The effects include infertility. Her periods will now cease, and ovary-produced hormones will stop circulating in her body. Replacement estrogen is often prescribed to ease the patient over the sudden hormone shortfall. More usually, the uterus alone is removed, or the uterus and one ovary. Then, although monthly bleeding stops, the woman's female sex hormones will go on being produced. (She will not, of course, be able to bear children.)

MENSTRUAL CHANGES

The first sign of the menopause is often irregular menstrual bleeding. Periods may be lighter, or later, than usual, and a woman may miss a month altogether. Sometimes periods are light one month and heavy the next; this heaviness may be marked if the period is late. Gradually, months or perhaps years later, the periods cease completely. (In some women, however, they may stop abruptly.) Twelve months after the last period, a woman of 50 plus is estimated to be infertile. However, it is safest if she continues with contraception for 2 years after her last period, to avoid any risk of pregnancy - and a woman under 50 should certainly do this.

Because women on the pill appear to continue menstruating, doctors often recommend periodic switching to another method of contraception after the age of 42 to see if the menopause has begun.

PROBLEMS AT THE MENOPAUSE

In many women, loss of periods is the only sign of the menopause. However, some other effects are often experienced.

HOT FLASHES, or flushes, are most common. These are a response of the hypothalamus gland to the falling estrogen level in the body. They often start as a warm feeling in the chest, moving to the neck and face, which color up. The disorder may spread to the rest of the body, perhaps with a prickling sensation. Sweating sometimes follows. Hot flashes can last up to 15 minutes and may occur several times a day, or they may be transient and infrequent. Some coincide with the due dates of periods. Hot flashes may start before periods stop and recur over 2 or 3 years. The worst types cause discomfort and depression, and "night sweats" may even break up sleep. In such cases, treatment with estrogen or the drug "Clomiphene" may be prescribed.

GENITAL SYMPTOMS Hormonal change may cause itching in any part of the body, but especially in the genitals. Vaginal dryness is also typical. In both cases, creams and ointments can be prescribed.

OTHER PHYSICAL PROBLEMS Many physical symptoms have been blamed on the menopause, and some doctors talk of a "menopausal syndrome." Apart from symptoms already described, this might include: dizzy spells, headaches, and insomnia; fatigue and lack of energy; abdominal bloatedness; digestive troubles, including pain, flatulence, constipation, and/or diarrhea; and breathlessness and palpitations.

But all these symptoms can be very variable from day to day, and many women do not experience them at all. In fact, no direct link with the menopause has been proved. (Some, though, are sometimes signs of illness, so always check with a doctor.)

WEIGHT GAIN At the menopause, appetite often becomes variable and may increase, while the body's energy needs fall. Obesity results unless food intake is controlled.

EMOTIONAL PROBLEMS Moodiness is quite common, and sometimes also irritability, forgetfulness, anxiety, and depression. Causes may include:
a) possibly, the hormonal changes themselves - though no direct link has been proved;
b) the mental consequences of such physical symptoms as hot flashes, headaches, insomnia, etc; and
c) in some cases, such psychological factors as fears of aging or of an altered sex life, or maybe regret at unrealized motherhood.

In the occasional serious case tranquilizers and anti-depressants will help. Hormone treatment may also be valuable.

If a woman starts to develop, or better still has always had, interests beyond her family and home, emotional problems at the menopause may be less likely, and less distressing if they do occur.

AT THE MENOPAUSE

SIGNS AND SYMPTOMS

a) Fatigue, headaches, dizzy
spells, insomnia; moodiness,
forgetfulness, irritability,
difficulty in concentrating,
anxiety, depression.

b) Hot flashes and sweating

c) Palpitation
d) Breathlessness

e) Tendency to gain weight

f) Variable appetite, digestive
troubles, abdominal bloating
g) Ovaries stop producing eggs,
and, therefore, estrogen

h) Menstrual bleeding changes
character and eventually stops
i) Vaginal dryness and itching

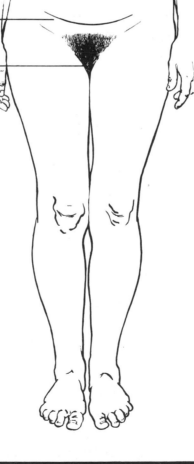

TREATMENT

a) Normally no treatment:
symptoms eventually disappear.
Short-term prescription of
tranquilizers or anti-depressants
in severe cases; possibly control
by hormone therapy in very severe
ones

b) Control by hormone treatment
in severe cases
c) None if only due to menopause
d) None if only due to menopause

e) Diet control

f) None if only due to menopause

g) An irreversible change

h) Normally no treatment needed.
If bleeding occurs after the
menopause, see a doctor
i) Use of creams and ointments,
possibly hormonal. Also lubricant
creams and jellies for intercourse

©DIAGRAM

WOMEN AND THE MENOPAUSE 3

CHECKING FOR CANCER

Certain forms of cancer occur more often at or after the menopause than before. They are: cancers of the breast, stomach, lower bowel, uterus, and genitals At this period in her life, a woman should be watchful for the warning signs described on p. 126 and especially careful in noting changes in her breasts (see p. 130). She should also have regular examinations of the neck of the uterus, or cervix, including a cervical smear (also called a "Pap test"--see p. 134). This test, which is painless and brief, involves taking a minute piece of tissue from the cervix, usually on a wooden spatula. From it, laboratory tests can tell whether cancer is likely to develop there or not. Over 80% of women operated on at an early stage of cancer of this sort (or of the breast) have no recurrence of the disease. For menopausal women, this examination also has another advantage, since it can reveal the nature and extent of any hormone lack and make treatment possible

SEX AND THE MENOPAUSE

One of the main factors often underlying menopausal gloom is a woman's fear that she will lose her attractiveness. She may feel that, because she can no longer bear children, she is no longer feminine. And the step from feeling unloveable to being embarrassed by physical expressions of love is a short one.

There is absolutely no reason, though, why lovemaking should stop at, during, or after the menopause. Occasionally, a woman does experience some temporary lessening of sexual desire during the menopause itself. But this is normally brief, and a post-menopausal woman has lost neither her sexuality nor, given reasonable care, her looks: all that has changed is her ability to conceive. Sexual desire is stimulated by the male, rather than the female, hormones, and a woman continues to have her share of these right through her life. Certainly no woman, however old she is, should feel ashamed of enjoying sexual intercourse.

Indeed, many find that there is greater enjoyment now that the risk of pregnancy is past.

A menopausal woman should, however, get a doctor's advice as to when contraceptives can be safely abandoned. A doctor can also suggest how to deal with vaginal dryness - usually by use of a lubricant.

Sexual response of women after the menopause as a percentage:

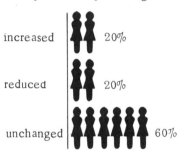

increased   20%

reduced   20%

unchanged   60%

POST-MENOPAUSAL CHANGES

Estrogen influences the growth and nourishment of breasts, uterus, vagina, smooth muscle, and skin. It also tends to protect women against circulatory diseases, and possibly against loss of calcium from the bones. Thus the fall in estrogen accompanying the menopause heralds aging changes in the body.

Estrogen loss largely explains why muscles lose tone and skin loses elasticity and becomes wrinkled. Deprived of estrogen, the breasts gradually flatten and droop. The womb and ovaries shrink, and the vaginal wall becomes thinner. The vagina is also likely to become drier and lose some of its natural protective acidity, making it more prone to infections. In addition, sexual intercourse may become difficult and even painful. The vulva atrophies. Tissues supporting the vagina and the muscles of the pelvic floor become less elastic and rather more flabby over the years. Prolapse of the uterus or vagina may follow; it is usually treatable without surgery.

Changes in secondary sexual characteristics include the loss of some pubic hair, and growth of hair on the upper lip and chin. More serious, the risk of heart attack increases, while obesity, if it occurs, increases the risk of arthritis. Finally, well after the menopause, calcium loss from the bones can produce a curve in the spine. Many of the problems are treatable.

NON-OVARIAN ESTROGEN Some women do not suffer hormone deficiency conditions as severely as others. These women make up for loss of ovarian estrogen by that produced in other parts of the body, probably including the liver. The women suffer few menopausal symptoms and they seem to avoid some of the aging and the disturbed sex life often linked with the menopause.

AFTER THE MENOPAUSE

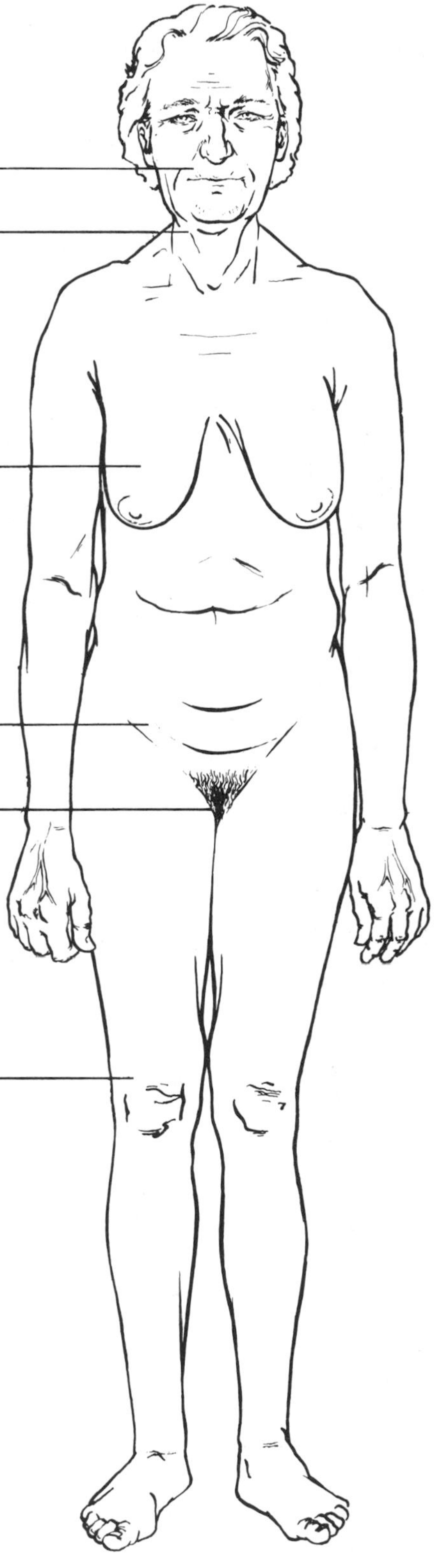

SIGNS

TREATMENT

a) Straggling hairs may begin to appear

b) Skin loses elasticity

c) Bones lose calcium; spine eventually affected

d) Breasts eventually flatten and droop

e) Increased risk of circulatory disease

f) Ovaries and uterus shrink. Uterine muscle becomes fibrous.

g) Vaginal wall grows thin and is liable to irritation (less so if intercourse continues)
h) Vulval walls atrophy
i) Urinary infection and/or incontinence may occur
j) Pubic hair gets scantier

k) Increased risk of arthritis

a) Facial hair removed by electrolysis or depilatory waxes

b) Skin care and massage help skin tone; estrogen tablets are a possible treatment but mistrusted by some doctors
c) Estrogen tablets may retard calcium loss from the bones; exercise and plenty of calcium in food is arguably a better prevention.
d) Hormone therapy still questioned by some doctors; support sagging breasts with a well-fitting brassiere
e) Circulatory disorders are best prevented by regular exercise and sensible diet

f) No treatment needed. Exercise helps muscle tone.

g) Application of estrogen cream helps, and use of lubricants in intercourse.
h) Hormonal cream may help.
i) Drugs for infection; treatment of incontinence depends on cause
j) No treatment needed

k) Arthritic pain is helped by anti-inflammatory drugs, such as aspirin. Surgery may be needed in severe cases.

© DIAGRAM

THE MENOPAUSE EXPERIENCE

HORMONE REPLACEMENT

Hot flashes and vaginal atrophy are treated by many doctors with prescriptions of estrogen. This, however, has usually been essentially short term: the aim being to cure specific disorders, and tide the woman over a difficult patch. Hormone replacement therapy (HRT) takes a longer view. According to supporters, menopausal women who are given estrogen for the rest of their lives will be protected from some types of disease and to some extent escape the aging process. The estrogen is prescribed in the form of tablets, creams, and injections. Therapy of this type is much more controversial than the short-term treatment. It is most widely accepted to combat surgical menopause (see F05), very severe "hot flash" symptoms, and, in some cases, loss of calcium from the bones (see M04). It may also delay the onset of arthritis (though not cure it), and promote clearer skin and general well-being. Local cream application can certainly help maintain the vaginal tissue and vulva.

Beyond this, the usefulness of replacement therapy is uncertain. Moreover, estrogen intake in post-menopausal women may cause breast tenderness, gastric upsets, and swelling of the ankles; and, more important, there is some evidence of a link with blood thrombosis and with breast and uterine cancer. Certainly estrogen helps any existing breast cancer to grow. It also may cause post-menopausal menstrual bleeding; this is significant simply because any such bleeding must always be investigated (usually by D & C) to establish that the cause is not uterine cancer.

DO MEN HAVE THE CHANGE?

Since men do not have women's equipment for child-bearing, they cannot have a "change of life" in the way women know it. Their hormonal decline is gradual. As shown in diagram (a) testosterone production in the typical male reaches a peak in early adulthood and declines into old age. Although the amount of testosterone produced by a man of 60 has dropped to the level of a pre-teens boy, it is usually adequate to allow sexual activity.

A man's sexual functioning, however, does slow down naturally with the onset of old age. Semen production, erection, and ejaculation all take more time and more stimulation, while the need and ability for orgasm declines. Again, poor nutrition, general bodily decline, depression, and a hidden relief at being able to drop an activity which has caused anxiety, may all play a part in reducing virility. But a healthy and fit man should nevertheless be capable of some sexual activity at any age. Diagram (b) shows that the percentage of men suffering from total impotence increases slightly through the middle years, and more rapidly into old age. But impotence is often caused by psychological rather than physical problems.

a) Testosterone produced in the life of a typical male

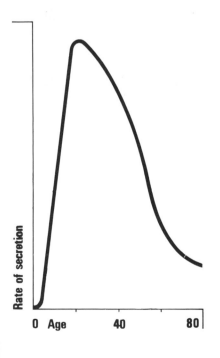

b) Percentage of men suffering from total impotence

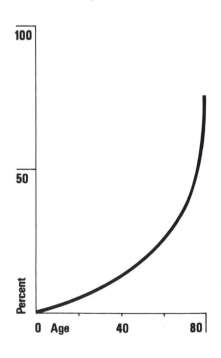

MENOPAUSE AND LIFE EXPECTANCY

The menopause is a highly charged time for many women. It is associated with aging but more significantly it is the time when a woman's reproductive abilities come to an end. Many women still seem to think of childbearing and home-keeping as their prime functions. Not surprisingly therefore their fears reveal an attitude that equates the menopause with the end of a woman's usefulness and productive life.

Nothing could be further from the truth. In the United States just over 20% of the female population is over 55. Life expectancy for the average American woman is 76. With the menopause occuring around 50, these women may well have one-third of their lives to come. Their life situation is very different from that at the turn of the century, when average life expectancy was 50 years and the menopause occurred in the mid or late 40s.

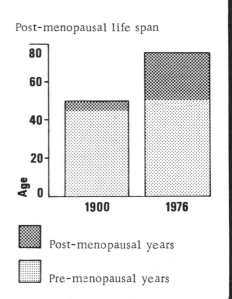

Post-menopausal life span

Post-menopausal years

Pre-menopausal years

THE MENOPAUSE EXPERIENCE

Symptoms associated with the menopause have been described on p.337. But they are not experienced by all women. A study was carried out in London and covered 638 women aged between 45-54. They were divided into groups ranging from those who were still menstruating to those whose periods had been over for 9 or more years.

The women were asked to report on 8 symptoms - hot flashes, night sweating, headaches, dizzy spells, palpitations, insomnia, depression and weight gain. The results are shown in the diagram. Nearly half (49.8%) reported hot flashes while 35-50% reported other symptoms. No symptoms were reported by 8.6%. Further investigation showed that, apart from hot flashes and night sweating, the occurrence of symptoms did not vary greatly from one group to another. Hot flashes, however, were reported by 75% of women whose periods had been over for 3-12 months as opposed to less than 30% of women whose periods had been over for 9 or more years.

DIRECT SYMPTOMS
The London survey and others seem to confirm that only hot flashes night sweats and vaginal atrophy are directly linked to the menopause. The other commonly associated symptoms such as depression and insomnia are probably indirect and may result from physical discomfort or emotional upset.
Interestingly an American sociologist, Pauline Bart, found

that many of these emotional symptoms are more likely to be reported by pre-menopausal women and by adolescents than by post-menopausal women.
SEVERITY OF SYMPTOMS
In general it seems that up to 90% of women may experience one or more menopausal "symptoms." Some experience great distress and discomfort. But it has been found that not more than 20% of women actually have symptoms severe enough to disrupt their lives.

Percentage of women reporting symptoms

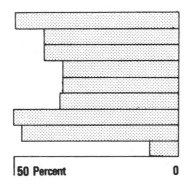

Hot flashes
Night sweats
Headaches
Dizzy spells
Palpitations
Insomnia
Depression
Weight increase
No symptoms

50 Percent   0

©DIAGRAM

THE PROCESS OF AGING

AGING

The process of aging begins in the middle to late twenties, and continues until death. No one escapes its effects; but there are often great differences in its degree of impact on people of the same age.

The efficiency of a man's body functions at the age of 75 is proportionately less than at 30. The extent of this decline is shown here.

Decline in body functions

Retained at 75   Lost by 75

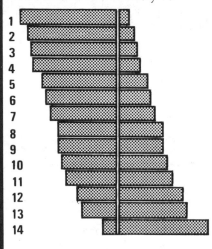

1 Body weight
2 Basal metabolic rate
3 Body water content
4 Blood flow to brain
5 Cardiac output at rest
6 Filtration rate of kidneys
7 Number of nerve trunk fibers
8 Brain weight
9 Number of kidney glomeruli
10 Maximum ventilation volume
11 Kidney plasma flow
12 Maximum oxygen uptake
13 Number of taste buds
14 Return of blood acidity to equilibrium

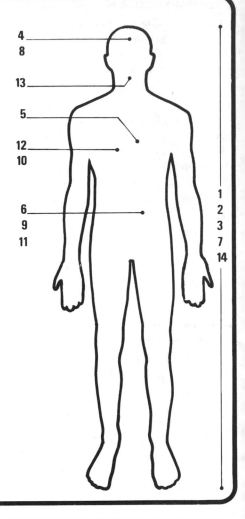

FACIAL APPEARANCE

Changes in facial appearance come about in adulthood through atrophy of the facial bones, recession of the gums, and loss of teeth. The skin becomes dry and loses its elasticity, taking on a wrinkled appearance. This deterioration is speeded by the loss of fat deposits from under the skin surface. Hair becomes grey as pigment ceases to be produced. Also both sexes undergo loss of hair, though actual balding is much more common in men.

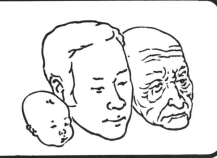

STATURE

Full stature is reached by the age of 20 or shortly afterwards. After that there is little change in the size of any individual bones. However, adult height does decline with the aging process. This is partly because the vertebral discs in the spine deteriorate, causing the spine to shorten slightly. But more important is the gradual weakening of the body muscles, so that the spine is not held so erect.

POSSIBLE CAUSES OF AGING

Medical science can prolong the life span - but it cannot yet prolong youth. Gerontology - the serious study of aging - is still a new science, and even the process and causes of ageing are not yet properly understood.

CELL MUTATION

This is currently thought of as the main cause of aging.

Most cells in the body reproduce to replace cells that have died. They do so by "somatic division," i.e. by dividing into two. In this way the exact characteristics of the original cell are preserved. However, it is possible for mutation to occur in a cell. This is any form of damage affecting the chromosomes, which are the code system built into the cell that decides how it operates.

Mutation can be caused by the gradual exposure over a lifetime to natural radiation (from the sun or from naturally occurring isotopes). Less normally, it may also be caused by: disease; chemical action; or radiation from nuclear activity, exposure to X rays, etc.

When mutation occurs, a cell may become inactive, or do its job badly, or be actively dangerous (as in the case of cancer). Moreover, because chromosomal damage is involved, the distortion is passed on whenever the original cell reproduces. Somatic division means that the number of mutated cells increases in geometric progression (1, 2, 4, 8, 16, 32, 64). In this way, areas of the body's activities become inefficient or disrupted. The effect is increased when the process occurs in certain organs, eg in the endocrine glands, which form the body's chemical control system.

NERVE CELL LOSS

From the age of about 25, there is a continuous loss of nerve cells ("neurons") from the brain and spinal cord. These cells cannot be replaced once lost; and the rate of neuron loss is accelerated in age by the onset of arteriosclerosis.

THE BODY

Muscles lose strength, shape, and size. Joints become worn and lose their ease of articulation. Combined with the degeneration of the nervous system, ease and confidence of movement is lost. Loss of bone tissue makes the skeleton brittle. As calcium is lost from the bone, it tends to be deposited in other areas - especially the walls of the arteries and the cartilage of the ribs. This causes loss of elasticity. One effect is a lowering of lung capacity. Most internal organs - such as the liver, heart, and kidney - become reduced in size and function.

The arteries harden and narrow ("arteriosclerosis"). This increases the normal rise in blood pressure, which goes up about 0.5mm Hg a year from the onset of aging. The speed of blood flow also rises - though usually not excessively. When arteriosclerosis is combined with atheroma (degeneration of the arteries' inner lining) the condition is known as atherosclerosis. These disorders of the vascular system speed up tissue decay, through inadequate blood and oxygen supply. This especially affects the heart and brain.

OTHER FACTORS

Some other factors have been seen to play a part in aging - but they are not "causes" in the same sense as cell mutation and nerve cell loss are thought to be.

STRESS

Psychological stress often has physical manifestations, and it has been noticed that stress of all sorts (physical danger, pain, mental strain, etc) can cause premature aging. However, the biological process whereby this happens is not known.

METABOLISM

As a person grows older, there is a drop in his basal metabolic rate: that is, the energy production of his body at its lowest waking level. For example, the body temperature of an old man is on average $2^{\circ}F$ less than that of a 25-year-old. Metabolic decline is a sign of the aging process, rather than a cause, but it has a wide impact on the body's functions and abilities.

HORMONE PRODUCTION

During the female menopause, the ovaries stop producing estrogen, and male production of testosterone also declines after the middle years (though it never reaches zero level). It was therefore natural for gerontologists to consider using injections of the appropriate hormone to make up the body's failing supply.

However, although injections of these hormones can reduce some physical signs of aging (smooth out wrinkled skin, for example), they do not seem to prevent the basic physiological process of aging going on.

In general, hormonal decline seems to be one of the ways in which aging expresses itself, but not a basic cause.

©DIAGRAM

AGING AND ABILITIES

DECLINE IN ABILITIES
Physical abilities

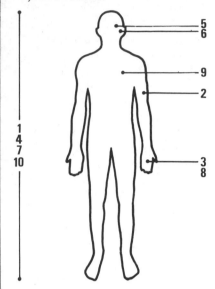

The decline in the efficiency of the body also affects an aging person's physical strength, conscious capabilities, nervous control, and mental powers (maximum measures in these are mostly reached at different ages from 20 to 30).

1 Nerve conducting velocity
2 Circumference of biceps
3 Persistence of grip
4 Maximum work rate
5 Visual acuity at 20 ft
6 Hearing
7 Reaction time
8 Hand grip
9 Maximum breath rate
10 Maximum work rate
(for short periods)

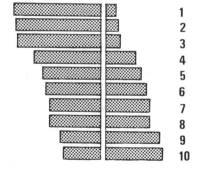

Retained at 75   **Lost at 75**

Mental abilities
1 Vocabulary
2 Information
3 Verbal intelligence
4 Comprehension
5 Arithmetic
6 Non-verbal intelligence
7 Intellectual efficiency

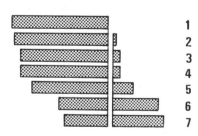

Retained at 75   **Lost at 75**

SELF-IMAGE

Changes in temperament and behavior in old people may be accepted as inevitable. But how far they are really due to neurological and mental deterioration is often hard to judge. The changes may rather be a psychic reaction to the person's social, psychological, and physical situation. Old age often brings with it a dramatic change in a person's experience of life. Declining physical ability and efficiency, perhaps involving being looked after by others; the end of the working life; and isolation, due to the disappearance of work contacts, family mobility, and death of friends - all these can affect an old person's self-esteem, and lead to depression and melancholia. Old people find that they have no role to fulfil, and no social label or way of identifying themselves other than by that term "old person" - the associated stereotypes of which are not inspiring. Society's subtle message can seem the same: you are no longer really useful, and though you are enjoying the deserved fruits of your labor, your difficulties and incapabilities are something of a problem for us. Of course, many old people keep up a wide range of active interests - but for others it is difficult, due to lack of finance, isolation, physical incapacity, and lack of mental stimulation. The rate of change in modern society adds to their disorientation; and the way of life in many old people's homes does little to help. All this can result in apathy, listlessness, resentment and mental stagnation, which others then dismiss as inevitable senility.

THE NERVOUS SYSTEM

In man, as in all vertebrates, different levels of organization can be seen in the nervous system. The reflexes of the lowest level (the spinal cord and oldest parts of the brain) are present in the fetus. Those of the forebrain and cerebral hemispheres develop after birth. With the onset of aging, the course of development is reversed. First, the higher levels are affected: memory, thought, and complex mental functions become slower and less reliable. Eventually the individual may pass through a second childhood, with, finally, only basic reflexes remaining, such as eating, walking, coughing. On average, by the age of 70, the brain has lost 50% of its weight.

CHARACTER CHANGES

The frontal lobes of the brain - the first part to deteriorate - are less concerned with intelligence and intellectuality than with general personality, interest in life, deliberation, and consideration. Moreover, these higher functions often bear a repressive relationship to the lower, so that as the higher deteriorate the lower are released, in what appears to be an exaggerated form. Social inhibitions are removed in the same way as they were originally formed: the person becomes increasingly selfish, inconsiderate, obstinate, and emotional.

MENTAL ILLNESS

About 10% of people over 65 show some signs of organic brain disorder. Such mental illnesses due to old age were originally undifferentiated under the term "senility" - the loss of mental faculties with age. However, four main conditions are now recognized.

ACUTE CONFUSION

This is one of the commonest mental disorders in old people. It is a disturbance of the brain due to physical illness elsewhere in the body, and is also known as "acute brain syndrome". Strange surroundings and other psychological factors may also play a part.

The symptoms are confusion, delirium, and disorientation in time and place. Perception is dulled, and the sufferer is frightened, often reacting violently to situations that he has completely misinterpreted. Speech becomes incoherent and rambling. The outcome depends on the pre-existing mental state and the original causal illness, but complete recovery is rare.

SENILE DEMENTIA

This is the disorder that links most directly with the slow process of natural nerve cell loss. It usually begins to be noticeable between the ages of 70 and 80, and primarily affects the memory. It begins gradually with recent memory, and may proceed to the point at which the patient forgets his relations and even his own name.

This forgetfulness leads to incompetence in personal care and management: the person needs more and more attention as time goes on.

Disorientation in time and place also occur. Emotions are blunted, and there is an increasing lack of consideration for others. Whether the course of the illness is rapid or slow, it is irreversible, and deterioration continues until death. As the numbers of old people rise, so inevitably do the cases of senile dementia. At present it is more common in women than men.

ARTERIOSCLEROTIC DEMENTIA

This is also due to the death of brain cells, though in this case the cells die because blockage in the arteries impairs their blood supply. The onset may be gradual, or follow suddenly upon a major stroke . In either case the effects, in an old person, are irreversible. Because the brain damage is restricted to the areas affected by the blockage, the basic personality may remain more intact than in senile dementia. The person usually retains more awareness and insight into his condition - though this can result in depression and fear.

DEPRESSIVE ILLNESS

This is a mental illness, but organic nervous decline plays a part in its appearance. It is characterized by acute feelings of sadness, inadequacy, anxiety, apathy, guilt, and fear. It may be triggered off by internal factors of mental make-up (endogenous depression), or by external events such as bereavement or knowledge of an incurable disease (reactive depression). It often results from a combination of both.

The illness manifests itself in moods ranging from apathy to despair, and in delusions, loss of appetite and weight, and a preoccupation with thoughts of suicide. It is, in fact, a major cause of suicide in the elderly. Although treatment may produce a cure, the chances of relapse are very great.

©DIAGRAM

AGING AND THE SENSES

THE SENSES IN AGING

The physical atrophy that occurs with age also affects the five senses. As the nervous system degenerates through neuron loss, touch, taste, and smell become less sensitive. This does not affect the body's physiological functioning, but may have dangerous consequences (eg an impaired sense of smell may mean that escaping gas is not detected). More serious sensory loss, though, comes from some illnesses brought about by, or occurring in, old age, and affecting the more complicated sense organs such as eye and ear.

HEARING

PRESBYACUSIS is the term for hearing loss as a direct result of old age. It is due to a combination of loss of elasticity and efficiency in the actual hearing mechanisms, and the slowing down of mental activities that old age brings. Mechanical losses especially affect the ability to detect high frequency sounds. By the age of 60 years, hearing of these has been reduced by 75%, though normal conversation and most other everyday sounds can still be distinguished with only slight distortion.

OTITIS EXTERNA (infection of the outer ear) is especially common in the aged, due partly to hearing aid earpieces that fit badly and are too infrequently cleaned.

SEEING

PRESBYOPIA is the term for the natural changes in the eye that result from old age. The most common (apart from the general loss of efficiency with age) is hardening of the lens. From the age of about 10 years onward the lens gradually loses its elasticity, and so its ability to adjust focus. As a result, by the age of 60 years the eye is often unable to focus on objects close at hand. Spectacles with convex lenses are then needed for close work such as reading.

CATARACT This is an opacity in the lens of the eye. It is caused by deficiencies in the lens proteins, resulting in a special type of hardening and shrinkage at the center of the lens. The lens cracks and disintegrates, so losing its transparency. The process spreads from the center outwards, and impairment of vision increases as the opaque area (the cataract) spreads. Cataracts occur in most people over 60 years of age, though they are usually too small to have a significant effect on sight.

GLAUCOMA occurs most frequently after the age of 50 years. It is caused by a build up of the fluid pressure in the eye. The fluid that fills the eye (the aqueous humor) is normally being continually drained away and replaced by fresh. Drainage takes place along the "canal of Schlemm", which lies at the junction of the iris and cornea.

But sometimes the canal becomes blocked, through inflammation of the eye or swelling of the lens pushing the iris forward. The amount of fluid in the eye then increases as secretion of fresh fluid continues. Pressure builds up, damaging the optical disk and the visual fibers of the retina. Loss of vision is first experienced in the peripheral field, and it spreads gradually to the whole visual field. There is no pain in the early stages, and the process is so slow it goes undetected. But early treatment is vital, as lost vision cannot be replaced. Yearly pressure check-ups for those over 40 are recommended. (Note that the usual watering of the eye with tears has nothing to do with the drainage of aqueous humor; so continued tear production does not prove that all is well.) Treatment is largely with drugs, to control aqueous humor production and/or drainage. If this fails, surgery may be used.

SENILE MACULAR DYSTROPHY The macular lutea is a small yellowish area of the retina, and the fovea lies at its center. Here visual perception is most perfect, and differentiation of minute objects takes place (for reading, etc). Senile macular dystrophy is a degeneration of this area due to impaired blood supply as a result of age. It is the most common eye disorder in the old. But affected people can be taught to use their remaining vision so as to read and function almost normally.

DISORDERS OF AGE

The atrophy of age reduces the body's efficiency, creating greater vulnerability and likelihood of malfunction. The body is still susceptible to all the usual disorders, while its maintenance, defense, and repair processes are all much weaker. Respiratory and heart disorders, for example, occur with much more frequency and intensity, skin wounds are more liable to infection, and bone fractures more difficult to heal. Cancer becomes more likely. But in addition the deterioration of the body and its functions produces ailments rarely found at a younger age.

INCONTINENCE

This is the inability to control the emptying of bladder and bowels. Incontinence of the bladder is usually due to infection of the urinary tract, or to damage to the controlling nerves. Atrophy of the muscles concerned also contributes. As senility and mental damage associated with old age progress, a woman may lose awareness of her bladder, so conscious control is finally lost and it empties of its own volition. Restricted mobility, emotional insecurity, and abdominal stress (caused by laughing, coughing, lifting, etc) also play their part in precipitating incontinence.

Fecal incontinence is most often due to fecal impaction - the accumulation of a mass of feces in the rectum too bulky to be passed. This mass then acts as a ball valve, with fresh feces trickling round it and escaping in a continuous flow. In other cases it is due to the continual overflow of the original mass. It is occasionally due to diarrhoea, as might occur in gastro-enteritis or rectal prolapse (collapse of the wall of the rectum often due to over-straining).

NERVOUS DISORDERS

The nervous system of an old person is likely to be affected by degeneration, since nerve cells cannot be replaced. Also the person is more susceptible to strokes, and liable to falls which may damage spinal cord or brain. All these may impair movement, and ability to think, see, hear, and express onself.
PARKINSONISM occurs fairly often in old age, though less often in women than in men. It is due to degeneration of the nerve cells in one part of the brain, usually as a result of arteriosclerosis. It manifests itself in trembling and muscular rigidity, and often begins in one hand and then spreads to other parts of the body. The body's rigidity interferes with all movement, from facial expression to locomotion, and brings increasing discomfort as the disease progresses. Treatment involves drugs, exercise, physiotherapy, and possibly surgery.

SPINAL DISORDERS

SLIPPED DISK is the common name for a prolapsed intervertebral disk. The spine is built up of a column of bones (vertebrae) separated by disks of cartilage which act as shock absorbers. (The inside of the disk is made of spongy but firm elastic tissue, held in place by the strong fibrous tissue of the outer layer.) In a prolapse, one of these disks, in the lower part of the spine, slips out of position, impairing mobility and exerting painful pressure on the nerves of the spinal cord.
CERVICAL SPONDYLOSIS affects the upper part of the spinal column. Degeneration of the spine shortens the neck, forcing the vertebral artery to concertina, and so impairing the blood supply to the spinal cord and brain. Also pressure is exerted on the nerves of the spinal cord, so their functioning, and that of all connected nerves, is affected.

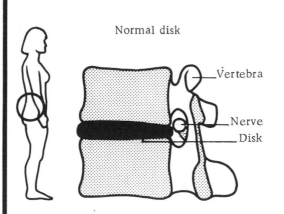

Normal disk

Vertebra
Nerve
Disk

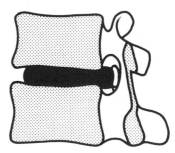

Prolapsed disk, pressing on nerve

DISORDERS OF AGE 2

HYPOTHERMIA

This is when body temperature falls below 95°F (35°C). It mainly occurs in old people who cannot afford adequate heating, food, or winter living conditions. The person becomes increasingly apathetic and lethargic, and below 90°F (32°C) coma usually occurs.

Direct heat must not be applied. The patient's surroundings should be warmed and a blanket wrapped around her to prevent heat loss. Because of hypothermia, deaths among old people may rise by 30 to 50% in winter, compared with summer months. The harder the winter, the worse the situation is. The trend begins even in the 50 to 54 age group, and becomes more marked over 60. Studies show that many elderly people live in temperatures that would not be allowed in factories or offices.

DISORDERS OF THE JOINTS

Degeneration of the joints is largely due to the constant wear and tear they receive throughout life, but injuries may accentuate and accelerate any disorders. OSTEOARTHRITIS occurs to some degree in 80 to 90% of people over the age of 60, though less in women than men. It originates from loss of elasticity in the cartilage of the joint. The cartilage breaks up with the joint's movement, and loose bits of cartilage may be deposited in the joint itself. These may grow and become calcified, increasing discomfort. The bone around the joint hardens, and cysts may develop, with spurs of bone around the joint's edges.

The knee, hip, and hand are most commonly affected. The process cannot be reversed, but can be delayed by gentle, regular exercise, to loosen the joint and strengthen the muscles.

RHEUMATOID ARTHRITIS usually begins in middle age, but its severity increases with time. It affects more women than men. The tissues of the joints thicken, so the cartilage becomes ulcerated and is eventually destroyed. There is overproduction of connective tissue, and ultimately the joint is swollen and may be fused solid.

The muscles waste with disuse. Special exercises, and rest in serious cases, are the main forms of treatment. Use of the drug cortisone is now thought to cause many problems, though it does give temporary relief.

CONTRACTURES These are deformities of the joints due to shortening (contraction) of the surrounding muscles and ligaments. They are caused by arthritic or neurological disorders, or simply by prolonged inactivity. If untreated they become permanent and cause severe disablement. Treatment is with muscle-relaxing drugs, and physical manipulation.

OSTEOPOROSIS is increased porousness of the bone, usually from unknown causes, but sometimes due to severe nutritional deficiency of calcium salts. It can also be due to hormonal changes following the menopause. The skeleton becomes brittle and prone to fracture. The vertebrae are most often affected. They may collapse as they become weaker, and as they lose weight and size the vertebral disks expand, producing increasing curvature of the spine. The condition may be triggered off by prolonged immobilization in bed. It is more common in women than in men.

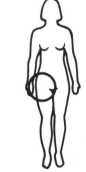

Normal hip

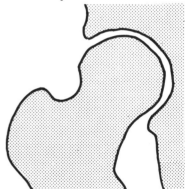

Osteoarthritic hip

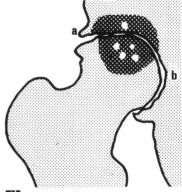

▨ Hardened bone
▨ Cyst
a Spur
b Reduced joint space

©DIAGRA

MALE SEXUAL DECLINE

PHYSIOLOGICAL FACTORS

Sexual activity declines with old age. But is this because of lack of ability, or lack of opportunity?

HORMONE PRODUCTION

The decline in testosterone production is quite steady, from the age of 20 onwards - there is no sudden point at which old age takes effect, and in fact after the age of 60 the rate of decline eases off.

The decline of sexual activity does seem to follow the same kind of pattern. But there are so many other factors affecting sexual performance in the older male that it is unlikely that the testosterone level exercises a major influence. The amount produced, even right to the end of life, is usually adequate for sexual activity.

In fact, at any time after puberty, sexual response and activity probably come to depend more and more on the higher centers of the brain (the hypothalamus) and on the automatic nervous system, rather than on testosterone production.

GENERAL PHYSICAL DECLINE

The physical degeneration of aging naturally slows down the processes of sex. Semen production, erection, and ejaculation all take more time and more stimulation. The need and ability for orgasm declines. All this continues the decline observable in men from an early age. But in old age, some new factors appear. Poverty may lead to poor nutrition - with a resulting loss of energy, and specifically sexual energy. And decline or disease in non-sexual bodily functions may make it difficult to perform sexual roles. Nevertheless, a healthy and fit person should be capable of some sexual activity at any age.

PSYCHOLOGICAL FACTORS

The main causes of declining sexual activity may often be psychological.

PSYCHOLOGICAL IMPOTENCE

The percentage of men suffering from total impotence increases slightly through the middle years, and more rapidly as old age is reached. In old age, physical factors are certainly involved. Yet one major cause of total impotence, at any stage of life, seems to be psychological. It occurs more often in those who have been sexually anxious, and for whom age provides a safe and valid excuse for ceasing sexual activity. In addition, the psychological trauma experienced by a male in middle age, when coming to terms with his life's work and expectations, may lead to depression, disappointment and self doubt, and this can sometimes affect future sexuality.

PARTNER PROBLEMS

The availability and attitude of, and the feeling for, a sexual partner will also increase or lessen desire. Again, age may give an excuse for ending sexual relations that have been merely dutiful.

SOCIAL NORMS

The prevalent attitudes of an individual's social surroundings - his national culture, religion, social contacts and intimate friends - can all impinge on the individual's attitude and thereby his performance. Many people, for example, end their lives in institutions in which any continued interest in sexuality is looked on as obscene.

EVIDENCE OF DECLINE

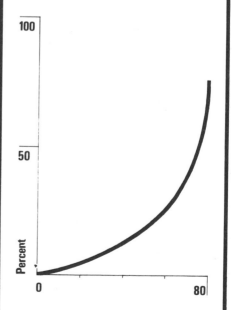

Above we show the pattern of testosterone production in the life of a typical male. It rises steeply until age 20, and declines thereafter. By 60, it has fallen to the level of a 9 or 10 year old.

The above graph shows the percentage of the male population impotent at each age.

© DIAGRAM

THE EXPERIENCE OF AGING

BODY CARE

EXERCISE Lack of exercise will speed physical decline, just when maintaining the ability to get out and about is vital for self-esteem. Any exercise chosen should be started gradually, without strain, and kept up regularly. Gardening, walking, and golf are ideal. Keep-fit programs can be followed, but only after expert medical advice.

SLEEP Recent evidence suggests that sleep needs do not decline with age. Still, during old age, sleep at night is often fitful and shallow. Daytime naps, illness, anxiety, loneliness, and discomforts such as stiffness of the limbs, can all contribute. An aging person may well try to accept the changed pattern of her sleep, and try to use waking hours whenever they occur. But a little exercise each day, avoidance of afternoon naps, a warm bedroom, a warm and comfortable bed, and a hot milky drink, can all help in achieving a sound night's sleep. If lack of sleep causes anxiety or depression, a doctor should be consulted.

WARMTH Adequate home heating is essential. Central heating is ideal. Open gas or electric bar heaters, or freestanding stoves, can be fire risks, while solid-fuel fires (coal, etc) can be both dangerous and difficult to maintain. Also, diminished awareness of pain may lead to self-scorching through sitting too near. With gas and oil heaters, good ventilation and regular safety checks are important. In bed, electric overblankets are safer than underblankets. Hot water bottles should always be cloth covered, never overfilled, and never used with electric blankets. Hot water bottles and blankets can also be a tremendous daytime help.

MOVING ABOUT THE HOUSE Reaction speed and balance decline with age, and bones and muscles weaken, so falls become more of a problem. Highly polished floors, loose mats, trailing appliance wires, frayed carpet or linoleum edges, and poor or uneven lighting, are all dangerous. Strong banisters, and good bath and toilet hand supports, are important. Awkward steps can be outlined in white paint. A light switch by the bed is also important.

TESTS AND CHECK-UPS Eyes should be tested yearly, and glasses changed if necessary: increasing farsightedness is typical with age. A doctor should be consulted over any hearing difficulties. The ears can be checked for wax, and the help of a hearing aid should not be ruled out. Elderly people who still have their own teeth should brush them regularly and visit a dentist every 6 months. Denture wearers should have a check-up every 5 years, or if the dentures cause discomfort or difficulty in eating or speaking.

FEET Shoes should be fitted with care; low heels are safest. Stockings should not be too tight. A chiropodist will help with problems caused by corns, calluses, bunions, or toenails too thick to cut.

ILLNESS A doctor should always be told of any symptoms appearing, such as poor appetite, loss of weight, blood in urine, etc; also of any discomfort, for in an old person even a bone fracture may feel no more than troublesome. Medicines should be clearly labeled. Sleeping pills should not be kept by the bedside, in case of mistakes.

BLADDER AND BOWEL CONTROL The causes of incontinence are often temporary, and control is regained naturally. Otherwise, medical treatment can often help. Pads, special sheets, and mattress covers can be used. If suitable help is available, it is usually best for the sufferer to remain at home.

DIET

NUTRITIONAL NEEDS The main change is simply in quantity. With age, decreasing physical activity and falling metabolic rate lower food energy needs; calorific intake should be gradually reduced. Otherwise, the kind of food needed stays basically the same at any age, though in old age there is reduced need for protein, fat, thiamin, glucose, and calcium (but see "Osteoporosis," p. 348). However, more foods may cause digestive problems, because of slower and therefore incomplete digestion and absorption.

DAILY DIET Ideally, two portions of meat, fish, cheese, or eggs should be taken each day, and some milk as well. Vegetables and fruit, wholemeal bread, and butter or margarine, are also important as sources of vitamins, minerals, and roughage. Several small meals during the day may be better than one or two large ones; breakfast should never be missed. Large meals late at night are best avoided, since they can interfere with sleep. Liquid intake should be at least 3 pints daily. Also, when catering for an old person, variety and attractiveness of food counts as much as good nutritional balance.

POOR NUTRITION Many old people eat badly. The lonely, impoverished, and neglected may do so through apathy, poverty, lack of facilities, lack of judgment, or general physical or mental disability. Others, unrestricted economically and practically, may overeat such foods as pastries and cakes, with a resulting weight gain that impairs health. Finally, unhealthy teeth and badly fitting dentures can encourage the old to choose comfortable rather than nutritious foods.

AGE AND SOCIETY

Life expectancy has risen steadily in industrialized countries in this century. So, with declining birth rates, their populations are increasingly older ones. In the USA, over 10% of the population are 65 or over: almost 22 million people. Of these, almost 60% are women. It is important to realize the variety and naturalness of old age. For a working person, post-retirement can be up to one-third of the life span and the majority of old people are not lonely, poor, incapacitated, neglected, or ignored. Nevertheless, they can count it as individual good fortune or good planning if they are not. Our society does not have a good record for its treatment of the old. There is insufficient provision for their physical welfare; there is little or none for their self-esteem. After a lifetime of work, elderly people too often find they have no role to fulfill, and no social label but that of "old person." Physical and financial difficulties can reinforce this. Society's subtle message can often seem to be: you are no longer really useful, and though you are enjoying the deserved fruits of your labor, your difficulties and incapabilities are something of a problem for us. Because they live longer, there are more retired women than men - and more living alone. Still, elderly women often keep their self-respect better than men. A man's identity faces a severe crisis when he retires from work. For a woman, even if she has had a job, the home has usually remained an important sphere of activity. Her crisis of identity - the menopause, and the departure of her children - comes earlier, when her mental resources are stronger.

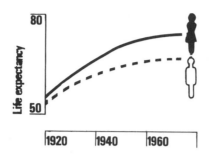

Figures for the USA show that life expectancy at birth has risen steadily in this century and that woman's advantage over man has increased.

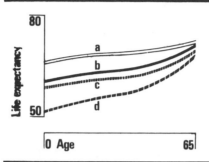

Life expectancy also increases with age, since at each age the fittest survive. Figures shown are for the USA now, for: white women (a); black women (b); white men (c); and black men (d). Expectancy for a white woman born now is 76 years; for one aged 65, over 82.

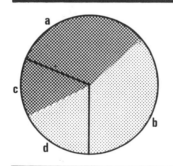

Because of female life expectancy, the retired population in a developed country consists overwhelmingly of women:
a) women living alone, 30%;
b) other women, 39%;
c) men living alone, 13%;
d) other men, 18%.
(Figures for Great Britain.)

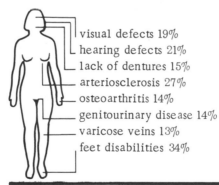

visual defects 19%
hearing defects 21%
lack of dentures 15%
arteriosclerosis 27%
osteoarthritis 14%
genitourinary disease 14%
varicose veins 13%
feet disabilities 34%

The diagram shows the estimated extent of unrecognized medical needs in those aged 65 and over, in a typically developed country. For example 34% of the group have unattended feet disabilities.

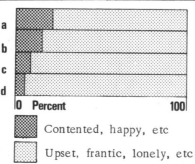

Contented, happy, etc

Upset, frantic, lonely, etc

The diagram illustrates how negative the reactions of most people are to retirement. Only 23% of people retire with any positive feelings (a); within 6 months (b) this percentage has dropped to 16% and within a year (c) to 10%. Of those who have been retired for more than 1 year (d) only 6% remain content.

©DIAGRAM

351

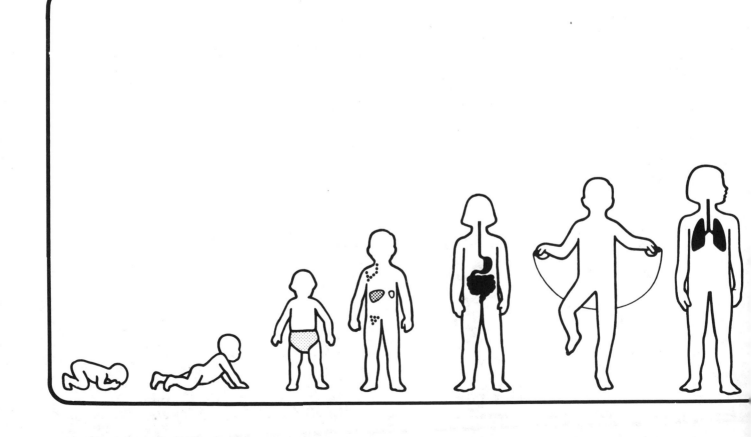

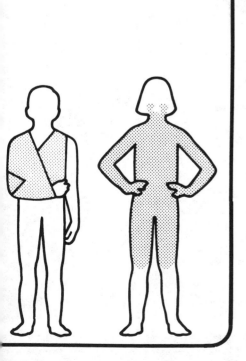

PART 2

CHILD'S BODY

10 Development year by year

Based on the work of specialists in child development — Gesell, Ilg, Ames, Illingworth, Sheridan, Griffiths — this chapter describes milestones of development typically reached at different ages. Many normal variations occur, however, and professional investigation is needed only if a child's progress is significantly out of line.

Left: Peeping through the windows of a candy store in the country (The Mansell Collection)

THE NEWBORN CHILD – REFLEXES

REFLEXES OF A NEWBORN

Described on these pages are a variety of reflexes - automatic reactions to particular changes in surroundings - that are present in a newborn baby. These "primitive" reflexes are thought to be a legacy from man's earliest ancestors, when such actions were vital for an infant's survival.

Several important reflexes are usually tested soon after birth, as they give an indication of the baby's general condition and the normality of his central nervous system. Parents are warned of the potential dangers of testing some of these reflexes for themselves. By three months, a baby's

primitive reflexes have generally been replaced by voluntary movements. This can only happen when a baby learns to associate a particular action with the fulfillment of a certain need; he learns, for instance, that hunger is satisfied by food, which is obtained by sucking.

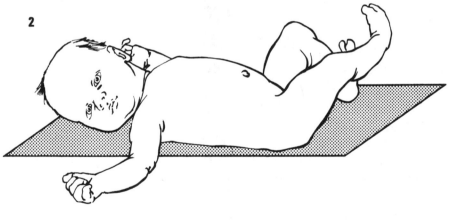

1 ROOTING REFLEX
If one side of a baby's cheek or mouth is gently touched, he will turn his head in the direction of the touch. This ensures that he will seek out the nipple when his cheek is brushed by his mother's breast. With the sucking and swallowing reflexes, the rooting reflex is essential for successful feeding in the newborn.

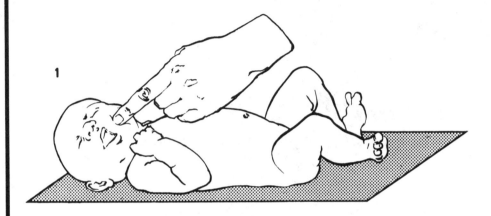

2 ASYMMETRIC TONIC NECK REFLEX
When on his back and not crying, a newborn baby lies with his head to one side, the arm on that side extended as shown, and the opposite leg bent at the knee.

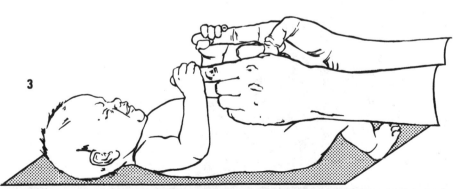

3 GRASP REFLEX
This reflex makes a baby automatically clench his fist if an object is placed in his palm. If a finger is slipped into each of his palms, a baby will grasp them so tightly that he can support his own weight. The grasp reflex can also be evoked from the toes.

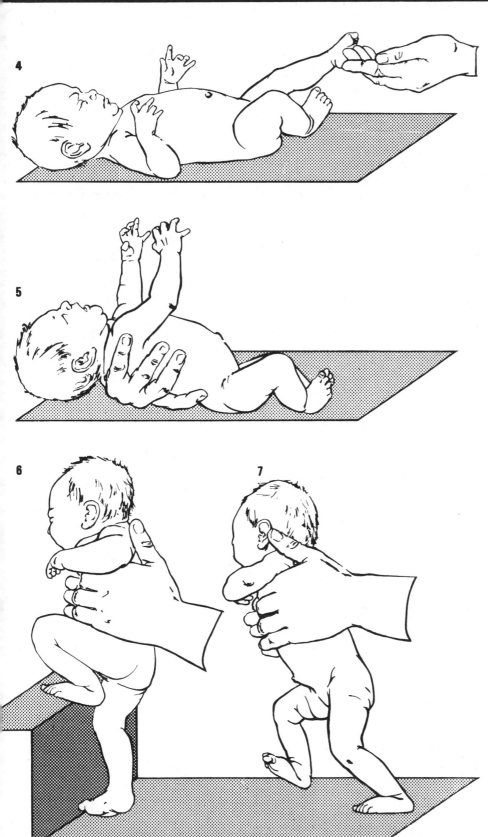

4 CROSSED EXTENSION REFLEX
If one of the baby's legs is held
in an extended position and the
sole of that foot stroked, the
other leg flexes, draws toward the
body, and then extends.

5 MORO REFLEX
This reflex is seen when the baby
is startled. The arms and legs are
outstretched and then drawn
inward with the fingers curled as
if ready to clutch at something.
A baby's Moro reflex is tested to
check muscle tone; if the limbs
respond asymmetrically, there
may be weakness or injury of a
particular limb.

6 "STEPPING" REFLEX
If the front of his leg is gently
brought into contact with the edge
of a table, a baby will raise his
leg and take a "step" up on to
the table.

7 "WALKING" REFLEX
If a baby is held upright with the
soles of his feet on a flat surface,
and is then moved forward slowly,
he will respond with "walking"
steps.

BLINK REFLEX
A baby blinks in response to
various stimuli - for instance, if
the bridge of his nose is touched.

DOLL'S EYE REFLEX
If a baby's head is turned to the
left or right, there is a delay in
the following of the eyes.

©DIAGRAM

NEWBORN / ONE MONTH

NEWBORN: PROFILE

A newborn baby depends on others for almost every need. With his large head, tiny legs, soft bones, largely "unwired" nervous system, and flabby muscles, he appears unlikely ever to achieve an independent life. Yet at least he breathes, feeds, excretes waste, and cries when hungry, and his brain is richer in promise than that of any other infant animal.

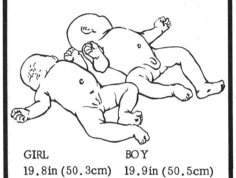

GIRL	BOY
19.8in (50.3cm)	19.9in (50.5cm)
7.4lb (3.35kg)	7.5lb (3.40kg)

ONE MONTH: PROFILE

The month-old baby cannot yet hold up his head, let alone sit up, crawl, walk, or talk. He cannot even pick up an object or smile responsively. Yet his body has made tremendous progress in the first four weeks. Billions of nerve fibers effecting countless new interconnections produce much improved muscle tone, breathing, swallowing, and temperature control.

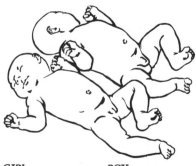

GIRL	BOY
21.01in (53.3cm)	21.2in (53.8cm)
8.5lb (3.86kg)	8.6lb (3.90kg)

NEWBORN: DEVELOPMENT

MOTOR DEVELOPMENT

Head and limbs hang down when body held face down, horizontally. Placed face down, baby lies with knees drawn up, arms bent across chest, head turned sideways (a). Releasing head of supported baby evokes Moro reflex.
Head lags if baby pulled to sit.
Has grasp and walking reflexes.

a

EYE AND HAND DEVELOPMENT

Shuts eyes when suddenly dazzled by bright light.
Pupils respond to light.
Opens eyes if held erect.
Doll's eye reflex for first few days only - eyes lag as baby is swung in horizontal arc (b).
Turns toward diffused light after first week.

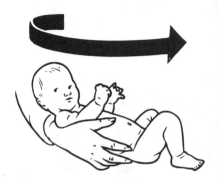

b

HEARING AND VOICE

Sudden sound produces blinking.
Visibly startled by sudden, loud sound.
"Freezes" in presence of steady, soft sounds.
Cries energetically (c), but stops in presence of fairly loud, steady adult voice.
Eyes may turn toward source of steady sound (but head does not).

c

PLAY AND SOCIAL DEVELOPMENT

Rooting (d) and sucking reflexes are present.

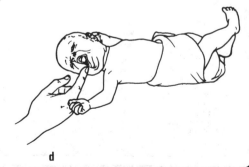

d

ONE MONTH: DEVELOPMENT

MOTOR DEVELOPMENT

Lies on back with head and limbs in positions illustrated (a). Makes big, jerky movements. Stretches limbs, fanning out toes and fingers. Rests with hands shut and thumbs turned in. Head flops when body lifted. Keeps head in line with body when held face down horizontally. Placed face down, lies with limbs bent, elbows out from body, and buttocks quite high. Momentarily holds head upright when pulled to sit. Presses feet down and adopts walking attitude if held standing.

a

EYE AND HAND DEVELOPMENT

Turns toward source of light. Stares at window or bright wall. Eyes briefly follow moving, nearby light of small flashlamp. Gazes at small, white ball moved 6-10in:15-25cm from face (b). Alertly watches mother's face when being fed by her. Blinks defensively by six weeks.

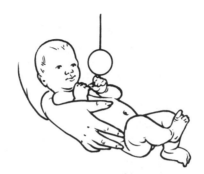

b

HEARING AND VOICE

Sudden noise may cause stiffening, blinking, quivering, stretching, and crying. Repeated, quiet, nearby sound causes "freezing." Eyes and head may turn toward source of sound. Soothing voice stops whimpering. Cries in hunger or discomfort. Gurgles if contented (c).

c

PLAY AND SOCIAL DEVELOPMENT

Spends most of time sleeping. Smiles socially and makes responsive sounds by six weeks. Grasps adult's finger when adult opens hand (d). Ceases crying if picked up and talked to. May turn to look at face of nearby speaker. Responsive awareness of bathing.

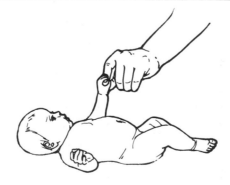

d

©DIAGRAM

THREE MONTHS / SIX MONTHS

THREE MONTHS: PROFILE

At three months a baby eyes his immediate world alertly and begins to respond in a lively way. Laid face down, he struggles futilely: on his back he brings his hands above his chest. He is becoming aware of the feel of his fingers and some objects.

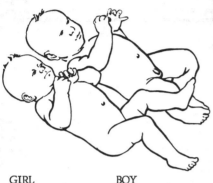

GIRL	BOY
23.4in (59.4cm)	23.8in (60.4cm)
12.41b (5.46kg)	12.61b (5.55kg)

SIX MONTHS: PROFILE

Now learning the meaning of people's gestures and expressions, he reacts with two-syllabled and other sounds. His improved grasp encourages continual handling, mouthing, and banging which help him learn about objects within his reach. He still cannot sit or stand unaided but has begun to master parts of these operations.

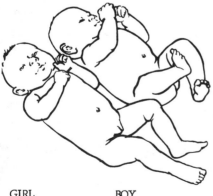

GIRL	BOY
25.7in (65.3cm)	26.1in (66.3cm)
16.01b (7.26kg)	16.71b (7.58kg)

THREE MONTHS: DEVELOPMENT

MOTOR DEVELOPMENT

Lies on back, face upward (a).
Brings hands together above body.
Kicks, and waves arms.
Placed on stomach, uses forearms as props to lift head and chest.
Held horizontally, face down, raises head and extends hips and shoulders.
Little head lag if pulled to sit.

a

EYE AND HAND DEVELOPMENT

Moves head to gaze around.
Alertly watches any nearby face.
Watches own hands clasping and unclasping, and in finger play.
Briefly fixes eyes on small objects less than 1ft away.
Briefly holds rattle (b).
Visibly excited by arrival of feeding bottle.
Tries to focus on small ball approaching face.

b

HEARING AND VOICE

Distressed by loud, sudden noise.
Quieted (unless screaming) by mother's voice.
Vocalizes happily when pleased.
Cries in annoyance or discomfort.
Head turns toward hidden source of nearby, intriguing sound.
Excited by sound of running tap, footsteps, or approaching voice.
Licks or sucks lips when hears food being prepared (c).

c

PLAY AND SOCIAL DEVELOPMENT

Gazes unblinkingly at face of person feeding him.
Baths and feeds evoke coos, smiles, and excited gestures.
Enjoys being bathed and cared for.
Responds happily to sympathetic handling, including tickling (d).

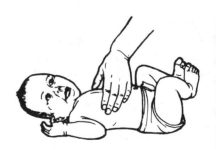

d

SIX MONTHS: DEVELOPMENT

MOTOR DEVELOPMENT

Lying on back, lifts head to regard feet (a).
Lifts legs and grabs feet.
Kicks legs vigorously.
Rolls over.
Holds up arms to be lifted.
Pulls self to sit if hands grasped.

Sits supported, with head erect and back straight. Briefly sits unaided.
If laid on stomach, extends elbows to lift head and chest.
Bounces, when held standing.
Exhibits "downward parachute" when quickly lowered.

a

EYE AND HAND DEVELOPMENT

If someone attracts his attention, eagerly moves head and eyes in all directions.
Focuses without squinting.
Focuses on nearby, small objects and puts out both hands to grasp (b).
Grasps a toy with whole hand and passes to other hand (c).
Watches toys dropped from hand until they land, but ignores toys that land out of sight.

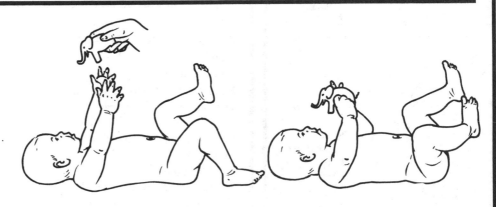

b

c

HEARING AND VOICE

Makes tuneful, sing-song, vowel and two-syllabled sounds ("goo," "adah," "a-a").
Turns at once toward sound of mother's voice heard across room.
Playfully squeals and chuckles (d).
Screams if annoyed.
Turns to very quiet sound from nearby, ear-level source.
Responds to varying emotional sounds in mother's voice.

d

PLAY AND SOCIAL DEVELOPMENT

Both hands reach out for and grab small toys.
Brings everything to his mouth (e).
Watches own feet moving.
Pats bottle when fed.
Shakes and watches rattle.
Passes objects from hand to hand.
Shy or anxious with strangers by about seven months.
Delighted if actively played with.

e

© DIAGRAM

NINE MONTHS / ONE YEAR

NINE MONTHS: PROFILE

He has now mastered sitting up, and may even pull himself to stand by holding on to the bars of his crib. Now, too, he can begin to roll or crawl along. Fingers are controlled with precision. He imitates simple actions, joins in some simple games, and babbles fluently.

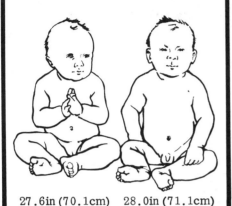

27.6in (70.1cm)   28.0in (71.1cm)
19.2lb (8.70kg)   20.0lb (9.07kg)

ONE YEAR: PROFILE

Rates of growth and development are slowing. But agile crawling now gives real mobility, and he can nearly stand and walk unaided. He repeatedly lets an object fall, and handles several objects one by one. He loves to imitate actions, speaks his first words, enjoys an audience, and shows jealousy, sympathy, and other emotions.

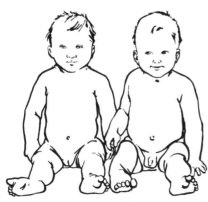

29.2in (74.1cm)   29.6in (75.2cm)
21.5lb (9.75kg)   22.2lb (10.06kg)

NINE MONTHS: DEVELOPMENT

MOTOR DEVELOPMENT

Sits alone on floor (15 minutes).
Keeps balance while bending forward to grasp toy (**a**).
Can turn sideways to look, while stretching forward for toy.
Rolls or wriggles along on floor, and tries to crawl.
Pulls himself to stand, briefly.
Extremely lively in bath and cot.

a

EYE AND HAND DEVELOPMENT

Watches nearby people or animals for several minutes.
Handles objects with interest.
Puts out one hand to grasp toy, but stares at strange toy first.
Uses first finger to poke and point at objects.
Grips string or piece of candy between thumb and finger (**b**).
Drops toy but cannot put it down.
Looks for toy fallen out of sight.

b

HEARING AND VOICE

Fascinated by ordinary sounds.
Comprehends "no-no," "bye-bye."
Cannot yet localize sound produced directly above and behind him (**c**).
Babbles loudly and repetitively ("ma-ma," "da-da," etc).
Shouts for attention.
Makes "friendly" and "annoyed" sounds at other people.
Copies cough, and playful vocal sounds made by adults.

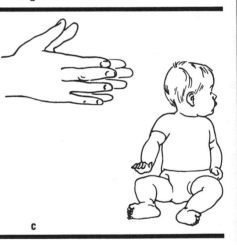

c

PLAY AND SOCIAL DEVELOPMENT

When feeding, uses both hands to grasp cup or bottle.
Can hold, bite, and chew cookie.
Hides face against familiar adult in the presence of strangers.
When annoyed, protests and stiffens body.
Holds bell in one hand and rings.

Offers toy to adult but still cannot release it.
Clasps hands and plays peek-a-boo in imitation of adult.
Retrieves a toy he has watched being partly or wholly hidden under a cover, or cannot discover it and cries.

ONE YEAR: DEVELOPMENT

MOTOR DEVELOPMENT
Can rise from lying to sitting, and sits well for a long time.
Crawls or shuffles quickly (**a**).
Perhaps crawls upstairs.
Grasps furniture to pull self up to stand (**b**), to lower self, and to walk sideways.
Walks forward, possibly unaided.
Briefly stands unaided.

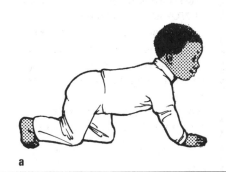

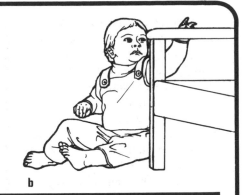

a

b

EYE AND HAND DEVELOPMENT
Gazes at moving people, animals, and vehicles out of doors.
Recognizes known people approaching 20ft (6.1m) away.
Begins to notice pictures.
Takes crumbs precisely between thumb and tip of first finger.
May favor use of one hand.
Clicks cubes together (**c**).
Deliberately drops and throws toys, watching them fall.
Can find a ball that has rolled out of sight.

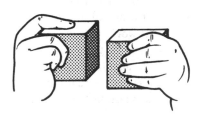

c

HEARING AND VOICE
Turns when called by name.
Soon ignores sound of hidden rattle after locating its source.
Understands some words in context.
Understands simple commands accompanied by gestures.
When requested, hands to an adult objects such as a cup or spoon (**d**).
Converses loudly in jargon.
Copies playfully made vocal sounds, and repeats some words.

d

PLAY AND SOCIAL DEVELOPMENT
Uses cup almost unaided (**e**).
Holds but cannot use spoon.
Offers arm and foot for dressing.
Bangs spoon in cup.
Rings bell with assurance.
Enjoys working sound-making toys.
Copying adult, inserts wooden blocks into cup and removes them.
Soon finds toys that he has watched being hidden.
Enjoys playing pat-a-cake.
Waves "bye-bye."
Shows affection to known adults.
Seeks their constant company.

e

©DIAGRAM

EIGHTEEN MONTHS

EIGHTEEN MONTHS: PROFILE

Immensely active, the eighteen-month-old child manipulates his body like a brash young driver trying out an automobile. Feet and arms somewhat apart, he runs about, exploring corners and clambering upstairs. He stops and starts but cannot turn corners easily or properly coordinate hands and feet. But he lugs big toys and even furniture around, learning what different places are. Lacking wrist control, he plays ball with movements of the whole arm. His world is one of here and now. His use of words is very limited. Self-willed, he takes but cannot give; and, unable to see other children as people like himself, he cannot share in their play.

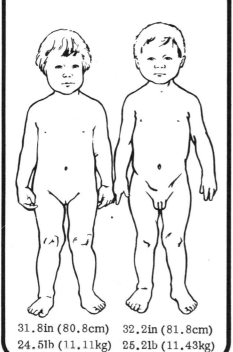

31.8in (80.8cm) 32.2in (81.8cm)
24.5lb (11.11kg) 25.2lb (11.43kg)

EIGHTEEN MONTHS: DEVELOPMENT

MOTOR DEVELOPMENT
Kneels upright, unsupported (a).
Squats to reach toy and stands up using hands as aids (b).
Enjoys pulling and pushing boxes and wheeled toys around (c).
Walks easily, with controlled starting and stopping, and feet fairly close together (d). No longer needs to stretch his arms to keep balance when walking.
Can run, but stares at ground just ahead (e), and stops for obstacles.
Likes clutching a big teddy bear or doll while walking.
Walks upstairs if hand is held,

EYE AND HAND DEVELOPMENT
Points to intriguing objects in the distance.
Absorbed by picture books; turns groups of pages (f), and points to brightly colored illustrations.
Becoming noticeably right-handed or left-handed.
Using precise pincer grip, swiftly picks up small objects such as beads.
Piles three cubes to build a tower when first shown how (g).
Grips middle of pencil in palm, and scribbles (h).

HEARING AND VOICE
Pays attention when spoken to.
Obeys requests to pass familiar named items to an adult; also understands and performs simple tasks like shutting a door.
Points to hair, nose, shoes (i), etc.
In play, jabbers continually.
Says 6-20 words and knows many more than that.
When spoken to, repeats last word of a short sentence.
Urgently vocalizes and points to an object that he wants (j).
Likes nursery rhymes and attempts to say them with an adult.
Tries singing.

PLAY AND SOCIAL DEVELOPMENT
Conveys food to mouth by spoon (k).
Drinks from cup, spilling little.
May control bowels but not bladder.
Fidgets and makes agitated sounds when urgently in need of toilet.
Removes own hat, socks, and shoes but usually cannot put them on.
Tries to open doors while energetically probing environment.
Plays alone on floor with toys.
Likes putting small articles into containers (l), and then taking them out again.
Remembers where familiar household objects are kept.
Copies everyday actions such as sweeping floor and reading.
Still needs much emotional support from mother or other familiar adult, but often resists authority.

and comes downstairs crawling
backward or bumping forward on
his buttocks.
Seats himself by backing or
slipping sideways into a child's
low chair, but climbs head-first
into a big chair before he turns
and sits.

a   b   c   d   e

f    g 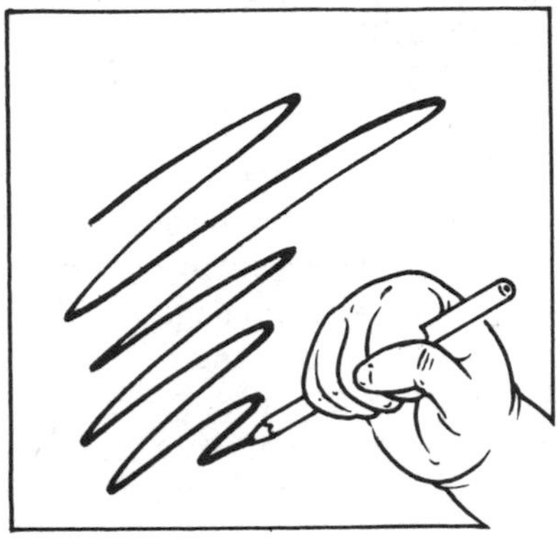   h

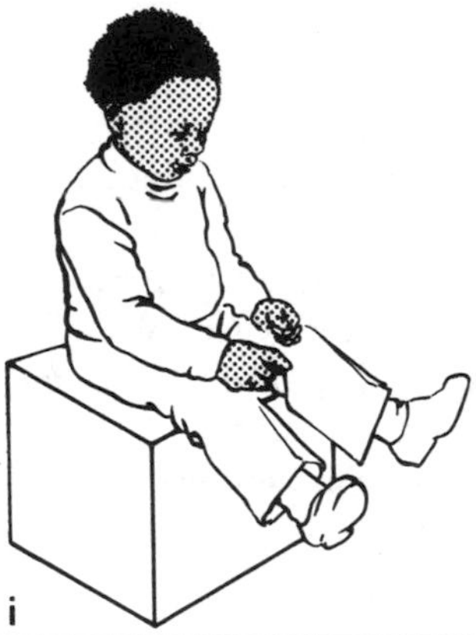

i    j

k 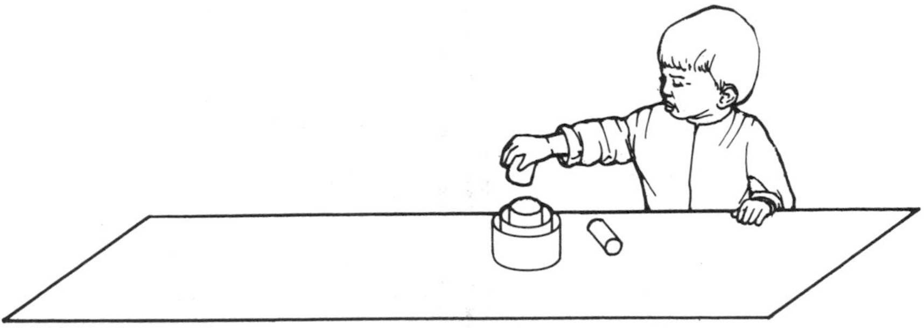   l

©DIAGRAM

TWO YEARS

TWO YEARS: PROFILE

No more an infant, the two-year-old gains control over bowels and bladder, cuts his last primary teeth, and rapidly builds up his vocabulary.

Surer now upon his feet, he loves to romp, chase, and be pursued. He seldom falls but walks with hunched shoulders and slightly bent knees and elbows, and, as he runs, leans forward. Rising and bending down are still done somewhat awkwardly.

Hand movements are now more varied and assured. A twist of the wrist and he can turn a door knob. He loves exploring objects: taking things apart; fitting things together; pushing in and pulling out; filling and emptying. He tests everything to hand by taste and touch. Daily chores intrigue him and he enjoys copying them.

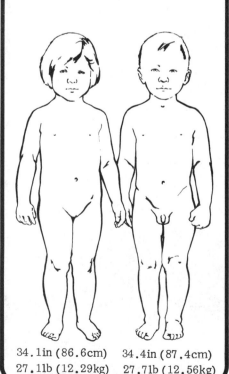

34.1in (86.6cm)   34.4in (87.4cm)
27.1lb (12.29kg)   27.7lb (12.56kg)

TWO YEARS: DEVELOPMENT

MOTOR DEVELOPMENT

Squats steadily and stands without using hands as props (a).
Runs easily (b), controlling stops and starts, and dodging obstacles.
Climbs on furniture to reach windows and door handles, and gets down unaided.
Increasingly aware of own size compared with objects around him.
Can throw a small ball forward while standing (c).
Walks up against a big ball in a bid to kick it (d).
Holding handrail or wall, walks up and down stairs, placing both feet on each step.
Steers tricycle, sitting astride, pushing along with his feet (e).

EYE AND HAND DEVELOPMENT

Identifies photographs of adults well known to him (but not of self) after one showing.
Identifies miniature toys of familiar objects.
Turns picture-book pages one at a time, and recognizes tiny details in the pictures.
Now definitely right-handed or left-handed.
Swiftly picks up tiny objects like crumbs and puts down carefully.
Can unwrap a small candy (f).
Uses six cubes to build a tower (g).
Grips pencil near point, between first two fingers and thumb, makes circular and to-and-fro scribbles and dots, and can copy a vertical line (h).

HEARING AND VOICE

Shows interest in conversation between other people (i).
Talks to self continually and largely unintelligibly as he plays.
Uses own name to refer to himself.
Says at least 50 words and knows many more.
Can say short sentences.
Asks the names of various people and things.
Names (after adult) and shows hand, eyes, mouth, shoes, etc.
Participates in songs and nursery rhymes.
Names and hands over familiar pictures and objects when asked.
Obeys simple commands, eg telling father a meal is ready.
Begins correctly assembling a miniature doll's house setting when asked to do so (j).

PLAY AND SOCIAL DEVELOPMENT

Competently spoon feeds and drinks from a cup.
Requests food, drink, and toilet.
May be dry by day.
Can put on shoes and hat (k).
Opens doors and runs outside, oblivious of danger.
Imitates mother as she performs household chores.
Demands her attention and often clings to her (l).
Resists authority and throws tantrums if frustrated.
Fiercely possessive over toys (m).
Does not play with other children and resents attention shown them by his parents.

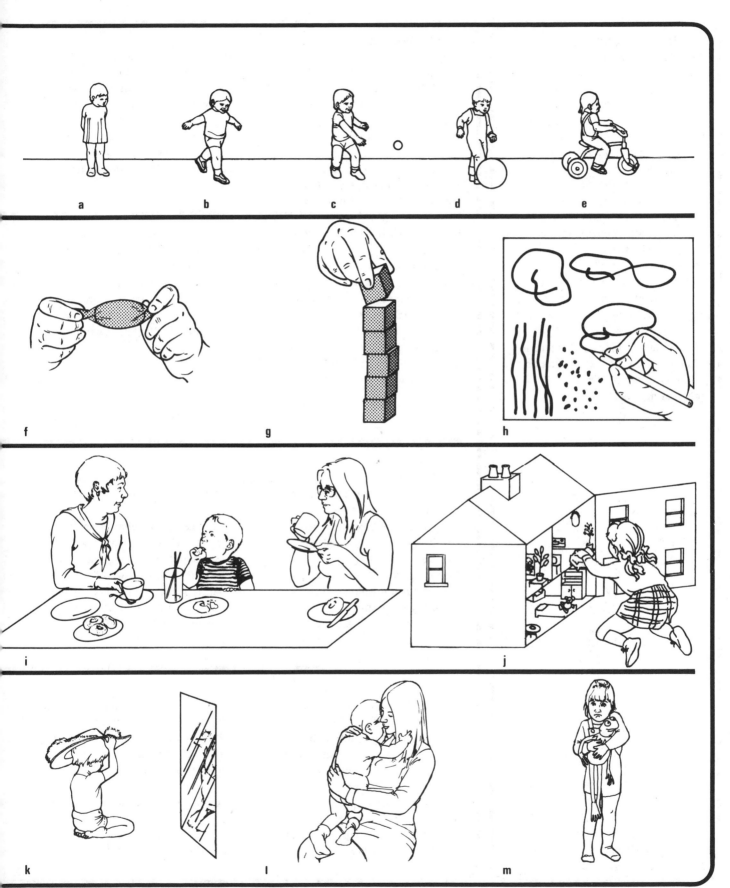

a   b   c   d   e

f   g   h

i   j

k   l   m

© DIAGRAM

THREE YEARS

THREE YEARS: PROFILE

The three-year-old has made great strides toward physical and psychological maturity and we glimpse in him the adult of the future. His motor mechanisms now mesh effectively; he walks erect, swings his arms in adult fashion, maneuvers around corners, and manages stairs with ease. He is also toilet trained. Hand-eye coordination are good enough for him to draw a recognizable copy of some simple shapes.

Increasing interest in words and numbers marches hand in hand with a growth of simple, verbal logic - comparing one object with another, for example.

Willingness to please, to share in play and wait his turn are major pointers to the psychological progress he has made.

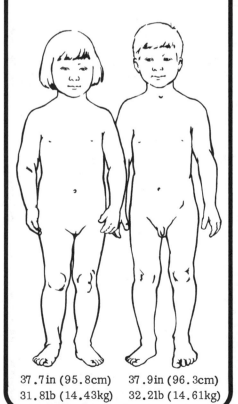

37.7in (95.8cm) 37.9in (96.3cm)
31.8lb (14.43kg) 32.2lb (14.61kg)

THREE YEARS: DEVELOPMENT

MOTOR DEVELOPMENT

Sits on chair, ankles crossed.
Stands briefly on one leg (a).
Can stand and walk on tiptoe.
Runs and hauls and shoves big toys around obstacles.
Walks forward, sideways or backward, pulling big toys.
Kicks ball hard (b), and catches a big ball between outstretched arms.
Places one foot on each step when walking upstairs, but both feet on each step when walking down.
May jump off bottom step.
Nimbly mounts children's furniture.
Pedals and steers a tricycle (c).
Now well aware of own size and movements in relation to objects around him.

EYE AND HAND DEVELOPMENT

Readily picks up crumbs, with one eye covered.
Cuts paper with scissors (d).
Builds a nine-brick tower, and a three-brick bridge (e).
Controls pencil well between thumb and first two fingers; copies circle and letters HTV; draws a man, showing head and maybe two other parts of body (f).
May name colors and match three primary colors.
Paints color wash all over paper; makes and names "pictures."
Matches up to seven letters with test letters 10ft (3m) away.

HEARING AND VOICE

Loves hearing favorite stories (g).
Cooperates in hearing test by carrying out required action.
Speaks with modulated pitch and volume.
Says many intelligible words, with childish mispronunciations and mangled grammar.
Accurately uses plurals and personal pronouns. Gives own name, sex, and maybe age.
Talks to self about actions during play (h).
Can briefly say what he is doing now and describe past experiences.
Asks countless questions starting "what," "where," "who."
Can repeat some nursery rhymes.
May count to 10 but is unlikely to understand quantities greater than three.

PLAY AND SOCIAL DEVELOPMENT

Uses spoon and fork at table (i).
Washes and dries hands with help.
Pulls panties up and down and can put on or take off some clothes with simple fastenings.
May now be dry day and night.
Affectionate and less rebellious.
Helps to shop, garden, etc, and tries to tidy own toys.
Imaginary people and things figure in make-believe games.
Plays with other children, sharing toys and candies (j).
Now affectionate toward sisters and brothers.
Begins to grasp differences between past, present, future.
No longer insists on instant gratification of desires.

a

b

c

d

e

f

g

h

i

j

© DIAGRAM

FOUR YEARS

FOUR YEARS: PROFILE

Exuberant of mind and body, the four-year-old breaks through constraints that held the three-year-old in check.

Well controlled motor muscles help him energetically climb, jump, hop, skip, and ride a tricycle. But he is also becoming good at tasks demanding careful hand-eye control, for instance sawing, lacing shoes, and scissoring along a line.

Boastful, bossy, and a smart aleck, he talks incessantly, trying out new words - inventing some, and often using adult terms incongruously out of proper context. Word play reflects the darting movements of his thoughts, which spawn inventive games and drawings. His mental life is blossoming.

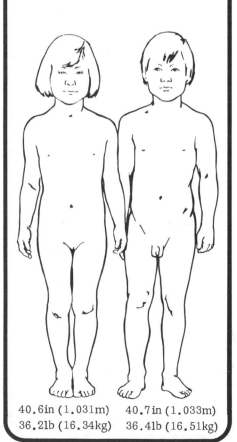

40.6in (1.031m)   40.7in (1.033m)
36.2lb (16.34kg)   36.4lb (16.51kg)

FOUR YEARS: DEVELOPMENT

MOTOR DEVELOPMENT
Sits on chair with knees crossed.
Can stand, walk, run on tiptoe.
Keeping legs straight, can bend at waist in order to pick up objects from the floor (a).
Can climb trees and ladders.
Turns sharp corners, running.
Walks or runs both upstairs and downstairs, placing only one foot on each step.
Can kick, catch, throw, and bounce a ball, and strike it with a bat (b).
Hops on favored foot and balances upon it for 3-5 seconds (c).
Can make sharp turns on a tricycle.

EYE AND HAND DEVELOPMENT
Names and matches colors.
Matches seven test-card letters with specimens 10ft (3m) away.
Readily picks up and replaces crumbs or other small objects with one eye covered.
Can thread beads (d), but still cannot thread a needle.
Builds a tower 10 cubes high (e).
When shown how, arranges six cubes to build three steps.
Shown how, presses thumb against each finger in turn.
Grasps pencil maturely; copies cross and letters HOTV; draws a man, showing head and legs plus probably arms and trunk (f).

HEARING AND VOICE
Can talk intelligibly, using correct grammar and few childish mispronunciations.
Can give own name in full as well as address and age.
Perpetually asks questions starting "why," "when," "how."
Asks what different words mean.
Loves hearing and telling long tales, often mixing up fantasy and fact.
Counts to 20 by rote, and counts actual objects up to five.
Likes jokes (g).
Accurately says or sings a few nursery rhymes.

PLAY AND SOCIAL DEVELOPMENT
Handles fork and spoon well.
Cleans teeth (h). Washes hands and dries them capably.
Can dress and undress except for managing difficult fastenings.
Sense of humor developing.
Independent - answers back.
Plays theatrical games involving dressing up.
Plays complex floor games (i).
Less tidy than at age three.
Builds things outdoors.
Argues with other children, but needs their companionship and learns to take turns.
Concerned for younger brothers and sisters, and sympathetic toward distressed playmates.
Understands differences between past, present, and future.

a

b

c

d

e

f

g

h

i

©DIAGRAM

FIVE YEARS

FIVE YEARS: PROFILE

Scatterbrained four develops into maturer five, whose mind works more precisely - less ruled by the mouth, as it were. He is more self-critical, and inclined to finish some project he has started. Well balanced and unflappable, the five-year-old has gained a clearer concept of himself and his role in the family and in a somewhat broader environment. He can now cope with most daily personal duties and some household tasks, and is ready for the wider world of school. But coordination of hand, eye, and brain are still developing and he is unlikely to be ready yet to learn to read and write with ease. He also remains unaware of many simple facts we take for granted.

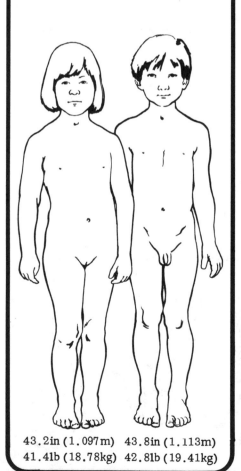

43.2in (1.097m)   43.8in (1.113m)
41.4lb (18.78kg)   42.8lb (19.41kg)

FIVE YEARS: DEVELOPMENT

MOTOR DEVELOPMENT

Stands on one foot 8-10 seconds.
Touches toes with legs straight.
Hops 2-3yd (2-3m) forward on either foot; skips on alternate feet.
Runs on toes (a), and moves to music.
Can walk along a thin line.
Climbs, digs, slides, and swings.
Plays ball games quite well.

a

EYE AND HAND DEVELOPMENT

Matches 10-12 colors.
Threads needle. Sews stitches.
Builds 3-4 steps from cubes.
Copies square (later triangle) and letters ACHLOTUVXY; writes some letters unprompted; draws man with trunk, head, arms, legs, and features (b); also house with roof, windows, door, chimney.
Colors pictures carefully.
Counts fingers of one hand.

b

HEARING AND VOICE

Speaks fluently and correctly but may confuse sounds f, s, and th.
Sings or says jingles and rhymes, and like riddles and jokes.
Enjoys hearing stories and later enacts them with friends (c).
Can quote full name, age, birthday, and address.
Explains meaning of concrete nouns by usage and often asks what various abstract words mean.

c

PLAY AND SOCIAL DEVELOPMENT

Uses knife and fork well (d).
Dresses and undresses self.
Increasingly independent and sensible, but rather untidy.
Chooses own friends and invents involved make-believe games in which shows sense of fair play.
Becoming aware of clock time.
Comforts distressed playmates and protects pets and young children.

d

SIX YEARS

SIX YEARS: PROFILE

Lively, expansive, eager to try out something new, the six-year-old has an insatiable appetite for fresh experiences. But he must succeed in everything he tackles; win every game he plays; have the most of anything that's going. Otherwise there will be tears and tantrums. Six is a demanding, stubborn, and unruly age.

By six years old the first adult tooth is pushing through. At school the child takes early steps along the paths that lead to an ability to read, write, and use numbers. But what he learns must still be firmly based on things that he can see and do. He cannot reason in an abstract adult way.

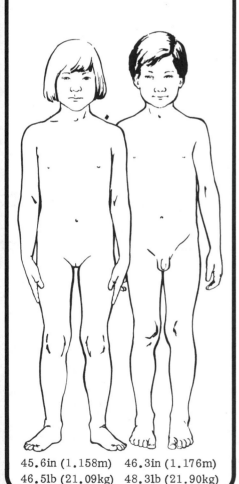

45.6in (1.158m) 46.3in (1.176m)
46.5lb (21.09kg) 48.3lb (21.90kg)

SIX YEARS: DEVELOPMENT

MOTOR DEVELOPMENT

Jumps over rope 10in:25cm high (a).
Hops 1-3 times on one foot so that it nudges a toy brick along.
Trots around a sizable open space such as a playground.
Skips at least 12 times indoors (more outdoors).
Holding a rope, skips three or more times.

a

EYE AND HAND DEVELOPMENT

Copies a square accurately.
Neatly copies a ladder (b).
Attempts to produce a copy of a diamond.
Draws a triangle more precisely than when aged five.
Can say in two cases how one object differs from another that in some way resembles it.

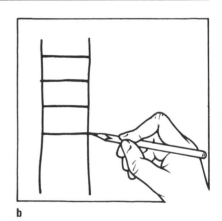

b

HEARING AND VOICE

Accurately repeats a sentence containing 16 syllables.
Uses at least three sentences to describe a test picture.
Names and recognizes at least 20 capital letters (c).
In a verbal test can say how one object resembles another.

c

PLAY AND SOCIAL DEVELOPMENT

Cuts own meat when using knife and fork at table (d).
Can now be expected to behave reasonably well at table.
Ties own shoelaces.

d

©DIAGRAM

SEVEN YEARS

SEVEN YEARS: PROFILE

Physical and mental progress march ahead. But broadening the mind's horizons predominates over physical activity. Many a seven-year-old prefers to watch instead of do. He reads and watches television, and often likes to do so on his own partly because he feels at odds with everyone around him. Moodiness is frequent.

Spells of intensive learning alternate with spasms of forgetfulness. He loves drawing and other tasks requiring precise hand-eye coordination. But he tends to overreach himself, and attempts tasks that he becomes too tired to finish.

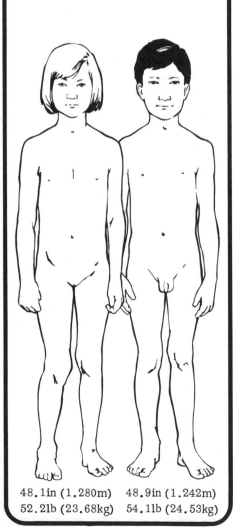

48.1in (1.280m)   48.9in (1.242m)
52.2lb (23.68kg)   54.1lb (24.53kg)

SEVEN YEARS: DEVELOPMENT

MOTOR DEVELOPMENT

Can jump off the bottom four steps of a staircase.
Rides a two-wheeled bicycle, but not very far.
Hops four times on one foot, so that it nudges a toy brick along (**a**).
Skips 12 times or more.

a

EYE AND HAND MOVEMENT

Writes 24-26 letters (**b**).
Fairly accurately draws a "window" (a cross inscribed in a square).
Draws a man with originality, eg clothed, side-face, seated.
Draws a diamond neatly.

b

HEARING AND VOICE

Can use at least four sentences to describe a test picture.
Correctly answers six questions from a comprehension test list.
Says in three cases how one object resembles a similar object.
Says in three cases how one object differs from another that in some way resembles it (**c**).
Gives the opposite meanings for four terms in a test list.

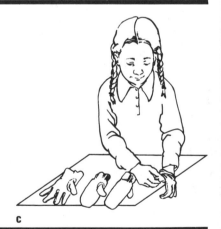

c

PLAY AND SOCIAL DEVELOPMENT

Has a special friend at school.
Sets a table, given some help.
Brushes own hair regularly (**d**).
Dresses and undresses unaided, managing fastenings well.

d

EIGHT YEARS

EIGHT YEARS: PROFILE

The eight-year-old actively explores his environment, believes no task too hard to tackle, forms new friendships, and is concerned over people's opinions.

His vocabulary is enriched by many adjectives because he now appreciates the qualities of objects and actions. But judgments involving generalizations and abstractions remain beyond him. Between now and his teens, the child stands somewhat on a plateau. He has a new, maturer independence of adults, and even at eight foreshadows the kind of person he will become. But full brain development, adult stature, and sexual maturity lie well ahead.

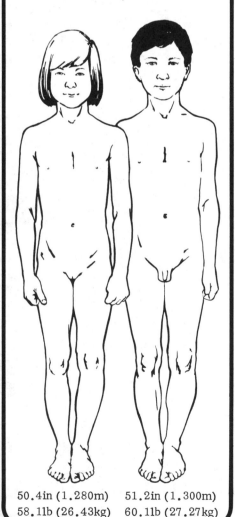

50.4in (1.280m)   51.2in (1.300m)
58.1lb (26.43kg)   60.1lb (27.27kg)

EIGHT YEARS: DEVELOPMENT

MOTOR DEVELOPMENT

Skips freely out of doors at least 20 times.
Rides a bicycle unaided with competence in any open space considered safe (**a**).

a

EYE AND HAND DEVELOPMENT

Draws a house embellished with a fair amount of detail (**b**).

b

HEARING AND VOICE

In four cases can say how one object differs from another that in some way resembles it.
Understands and tells the time in hours, half hours, and quarter hours (**c**).

c

PLAY AND SOCIAL DEVELOPMENT

Can set a table for everyday use with no help at all (**d**).

d

©DIAGRAM

11 Aspects in development

The development of
acquired abilities and
characteristics — from
standing up to solving
complex problems —
depends on the interaction
of biological maturation
and learning.

Left: Following the leader through
a concrete pipe (Photo: Tony
Roxall - Barnaby's Picture Library)

GROWTH 1

GROWTH TO THREE YEARS

A child grows rapidly in the first three years, but successive measurements within this period show that the rate of growth is slowing down. (Fastest growth in fact occurs in the three months before birth.)

WEIGHT In the first few days a new baby loses weight. Subsequent weight gain reduces individual differences: in the first six months a 5lb (2.3kg) baby trebles his birthweight while a 9lb (4.1kg) baby only doubles his. In the second six months weight gain is halved. Weight gain from one to two years equals that from six to 12 months. There is a further slowing down from two to three.

LENGTH Growth is fastest in the first year, but slowing down. From two to three years it is the same as in the first three months.

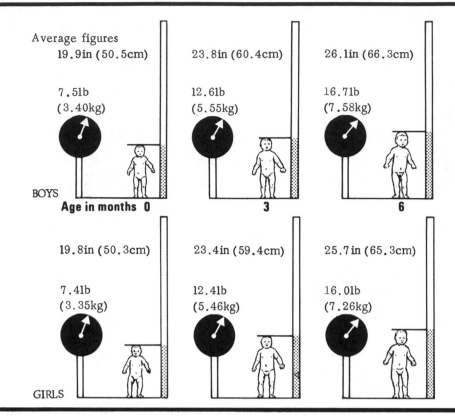

Average figures

19.9in (50.5cm) 23.8in (60.4cm) 26.1in (66.3cm)

7.5lb (3.40kg) 12.6lb (5.55kg) 16.7lb (7.58kg)

BOYS

Age in months 0 3 6

19.8in (50.3cm) 23.4in (59.4cm) 25.7in (65.3cm)

7.4lb (3.35kg) 12.4lb (5.46kg) 16.0lb (7.26kg)

GIRLS

BIRTHWEIGHT AND LENGTH

Birthweight and length vary greatly among healthy newborn infants. The graphs here show the percentage of full-term babies born with different crown-to-rump lengths and different weights. Most babies are born close to the middle length and weight figures. But some are a lot shorter than and weigh only half as much as others. Major factors that determine weight and length at birth are:

a) term of development inside the mother's uterus;

b) mother's health in pregnancy;

c) baby's nutrition before birth;

d) heredity;

e) race;

f) siblings (twins tend to be smaller than singletons, and first-born babies smaller than ones born subsequently).

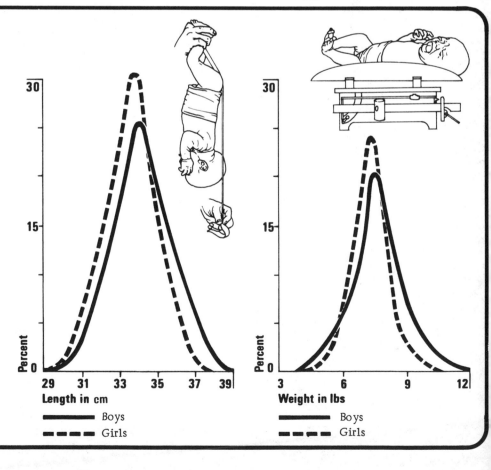

Length in cm — Boys ——— Girls ------

Weight in lbs — Boys ——— Girls ------

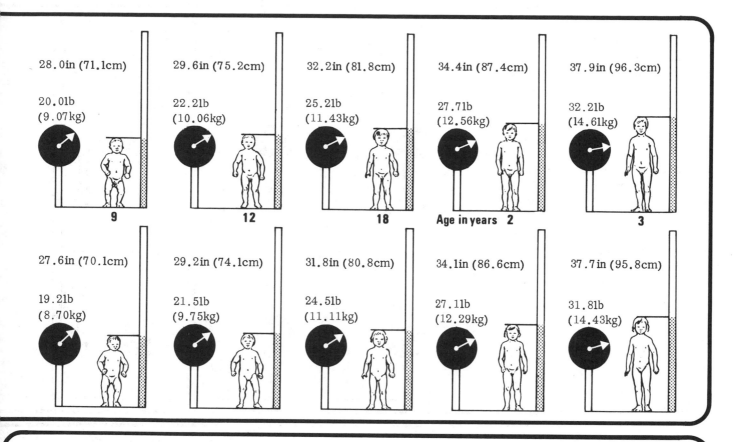

28.0in (71.1cm)

20.0lb
(9.07kg)

9

29.6in (75.2cm)

22.2lb
(10.06kg)

12

32.2in (81.8cm)

25.2lb
(11.43kg)

18

34.4in (87.4cm)

27.7lb
(12.56kg)

Age in years 2

37.9in (96.3cm)

32.2lb
(14.61kg)

3

27.6in (70.1cm)

19.2lb
(8.70kg)

29.2in (74.1cm)

21.5lb
(9.75kg)

31.8in (80.8cm)

24.5lb
(11.11kg)

34.1in (86.6cm)

27.1lb
(12.29kg)

37.7in (95.8cm)

31.8lb
(14.43kg)

BODY PROPORTIONS

From birth to adolescence body proportions are altered by changes in the growth rates of head, trunk, and limbs. While the head is one-quarter of the total length at birth it is only one-sixth by age six and one-eighth by adulthood. The legs start at only three-eighths of the total length but increase to one-half of total length by maturity. The trunk becomes relatively slimmer.

Changes in growth patterns for bone, muscle, and fat transform the chubby, short-limbed, relatively large-headed baby into the thinner, wiry schoolchild. Later, children broaden out - boys especially at the shoulders, girls at the hips. But eventual physique varies with individuals and may be detected early on, in some cases by the age of two.

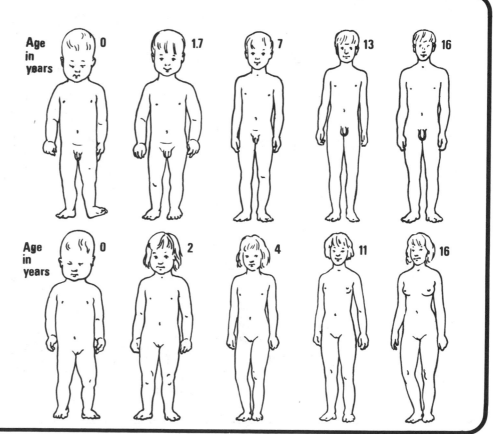

Age in years 0   1.7   7   13   16

Age in years 0   2   4   11   16

© DIAGRAM

GROWTH 2

GROWTH FROM FOUR YEARS

After the rapid growth of the first years, height increases more slowly in middle childhood. The onset of puberty is linked to a sharp spurt in growth that peaks just before the beginning of adolescence. Girls mature earlier than boys, and usually stop growing by about age 16. In males the growth spurt is later, and growth often continues after 18. Average patterns of weight increase broadly follow those for height gain, with biggest weight increases occurring from 10 to 14 in girls, and 12 to 16 in boys.

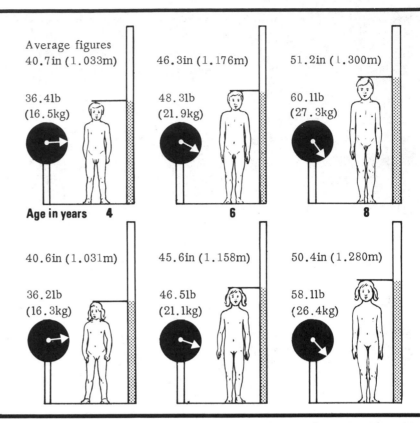

Average figures
40.7in (1.033m)   46.3in (1.176m)   51.2in (1.300m)

36.4lb (16.5kg)   48.3lb (21.9kg)   60.1lb (27.3kg)

Age in years 4   6   8

40.6in (1.031m)   45.6in (1.158m)   50.4in (1.280m)

36.2lb (16.3kg)   46.5lb (21.1kg)   58.1lb (26.4kg)

GROWTH: MALE AND FEMALE

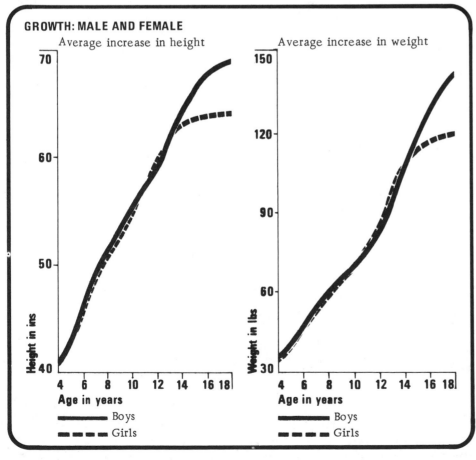

Average increase in height

Average increase in weight

Height in ins

Age in years

Weight in lbs

Age in years

—— Boys
- - - Girls

—— Boys
- - - Girls

GROWTH: PERCENTAGES

Boys: percentage of adult height and weight reached at different ages

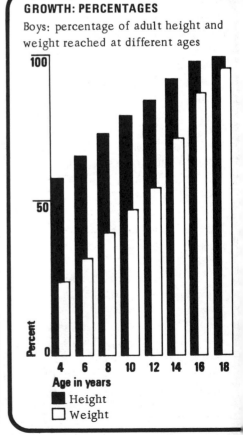

Percent

Age in years

■ Height
□ Weight

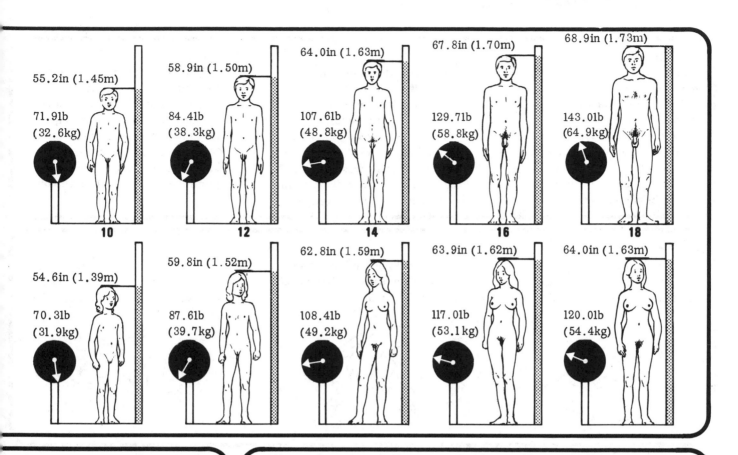

55.2in (1.45m)

71.9lb (32.6kg)

10

58.9in (1.50m)

84.4lb (38.3kg)

12

64.0in (1.63m)

107.6lb (48.8kg)

14

67.8in (1.70m)

129.7lb (58.8kg)

16

68.9in (1.73m)

143.0lb (64.9kg)

18

54.6in (1.39m)

70.3lb (31.9kg)

59.8in (1.52m)

87.6lb (39.7kg)

62.8in (1.59m)

108.4lb (49.2kg)

63.9in (1.62m)

117.0lb (53.1kg)

64.0in (1.63m)

120.0lb (54.4kg)

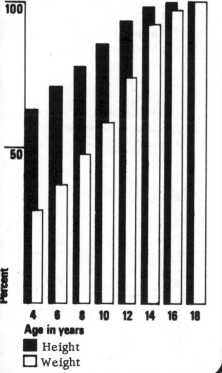

Girls: percentage of adult height and weight reached at different ages

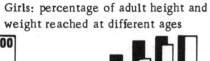

Percent

Age in years

■ Height
□ Weight

GROWTH OF BODY PARTS

Different parts of the body have different types of growth. This graph shows the percentage increase for each type from birth to adulthood.

a General type (affecting most of the body including muscles, lungs, blood volume, intestines) spurts in infancy, slows in childhood, spurts toward adolescence, then decelerates.

b Neural type (affecting brain, eyes, skull, spinal cord) spurts dramatically from birth to six years. By then 90% of neural growth is over.

c Genital type (testes, ovaries, etc) stays minimal, then spurts at puberty and during adolescence.

d Lymphoid type (thymus and lymph nodes) peaks from 10 to 12 years, then lymph and thymus tissue shrinks. Hence the graph shows lymphoid type reaching 200% before diminishing.

Percentage of different types of growth at different ages

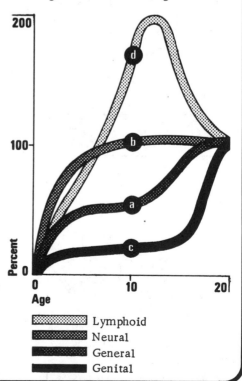

Percent

Age

▨ Lymphoid
▨ Neural
▬ General
▬ Genital

©DIAGRAM

381

MOTOR DEVELOPMENT 1

GENERAL TRENDS

There are three broad trends in motor development.

1) Outward: central body areas function before outer areas. Children control upper legs and upper arms earlier than lower legs, forearms, and hands.

2) Head to foot: head control comes before hauling the body along with the hands. Creeping on hands and knees follows. Walking comes last of all.

3) Big to small muscles: early movements are wild jerks of the whole body, trunk, or entire limbs. Control of small muscles is needed before a child can pick up a crumb with thumb and forefinger, and walk with an economical action. Each trend reflects increasing maturation of the nervous system. Walking practice is useless until the nervous system is ready. Moreover voluntary movements can come only after the newborn's primitive reflexes (B21) have gone.

HELD HORIZONTALLY

A newborn baby's head droops – he lacks the coordination and strength required to raise it.

At six weeks he can raise his head in line with his body, but can do so only for a moment.

By 12 weeks he can lift his head well above the general body line and hold it there for a time.

CRAWLING

Placed in a prone position, the newborn baby draws his knees up beneath his abdomen and lies with his pelvis high and his head directed to one side.

By six weeks old he tends to lie less tucked up. His pelvis is lower. Legs are more extended. Sometimes he kicks out. Now and then he lifts his chin well up.

By 12 weeks the baby lies with legs fully extended and pelvis flat upon the bed or couch. He raises chin and shoulders and may hold his head almost upright.

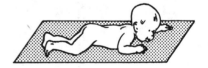

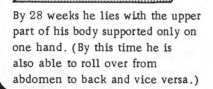

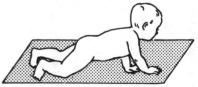

By 28 weeks he lies with the upper part of his body supported only on one hand. (By this time he is also able to roll over from abdomen to back and vice versa.)

The 36-week-old child learns that by thrusting down and forward on a surface he can move his body backward. He is well on the way to true crawling.

At 40 weeks he crawls forward on his belly, pulling himself along with his hands. At this stage his legs do no more than trail along unhelpfully behind.

SECONDARY RESPONSES

So-called secondary responses begin appearing at about four months when primitive reflexes decline. Some secondary responses may be absent in normal babies before one year. Vital in later life, secondary responses concern balancing, rolling, and protection of the falling body.

ROLLING RESPONSE From the fourth month, rolling the body over by the legs evokes a complex response involving the baby's head and unimpeded arm.

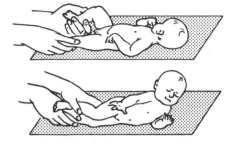

DOWNWARD PARACHUTE If someone holds a baby upright then swiftly lowers him, the legs stretch and move apart as though anticipating a hard landing.

FORWARD PARACHUTE If someone holds a baby upright and swiftly tilts him forward toward the ground, arms and fingers outstretch and spread protectively.

SIDEWAYS PROPPING REACTION This time a child sitting upright is tilted sideways. He extends an arm and hand, pressing on the ground to stop his body falling.

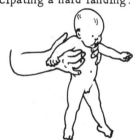

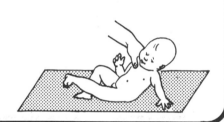

At 16 weeks he can press down with his forearms to lift his head and front part of the chest. He also stretches all his limbs, and "swims" on his abdomen.

By 20 weeks the infant has had some weeks of practice and uses forearms in an even surer manner than before to help him lift up head and upper chest.

By 24 weeks he can lift his entire head, chest, and upper abdomen. Thrusting down with outstretched arms, he bears the weight of his upper body on his hands.

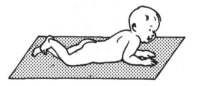

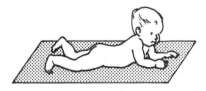

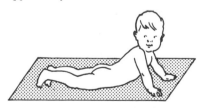

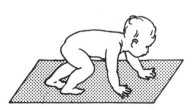

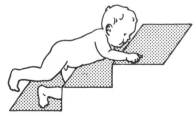

By 44 weeks he has learned a more agile mode of crawling. Arms and legs all work together as he now creeps along on hands and knees, with belly held up off the ground.

At one year old he rests some weight upon his soles and ambles bear-like on his hands and feet. He has reached the final stage before unaided walking.

At 15 months he sometimes still feels safest on all fours, for instance when he starts tackling that formidable stepped cliff known to adults as a staircase.

©DIAGRAM

MOTOR DEVELOPMENT 2

SITTING

The newborn baby cannot sit, and his head lags flaccidly if he is pulled into a sitting position.

At four weeks his head still lags but lifts briefly if he is held to sit. His back is rounded.

By 12 weeks a baby supported to sit holds his head up, but the head may still bob forward.

At 28 weeks he can sit alone on the floor, but uses arms and hands as props for his unsteady body.

By 32 weeks he is already able to sit on the floor, briefly unsupported by his arms.

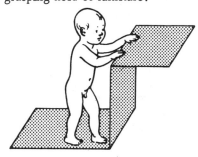

At 36 weeks he can sit for 10 minutes, regaining balance if he bends his body forward.

STANDING AND WALKING

A newborn baby held with the sole of the foot on a table moves his legs in a reflex walking action.

At eight weeks the baby briefly keeps his head up if he is held in a standing posture.

By 36 weeks he can pull himself up and remains standing by grasping hold of furniture.

At 18 months he can go up and down stairs without assistance.

By two he runs, walks backward, and picks things up without overbalancing.

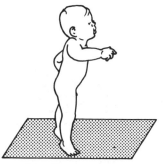

At 2½ he can balance on tiptoe, and jump with both feet.

At 16 weeks only the lower back of the supported baby is curved. The head wobbles if the body sways.

By 20 weeks there is no head lag or wobble and the baby pulled to sit keeps his back straight.

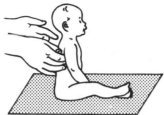

At 24 weeks he lifts his head to be pulled up, and sits supported in a high chair or baby carriage.

By 40 weeks the sitting child can lie down on his stomach and return to a sitting position.

By 48 weeks the sitting child can twist around to pick up something, yet keep his balance.

About 15 months he sits down in a low chair, climbing forward into it and then turning around.

By 48 weeks he can walk forward if both hands are held (or sideways, gripping furniture).

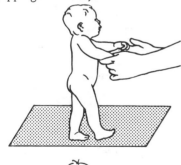

At one year the child walks forward if someone holds one of his hands.

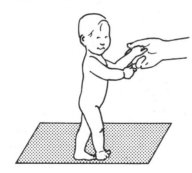

By 13 months the child has become capable of walking without help.

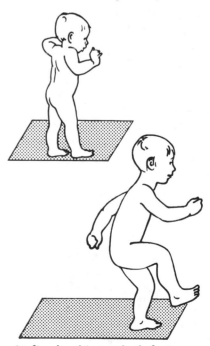

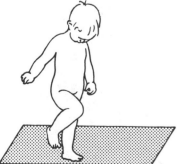

At three he can balance for some seconds while standing on one foot.

At four he walks downstairs by placing only one foot on each step.

At five he skips on both feet.

©DIAGRAM

EYE AND HAND DEVELOPMENT

FIELD OF VISION

At birth a baby's eyes move independently. At this early stage he finds it hard to fix on any object with his eyes.

Soon after birth (**a**) he can watch a dangling ball swung through a 45° arc - one-quarter of an adult's field of vision.

By six weeks (**b**) he moves his eyes to watch a ball through 90°. (This coincides with the development of binocular vision: visual fields now overlap so that most of the image formed on one retina duplicates most of the image on the other.)

By three months (**c**) he can watch a ball swung through 180° - the full adult field of vision.

This progress partly reflects the baby's increasing mastery of the six muscles attaching each eyeball to its socket.

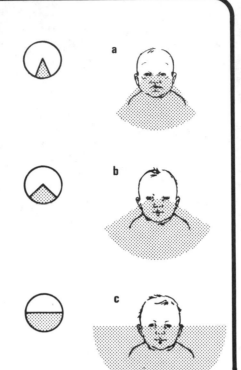

VISUAL PERCEPTION

Adults check their visual impressions against a memory bank stored in the brain. This helps instant recognition of the shapes, sizes, colors, positions, and other qualities of the objects that are seen.

To the newborn infant with no experience to draw upon, the visible world may seem no more than a muddle of blotches.

Perception dawns gradually. Tests devised by psychologists give an indication of the times at which visual perception of shape, size, and color appear.

By six months old, tested infants could perceive simple forms. When one of three test blocks - a circle, triangle, and square - was sweetened, the infants learned to recognize and suck it.

At six months children reached for a small nearby rattle rather than a larger one farther away, although

they appeared to be similar. (But a child is 10 years old before he can discount distance in a fully adult way to judge the size of an object.)

Color vision dawns at about three months, when the cones in the eye's retina become sufficiently developed. Tested children of that age gazed longer at colored paper than at gray paper of equal brightness.

DEPTH OF VISION

Depth of vision improves during the first year. A month-old baby gazes at a small white ball moved 6-10in (15-25cm) from his face, but not at one across the room. Tests with rolling balls of different diameters help to show the rate at which depth of vision improves. By six months old (**a**) a baby watches a ball ¼in (6mm) in diameter 10ft (3m) away. This suggests a visual acuity as measured by a standard eye-test card of 20/120: i.e. the baby can clearly focus an object 20ft away that someone with normal 20/20 vision would clearly see from 120ft away. By nine months (**b**) a baby watches a ball 1/8in (3mm) in diameter 10ft (3m) away, implying 20/60 vision. Such tests suggest 20/20 or normal visual acuity by one year old (**c**).

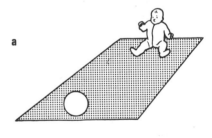

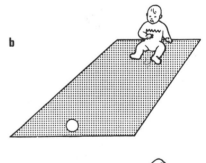

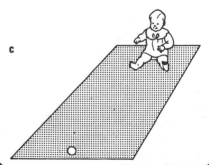

MANIPULATION

At birth the hands are closed; the reflex grasp action (B21) is present. The hands are still closed at one month, but are often opened by two months. Grasp reflex has gone by three months and a baby can hold a rattle. At four months the hands meet in play. By six months he can deliberately, but awkwardly, grasp a cube (a). By this time he passes it from hand to hand and bangs it on a table. By eight months he begins to grip the cube with his thumb opposite his fingers (b). By nine months thumb and finger grip is good enough to pick up a candy. By one year he holds a cube in a mature pincer grip of thumb and fingers (c). By two years he can turn a doorknob and unscrew lids. He is now clearly left- or right-handed. By $2\frac{1}{2}$ he holds a pencil in his hand, no longer in his fist, and starts to scribble and draw.

a

b

c

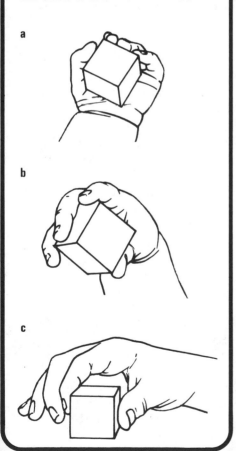

DRAWING

As hand-eye coordination and wrist control improves, children become capable of drawing an increasing number of shapes, always in the same sequence though sometimes earlier or later than the dates given below.
At two a child usually learns to copy a vertical line. At $2\frac{1}{2}$ he can copy a horizontal line. At three he can copy a circle. By then he is also drawing a man, but shows no more than the head and maybe two other parts of the body.
Only at age four does he manage to copy a cross by combining the vertical and horizontal strokes that he had mastered earlier. Another year may pass before he learns to rearrange these lines into a square. He can now draw a rather basic box-like type of house but quite likely cannot achieve the oblique strokes needed for showing a sloping roof. By five his man is more obviously human and may possess recognizable head, trunk, arms, and legs.
Between five and six the child becomes proficient at making oblique strokes and can now draw triangles. By seven he copies a diamond well, and fairly accurately inscribes a cross inside a square to make a "window."
He has now built up a repertory of vertical, horizontal, oblique, and curved strokes. Thus armed, he draws increasingly complex houses and people, often with some touches of originality. For instance his people wear distinctive clothes and may appear in profile. His house may have a smoking chimney and a garden containing flowers and bordered by a fence.
The illustrations included here show how a child's drawing of a man becomes increasingly developed between the ages of three and seven.

Development of drawing ability

Age two

Age three

Age three

Age five

Age seven

©DIAGRAM

HEARING AND SPEECH DEVELOPMENT

DEVELOPMENT OF HEARING

The hearing mechanism is fully formed at birth, but amniotic fluid in the eustachian tubes causes a few hours' deafness. Then a newborn reacts to harsh, sharp sounds. By 10 days he responds to a loudly ticking watch or voice, and soon responds to sounds of different pitch and loudness.

Sound localization improves over the first year. At three months a child turns his head vaguely toward a sound and his eyes seek it (**a**). At five months he turns his head, then inclines it, to locate a sound below ear level (**b**). At seven to eight months turning and inclination begin to merge (**c**). By nine to ten months the child swivels his head diagonally in a direct searching movement (**d**). By one year he localizes sound as well as any adult.

Localization of sound

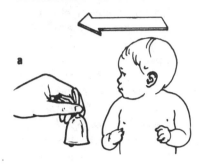
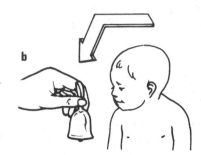
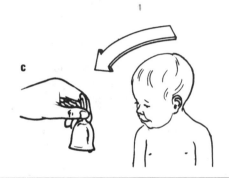
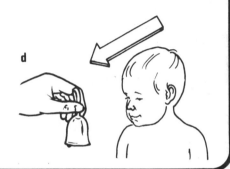

SOUNDS OF SPEECH

Speaking English involves mastering over 40 speech sounds, or phonemes. Children gain this skill in two ways. First, they learn to babble, and so find out how to make a wide variety of sounds. Second, adult responses to their babbling teach them to select for use the sounds they hear most often. Not all children learn phonemes in the same order. But phonemes do tend to appear in a pattern. Children usually say easily pronounced phonemes such as "a" before harder ones like "s." But the frequency with which people around them utter certain sounds also affects the learning sequence. A typical order for the emergence of sounds is described here. By eight weeks the child utters "a" and some other vowel sounds. By 16 weeks he mouths "m" as well as "b," "g," "k," "p." By 32 weeks he masters "t," "d," "w." But only later will he get his tongue around "s," "f," "h," "r," "th." Some sounds give trouble for considerable time - it may take three years to learn all 20 vowel sounds. The phonemic system is usually well established when a child is aged five to seven years.

Formation of speech sounds

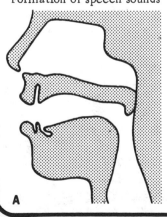
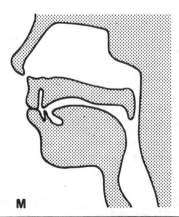
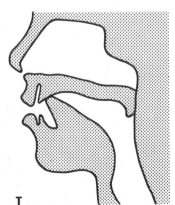
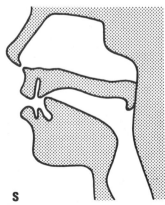

A   M   T   S

LEARNING TO SPEAK

Learning to speak involves making and monitoring sounds.

Speech occurs when motor nerves bearing signals from the brain operate the larynx and its vocal cords, also the pharynx, soft palate, tongue, and lips. Monitoring speech involves feedback to the brain. Sensory nerves bring the brain signals from the speech muscles and from the ears which have picked up sound waves pushed out by the voice. Thanks to this feedback system a child learns to modify the sounds he makes to match words that he has heard others speaking. It becomes markedly more difficult to learn to speak after age three. Thus deafness must be diagnosed and treated as soon as possible.

Speech   Feedback

1 Brain
2 Speech muscles
3 Ear

☐ Motor nerves
▨ Sensory nerves
∿ Sound waves

SPEECH DEVELOPMENT

Rates of vocabulary growth vary with a child's aptitude, parents, presence or absence of brothers and sisters, and general living patterns.

Speech development usually starts with vocalized vowels at about seven weeks. By 16 weeks a child utters some consonants, and produces syllables by 20 weeks. The first meaningful word appears at 44-48 weeks. By one year he says two or three words, and by 21-24 months is using two-word phrases. By three years he talks incessantly.

Included here are two graphs showing speech development in tested groups of normal children. The first shows the percentage of children uttering words, phrases, and intelligible speech by different ages. The second shows the average number of words understood at ages up to six.

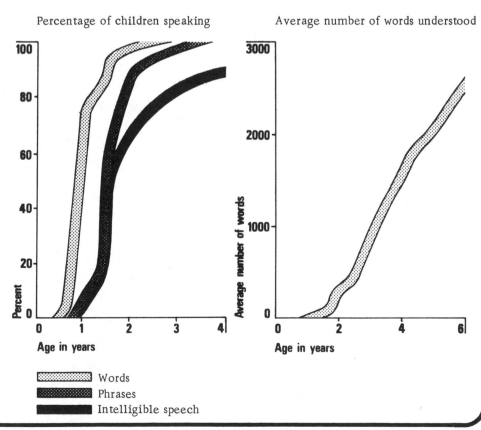

Percentage of children speaking

Average number of words understood

▨ Words
▨ Phrases
■ Intelligible speech

©DIAGRAM

CONCEPTUAL DEVELOPMENT

PIAGET'S THEORIES

Theories evolved by the influential Swiss psychologist Jean Piaget hold that a child's intellectual development moves through regular stages, where cognitive and emotional factors interact.

1) Sensorimotor stage (0-2 years). Intelligence is empirical, and largely non-verbal. Muscles and senses help children deal with external objects and events. They experiment with objects and graft new experiences onto old. They grasp that objects persist out of sight and touch. They start symbolizing - representing by words and gestures.

2) Pre-operational stage (2-7 years). Children use words for perceived objects and inner feelings, and experimentally manipulate them in the mind as they physically handled objects earlier. They act by trial and error, intuition, and experience.

3) Concrete-operational stage (7-12 years). Logical operations begin. Objects classified by similarities and differences.

4) Formal (adult) operations (12 years onward). Children use thought more flexibly, to handle hypothetical issues.

Grasping the idea of conservation of amount and number is a major advance in mental development as assessed by Piaget. In the test shown below a child under seven watching equal amounts of water poured into containers of unequal height may say that a tall, narrow container (a) receives more than a short, broad container (b).

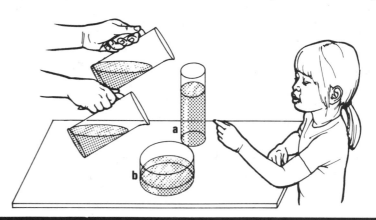

STAGES IN MENTAL DEVELOPMENT

6 mth Recognizes familiar faces. May perceive differences between some shapes.

1 Says first words. Defines some common objects by use.

0-2 Manipulates objects in trial and error fashion. May use observation of cause and effect to solve some simple, practical problems by action.

2 Gives own first name. Knows difference between "one" and "many." Understands simple language.

2½ Knows many more words than he uses. Knows own sex.

2-4 Believes that changing an object's shape also alters its size, weight, and volume.
Imagination developing: uses signs and symbols (words) to stand for absent objects and events but often confuses the sign or symbol with the thing signified or symbolized.

3 Knows own age in years. Now gaining accurate visual judgment of depth. Knows difference between "big" and "little."

4 Speaks fully intelligibly and now uses many words.
Knows "yesterday," "today," and "tomorrow."
Attempts to reason often involve confusing cause and effect.
Understands "higher," "longer," and "heavier."

4-7 Intelligent behavior largely limited to actions and intuitive thought based on incomplete perceptions.

5 Perceives relative sizes of objects well by now.
Counts up to 15 bricks.
May be learning to read.
Can write some letters.

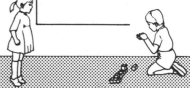

HOW CHILDREN LEARN

At age two a child given a form-board puzzle tries to force a block into the hole opposite irrespective of their shapes (**a**).

At 2½ years, if the block does not fit one hole the child tries to push it into another (**b**).

By age three the child matches shapes by eye before placing a block in a hole (**c**).

Such tests suggest that a child's early steps in learning are based on simple trial and error association. The puzzle is a stimulus evoking a response of effort pleasurably rewarded by solving the puzzle. Thus effort and reward become associated in the child's brain, and next time the puzzle is presented he will complete it

faster.

By age three he solves some problems in the mind. Language facilitates this process.

Some psychologists discount cognition ("the act of knowing") and consider that throughout life learning consists of mental habits formed by stimulus-response relationships.

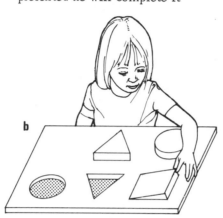
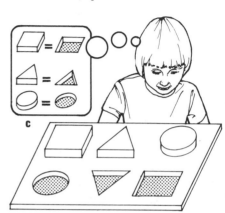

a   b   c

5-8 Intersensory perception developing rapidly.

6 May be learning to write joined-up letters by now.
Can say days of the week.
Counts up to 30 by rote.
Knows "right" and "left."

7 Tells time from clock.
Knows own birthday.
Realizes that changing an object's shape need not mean changing the amount of the substance.

8 Now reasons well about data he can see and touch.
Solves simple mathematical problems.
Enjoys reading children's books with more text than pictures.

8-11 Becomes able to group objects in classes and series. Hence concepts of space, time, number, and realization of an ordered material world first fully emerge.

9 Abstract response to the concept of weight is developing.

10 Mental problem solving improves as he draws conclusions from less concrete situations than before.

11 Begins to reason logically about statements instead of just about concrete objects and events.
Perception of physical volume well developed by now.

12 Historical time sense developing. Well able to formulate hypotheses and theories, make assumptions, and draw conclusions. Ability to reason increases through teens.

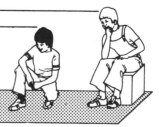

©DIAGRAM

SOCIAL DEVELOPMENT

UP TO EIGHTEEN MONTHS
By three months a child recognizes his mother (**a**). By eight months he is shy with strangers. By one year he is developing a deeper relationship with his father (**b**). Close, continuous physical and emotional contact with one person in this period is vital for developing a stable personality in adulthood.

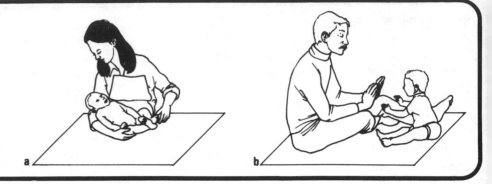

EIGHTEEN MONTHS TO FIVE YEARS
At first children closely identify with their parents, copying household actions (**a**). Parents discourage overdependency and outbursts of aggression. At two children cannot yet join in games with others, but at three do so (**b**) and show affection for younger brothers and sisters. Speech increasingly helps the promotion of new social relationships.

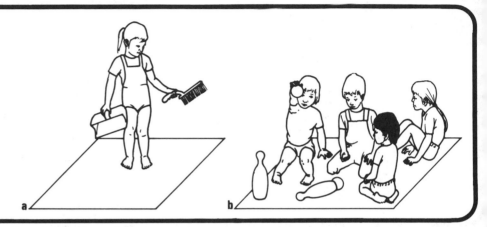

CHANGES IN BEHAVIOR
Sometimes a child gets on well with those around him. At other times he seems at odds with everyone. Psychologists have shown that, through childhood, phases when a child seems in balance with his world alternate with phases when he is unhappy and difficult. A child's personality pendulum swings between equilibrium and disequilibrium so regularly that psychologists have compiled timetables indicating when each swing is likely to occur. Behavior fluctuating from "focal" (eg clinging to the mother) to "peripheral" (eg expansive and exploratory) creates another rhythm. Growth usually strikes a balance, yielding a stable, socially well adjusted adult.

Diagrammatic representation of typical swings in behavior

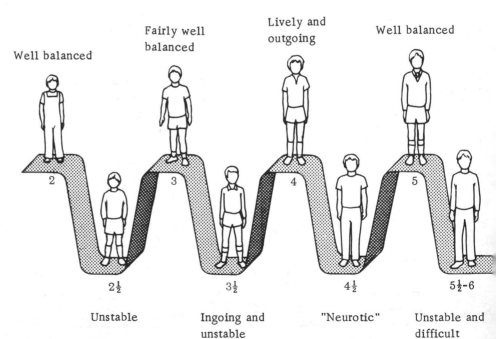

FIVE TO TWELVE YEARS

A five-year-old has a best friend (**a**), but group games and sports (**b**) gain in importance. He seeks to identify with his contemporaries and to conform with his own school, class, and friends. Less under his parents' influence than formerly, he questions parental values that he once unthinkingly accepted.

ADOLESCENCE

Young teenagers are disoriented by rapid growth and sexual development. They often gain reassurance from a stable home and from conforming with their age group in fads of fashion and behavior (**a**). By 16, adolescents are adjusting to their future role as adults. They begin heterosexual activity (**b**) and may develop a more mature relationship with their parents.

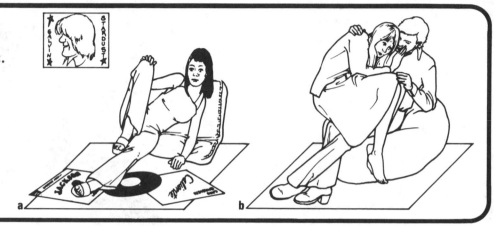

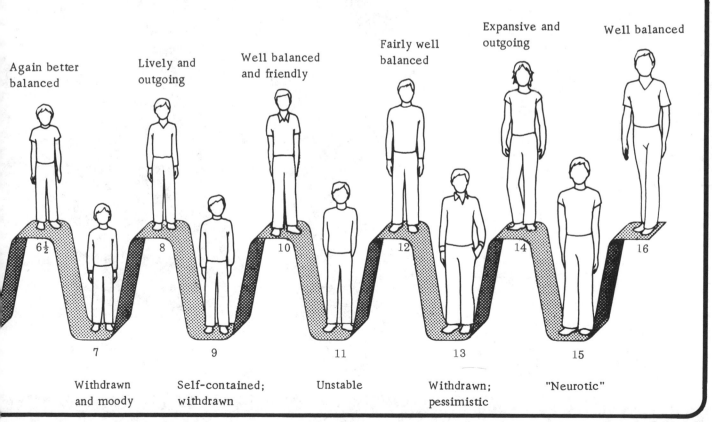

Again better balanced

Lively and outgoing

Well balanced and friendly

Fairly well balanced

Expansive and outgoing

Well balanced

6½ 8 10 12 14 16

7 9 11 13 15

Withdrawn and moody

Self-contained; withdrawn

Unstable

Withdrawn; pessimistic

"Neurotic"

©DIAGRAM

12 Practical child care

Once confidence grows and certain tasks become routine, a pattern can usually be established in which parents have time to enjoy their children as well as care for them.

Left: A vacation at the coast (The Mansell Collection)

THE NEW BABY

COPING WITH THE BABY

A new baby brings many changes to the parents' lives.

Feelings of anxiety in the early days are quite normal. Many mothers suffer from a short period of depression after giving birth to a child - "postpartum blues" - probably caused by hormonal changes in the body. If depression lasts more than a few days, the mother should consult her doctor. Anxiety about coping with the baby's needs is very common - particularly for new mothers or mothers whose lives were busy even before the new arrival. It is in such cases that understanding relatives and friends, baby-sitters, play groups for older children, and groups where mothers can discuss their problems can be so useful.

A major problem at first is that the parents do not know the baby's typical behavior patterns. In time they will learn what to expect, but meanwhile the baby will make his needs known by crying.

Specific aspects of care such as feeding, changing, and bathing may seem problematic to new parents - but will soon become second nature.

Some fathers feel neglected after the arrival of a new and demanding baby. Understanding is needed on both sides in the partnership of parenthood, and a mother should encourage her partner to take an active interest in the care of their child.

Parents should avoid neglecting their own interests; self-imposed imprisonment can lead to harmful resentment. Babies do not ask for total sacrifice - only for love, food, and warmth.

EQUIPMENT FOR A BABY

Preparations for the arrival of a new baby include the assembly of a wide range of equipment. This should be done well in advance, as shopping trips become more difficult in late pregnancy and after the baby's birth.

Many items need not be bought new; buying used furniture and nursery equipment can save a considerable amount of money. (It is important to remember that any repainting must be done with leadless paint.)

Large items, such as the crib and baby carriage, can sometimes be borrowed from friends.

There is no need for the layette to include large numbers of first-size garments - babies grow extremely quickly in the early months.

EARLY ROUTINE

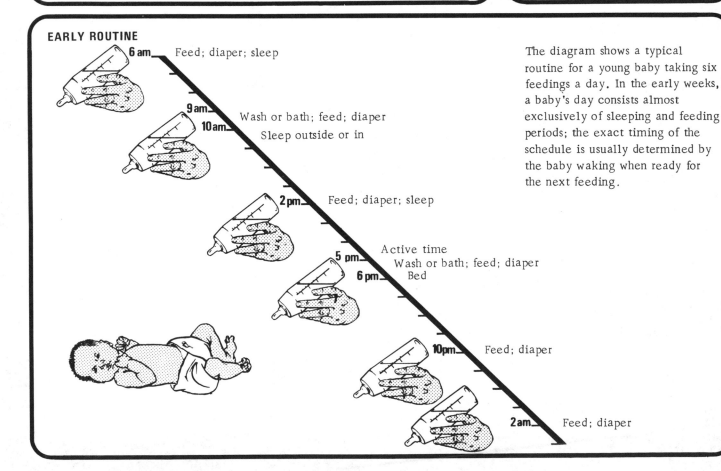

6 am — Feed; diaper; sleep

9 am
10 am — Wash or bath; feed; diaper
Sleep outside or in

2 pm — Feed; diaper; sleep

Active time
5 pm — Wash or bath; feed; diaper
6 pm — Bed

10 pm — Feed; diaper

2 am — Feed; diaper

The diagram shows a typical routine for a young baby taking six feedings a day. In the early weeks, a baby's day consists almost exclusively of sleeping and feeding periods; the exact timing of the schedule is usually determined by the baby waking when ready for the next feeding.

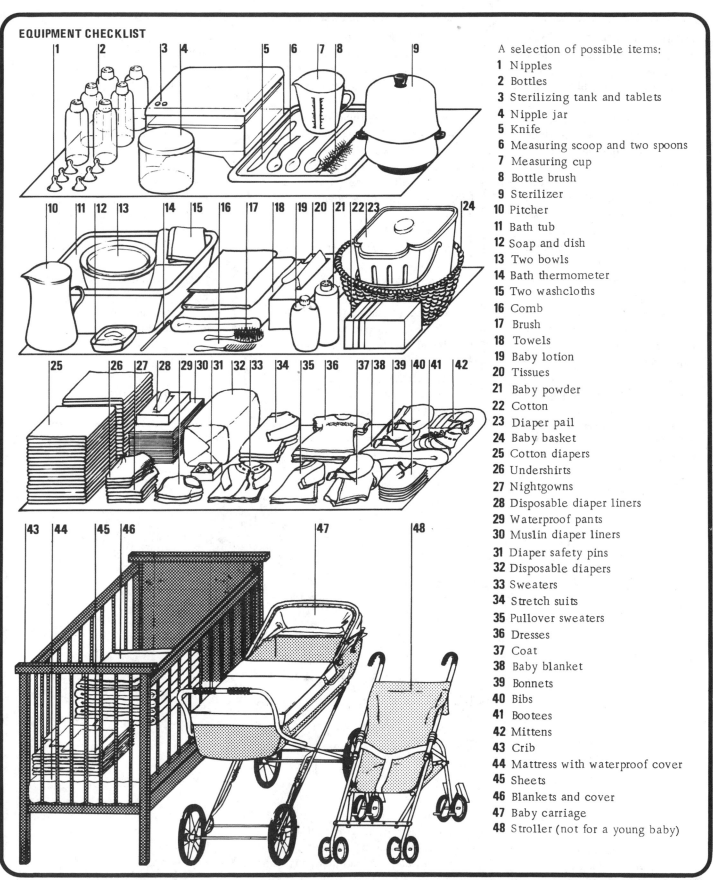

EQUIPMENT CHECKLIST

A selection of possible items:

1 Nipples
2 Bottles
3 Sterilizing tank and tablets
4 Nipple jar
5 Knife
6 Measuring scoop and two spoons
7 Measuring cup
8 Bottle brush
9 Sterilizer
10 Pitcher
11 Bath tub
12 Soap and dish
13 Two bowls
14 Bath thermometer
15 Two washcloths
16 Comb
17 Brush
18 Towels
19 Baby lotion
20 Tissues
21 Baby powder
22 Cotton
23 Diaper pail
24 Baby basket
25 Cotton diapers
26 Undershirts
27 Nightgowns
28 Disposable diaper liners
29 Waterproof pants
30 Muslin diaper liners
31 Diaper safety pins
32 Disposable diapers
33 Sweaters
34 Stretch suits
35 Pullover sweaters
36 Dresses
37 Coat
38 Baby blanket
39 Bonnets
40 Bibs
41 Bootees
42 Mittens
43 Crib
44 Mattress with waterproof cover
45 Sheets
46 Blankets and cover
47 Baby carriage
48 Stroller (not for a young baby)

©DIAGRAM

CONTACT AND COMFORT

PHYSICAL CONTACT

A reliable and continuous loving relationship experienced from birth on provides a baby with a firm basis for future development. Research suggests that close physical contact, and not food, is the most important factor in the formation of a baby's first emotional attachments - with parents or parent substitutes.

Babies like to be held in close body contact - being particularly happy in positions that simulate clinging - and a reassuring cuddle will often work wonders when a child is upset.

Aids such as baby carriers and baby carriages may be a blessing to mothers, but if used to excess can deprive a baby of valuable physical contact. Baby slings or backpacks worn by the parent can be useful alternatives.

THUMB OR PACIFIER

Young babies have an instinctive need to suck, and many like to suck their thumbs in addition to sucking during feedings. Thumb sucking may continue later, by which time it is essentially a comforting device.

Only if thumb sucking continues after a child has his adult teeth is there a risk of permanent damage - and by this time even the most enthusiastic thumb sucker is likely to have given up the habit.

Some parents prefer their baby to suck a pacifier or dummy, and it does seem to be easier for a child to abandon his pacifier than to give up sucking his thumb.

If a pacifier is used it must be kept scrupulously clean - and never dipped in tooth-decaying sweet substances.

HANDLING A BABY

Handling a baby may seem a daunting prospect to the new parent. In practice, however, most people soon master all the procedures they will need.

When lifting or carrying a young baby it is important to support his neck to prevent it from jerking back; supporting the neck with your hand or arm is all that is needed to prevent an alarming shock reaction.

A baby is likely to be startled if he is picked up suddenly, especially if he is unaware of the handler's approach. Talking to him as you approach him will remove the element of surprise. Illustrated here are two basic handling techniques - a method for turning and lifting the baby, and for raising him to a sitting position.

CARRYING A BABY

Illustrated here are four ways of holding and carrying a baby.

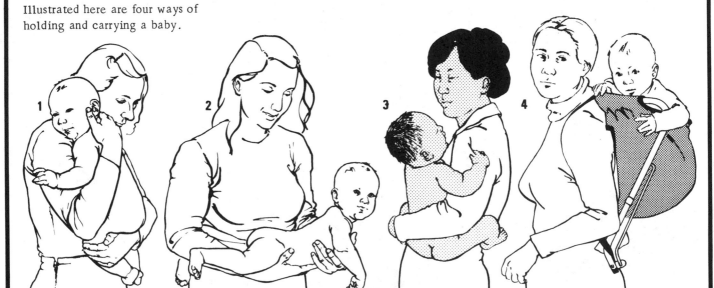

1 A traditional "comforting" position with the baby's head against the parent's shoulder.
2 A forward-carrying position that allows the older baby free movement of his arms and legs.
3 Hip-carrying position - a good way of carrying an older baby who can control his head movements.
4 Baby backpack - also for an older baby who can hold his head up - gives everyone mobility.

TURNING AND LIFTING

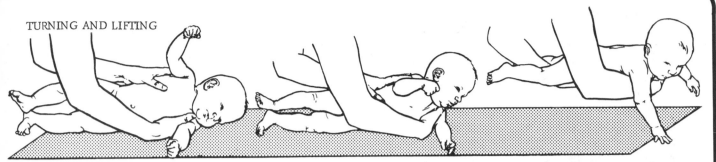

Grip upper part of baby's body.

Turn baby to rest against your arm.

Using your free arm, reach through baby's legs and lift.

RAISING TO A SITTING POSITION

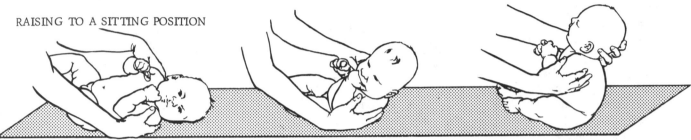

Grip baby by his upper arms and shoulders.

Turn baby slightly to one side.

Supporting head, raise baby to sitting position.

COLD AND HEAT

It is possible for a baby, from birth, to regulate his body temperature in response to changes in temperature around him. But the body mechanisms that allow him to do this are still inefficient.

COLD

Cold is generally more of a problem for babies than heat. A baby who is inadequately dressed in a cold environment needs a lot of energy to keep himself warm. A baby who has become chilled normally responds to the cold by using energy to create additional body heat - but he is unable to store this heat, and so is forced to continue his efforts until relieved by outside warmth.

Serious problems can occur if the air temperature drops suddenly and markedly when the baby is asleep. It seems that a baby's temperature control mechanisms start to function only when he is almost awake. A controlled bedroom temperature, and swaddling or a sleeping bag are therefore strongly recommended. If a baby has become chilled it is important to get him warm before adding clothing - otherwise the extra clothing will merely keep in the cold. If the chilled baby is lethargic, with reduced respiration and pulse rates, medical attention should be sought at once.

HEAT

High environmental temperatures are less likely to cause serious problems than low ones. A baby will soon cry if he is too hot - and the problem is easily identifiable. Removing a layer of clothing, wiping away the perspiration, and perhaps a drink of water are usually all that are needed to restore his comfort.

FRESH AIR AND SUN

Babies and children generally benefit from spending part of the day outdoors.

Fresh air improves the appetite, brings color to the cheeks, and probably reduces the risk of infection by preventing the air passages from becoming too dry. It is normally quite safe for even a young baby to be taken outside in his baby carriage - provided that he is suitably dressed, well sheltered from the wind, shaded from the sun, and that there is no fog. Most older children enjoy a daily walk or a period of outdoor play.

Babies and children have sensitive skin that burns easily, and care must be taken to avoid sunburn. A short sunbath - two minutes maximum on the first day - can be beneficial. A light sunhat is a must even for older children when the sun is hot.

©DIAGRAM

CRYING

COPING WITH CRYING

Crying is the only way a young baby can communicate with his parents; if he cries it may be an indication that something is wrong. Babies have characteristic ways of crying depending on what is troubling them, and most parents soon learn to distinguish between cries for different reasons.

Normally the causes of crying are straightforward, and once they have been dealt with the crying soon stops. Frequent causes include hunger, gas (wind), teething, general discomfort, and loneliness. Crying may also be caused by disturbing a child to undress or change him. A baby crying for no apparent reason is probably only asking to be cuddled.

Some reasons for crying are, however, more difficult to remedy. Notable among these is colic, which is painful for the baby and can cause great distress to the helpless parents.

Another difficult problem is that of dealing with what are sometimes called "hypertonic" babies. These babies are particularly tense, and start at even the slightest noise or handling. Many suffer from colic, or are prone to long periods of irritable crying. In time a hypertonic baby will calm down, but for the first few months it is best to disturb him as little as possible.

If crying is in any way unusual, particularly persistent, or if the child appears in any way unwell, medical advice should be sought without delay.

In general, most parents find that coping with crying becomes much easier once they have become accustomed to their own baby's individual behavior patterns.

DISCOMFORT causes babies to cry. The baby should be dressed in light clothes, and kept in a warm draft-free room away from noises and bright lights.

GAS is often relieved by cuddling the baby over the shoulder. Vigorous back-slapping is usually a waste of time.

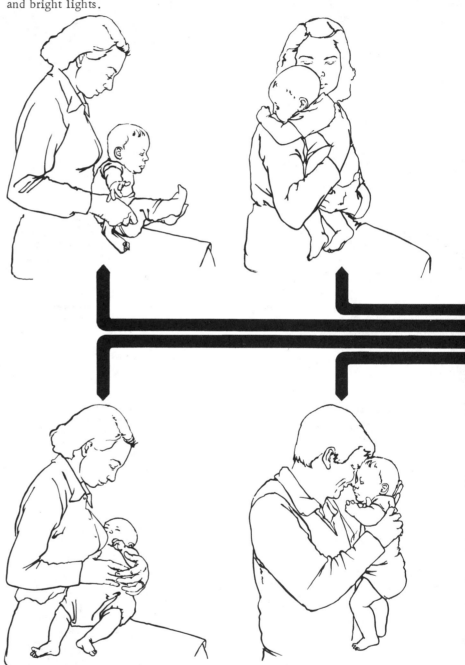

HUNGER will make a baby cry. A baby who has taken insufficient milk at a feeding typically wakes after about an hour and cries for another feeding.

LONELINESS Babies cry for attention and love. They need to be cuddled and cannot be "spoiled" in the first months of life.

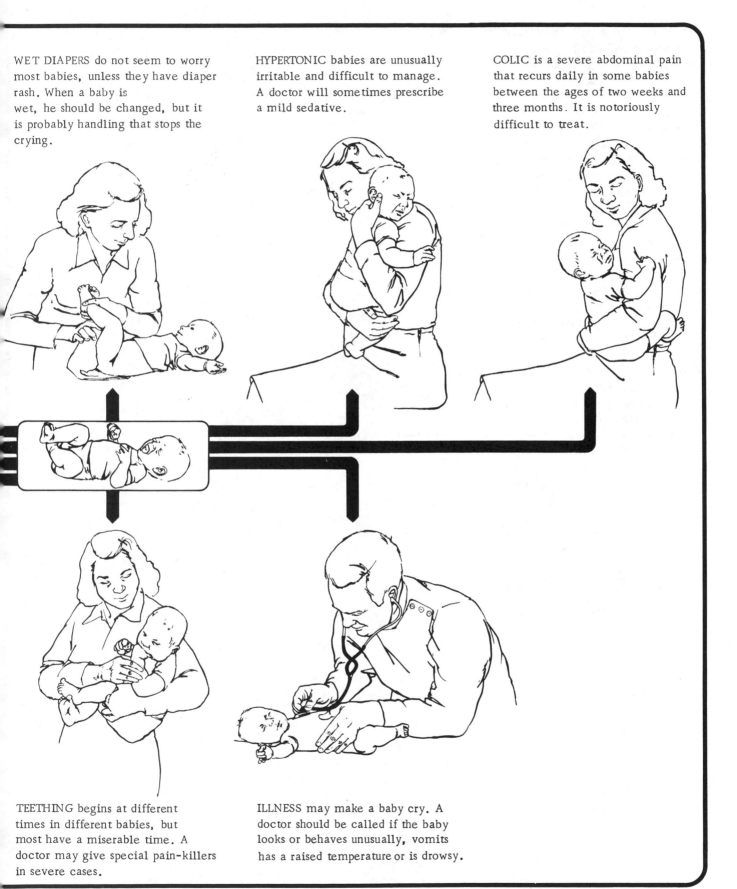

WET DIAPERS do not seem to worry most babies, unless they have diaper rash. When a baby is wet, he should be changed, but it is probably handling that stops the crying.

HYPERTONIC babies are unusually irritable and difficult to manage. A doctor will sometimes prescribe a mild sedative.

COLIC is a severe abdominal pain that recurs daily in some babies between the ages of two weeks and three months. It is notoriously difficult to treat.

TEETHING begins at different times in different babies, but most have a miserable time. A doctor may give special pain-killers in severe cases.

ILLNESS may make a baby cry. A doctor should be called if the baby looks or behaves unusually, vomits has a raised temperature or is drowsy.

©DIAGRAM

DIAPER CHANGING

DIAPERING EQUIPMENT

Various styles of diaper are currently available.

Traditional cotton diapers are usually square, but other shapes can also be bought.

To make laundering easier many people favor using muslin or disposable diaper liners along with cotton outer diapers.

Plastic pants are usually used over the diaper, but should be avoided if the baby is suffering from diaper rash.

Disposable diapers either have a plastic outer lining, or are worn under specially designed plastic pants. Although comparatively expensive, disposable diapers are often preferred by people who are short of time or away from home. Constant use of disposables, however, can cause chafing or diaper rash in some babies.

1 Cotton diapers and pins
2 Muslin liners
3 Pants for disposable diapers
4 Cotton for cleansing
5 Tissues
6 Baby lotion or cream
7 Baby powder
8 Disposable liners
9 Disposable diapers
10 Waterproof pants for fabric diapers
11 Diaper pail

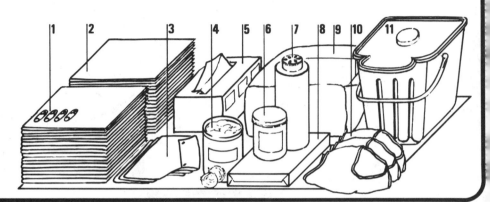

CHANGING A DIAPER

It is useful to have a flat surface on which to change a baby's diaper. All the equipment needed should be directly at hand as a baby must never be left alone where he may fall.

The method illustrated here is for a triangular-fold diaper but stages 1 and 2 apply whichever folding method is used. (Also see E13 for diaper folding methods.)

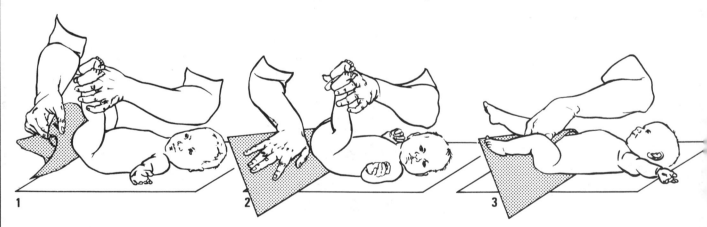

1 Raise the baby's body by grasping his legs with one hand, the index finger between his ankles. Remove the soiled or wet diaper and place in pail.

2 After a bowel movement, wipe the diaper area with tissues. Wash with cotton moistened with warm water, working from front to back. Apply lotion or powder, if used. Slide folded diaper under baby.

3 Fold one corner of the diaper across the baby's stomach and tuck it between his legs.

FOLDING A DIAPER

Illustrated here are three popular ways of folding a regular square diaper.

TRIANGULAR FOLD

Fold the diaper into a triangle and fasten the three corners with a single pin.

KITE FOLD

Fold in two sides to give a long, pointed shape. Fold over top and bottom flaps; secure with two pins. This method gives a thick center panel.

TRIPLE FOLD

Fold over one-third of the diaper and then fold the rectangle into three. Fasten the diaper with two pins. The extra thickness should go at the back for girls and in front for boys.

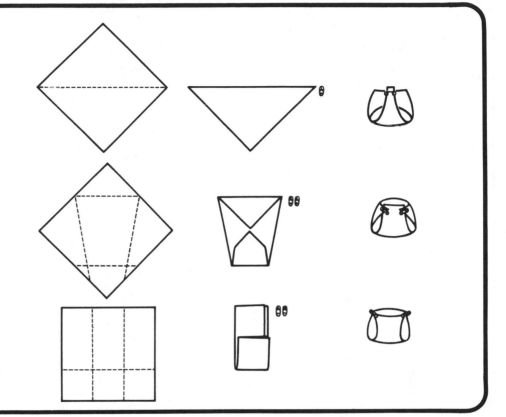

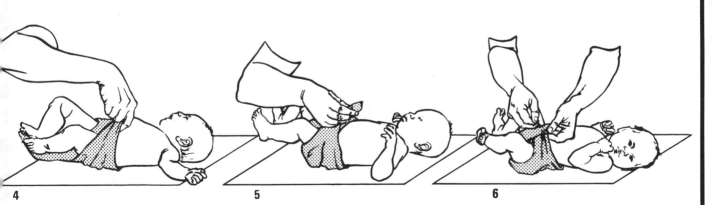

4 Lay the second corner of the diaper across the baby's stomach.

5 Bring up the third corner of the diaper between the baby's legs.

6 With the fingers beneath the diaper to protect the baby's stomach, pin all three thicknesses together, placing the pin horizontally.

©DIAGRAM

DIAPERS AND HEALTH

DIAPER RASH

Diaper rash - patches of redness and spots in the diaper area - is a cause of great discomfort. It can result from sensitivity to soap, bleaches, or fabric rinses used on the diapers; from powders or lotions used directly on the skin; or from irritation caused by wet or dirty diapers. In older babies, it is usually caused by the reaction of the skin to ammonia formed by bacteria in the urine.

To treat the condition: boil diapers for 10 minutes and rinse with an antiseptic rinse; try using one-way liners or disposable diapers; change the diaper as frequently as possible; do not use waterproof pants; use a cream on the diaper area; and expose the area to the air as much as possible.

NORMAL STOOLS

The stools (bowel movements) of babies vary in appearance and frequency according to age and method of feeding.

During the first two or three days of life a baby normally passes "meconium stools" which are sticky and green-black in color. These stools consist of the waste products that accumulated in the intestines during the last weeks before delivery. For many babies, the first of these stools is gray in color and is called the "meconium plug."

As the baby adapts to milk feeding the stools become green-brown and are known as "changing stools." When milk feeding is established the stools become mustard yellow. The transition from "changing stools" can take just a few days or as long as four weeks.

The frequency of passing stools can vary enormously. Some breast-fed babies produce stools after every feeding during the first few months; others produce as many as 15 daily. Bottle-fed babies tend to produce fewer stools at this time - usually between one and four a day.

In general the stools of a breast-fed baby are softer and less well formed than those of a bottle-fed baby.

As they grow older, both breast-fed and bottle-fed babies tend to produce stools less frequently. Breast-fed babies in particular are likely to have very irregular bowel movements at times; an interval of seven days between stools is not rare, but the doctor should be called if the baby appears unwell.

CLEANING DIAPERS

The traditional method of cleaning diapers is by washing. The stools are first scraped from the diaper into the toilet and then the diaper is rinsed and washed in very hot, soapy water. Diapers should also be boiled for a few minutes to sanitize them.

The task of cleaning diapers has recently been made easier with the development of special sanitizing products. In this method the diaper is scraped clean, immersed in a sanitizing solution in a plastic pail for two or more hours, and then rinsed several times in clean water.

Diapers are ideally dried in the sun or in an automatic drier; both methods destroy many bacteria. Many people now decide to use labor-saving disposable diapers or liners, or to make use of a local diaper-cleaning service.

PROBLEMS WITH STOOLS

DIARRHEA in infants is a serious complaint that needs medical attention.

Sudden onset is usually a result of infection. Gradual onset can be a result of careless formula mixing - too much sugar is often responsible - or the introduction of a new substance to the diet.

CONSTIPATION is most common in bottle-fed babies. Stools are hard and difficult to pass.

Adding brown sugar or maltose to formula can help. Laxatives should only be given on medical advice.

BLOOD IN STOOLS

Blood in stools of newborn babies results from swallowing maternal blood at delivery. Streaks can result from fissure in anus caused by constipation. Large quantities indicate a serious disorder needing urgent medical attention.

STOOLS OF UNUSUAL COLOR

Introduction of any new substance to diet can cause change of color. Bulky gray stools result from over-concentrated formula mix. Black stools can be caused be medicinal iron prescribed for anemia.

TOILET TRAINING

TOILET TRAINING

The paragraphs below describe a typical pattern of toilet training, but obviously there will be considerable individual variations. Some babies are put on a pot from a very early age in the hope of "catching" urine or stools. It is now generally agreed, however, that true toilet training cannot begin until the child is getting near the age at which he is physically able to control his bladder and bowels.

Between 15 and 18 months, some children give a signal that the diaper has been wetted or soiled. This should be encouraged and most children then progress quickly to indicate that a bowel movement or urine is imminent. At this stage, the notification and the action of the bowels or bladder may be simultaneous so that on many occasions it is too late to get the child to the pot in time. Nevertheless, praise should still be given. Later, the child will be sufficiently familiar with the sensations of emptying his bladder or bowels to give plenty of warning on most occasions.

Some young children are frightened of falling from the toilet seat, or by the sound of the flush, and should always be accompanied by an adult or older child. Graduating to the toilet from a pot can be made easier if a special infant seat is used.

Accidents will occur, especially with urine, when the child is tired, ill, or excited. But no reprimand should be given.

Night dryness is hardest to achieve, although taking a child to the toilet or pot at around 10pm can often prevent wet bedclothes.

TRAINING EQUIPMENT

The most basic item of equipment used during toilet training is a pot. Most pots are made of plastic and some have a shield that makes them suitable for boys and girls. It is important to wash and sanitize the pot thoroughly after use. Training pants - made of terry cloth with a waterproof backing - are useful in the early stages of training. Later, a special seat and step for the adult toilet may be used. Some infants prefer a commode chair.

1 Commode chair   4 Toilet seat
2 Training pants   5 Toilet step
3 Pot

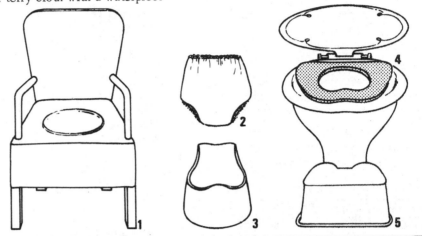

STAGES IN TRAINING

Children go through the same stages in toilet training but timing varies from child to child.

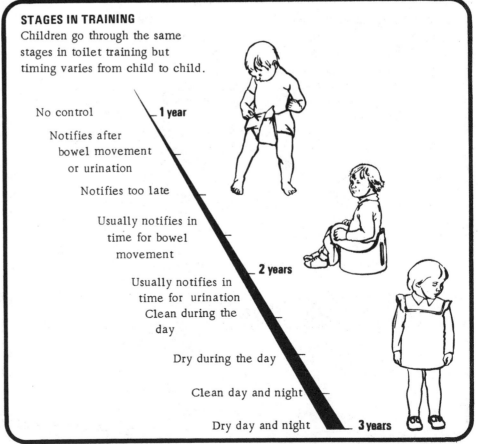

No control — 1 year
Notifies after bowel movement or urination
Notifies too late
Usually notifies in time for bowel movement
2 years
Usually notifies in time for urination
Clean during the day
Dry during the day
Clean day and night
Dry day and night — 3 years

©DIAGRAM

BATHING 1

BATHING A BABY

A bath is part of most babies'
daily routine, and is usually given
just before either the mid-morning
or early-evening feeding.

It is usually possible to give a new
baby a tub bath, but an all-over
wash or "sponge bath" may be
recommended for the first few
weeks.

Whatever bathing method is used,
it is important to have all the
necessary equipment at hand.

This includes all items used during
the bath itself, and also towels, a
clean diaper, and fresh clothing.
For tub baths a plastic tub on a
stand or table is useful, but the
sink is a good substitute.

A bath thermometer is useful if it
provides reassurance, but is not
really necessary.

1 Basin
2 Pitchers of hot water
3 Baby shampoo
4 Soap in dish
5 Bath towel
6 Cotton
7 Washcloth

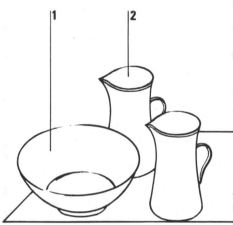

BATHING CHILDREN

Many older babies and children
like being bathed together, and
simple bath toys can add to their
enjoyment. It is important never
to leave young children unattended
in the bath tub.

NAIL CARE

A baby's nails may be cleaned
very gently with the blunt end of
a toothpick.

Nails may be trimmed with blunt-
ended scissors, but nail clippers
should not be used as their action
may cause damage to a baby's
nails.

It is a good idea to trim nails
when the baby is asleep.

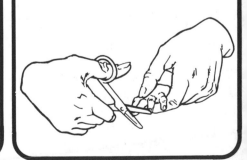

8 Bath thermometer
9 Toothpicks; cotton-tipped sticks
10 Baby powder or cornstarch
11 Baby lotion
12 Diaper

13 Clean clothes
14 Safety pins
15 Bottle for after the bath
16 Tub
17 Waterproof bath apron

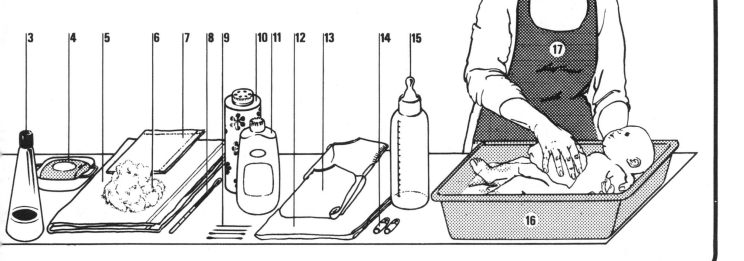

EAR AND NOSE CARE

A baby's ears and nose are very delicate and should be cleaned only when they become clogged. For this, moistened cotton swabs or cotton-tipped sticks are usually used. It is important to clean only the outer areas, as deeper penetration may cause damage.

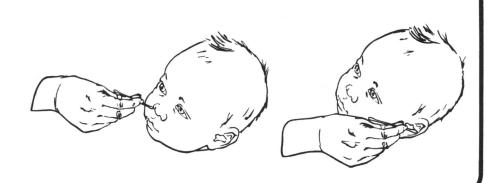

HAIR CARE

A young baby's hair and scalp should be washed with gentle soap or shampoo up to three times a week, and with warm water at other bathtimes. The baby's head should be supported over the tub or basin, as shown. Care must be taken to rinse the head carefully if soap or shampoo is used.

Cradle cap - a yellowish, waxy crust that commonly forms on the scalp at about six weeks - should be treated by massaging with baby oil before washing. An older baby need not have his scalp rinsed so often, but should have his hair washed regularly. Clean, good-quality soft brushes and combs should always be used, and care taken not to tug the hair when removing tangles.

©DIAGRAM

BATHING 2

BATHING PROCEDURES

Bathe the baby in a warm room, using water comfortable to your elbow (95-100°F, 35-38°C). Use mild soap or baby cleansing lotion. The baby may be given a sponge bath or a tub bath. If a tub is used, the baby may be soaped before or after placing in the tub.

SPONGE BATH
Take off clothes above waist. Apply soap to neck, chest, arms, and hands. Pay special attention to folds and creases. Carefully rinse with clean, warm water. Pat dry with soft towel. Unpin diaper but do not remove.

Turn the baby over, supporting neck with one hand. Apply soap to back and, lowering diaper, also to buttocks. As before, pay attention to all folds and creases. Rinse carefully with clean, warm water. Pat dry. Return baby onto back.

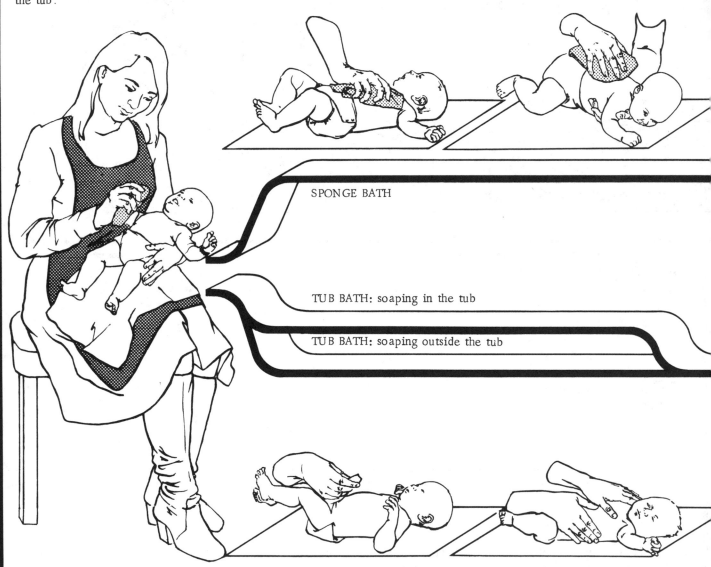

SPONGE BATH

TUB BATH: soaping in the tub

TUB BATH: soaping outside the tub

Collect everything you will need and wash your hands. Sit down with the baby on a big towel on your lap. Clean ears and nose. Gently clean face with washcloth or cotton dipped in warm water. Wash hair.

TUB BATH
Remove the baby's clothes - leaving his diaper until last. The baby is now ready to be soaped. This may be done either before or after the baby is placed in the tub.

If the baby is to be soaped outside the tub, this should be done quickly to prevent lather drying on the skin. Using hands or a cloth, apply soap to baby's body and limbs. Use front to back movements in the genital area.

Remove diaper. Wash, rinse, and dry the baby's abdomen, being very gentle with the navel. Wash, rinse, and dry the legs, feet, and genitals, being careful always to use a front to back movement in the genital area.

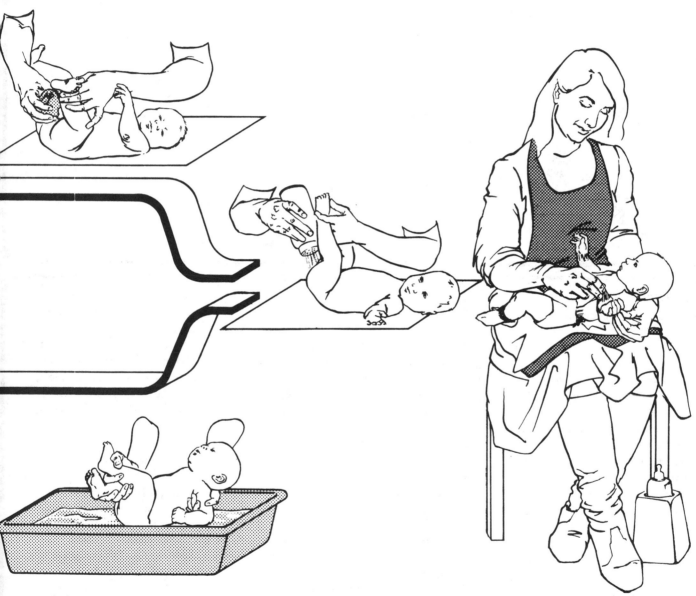

Lower baby into tub - supporting his head and back, and grasping his ankles as illustrated (forefinger between them). Continue to support the baby's head and back with one hand, while soaping or rinsing with the other. Lift out.

To prevent any chafing always make sure that the baby is patted completely dry, particularly around the navel and wherever the skin is folded. Baby powder, lotion, or cream may then be applied as required.

The baby is now ready for a fresh diaper and clean clothes. Babies should never be fed just before a bath, but a bottle or breast feeding should be very welcome once the bath is finished.

©DIAGRAM

EVERYDAY ROUTINE

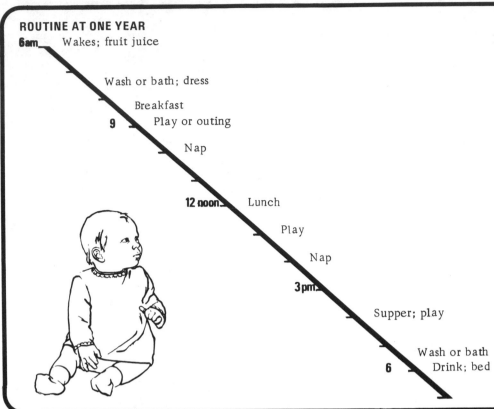

ROUTINE AT ONE YEAR

6am — Wakes; fruit juice

Wash or bath; dress

Breakfast

9 — Play or outing

Nap

12 noon — Lunch

Play

Nap

3pm —

Supper; play

Wash or bath

6 — Drink; bed

Illustrated here is an outline routine for a one-year-old. Rigid adherence to a routine is probably unrealistic for most parents and children - but a measure of planning makes life easier for everyone.

ROUTINE FOR A BABY

A regular daily routine can make life easier for new parents and help give a young baby a valuable sense of security. In addition, performing the same set of actions every day can give confidence to an anxious mother in the few weeks following her baby's birth. In the early months, the routine will be largely dictated by the baby's physical demands. These demands, and their timing, may vary slightly from day to day and it is important that the routine should be flexible enough to accommodate them. It is not, for instance, essential to bathe a baby at the same time each day, nor is it necessary to diaper him before each feeding if his hunger seems to outweigh his discomfort.

PRE-SCHOOL ROUTINE

After the first few months the daily routine for a baby becomes more varied. The number of fixed points in the day decreases, as less of the infant's time is spent asleep or feeding. By 12 months a baby is typically taking three meals a day and having a nap in the morning and afternoon. (An outline routine for a one-year-old is given above).

During the next few years much of the parents' time and energy will be spent in guiding the child's early exploration of the world about him. It is an exhausting and demanding time for parents - particularly if it is accompanied by the addition of a new baby to the family.

A day with some degree of structure is recommended throughout the pre-school years. (An outline routine for a three-year-old is

given in E31.) Careful planning will make life easier for all the family; children will know what to expect, and parents should gain time to pursue interests of their own. Realistically, however, a considerable element of flexibility is going to be required.

Except when the weather is bad, children generally benefit from spending some part of the day outdoors. Also recommended in the daily routine is a period of play with other children. The very young child plays alongside rather than with his contemporaries, but from about two-and-a-half he will enjoy joining others in their games. Organized groups for pre-school children have become increasingly popular in recent years, and a well-run group has advantages for both child and parent.

ROUTINE AT THREE YEARS

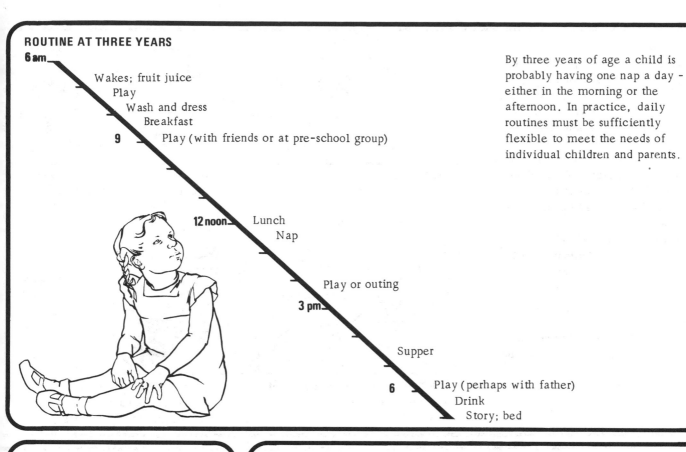

6 am — Wakes; fruit juice
Play
Wash and dress
Breakfast
9 — Play (with friends or at pre-school group)

12 noon — Lunch
Nap

Play or outing
3 pm —

Supper

6 — Play (perhaps with father)
Drink
Story; bed

By three years of age a child is probably having one nap a day - either in the morning or the afternoon. In practice, daily routines must be sufficiently flexible to meet the needs of individual children and parents.

DISTURBED ROUTINE

Children vary in their ability to cope with a disturbed routine, but parents should be reassured that an unfavorable reaction is not unusual at some stage. In general, understanding handling can usually minimize the trauma.

A visit to the home of relatives or friends can impose extra strain on even the most placid and easy-going child. Good behavior is expected from him in unfamiliar surroundings and he may become exhausted and react with tears or bad temper. A similar reaction can occur after or during the visit of an indulgent relative - perhaps a grandparent.

A family outing, planned as a special treat, can end disappointingly. The excitement of the occasion can quickly tire a child and make him balky and irritable.

SCHOOL-AGE ROUTINE

A child's daily routine changes radically when he first starts attending school, and most children need a good deal of parental support to help them in this new situation.

A school-age child spends much of his day away from home, but a fair amount of routine planning is still required.

In the morning, different members of the family must try to coordinate their times of waking, using the bathroom, eating breakfast, and leaving home. Schoolchildren should always be given a nourishing breakfast, with ample time allowed to eat it. Preparing for school should also be as unhurried as possible, and parents should ensure that their children remember any special equipment that may be needed. Whether or not a parent actually accompanies a child to school, it is ultimately the parents' responsibility to ensure that children arrive at school on time. At the end of the school day, a parent picking up a child should make every effort to arrive on time. If the child is old enough to come home alone, ensuring that someone is there to greet him will increase his sense of security.

If a child has schoolwork to do at home parents should try and see that it is done, providing a quiet work place if possible.

Evening activities and eating times vary a great deal from home to home. Each family must work out a pattern that suits it, and children should learn from an early age the need to fit in with other people's plans.

Views on bedtimes also vary. In general, children should go to bed early enough to ensure that they will not be tired at school.

©DIAGRAM

CLOTHING AND SHOES

DRESSING A BABY

Babies grow quickly, so careful planning is needed when organizing baby clothes. Seasonal requirements vary and should be remembered when selecting baby clothes of different sizes.

Garments should be easy to launder and dry, and preferably need no ironing. Deeper colors look fresh longer than the traditional pastel shades.

Dressing a baby is simplified if clothes have few buttons or frills. Lacy-knit or fluffy garments are best avoided, as the baby may catch his fingers or inhale some of the wool.

In the early days, many people favor simple gowns that tie at the neck. Also very popular are one-piece stretch suits.

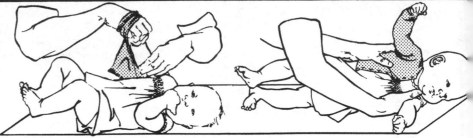

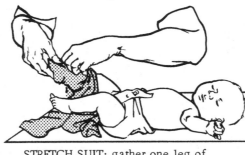

SWEATER: gather sleeve into loop with one hand; with the same hand grasp baby's hand and draw arm down through sleeve;

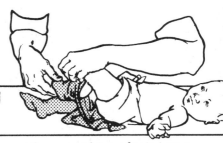

rock or lift baby onto other side; ease the sweater around baby's neck; draw other arm down through sleeve as described.

STRETCH SUIT: gather one leg of suit into a loop and ease baby's foot and leg into it;

gather second leg of suit into a loop and ease baby's other leg into it; grasp baby's ankles;

CLOTHES FOR CHILDREN

Many factors affect the choice of clothes for children. In general, clothes should be selected that are appropriate to a child's varying activities.

Clothes for play should be in a hard-wearing, easy to launder fabric, and preferably dark in color. Constant nagging about getting dirty is bad for both parent and child - much better to dress the child in play clothes that need little attention.

School clothes should be easy for the child to manage by himself; all unnecessarily difficult fastenings should be avoided. This is particularly important on days when the timetable includes a sporting activity for which the child may have to change. "Best" clothes are worn comparatively little and since

they are soon outgrown it is unnecessary to pay large sums for them. (Shoes must always fit well, and the speed at which children's feet grow makes it particularly uneconomical to keep any for special occasions.)

At certain ages children have very definite ideas about clothes and fashion. When this is the case the parents must decide the extent to which they are prepared, or financially able, to let their children have their own way. Hand-me-downs are extremely useful, but to avoid jealousy parents should ensure that younger members of the family receive at least some new garments. Clothes will last longer if they are well taken care of - with regular laundering, prompt mending, and appropriate storage.

DRESSING THEMSELVES

From about age two, most children show an interest in dressing themselves. By school age, most can manage all but the most

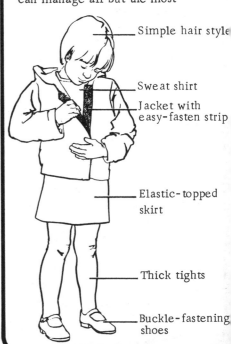

Simple hair style

Sweat shirt

Jacket with easy-fasten strip

Elastic-topped skirt

Thick tights

Buckle-fastening shoes

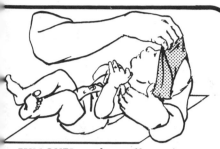

PULLOVER: gather pullover into loop; slip opening over back of head and then forward over front, stretching over forehead and nose.

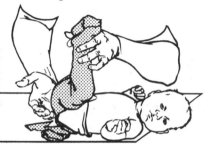

raise baby's legs and ease suit up to waist; put on top of suit in same way as sweater.

CARE OF A BABY'S FEET

A baby's feet are soft and pliable and easily distorted.
In the early months, damage can be caused by bedclothes tucked too tightly around the feet when the baby is lying on his back.
It is important to check that socks and the feet of stretch suits are big enough to allow free movement. Soft shoes for non-walking babies are best avoided as they may hinder the normal development of the feet.
A baby learning to walk should go barefoot as much as possible; shoes worn for outings must fit correctly.

CHILDREN'S SHOES

The fit of children's shoes matters much more than their appearance. Children's feet grow fast and should be measured regularly for width and length with the special gauges used in shoe stores.
The structure of the shoes is also important. The soles should be flexible to allow a springy step. The uppers should be supple where the toe joints bend but firmer in the arch for support. Shoes that fasten with adjustable straps or laces are recommended as they hold the foot at the back of the shoe and prevent it from sliding forward and cramping the toes. Though it is important to keep the feet dry in wet weather, it is unwise to wear rubber boots for long periods, as they restrict the evaporation of perspiration.

difficult fastenings. It is sensible to encourage independence by providing simple clothes such as those illustrated here.

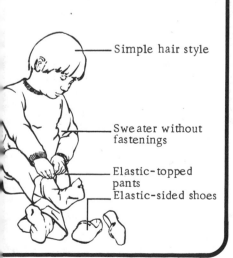

- Simple hair style
- Sweater without fastenings
- Elastic-topped pants
- Elastic-sided shoes

KEEPING TIDY

Accessible storage areas will help children to keep clothes tidy. Ideally there should be enough closet space for hanging up clothes and keeping shoes - with lowered rails for easy reach. Dressers can be used to store underwear and sweaters, and the flat surface on top is a good place to keep combs and brushes.

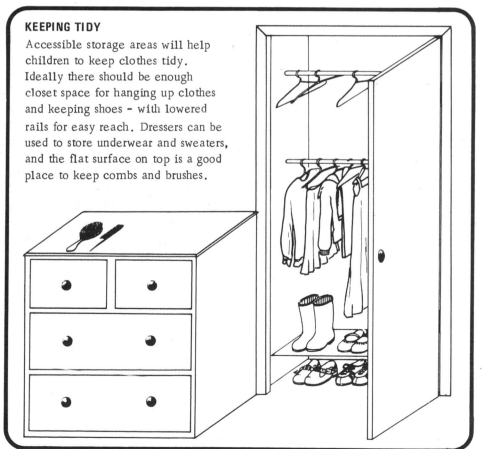

©DIAGRAM

TRAVEL HINTS

TRAVEL WITH CHILDREN

Trouble-free travel with children relies to a great extent on careful planning by the parents. An important aspect of this is the provision of all the equipment needed for a particular trip.

For babies it is important that the normal routine remains unchanged as far as possible, and it is sensible to carry in a large bag all the equipment needed for feeding and diapering.

A damp washcloth, a towel, and tissues are useful for children of all ages.

Extra garments and a blanket should be carried for changes in temperature, and some form of entertainment - toys, books, or games - should also be included.

For car travel, safety equipment such as a junior car seat or seat belt is essential.

EQUIPMENT FOR BABY

1 Baby carrier and restraining straps
2 Diapers, possibly disposable
3 Prepared bottles
4 Pacifier
5 Baby food
6 Tissues
7 Extra garments
8 Pot
9 Toys
10 Damp washcloth and towel in plastic bags

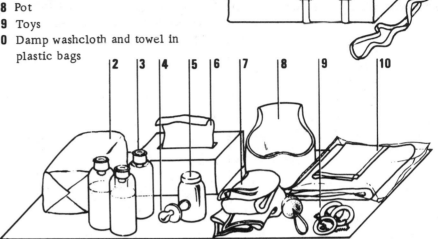

CAR TRAVEL

Car trips are a regular feature of modern family life, and, with planning, parents can ensure that as little upset as possible is caused to younger members of the family.

Two factors need consideration when a trip is planned - safety and comfort.

SAFETY

It is unwise to allow a child to travel in the front seat of a car.

A baby can travel in his carrier, which should be held on the back seat with restraining straps.

Children need special safety seats or junior seat belts (see E40).

"Childproof" locks should be fitted to doors.

COMFORT

The car should be warm and adequately ventilated; a small child should not be in a draft from an open window. Clothing should be easy to adjust or remove if the child becomes too warm, and an extra garment or blanket should be available if the child becomes cold.

Trip times should be planned to allow for frequent stops - for toilet visits and exercise.

On longer trips, provision must be made for feeding the family. A picnic meal is usually ideal, and a bottle of plain water should always be carried. For a bottle-fed baby, bottles of formula should be made up in advance and either given cold or heated in a pitcher of hot water from a thermos.

It is essential to provide some form of amusement or diversion for children on a trip - even a baby enjoys having a familiar toy close by.

BUS AND TRAIN TRIPS

Long-distance travel with young children on buses or trains is never easy, especially when they are crowded. For this reason, it is preferable to travel midweek and, if possible, to reserve seats in advance. Sleeping cars are recommended for overnight travel on trains.

The breast-fed baby presents few problems on a coach or train journey unless the mother is unhappy about feeding the child in anything but strict privacy. Ladies' rooms in stations and hotels often have special facilities. Bottles of formula should be made up in advance for the bottle-fed baby.

Older children should be given plain food to minimize the chance of sickness.

EQUIPMENT FOR OLDER CHILD

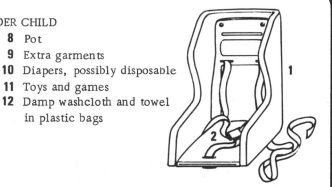

1	Car seat	**8**	Pot
2	Seat belt	**9**	Extra garments
3	Tissues	**10**	Diapers, possibly disposable
4	Pillow	**11**	Toys and games
5	Blanket	**12**	Damp washcloth and towel
6	Thermos		in plastic bags
7	Food		

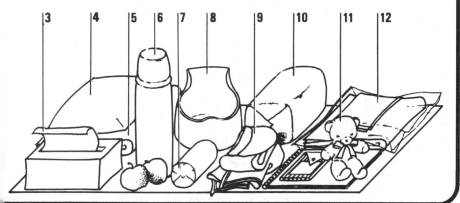

MOTION SICKNESS

If a young child is susceptible to motion sickness, the parents should never fuss or show concern before the trip.

Rich food should not be given before or during a trip. An apple or dry cracker may help relieve feelings of sickness.

Adequate ventilation and frequent stops can make car travel more pleasant, and amusements for the child may also help.

An anti-sickness medication taken in advance can also help ensure a comfortable trip.

During the trip it is useful to have close at hand plastic or paper bags, a damp washcloth, a towel, and a change of clothes; these can give a sense of security to an anxious child.

AIR TRAVEL

Most airlines are extremely helpful to parents traveling with children, especially if they are warned in advance. Some aircraft have cribs, facilities for heating bottles or canned baby food, and a supply of disposable diapers; this should be checked before the trip.

Takeoff and landing can sometimes be alarming for a child. A baby may be given a bottle of boiled water or a zwieback (rusk), and an older baby a candy; these will help equalize pressure in the ears. Children need some form of amusement during the flight and waiting periods.

Delays are sometimes unavoidable so provision should be made for extra time spent traveling. This means keeping extra clothing and food at hand and not packed away in baggage in the aircraft hold.

BOAT TRIPS

A short boat trip with children is often more difficult to manage than a long sea voyage.

If cabin accommodation is available, most of the problems associated with feeding and caring for a baby will be solved. Toddlers and older children should be carefully controlled at all times for their own safety. Rich meals should naturally be avoided.

For long trips, most large ships offer special facilities for babies and children, including supplies of formula milk, disposable diapers, and baby foods; special mealtimes; and a supervised nursery.

Before a long trip it is important to find out about the places visited en route so that the child may be dressed comfortably whatever the temperature.

TRAVEL ENTERTAINMENT

Travel can be easier and more pleasant for everyone if some effort is made to entertain the children during a trip.

For a special trip, reading in advance a book about boats, trains, or aircraft can greatly increase a child's enjoyment and interest.

On trains, aircraft, and boats reading or drawing books and play materials such as modeling clay can be useful, though this kind of activity can cause a headache or sickness in a car.

Spotting games like "I spy" are entertaining and can be instructive. Also popular are counting and guessing games.

Singing or reciting for short periods can amuse children of all ages, and stories, told by parents or children, can provide an enjoyable diversion.

©DIAGRAM

SLEEP 1

TYPES OF CRIB

A baby spends much of his time
asleep, so all the equipment must
be both safe and comfortable.
In the early months, many babies
sleep in a bassinette, or baby
carrier on a stand. These have
enclosed sides that reduce the
risk of chilling from drafts.
When buying a full-size crib there
are several points to consider. It
should be sturdily built with
vertical bars no more than 2½in
(6.4cm) apart, and a safe latch
on the side that can be raised and
lowered. The mattress, either
foam or horsehair, should have a
waterproof cover. A pillow is not
recommended for the first year.
Fitted sheets are timesaving and
easy to use. Blankets should be
light, warm, and easy to launder;
acrylic fiber is a good choice.

1 Bassinette
2 Baby carrier
3 Crib

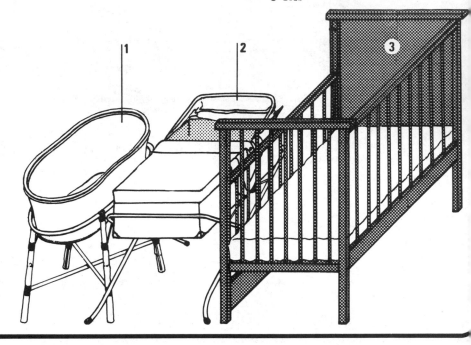

EQUIPPING A NURSERY

A room of his own is ideal, but a
baby can successfully share his
parents' room if necessary.
The crib is the only essential
item of equipment in a nursery.
A flat surface for diaper changing
and a storage unit for baby clothes
and equipment are, however,
strongly recommended. Also
useful is a comfortable nursing
chair for feeding.
A number of safety and health
factors need careful consideration.
Walls and furniture should be
painted with leadless paint.
Drapes and blinds should be
made of flameproof fabric.
Flooring should be easy to clean
and non-slip. Scatter rugs may
cause a serious fall.
The room should be kept warm -
not lower than 65°F (18°C) - and
the crib should be placed in a
draft-free position.

1 Crib
2 Changing surface
3 Storage unit
4 Nursing chair

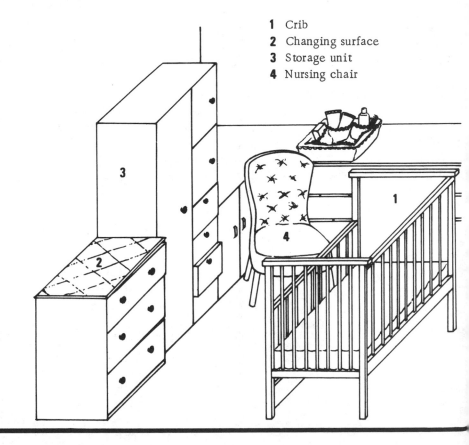

BABIES' SLEEPWEAR

A simple cotton gown (**1**) is the traditional sleepwear for young babies. Many people, however, now favor dressing their babies for sleep in an easy-to-wash terry cloth stretch suit (**2**). As babies become more active and are likely to kick off their cot covers it is a good idea to use a sleeping bag (**3**).

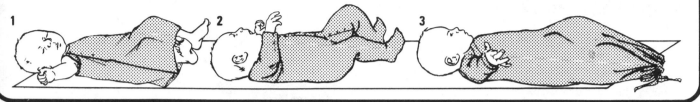

SLEEPING POSITIONS

For the first three months a baby sleeps in the position in which he is placed. The front position (**1**) is sometimes recommended, as any regurgitated feeding can easily trickle out of his mouth. The back position (**2**) allows the baby to look around when he wakes. Only from about six months old are babies able to sleep unsupported on their sides (**3**); a younger baby can be kept in a side sleeping position by supporting his back with a rolled towel or sheet.

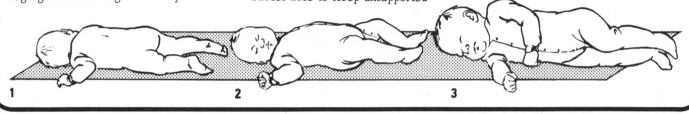

HOW MUCH SLEEP?

Individual sleep needs vary enormously. Some young babies sleep as much as 80% of the time, while others seem to be wakeful from birth. Children and adults, too, appear to have very different personal sleep requirements. In general, however, sleep needs decrease with age. An indication of sleep needs at different ages is given here in the diagram (based on the findings of H. P. Roffwarg, J. N. Muzio, W. C. Dement). Researchers have discovered that there are two basic types of sleep: light, or REM, sleep (when rapid eye movements can be observed), and deep, or NREM, sleep (without rapid eye movements). In older children and adults, dreams are most likely to occur during REM sleep. In infants, REM sleep is thought to be not necessarily related to dreaming.

Typical sleep needs and types of sleep at different ages

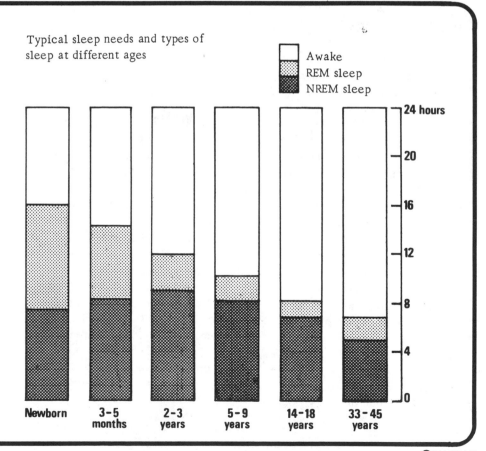

© DIAGRAM

417

SLEEP 2

SLEEPING PATTERNS

The distribution of sleep during a 24-hour period changes with age. A young baby typically has five or six sleep periods a day, waking when hungry and sleeping when fed. As the baby gets older he spends more of the time awake, with sleep concentrated into a long night sleep and a number of daytime naps. A two-nap-a-day pattern is usually established in the second half of the first year. This is followed, usually early in the second year, by a one-nap-a-day pattern that persists through early childhood. From about age four children usually sleep only at night, with total sleep needs decreasing as they get older.

Typical sleeping patterns at different ages

□ Awake
▨ Asleep

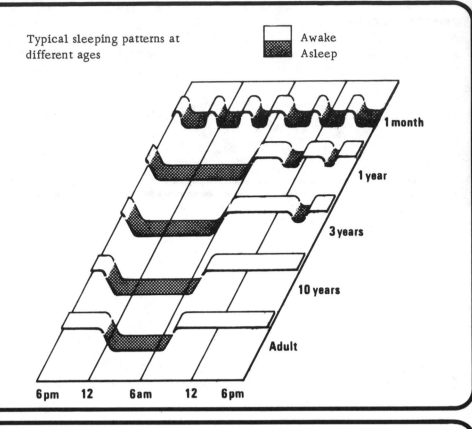

6 pm　12　6 am　12　6 pm

1 month
1 year
3 years
10 years
Adult

BEDTIME ROUTINE

The establishment of a regular bedtime routine can be comforting and reassuring for a young child, and may in some cases prevent the formation of severe sleep problems. Though exact procedures vary with the age and temperament of the child and with the particular family situation, the observation of a few general points can help make bedtime easier for both child and parent.

The period before bed should be as calm as possible and a story can be an ideal way of rounding off a child's day. Most children thoroughly enjoy a story and derive considerable satisfaction and a valuable sense of security from receiving the undivided attention of a parent – even for just a few minutes.

It can also be helpful to encourage a child to take an active part in bedtime preparations. He can be allowed to turn on the bath tap (under supervision), undress himself, help dry himself, and put on his nightclothes.

Ideally, the child should go to bed at approximately the same time each evening. The parents' attitude to bedtime should be pleasant but firm; if the pre-bed routine is tackled with an air of certainty the child will quickly learn that he cannot successfully delay going to bed by discussion, argument, or pleading.

A certain amount of bedtime ritual is comforting - indeed necessary - for children. Many insist on having particular toys or playthings in bed with them; such objects can help give a child a feeling of security when he is left alone and may help him not to disturb his parents if he wakes in the night or early in the morning. Others may ask for a drink or an extra kiss, or that the bedroom door be left open a certain distance. As long as bedtime requests remain reasonable it is probably sensible to grant them. Some children, however, may use them as a deliberate delaying tactic, and parents should guard against the establishment of long and involved bedtime rituals.

In general, the parent should settle the child and leave him with a firm but cheerful farewell. Some children will fall asleep almost immediately, or at least lie quietly until they are ready to sleep. Others are more troublesome, beginning to cry as soon as the parent has left the room.

CHILDREN'S NIGHTWEAR

Nightwear should be carefully chosen to meet a child's needs. A sleeping suit is a good idea for young children. Simple, short nightdresses may be worn by young girls, but long ones should be reserved for older girls for safety reasons. All children's nightwear should be made from flameproof fabric.

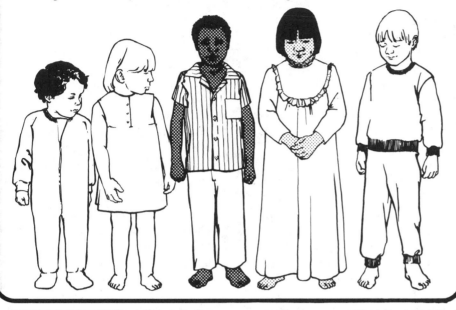

REFUSING TO SETTLE DOWN

Almost every child goes through phases of refusing to settle down when put to bed. There are a number of common reasons. Childhood is interspersed with countless new experiences, and at bedtime the child may be left too excited to sleep. Children, like adults, need time to unwind before sleep is possible.

A young child's growing enjoyment of the company of others can make him reluctant to leave the family, and his unwillingness to sleep can be reinforced by interesting sounds from elsewhere in the home. Alternatively, he may simply not be tired. Afternoon naps continued too long or taken too late in the day may lead to problems at bedtime. Ajdusting the daily routine as the child gets older should put matters right.

Sometimes a child's refusal to settle down is due to anxiety or a genuine fear of being left alone in the dark. Parental reassurance, together with a comforting toy, a nightlight, or leaving the bedroom door open and another light on should help the phase to pass more quickly.

Some toddlers refuse to lie down in their cribs at bedtime. They should be left sitting or standing; the parent can return later to cover the child.

Many young children habitually cry when left for the night. In most cases this is merely a "testing cry;" it rarely warrants the parents' return as the child usually drifts to sleep quite quickly. More persistent bedtime crying is more of a problem, and in severe cases a a doctor may prescribe a very mild sedative to help reestablish the habit of falling asleep.

NAPS

Babies and young children need to supplement their night sleep with daytime naps. The timing of a child's naps is influenced by sleep needs at different ages and by general family routine.

Problems sometimes occur when the child is making the change from two naps a day to one. If he does not sleep in the morning, he is too tired to eat lunch. If he naps in the morning, he becomes overtired in the late afternoon. A very early lunch followed by one long nap is probably the best solution at this stage.

Similar flexibility is often needed when the child is almost ready to give up naps altogether. A quiet period in the afternoon or a short sleep somewhere other than in his bed may help a child over this period.

EARLY WAKING

Many children regularly wake very early in the morning. This is perfectly normal but can cause an unwelcome disturbance of the parents' sleep, especially in the case of a young child who is at his brightest and most sociable after a night's rest.

It is often possible to encourage a young child to play quietly in his bed in the early morning; a mobile to watch or toys to play with may delay any disturbance.

In some cases a "bribe" of a drink or cookie set out the night before may also be useful.

Older children can sometimes be taught not to make too much noise until they hear a certain sign - for instance, the ringing of their parents' alarm clock.

Sharing a room may also help solve the problem as the children may amuse each other and play quietly until a reasonable hour.

©DIAGRAM

DISTURBED NIGHTS

Young babies, provided that they are tired, comfortable, and not hungry, sleep very soundly. From the age of about six months, however, children wake more easily, and may do so frequently for a variety of reasons. Disturbance by other people is a common cause of waking at night. Too many visits by over-anxious parents can cause the child to wake up and demand attention. Sharing a room - either with the parents or another child - can also cause problems. Bedtime preparations must be made with the minimum of noise and fuss if a child is already asleep in the room. If two children are sharing a room and one of them wakes in the night it is extremely likely that he will disturb the other - either deliberately or not.

Accustoming a child to complete silence when he is in bed is not necessarily a good idea - making it difficult for him to sleep away from home or if there is any unusual noise in the night. It is not, of course, reasonable to expect children to sleep through high levels of noise, for example, from a loud television or party. Some children wake at night simply because they are no longer tired - a later bedtime, or cutting out a nap, may be the solution. Many children need to visit the toilet during the night - leaving a convenient light on can persuade an older child to manage alone. Nightmares are another common cause of disturbed nights. Gentle reassurance from the parent will usually get the child to go quickly back to sleep.

MOVING TO A BED

The age at which a child is moved into a bed can vary considerably - some children are ready at two while others are happy in a crib until they are four. When a toddler begins to climb out of his crib regularly, he is probably safer in a bed. Although it represents "growing up" some children may be anxious about moving from a crib to a bed. The transition need not be made suddenly; a child can become accustomed to a new bed by taking naps there, and only moving on to sleeping in it at night when he feels ready.

If the crib is needed for another child, it is wise to establish the toddler in his bed well before the new baby's arrival in order to avoid giving him the impression that he is being "pushed out."

CHOOSING A BED

The first essential for any bed is a firm mattress to support growing bones and encourage good posture. In the early months it is wise to place the bed against a wall, and to have a special guard rail or a chair against the other side of the bed for extra reassurance. Junior beds, smaller than the regular single version, are widely available, but have the disadvantages of needing special bedding and being soon outgrown. Bunk beds save floor space, and their play potential makes them popular with many children.

It is a good idea to store a camp bed under the child's bed, for use by visiting friends.

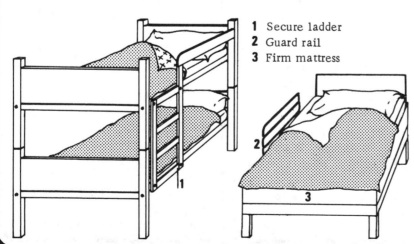

1 Secure ladder
2 Guard rail
3 Firm mattress

SHARING A ROOM

In many families it is necessary at some stage for two or more children to share a bedroom - a situation that has both advantages and disadvantages.

Young children often enjoy sharing a room and perhaps will amuse each other in the early morning, so allowing the parents a little extra sleep. At the other end of the day, however, it can sometimes be a problem to persuade children sharing a room to go to sleep. Also, a disturbed night for one child can often mean a disturbed night for both.

As they get older, children may resent the lack of privacy in a shared room. It is a good idea at this stage to allocate specific areas of the room to each child, either by using some form of room divider or by giving each child his own storage and display areas.

BABY·SITTERS

GENERAL PROCEDURE
1 Agree in advance the hours, pay, and duties expected of the sitter.
2 Indicate whether you allow the use of the TV, radio, and telephone; whether you allow the sitter to bring a friend; and whether you will provide food.
3 Show the sitter around the house.
4 Pick up the sitter and take home if necessary.

CARING FOR THE CHILD
1 Choose a person who can cope with the needs of a particular child, and preferably someone the child knows.
2 Check that the sitter can, if necessary, change a diaper and give a bottle. Indicate where the equipment is kept.
3 Warn the sitter of any point of routine such as leaving a door ajar, or any favorite toy, that the child finds comforting.
4 Establish, in the presence of an older child, details about snacks, drinks, play, bedtime, etc.

EMERGENCIES
Leave a list of telephone numbers, including the number at which you can be contacted, and the numbers of a reliable neighbor, the doctor, and the police.

BRIEFING THE CHILD
If the child is old enough, brief him as follows:
1 Warn him in advance that you are going out and that the sitter will be taking care of him.
2 Remind him to behave well.
3 Introduce him to the sitter.
4 Establish, in the presence of the sitter, details concerning snacks, play, bedtime, etc.

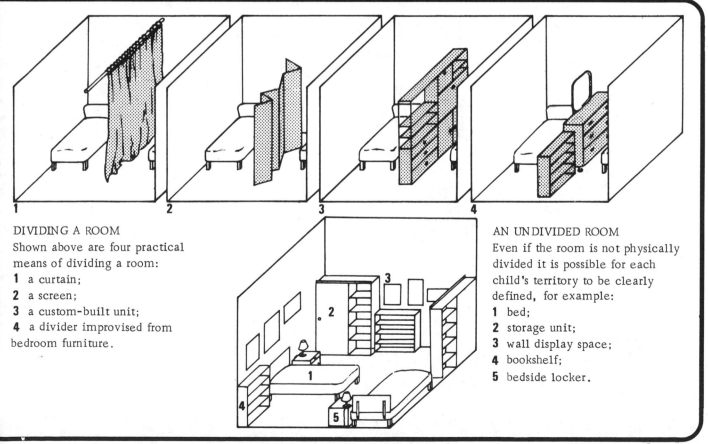

DIVIDING A ROOM
Shown above are four practical means of dividing a room:
1 a curtain;
2 a screen;
3 a custom-built unit;
4 a divider improvised from bedroom furniture.

AN UNDIVIDED ROOM
Even if the room is not physically divided it is possible for each child's territory to be clearly defined, for example:
1 bed;
2 storage unit;
3 wall display space;
4 bookshelf;
5 bedside locker.

©DIAGRAM

421

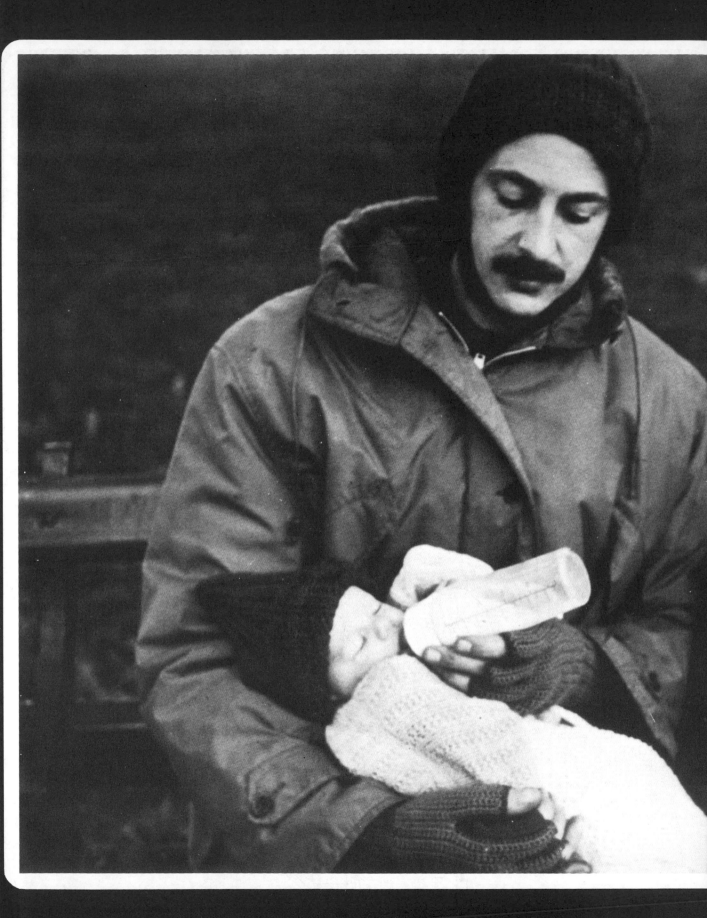

13 Feeding

Few aspects of child care
cause more anxiety than
feeding. From the birth of
the baby onward there are
numerous questions to be
answered.

Left: A park bench feeding (Photo
Richard and Sally Greenhill)

MILKS

BREAST MILK CONSTITUENTS

For the first week of a baby's life all his nutritional needs are supplied by milk - either his mother's breast milk or a specially developed formula milk.

The exact composition of breast milk varies slightly from mother to mother - and there is a similar variation in the precise requirements of different babies. The diagram shows a detailed analysis of 100gm of breast milk. Water constitutes a very large proportion of the total, and carbohydrate, protein, and fat account for most of the remainder. Minerals and vitamins have an importance out of all proportion to their weight.

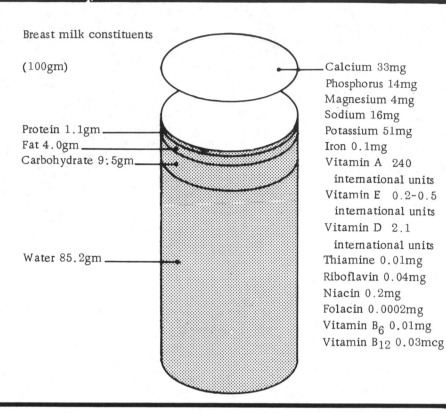

Breast milk constituents

(100gm)

Protein 1.1gm
Fat 4.0gm
Carbohydrate 9:5gm

Water 85.2gm

Calcium 33mg
Phosphorus 14mg
Magnesium 4mg
Sodium 16mg
Potassium 51mg
Iron 0.1mg
Vitamin A 240
    international units
Vitamin E 0.2-0.5
    international units
Vitamin D 2.1
    international units
Thiamine 0.01mg
Riboflavin 0.04mg
Niacin 0.2mg
Folacin 0.0002mg
Vitamin B_6 0.01mg
Vitamin B_{12} 0.03mcg

COMPARISON OF MILKS

Human milk is ideally suited to the developmental needs of a human baby. The needs of the offspring of different types of animal are similarly met by their own mothers milk.

Obviously the needs of human babies and the needs of different young animals vary - and so do different types of milk.

The diagram illustrates the most obvious difference between human milk and various types of animal milk - all milks contain carbohydrate, protein, and fat, but the relative proportions of these different nutrients vary considerably from milk to milk. Human milk is seen to be comparatively rich in carbohydrate and low in protein.

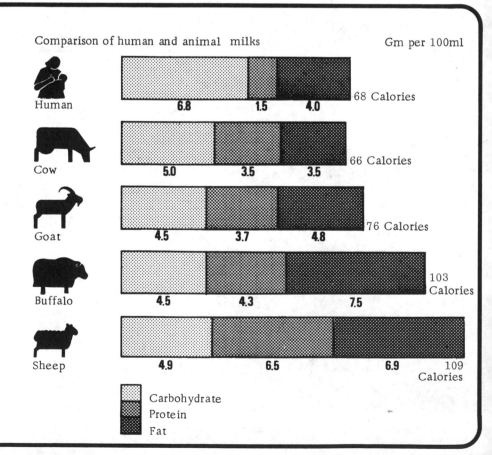

Comparison of human and animal milks

Gm per 100ml

Human
6.8   1.5   4.0   68 Calories

Cow
5.0   3.5   3.5   66 Calories

Goat
4.5   3.7   4.8   76 Calories

Buffalo
4.5   4.3   7.5   103 Calories

Sheep
4.9   6.5   6.9   109 Calories

Carbohydrate
Protein
Fat

HOW MUCH MILK?

The quantity of milk required by a baby varies according to age and weight. In the first few days after birth babies take tiny feedings at irregular intervals. By the end of the first week a more regular feeding pattern is usually established. From this point food requirements increase fairly steadily and are reflected in a correspondingly steady gain in weight.

It is possible to calculate a baby's approximate daily food needs in ounces by multiplying his weight to the nearest pound by $2\frac{1}{2}$. Each baby's requirements are slightly different, and a particular baby's intake may vary from day to day. The diagram is a guide to the amount of milk required by typical babies at different ages.

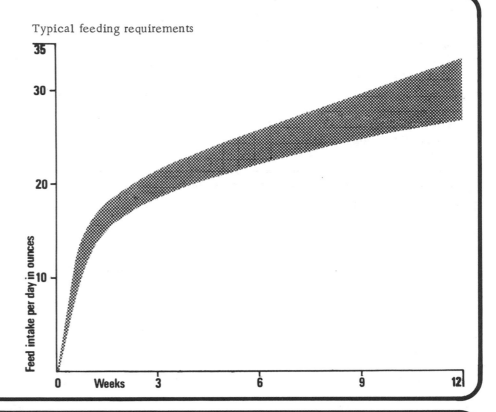

Typical feeding requirements

FORMULA MILKS

A wide range of commercially produced milk formulas is now available for bottle-fed babies. Scientific research into the constituents of human and animal milks has enabled manufacturers to produce milk formulas that are well suited to babies' needs. Baby milk formulas are generally made by modifying cow's milk to make it more like human milk.

One important change is illustrated in the diagram – an adjustment in the relative proportions of carbohydrate, protein, and fat. The actual nature of these nutrients is also modified – as are the proportions of different minerals and vitamins.

A very few babies are unable to take milk of any kind – and special non-milk formulas have been developed to meet their needs.

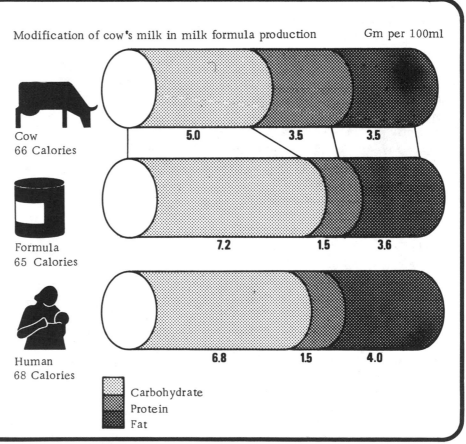

Modification of cow's milk in milk formula production   Gm per 100ml

Cow
66 Calories   5.0   3.5   3.5

Formula
65 Calories   7.2   1.5   3.6

Human
68 Calories   6.8   1.5   4.0

Carbohydrate
Protein
Fat

©DIAGRAM

FEEDING CHOICES

SCHEDULE OR DEMAND

Experts used to recommend a strict four-hour feeding schedule.
At present, however, a more flexible approach is generally in favor - with many babies now being fed on a "self-demand" basis. Most babies, whether fed on schedule or demand, have dropped the late-night sixth feeding by eight weeks.

SCHEDULE
The most common timetable for a schedule-fed baby receiving five feedings a day.

SELF-DEMAND
A typical self-demand pattern for an eight-week-old baby on five feedings a day.

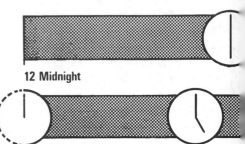

12 Midnight

BREAST OR BOTTLE?

It has been estimated that in recent years up to 75% of babies in the United States have never been breast-fed at all.

There is now evidence that this pattern is changing. More and more women are choosing to breast-feed their babies, and most doctors welcome this as they regard breast-feeding as the safest and most natural method of infant feeding. Mothers who have successfully breast-fed reinforce this view, claiming that breast-feeding is also a unique and rewarding experience.

Some women, however, are uncertain about breast-feeding. Perhaps they have commitments that would make it impossible, or they may find the whole idea unpleasant. And for some women who had planned to breast-feed, illness or problems arising after the birth force them to turn to bottle-feeding.

Current promotion of breast-feeding may make bottle-feeding mothers fear that they are somehow inadequate or uncaring. Such worries are unnecessary - although breast-feeding is usually desirable, the vital physical contact between mother and baby can be as intimate, warm, and loving whether the baby is fed by breast or bottle.

Some of the main advantages and disadvantages of each feeding method are presented here.

QUALITY
The constituents of breast milk are present in near perfect proportions for the baby's optimal growth and development throughout the breast-feeding period.
Colostrum, secreted by the mother in the first few days after the birth, is rich in antibodies and gives invaluable protection against some gastric infections.
Breast-fed babies are less likely to suffer from diaper rash, and less likely to be overweight.

QUALITY
Formula milks have been developed that are very similar in composition to breast milk.
But finding the formula that is most suitable for a particular infant can be a matter of trial and error.
There is no formula equivalent to colostrum.
Bottle-fed babies are more likely to suffer from diaper rash and more likely to be overweight.

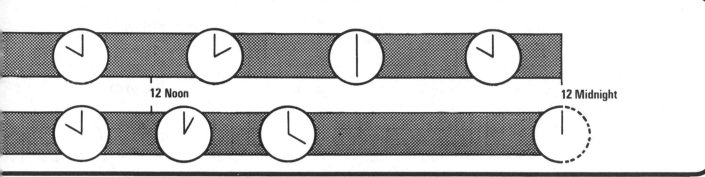

12 Noon   12 Midnight

QUANTITY

Only by test weighing the baby is it possible to tell how much milk has been taken at a feeding. Milk supply generally adjusts itself to cope with the baby's demands. But sometimes the amount of milk produced at a particular feeding does not coincide with the baby's needs at the time.

EXPENSE

Breast milk is virtually free. But the breast-feeding mother may need to spend rather more money than usual on her own diet, which must be rich in protein, calcium, and vitamins.

CONVENIENCE

Breast milk is instantly available - at the correct temperature and usually sterile. But only the mother can feed the baby, and her health and well-being affect the milk supply. Any medication or alcohol taken by the mother may also affect the breast-fed infant.

QUANTITY

It is possible to see at a glance how much milk the baby has taken. The quantity of milk offered can be easily adjusted to meet the baby's needs at any time.

EXPENSE

Initial expense is involved in purchasing the various types of equipment needed for bottle-feeding. Milk-formula must be purchased over a period of several months.

CONVENIENCE

Formulas need mixing, and strict sterilization procedures must be followed. The mother can delegate feedings - which allows her more freedom and gives others, particularly the father, an opportunity to help in the care of the baby.

©DIAGRAM

BREAST-FEEDING 1

SUCCESSFUL FEEDING

For successful breast-feeding, a nursing mother needs a well balanced diet and adequate rest. She must also ensure that she is relaxed and comfortable during the feeding itself, as tension and anxiety affect the supply of milk from the breasts.

DIET

Although most nursing mothers can continue to eat all the foods that they normally enjoy, it is advisable to try and include the following in the daily diet:

a) 1 US quart milk (just over 1.6 Imperial pints: 0.9 liters);

b) meat, poultry, or fish;
c) 1 egg;
d) whole grain bread or cereal;
e) butter or margarine;
f) fruit and vegetables.

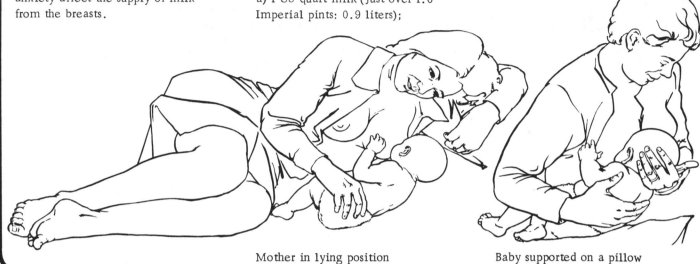

Mother in lying position

Baby supported on a pillow

HOW MILK IS PRODUCED

The breasts prepare for milk production just before childbirth as the estrogen and progesterone output from the ovaries decreases. The reduction of the level of these hormones in the bloodstream affects the hypothalamus which then causes the pituitary to produce prolactin. It is this hormone that sparks off the secretion of colostrum, and later milk, from the breasts. (Milk yield does not start until several days after childbirth.) The flow is stimulated by the baby sucking the nipple. This sends nerve impulses to the hypothalamus which releases oxytocin that travels via nerve fibers to the pituitary. From there, oxytocin flows through the bloodstream causing the alveoli to contract, thus forcing liquid through the ducts to the nipples.
Milk flow usually starts about 30 seconds after nursing begins.

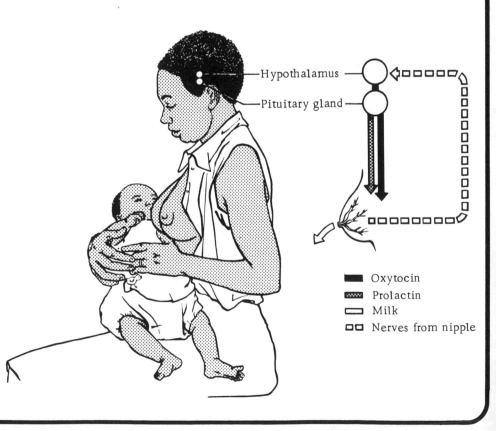

Hypothalamus

Pituitary gland

■ Oxytocin
▨ Prolactin
▢ Milk
▢▢ Nerves from nipple

428

NURSING POSITION

The mother should adopt any comfortable nursing position. Three examples are shown below.

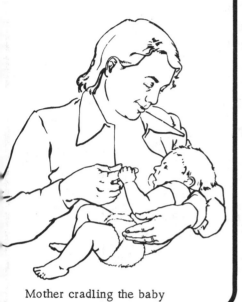

Mother cradling the baby

STARTING FEEDING

As soon as the baby is put to the breast, his rooting reflex will help him find the nipple. The mother should ensure that the baby takes into his mouth the entire areola (the dark area around the nipple), and she may also need to hold the breast away from his nose to help him breathe.

At the end of the feeding, the mother can release the nipple by gently inserting her finger into the corner of the baby's mouth. The first few feedings will be very short - perhaps only two to five minutes - but after the first few days the baby will demand longer feedings at less frequent intervals. The changing composition of breast milk during the first ten days is shown in the diagram. The milk itself is preceded by colostrum - a yellow liquid rich in antibodies.

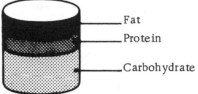

Colostrum (birth to 5 days)
— Fat
— Protein
— Carbohydrate

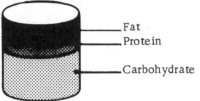

Transitional breast milk (6-10 days)
— Fat
— Protein
— Carbohydrate

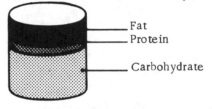

Mature breast milk (after 10 days)
— Fat
— Protein
— Carbohydrate

CARE OF THE BREASTS

Care of the breasts and nipples is extremely important for the nursing mother.

Some doctors advise regular massage of the nipples during the last weeks of pregnancy to prepare for breast-feeding.

The enlarged breasts should be well supported in the later stages of pregnancy and in lactation. Many women find that a special nursing bra is also ideal for the last weeks of pregnancy.

Nursing bras are designed to allow the mother to expose her nipples for nursing without removing the whole garment. Most styles also have an adjustable back fastening. Because the nipples may leak a little milk, some nursing bras have special waterproof cups (alternatively, absorbent cotton pads may be placed in the cups).

Front-fastening nursing bra: this style provides the breasts with good support and unfastens at the front to allow easy access to the breasts for feeding.

Nursing bra with flaps: this style fastens at the back, but the flaps on the cups allow the nipples to be exposed without unfastening the whole of the bra.

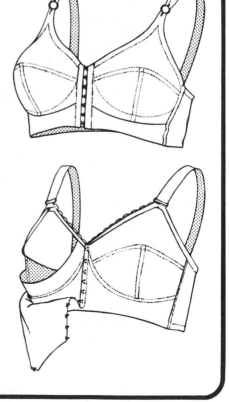

©DIAGRAM

TEST WEIGHING

It is possible to calculate the exact quantity of milk taken by a breast-fed baby by a series of test weighings carried out over a period of 24 hours. It is important to calculate a whole day's intake as the intake at individual feedings varies considerably.

Before receiving the first feeding of the day the baby is weighed in his clothes. Immediately after the feeding he is weighed again in the same clothes without having his diaper changed. The difference between the two weights gives the total milk intake at one feeding. By repeating the procedure at each feeding and totaling the separate milk intakes, the day's total intake can be calculated.

The printed table (right) gives typical results obtained from weighing an infant before and after each feeding. Illustrated below is the calculation of the infant's daily intake based on these results.

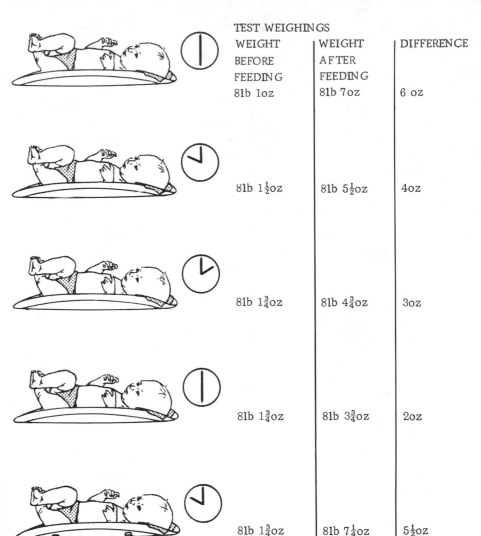

TEST WEIGHINGS		
WEIGHT BEFORE FEEDING	WEIGHT AFTER FEEDING	DIFFERENCE
8lb 1oz	8lb 7oz	6 oz
8lb 1½oz	8lb 5½oz	4oz
8lb 1¾oz	8lb 4¾oz	3oz
8lb 1¾oz	8lb 3¾oz	2oz
8lb 1¾oz	8lb 7¼oz	5½oz

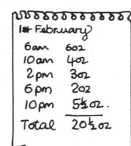

1st February

6am	6oz
10am	4oz
2pm	3oz
6pm	2oz
10pm	5½oz
Total	20½oz

PROVIDING EXTRA MILK

If the supply of breast milk is insufficient, the mother can try to increase it by giving feedings more often, being more relaxed, and increasing her own fluid intake. If these measures are unsuccessful and the mother is determined to continue breast-feeding she will have to increase the baby's milk intake by adding bottle feedings. These may be given in two forms.

COMPLEMENTARY BOTTLES
Feedings given after breast-feeding to "top up" the milk intake are known as complementary feedings. Some mothers find that they need to give their baby a 2-3oz bottle after each breast-feeding. Others find that a complementary bottle is necessary only after the day's scantiest feeding - usually in the afternoon or early evening.

SUPPLEMENTARY BOTTLES
Feedings given instead of a breast-feeding are called supplementary feedings. A supplementary bottle is most commonly given instead of the afternoon feeding. The breasts should then be full enough by early evening to provide the baby with a substantial feeding.

MOTHERS' PROBLEMS

Breast-feeding mothers may have to face one or more of the following problems.

SORE OR CRACKED NIPPLES

These conditions may be caused by not allowing the nipples to dry properly before covering them after a feeding, or by the baby "chewing" on the nipple rather than suckling the entire areola area. The mother is often advised to stop or restrict nursing until the nipples heal, during which time she must express her milk for feeding in a bottle. The doctor may prescribe an ointment to speed recovery. In some cases the use of a nipple shield may prove helpful.

RETRACTED NIPPLES

If the nipples are retracted or inverted the baby is unable to take them into his mouth for satisfactory suckling. Gentle massage of the nipples before a feeding may make them stand out a little more; or a nipple shield may be used during feeding.

ABSCESSES

The presence of an abscess is indicated by an extremely painful red patch on the surface of the breast. The condition is often accompanied by a high temperature. It is sometimes possible to continue breast-feeding but it is essential to seek medical advice.

ENGORGEMENT

This is the medical term used to describe the overfilling of the breasts with milk. It is most likely to occur toward the end of the first week, as the mother's milk supply gets under way.

In areolar engorgement only the sinuses behind the nipple are affected, making the areola too firm for the baby to take into his mouth. Expressing a little milk (see below) before the feeding usually solves this problem.

Total engorgement, affecting the whole breast, is also treated by expressing milk and sometimes by massage and hot and cold compresses.

NIPPLE SHIELDS of different styles are sometimes recommended if the nipples are sore or retracted. The baby sucks on the rubber nipple, which creates a vacuum and draws milk from the breast.

EXPRESSING MILK It may be necessary for a mother to express milk from her breasts - to maintain her milk supply if the baby cannot suckle, or if her breasts become engorged. In manual expression (**a**)

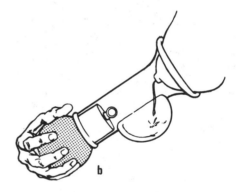

the mother squeezes the areola between her finger and thumb. Milk may also be expressed with a hand pump (**b**) or with a more sophisticated electrically operated breast pump.

GIVING EXTRA NUTRIENTS

Breast or formula milk supplies almost all the nutrients a baby needs during the first few months of life.

Medical opinion, however, now favors the giving of some form of vitamin supplement. The most common of these is a commercial preparation in drop form containing vitamins A, C, and D. The quantity needed by each baby varies according to his age and how he is fed, and because excessive doses of vitamin D can be harmful, it is essential always to follow medical instructions. For bottle-fed babies, the drops are mixed with the formula; for breast-fed babies, they are placed inside the baby's cheek or on his tongue, or mixed with a bottle of cooled, boiled water.

From about four weeks babies can be given diluted, or enriched fruit juices either on a spoon or in a bottle. These are valuable sources of vitamin C and are also thirst-quenching. Fruit juices should not be given just before a feeding, and because of the risk of tooth decay must never be given undiluted on a pacifier.

©DIAGRAM

BOTTLE-FEEDING 1

FEEDING EQUIPMENT

The mother who intends to bottle-feed her baby will need to have all the feeding equipment ready before the birth. Most mothers who breast-feed their babies also keep a small supply of bottles and nipples for the occasional supplementary feeding or fruit drink.

1

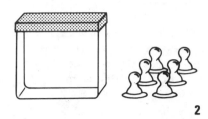

2

WORKING AREA

It is a good idea to keep close together the different items needed for preparing feedings. Most people find it convenient to store equipment and formula in an area of the kitchen where they are near both a sink and a stove, and where there is a flat working surface. Spare formula and items not in regular use are best stored in in a cupboard or on a nearby shelf.

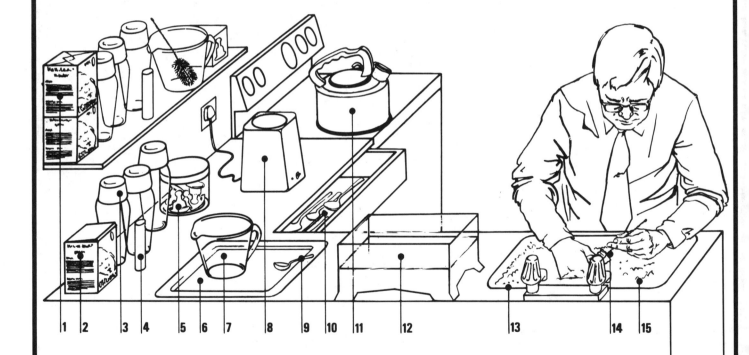

1 2   3 4   5 6 7   8   9 10 11   12   13   14 15

1 Spare equipment:
  Formula
  Bottles
  Sterilizing tablets
  Measuring cup
  Bottle brush
  Nipples
2 Formula
3 Bottles
4 Sterilizing tablets
5 Nipples in a jar

6 Tray
7 Measuring cup
8 Bottle warmer
9 Scoop
10 Cutlery
11 Kettle
12 Sterilizing unit
13 Sink
14 Bottle brush
15 Hot, soapy water

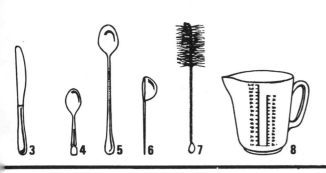

1 6 bottles and caps
2 Jar and 6 nipples
3 Knife
4 Spoon for adding sugar (if used)
5 Long-handled spoon for mixing
6 Measuring scoop for formula
7 Bottle brush
8 Measuring cup
9 Tray for storage

FEEDING BOTTLES

Feeding bottles are available in a variety of styles. They are usually made from heat-resistant glass or plastic. A wide-necked bottle (1) has the advantage of being easy to clean with a bottle brush. But a narrow-necked bottle (2) helps prevent the baby from taking in air with the milk. Small bottles like (3 and 4) are useful for using in addition to a breast-feeding or for giving water or fruit drinks. The straight-sided bottle (5) is fitted for convenience with disposable, ready-sterilized, plastic lining sacs.

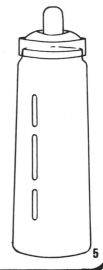

NIPPLES (TEATS)

The size of hole in the nipple is extremely important. Nipples with large holes should not be used because they allow too fast a flow, causing indigestion or even choking. Small holes restrict the milk flow but can be enlarged - either with a hot needle held in a cork (1), or by using a razor blade to make a cross-cut in the end of the nipple (2).

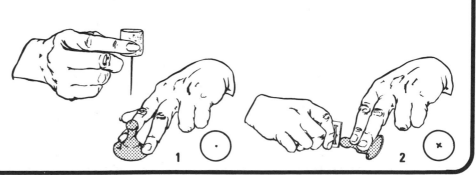

©DIAGRAM

BOTTLE-FEEDING 2

MIXING A FEEDING

There are two basic types of formula milk - powder and liquid. Both are made from modified cow's milk and are reconstituted with water for feeding. Some brands are unsuitable for young babies so it is important to follow the instructions of the nurse or doctor. It is also wise to seek medical advice before changing from one formula to another.

Pour water into sterilized bottle to half the final feeding quantity required.

Scoop powder from package into bottle, leveling each scoop with the back of a knife. Count the scoops carefully. (Sweetening is now added only if specified in manufacturer's instructions.)

METHOD OF PREPARING POWDER FORMULA

Check that the baby has a dry diaper; change if necessary. Wash hands. Boil water in kettle; leave to cool slightly.

METHOD OF PREPARING LIQUID FORMULA

Scrub and scald top of can. Open can and pour prescribed quantity of formula into sterilized bottle.

Add required amount of cooled, boiled water from kettle.

Fix sterilized cap to top of bottle and shake. Remove cap and top up to desired level with cooled, boiled water.

Fix on sterilized cap and shake bottle.

Attach sterile nipple. Check temperature and rate of flow by shaking a few drops onto the inside of the wrist. The milk must not feel hot and should run in a rapid stream of drops.

Feed, and discard any formula left in the bottle.

WARNING
1) Never make a bottle too strong by overloading the scoop or adding extra formula. Over-concentrated mixture can cause stomach upsets or even kidney problems and can result in overweight.
2) Never add extra sweetening.

©DIAGRAM

BOTTLE-FEEDING 3

MIXING IN A MEASURING CUP

Some people prefer to mix the formula in a measuring cup before transferring it to the bottle. This method is particularly useful for formulas to which sweetening must be added, or for formulas that need stirring with a spoon or fork. It is also recommended when preparing several feedings at the same time.

With this method, formula for one or more feedings is measured into a large cup where it is mixed to a paste with a little cooled, boiled water. More cooled, boiled water is then added to bring the feeding up to the right amount. The formula is then poured into one or more bottles for use immediately or after storage in a refrigerator.

1 Formula is mixed in a large measuring cup

2 Bottles are filled with formula for immediate or future use.

FEEDINGS FOR THE DAY

If a refrigerator is available it may be preferred to make up enough feedings for the whole day.

The bottles can be made up one by one or the entire quantity can first be mixed in a large measuring cup.

After making up, the bottles not needed for an immediate feeding should be placed in the refrigerator. Made-up bottles must never be stored for more than 24 hours. When a bottle is required, it is not necessary to reheat it: most babies find cold milk quite acceptable. Many people, however, prefer to warm the milk. This can be done by: standing the bottle in in a pitcher of hot water; putting the bottle in a saucepan of water while the water is brought to the boil; or by means of an electric bottle warmer.

ASEPTIC PREPARATION

When using the aseptic method of food preparation, all equipment is sterilized beforehand and the milk formula is mixed with cooled, boiled water.

1 All equipment - bottles, nipples, spoons, etc - must be sterilized before mixing begins. Many people now do this with a commercially obtainable sterilizing product. Sterilizing liquid or quick-dissolving tablets are added to water in a special container designed to keep the equipment completely immersed. (Note the importance of following the manufacturers' instructions when using these products.) Alternatively, equipment may be sterilized by boiling for 25 minutes in a sterilizer unit or in a covered saucepan.

2 The formula is mixed in the bottles, using cooled, boiled water, and handling the sterilized equipment as little as possible. Alternatively, the formula may first be mixed in a sterilized measuring cup.

3 Any bottle not intended for immediate use should have its nipple inverted in the neck.

4 Caps are tightly screwed on to all bottles prepared for future use.

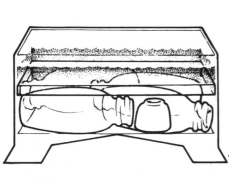

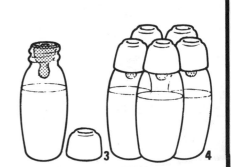

GIVING A BOTTLE FEEDING

It is important that a bottle-fed baby should enjoy the same sort of warm and intimate relationship that a breast-fed baby enjoys with his mother. To achieve this, the baby must be held close and cuddled throughout the feeding. In this way he will come to associate the pleasure of feeding with the people who love and care for him. On some busy occasions it may be tempting to "prop" the bottle on a folded diaper or by means of a bottle-holder. But this is much less satisfactory from the baby's point of view. Very young babies, in particular, must never be left alone with a bottle because of the risk of choking. From the age of about six months a baby may be allowed to hold his own bottle under supervision.

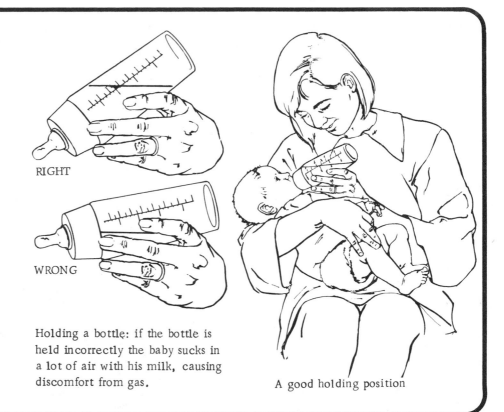

RIGHT

WRONG

Holding a bottle: if the bottle is held incorrectly the baby sucks in a lot of air with his milk, causing discomfort from gas.

A good holding position

TERMINAL STERILIZATION

In the terminal method of sterilization the prepared milk is sterilized along with the bottles and nipples.

1 Milk formula is measured into the bottles, which have previously been washed but not sterilized. (Hot, soapy water is used for washing the bottles and a bottle brush is recommended to get them really clean. Care should always be taken to rinse the bottles thoroughly with clean water.)

2 Cold water from the tap is used for mixing the formula.
3 A clean nipple is inverted in the neck of the bottle. (Rubbing nipples with a little salt before washing will remove the film of milk that collects inside them.)
4 The bottle caps are put on - loosely so that steam can pass under them - and the bottles placed upright in 3in of water in a sterilizer unit or saucepan.

5 The sterilizer or saucepan is covered, and the water brought to the boil. After 25 minutes' boiling the sterilizer should be left to cool with the lid on for about 2 hours. (Gradual cooling prevents skin from forming on the milk.) The bottles are then taken out of the sterilizer, and those not intended for immediate use should have their caps screwed on tightly before storage.

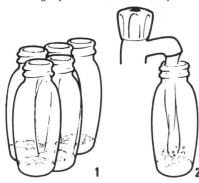
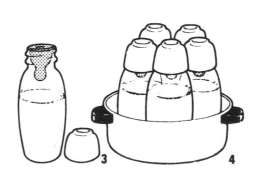

1   2   3   4   5

©DIAGRAM

PROBLEMS WITH FEEDING

BURPING (WINDING)

During a feeding a baby swallows air which forms a bubble in his stomach. If discomfort is to be prevented it is important to get up this air bubble at least once during a feeding and again at the end. Three ways of burping a baby are illustrated here.

1 The baby is held against his mother's shoulder while she gently rubs his back.

2 The baby is held upright in his mother's lap while she gently rubs his back.

3 The baby lies across his mother's lap while she rubs his back.

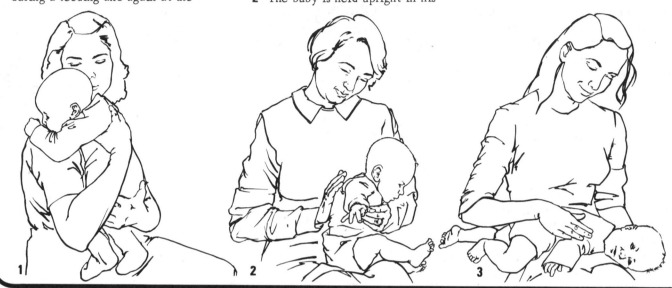

PREMATURE BABIES

Feeding a premature baby during the first few days, or even weeks, of life, demands the expert attention of hospital staff.

In a premature infant the control of vital functions such as breathing, sucking, and swallowing is restricted and, in many cases, the digestive system is still immature and unable to cope easily with the normal processes of digestion.

One of the major hazards of feeding premature infants is that improperly digested food may be regurgitated and breathed into the lungs, causing choking.

Most premature babies are not fed until 48 hours after birth. They are fed very small quantities of milk, usually every three hours, from a dropper, or by means of a tube passed down the throat into the stomach. The feeding consists of formula milk or breast milk expressed by the mother. When the baby's weight has increased and his sucking ability is more fully developed, he can be fed normally from breast or bottle. The baby will still require very small quantities at frequent intervals.

The premature baby has few nutritional reserves and is often so short of iron that slight anemia results. This can be corrected by a prescription of medicinal iron from about the fourth week. Vitamin supplements are usually given from the second week.

MILK ALLERGIC BABIES

A very few babies who fail to thrive are found to be allergic to milk. This allergy is identified by a series of hospital tests, after which the doctor will prescribe the use of a special non-milk formula.

These specially developed milk-free formulas typically contain such ingredients as soybeans, corn and coconut oils, sucrose, and corn sugar. Other formulas have been developed using modified meat protein.

A baby fed on a special non-milk formula can be expected to have normal rates of growth.

COMMON FEEDING PROBLEMS

Some babies have feeding problems that can cause considerable distress and anxiety to those around them.

Very young babies sometimes feed for a very short time and then fall asleep, only to wake a few minutes later for more milk. This is frustrating and exhausting for everyone but it is almost certainly only a passing phase. The nervous and digestive systems of some babies do not at first work properly together, so that the baby does not make the connection between sucking and the relief of hunger. He will soon stay awake long enough to satisfy his hunger. Meanwhile the mother could try changing the baby's feeding position, or could move him frequently from breast to breast. Feeding problems may also arise because babies are sensitive to the moods of those who feed them. If the person giving the feeding is tired, rushed, or anxious, the baby may sense this and become restless and difficult to feed. A calm and relaxed atmosphere at feeding times should help reduce problems.

Older babies may feed badly when they are teething or if they have an ear infection that makes jaw movements painful. If discomfort temporarily prevents a baby from sucking, expressed breast milk or formula milk may be taken from a cup or spoon.

BREAST-FED BABIES

In the case of a breast-fed baby, problems affecting the mother, such as retracted nipples or engorgement, can mean that the baby has considerable difficulty in taking the areola into his mouth to begin feeding. Occasionally a baby refuses to nurse during the mother's menstrual period. If this happens, the mother should express her milk and substitute a formula bottle-feeding.

BOTTLE-FED BABIES

A bottle-fed baby may lose interest in his feeding if the milk does not flow quickly enough through the nipple. This can be remedied by enlarging the nipple holes as described on p.433.

HICCUPS AND SICKNESS

Many young babies suffer from hiccups after feedings, but they are rarely upset by them. Some mothers find that they can stop the hiccups by burping the baby or by giving a drink of cooled, boiled water.

Some gentle regurgitation or "spitting" of milk is quite normal after a feeding, particularly during the early months. As long as the baby's weight gain is satisfactory there is rarely anything to worry about.

True vomiting - the regurgitation and ejection of the stomach's contents with some force - is much less common. It may be a symptom of infection or obstruction, and cases of persistent vomiting always need medical attention.

COLIC

Colic is a severe abdominal pain that affects many babies between the ages of two weeks and three months. Attacks usually occur daily, often in the evening, and can last up to four hours.

A baby with colic typically refuses to settle down after the late afternoon or early evening feeding. He begins to scream - often adopting the position illustrated. During an attack the baby can be comforted temporarily if he is cuddled or wrapped up tightly but his distress returns as soon as he is left again.

The causes of colic are uncertain and medical opinion is still in disagreement. Some doctors, however, suggest some precautions that may possibly prevent an attack. The baby should not be allowed to feed too quickly; he should be thoroughly burped after a feeding; and he should be kept warm.

The illustration shows a baby with characteristic signs of colic:

a) screaming;
b) legs drawn up to the stomach;
c) tightly clenched fists.

©DIAGRAM

WEANING

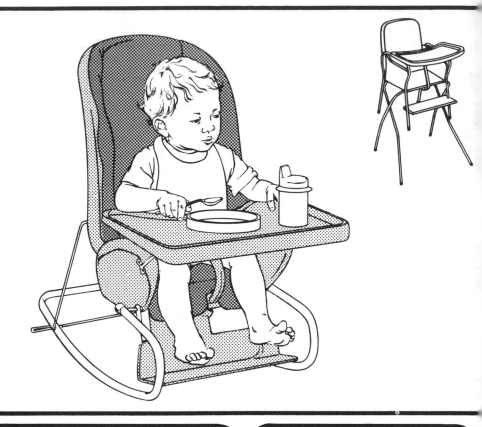

EQUIPMENT FOR WEANING
From the time a baby first begins to take solid food a considerable variety of equipment will be needed at mealtimes.

Learning to eat is always a messy process - particularly when a baby starts to feed himself. For this reason, it is advisable to have terry cloth or plastic bibs to protect clothing, and easy-wipe surfaces or mats to protect the furniture. Some feedings may be given on the parent's lap, but the baby will need a rigid seat with a table when he begins to feed himself. Many baby chairs - like the one shown, right - can be used as a low chair or fitted on a stand for conversion to a high chair.

Cups and dishes should be unbreakable and designed for easy use. Cutlery should be small enough for the baby to handle.

WEANING
Weaning is the term used for the gradual substitution of solid foods for milk in an infant's diet. Exact procedures and schedules for weaning vary considerably - and the advice of a doctor or nurse is the best guide for the optimal growth and development of a particular baby. Typically, weaning begins at three to four months, and by 12 months the infant should be taking a wide variety of solid foods.

PHASING OUT MILK FEEDS
During the first stages of weaning the infant continues to receive five breast or bottle feedings daily, but the amount of milk is reduced at any feeding at which solid food is also given. From about six months an infant may safely be given undiluted cow's milk in place of breast or formula milk, and milk and milk products remain an important aspect of diet throughout childhood.

INTRODUCING SOLIDS
Traditionally the first solid food consists of cereal or a zwieback mixed to a creamy consistency with milk. Some people, however, prefer to start weaning with pureed apple or a small amount of canned baby food.

Whichever solid food is chosen, it is first given at one feeding daily and is given in addition to breast or formula milk. After a week or so solids are given at two of the five milk feedings, and after a few more weeks the infant should be receiving a selection of different solid foods at three of his feedings. Gradually the infant's food pattern becomes more like that of the rest of the family - with three meals a day and drinks on waking and at bedtime.

FEEDING PROCEDURE
In the early stages of weaning, small tastes of solid food should be given on a teaspoon. The spoon is placed in the baby's mouth so that he can eat the food and swallow it. He is likely to fuss at the first few feedings and it is often worth giving at least part of the milk feeding before offering any solids.

A baby should also be encouraged to drink from a cup. An eggcup or a double-handled cup with a spout is suitable for this.

After a few months, many babies begin to pick up food in their fingers and show signs of wanting to feed themselves. This is the start of a messy and lengthy process but it is vital to allow the baby to persevere. The feeding can be speeded up if the parent slips an occasional spoonful of food into the baby's mouth when he becomes tired or fretful.

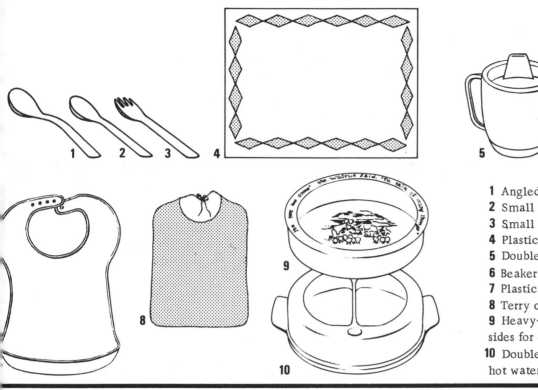

1 Angled spoon
2 Small spoon
3 Small fork
4 Plastic mat
5 Double-handled cup with spout
6 Beaker with spout
7 Plastic bib with trough
8 Terry cloth bib
9 Heavy-based dish with straight sides for easy use
10 Double dish with hollow base for hot water to keep food warm

HOME-PREPARED FOODS

Some people like to prepare their own baby foods, and with a little time and trouble many items from the family diet can be adapted to a baby's needs.

Food for the young baby must be in puree form, so a sieve and food mill or blender are essential. As the baby becomes accustomed to solids, mashing, mincing, or chopping is usually sufficient.

Too much salt and sugar is harmful to babies, and seasoning and sweetening of baby foods should be kept to a minimum.

Vegetables - particularly carrots, tomatoes, peas, string beans, and potatoes - are good weaning foods. If they are offered one at a time the baby's likes and dislikes can be easily identified.

Egg yolk is a valuable source of iron, and can first be introduced with a little cereal. (Whole egg may cause an allergic reaction and should be avoided during the early stages of weaning.)

Meat broth can be used as an early weaning food and the tasty juices from roast meat can be mixed with vegetables. Beef, lamb, chicken, and liver - pureed or minced - are all suitable for the slightly older baby.

Non-oily fish can be served if all the bones are removed.

Cheese is another nourishing weaning food. Grated cheese can be mixed with cereal as an alternative to sweetening. Later, many babies like to hold and eat a piece of cheese.

Firm foods should be given to babies to encourage chewing and teething. Suitable foods include small pieces of toast, crusts of bread, raw carrot, and apple.

Fruit is popular with many babies - particularly apples, bananas, and apricot puree.

COMMERCIAL BABY FOODS

It is often convenient to use commercial baby foods, and manufacturers have undertaken a great deal of research to produce foods to meet the nutritional needs of infants.

There are two basic types - ready to serve in cans or jars, and powdered foods for mixing with milk. Ranges include foods suitable for the early stages of weaning - broths, and purees of fruits, vegetables, and meats - and a great variety of foods suitable for the slightly older child.

Particularly in the early stages of weaning, powdered foods may prove most economical as tiny quantities can be prepared without affecting the quality of the remainder of the package.

Food from a can or jar, once opened, can be stored for re-use provided that certain precautions are taken.

© DIAGRAM

WEANING AND AFTER

USING COMMERCIAL FOODS

Small quantities of baby food can be heated in a cup over hot water. The food can then be fed directly from the cup or may be transferred to a dish. If the baby has not been fed from the jar or can, and if it has not been heated up, any remaining food can be stored in a refrigerator for up to two days.

Sufficient food for one meal is transferred to a cup **1**, heated by placing the cup in hot water **2**, and then transferred to a dish for serving **3**.

BABY FOODS COMPARED

Details of the contents of three popular varieties of canned baby food are given here as an indication of the wide range of commercially produced baby foods.

Protein
Carbohydrate
Fat

Analysis per 100gm

BEEF AND CARROT CASSEROLE

Contents: beef, carrots, potatoes, tomatoes, flour, soya flour, cornflour, herbs, hydrolyzed vegetable protein, iron sulfate.

Calories: 80

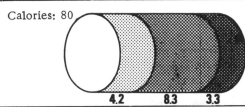

4.2   8.3   3.3

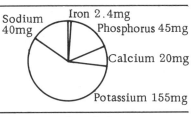

Sodium 40mg
Iron 2.4mg
Phosphorus 45mg
Calcium 20mg
Potassium 155mg

EGG AND CHEESE

Contents: eggs, milk, cheese, semolina, modified cornstarch, vegetable oil, spice.

Calories: 88

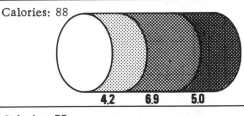

4.2   6.9   5.0

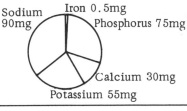

Sodium 90mg
Iron 0.5mg
Phosphorus 75mg
Calcium 30mg
Potassium 55mg

BANANA DESSERT

Contents: banana, sugar, modified cornflour, orange juice, lemon juice, vitamin C.

Calories: 77

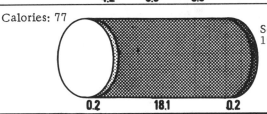

0.2   18.1   0.2

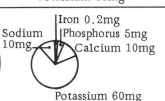

Sodium 10mg
Iron 0.2mg
Phosphorus 5mg
Calcium 10mg
Potassium 60mg

NUTRIENTS

There are five basic types of nutrient found in foods - proteins, carbohydrates, fats, minerals, and vitamins. All play a part in a healthy diet and are especially important for a growing child.

PROTEINS are complex substances vital for the growth and repair of the body.
The richest sources of protein are animal products such as meat, fish, eggs, milk, and cheese. Protein is also found in some vegetables such as peas and beans, and in grain and grain products like wholemeal bread.

CARBOHYDRATES are the main source of energy for immediate use, and play an important part in the functioning of vital organs. Bread, rice, potatoes, and almost all sweet foods are rich in carbohydrates. An excess of these foods in the diet - particularly of sweet foods that contain few other nutrients - can lead to an undesirable weight gain.

EATING SENSIBLY

Most parents appreciate the importance of establishing sensible eating habits in their children. This is not always easy, however, as a variety of feeding difficulties can cause parents considerable anxiety.

Among the most common problems are: refusal to eat particular foods; refusal to eat properly at family mealtimes; and a preference for sweet foods and snacks.

Refusing to eat is rarely serious, and parental anxiety will only communicate itself to the child and make the problem worse. Food should be presented as attractively as possible, and after a certain time removed without fuss if the child is not eating.

Cookies, cake, and candies are best avoided except at the end of a meal. Recommended snacks include fresh fruit, fruit juices, and cheese.

DAILY FOOD GUIDE

A properly balanced diet is essential for the health and well-being of even the youngest child. Assessing whether a child's nutritional needs are being met, however, can be confusing and difficult. A simple means of checking is to use the daily food guide described here.

Devised by a dietitian, the guide divides foods into four groups:
a) milk, cheese, and yogurt;
b) fruit and vegetables;
c) meat, fish, and eggs;
d) cereals and bread.

It is recommended that two helpings from each group should be served each day if possible, though an overweight child may need only one serving from the last group.

Children suffering from certain illnesses, such as diabetes, need diets planned under medical supervision.

MILK GROUP
Contains protein, carbohydrate, fat, calcium, vitamin A

VEGETABLE AND FRUIT GROUP
Contains carbohydrate, minerals, vitamins C and A

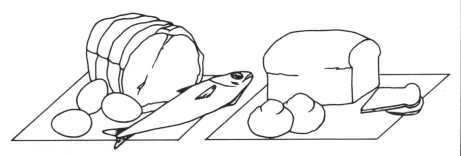

MEAT, FISH, AND EGG GROUP
Contains protein, fat, iron, vitamin A, vitamin B, vitamin C

CEREALS AND BREAD GROUP
Contains protein, carbohydrate, calcium, iron, vitamin B

FATS are a concentrated source of energy. They are found in two main forms.

Saturated fats are normally hard at room temperature. They occur in meat, and in dairy products such as butter and cheese. Polyunsaturated fats are usually consumed in the form of liquid vegetable oils, such as soybean or corn oil, or some margarines.

MINERALS play a vital role in the diet. Calcium, found in milk and cheese, is important for healthy bones and teeth; iron for making red blood cells is present in wheat germ and liver. Other valuable minerals - potassium, phosphorus, sodium, and iodine - are found in a variety of natural foods such as meat, milk, fish, eggs, fruits, vegetables, and whole grains.

VITAMINS are substances that the body cannot manufacture for itself but which are vital to health. Vitamin A is found in carrots, egg yolk, and butter, and vitamin C in fruit and vegetables. Vitamins of the B complex are present in wheat germ and yeast, while the major source of natural vitamin D is sunshine.

© DIAGRAM

CALORIES AND WEIGHT

CALORIES

The energy value of food and drink is measured in units called Calories.

People use up Calories every minute of the day - to maintain body functions and as fuel for physical exercise. Children have comparatively high Calorie requirements because energy is also needed for growth.

An individual's Calorie needs vary according to age, sex, size, physical activity, and climate.

If a person takes in as food more Calories than are needed, the surplus is converted into fat to be stored for future use.

If Calorie intake is below requirements, fat stores in the body are converted into energy and body weight decreases.

DAILY CALORIE NEEDS

Daily Calorie needs change with age. Exact estimates of needs vary, but different sets of statistics show the same general pattern. (Our diagram is based on the 1974 statistics of the US Food and Nutrition Board.) For the first 10 years or so, boys and girls of the same age have similar needs. Later, differences reflect changes at puberty, comparative growth rates and body size, and typical exercise levels.

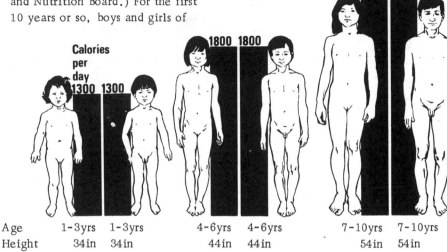

Age	1-3yrs	1-3yrs	4-6yrs	4-6yrs	7-10yrs	7-10yrs
Height	34in	34in	44in	44in	54in	54in
Weight	28lb	28lb	44lb	44lb	66lb	66lb

SNACKS AND MEALS

Eating too many snacks is a common bad habit among children and adults alike. As this diagram shows, the Calorie content of many popular snacks is comparatively high - making it all too easy to add a few hundred surplus Calories to the daily intake. In addition, snacks may also be responsible for spoiling the appetite for other foods needed to give a balanced diet.

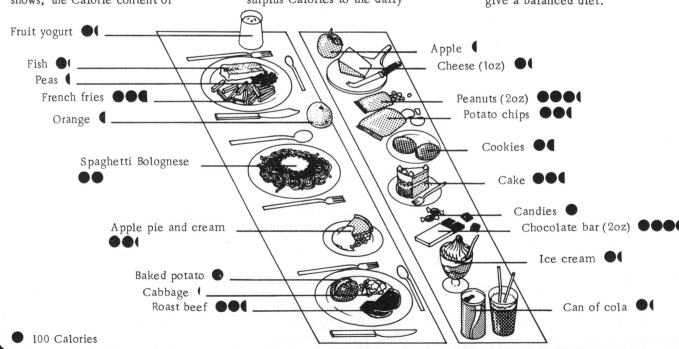

● 100 Calories

444

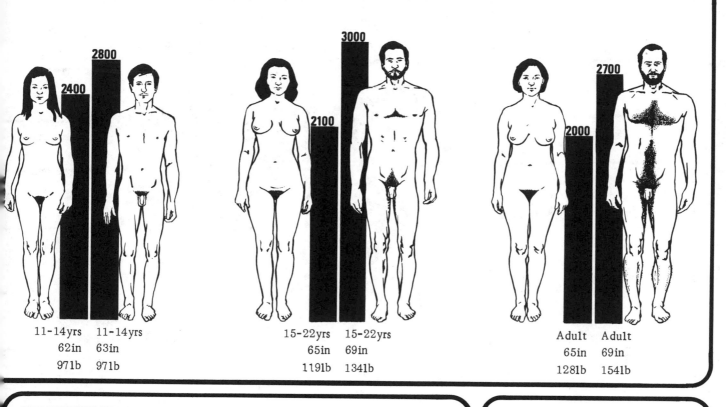

11-14yrs	11-14yrs		15-22yrs	15-22yrs		Adult	Adult
62in	63in		65in	69in		65in	69in
97lb	97lb		119lb	134lb		128lb	154lb

THE OVERWEIGHT CHILD

In all but a very few cases, overweight in infants and children is caused by overeating. All too often this is the parents' responsibility - bad eating habits are established at an early age, and continue through childhood and adolescence into adult life. Too many parents, either consciously or unconsciously, regard feeding their offspring as a means of expressing affection.

OVERWEIGHT INFANTS

Researchers have found that there is a marked tendency for overweight infants to become overweight children and adults. For this reason it is especially important to pay attention to an infant's diet and to stop him from becoming overweight. Common errors in feeding include: making formula feeds too concentrated or too sweet; introducing cereal into an infant's diet unnecessarily early; offering too many sweet foods so that the infant fails to acquire a taste for less fattening savories.

OVERWEIGHT CHILDREN

Excessive weight gain is commonly caused by too many between-meal snacks. Once a child has acquired the habit of eating fattening snacks it is hard to break him of it - much fairer to limit his intake from the start and try to be firm about when such "treats" are or are not allowed. Overweight in children is often made worse by lack of exercise. In many cases the child falls into a vicious circle in which he is too embarrassed to take part in sport or active play - and then makes matters worse by overeating for consolation.

HELPING A CHILD REDUCE

A plump child will very often slim down after puberty without special effort. An obviously overweight child is, however, likely to be healthier and happier if he is helped to lose weight. If the child is very overweight, it is recommended that medical advice be sought before embarking on any drastic reducing program. The parents of an overweight child will probably find it useful:
a) to plan menus so that all the family can eat the same thing - for a time excluding fattening foods from everybody's diet;
b) to avoid having fattening snacks in the house - using fresh fruit as a healthy alternative;
c) to check that the child is not overeating because he is unhappy;
d) to gain the child's cooperation so that he does not eat fattening foods away from home.

©DIAGRAM

14 Child, home, and community

Early experiences with parents and siblings help a child's emotional development and prepare him for life within a larger community.

Left: Fun and games in a popular cartoon strip (Archive K)

CHILD IN THE FAMILY

MEETING A CHILD'S NEEDS

From the moment of his birth a child has many needs that must be met by adults. Some of these needs, such as the provision of food and shelter, are obvious to everyone. Others, notably those involving a child's emotional development, are perhaps harder to recognize and therefore easier to neglect.

Recognizing and meeting their children's various physical and emotional needs is a major responsibility of parenthood. In doing so, the parent helps to prepare his child for physical and psychological independence in later years.

A secure and loving relationship is a basic requirement for the well-being of any child. A child who is deprived of loving contact in the early years is more likely to experience difficulties in forming close personal relationships in maturity.

Parents can show love for their children in many ways - by close physical contact in the form of a hug or a cuddle, or by giving them plenty of attention and showing real interest in their activities and childish chatter. Stimulation is a further need in childhood that may sometimes be ignored. A parent can do a great deal, quite simply, to encourage a young child's developmental progress. Providing interesting toys, reading to him, encouraging him to play with friends, taking him on outings, and, simplest of all, talking to him, all help to enrich a child's pre-school years and prepare him for the future.

THE FAMILY CONTEXT

Many of today's children grow up within a small two-generation family unit consisting of father, mother, and children. This "nuclear" family structure - generally resulting from increased population mobility in search of work or a new life - has in many instances replaced the "extended" family structure more typical of earlier generations.

In an extended family, a child is surrounded not only by his immediate family but also by aunts, uncles, cousins, and grandparents, most of whom live nearby. This has obvious advantages for a child. Against a general background of care and affection, he has a first-hand opportunity to become acquainted with members of different generations. From the parents'

PARENTS AND CHILDREN

1 A close and loving relationship with his parents is important for the healthy emotional development of a child, and gives him a firm basis on which to build personal relationships in the future.

2 Parents can help the development of a pre-school child by providing stimulation in the form of toys and playthings that are appropriate to the child's particular age and ability.

3 Reading stories to a child and sharing a picture book with him can usefully begin before he is 12 months old. Language development is encouraged, and stimulus given to his imagination.

viewpoint, an extended family can mean a great deal of on-the-spot support, both emotional and material, and advice - both welcome and unwelcome.

By contrast, the nuclear family requires a greater degree of self-reliance. In many cases this has been gained by a new sharing of responsibilities - with fathers participating more fully than ever before in all aspects of domestic and family affairs.

Support for the nuclear family can also come from the community as a whole. Friends and neighbors can fulfill many of the roles traditionally played by other family members. In consequence, it is essential that the value of a flourishing community is not lost sight of in these present days of rapid social change.

ROLE REINFORCEMENT

In recent years, there has been a widespread questioning of traditional male and female roles. Women are no longer viewed necessarily as child rearers, nor men as breadwinners, and this has also led to changing attitudes to parental roles in the family.

As a result, a child's imitative play is no longer clearly "masculine" or "feminine" in the traditional sense. Many children still prefer the activities and toys traditionally associated with their sex, but others prefer to spend their time playing in a style once thought more appropriate to members of the opposite sex.

4 Drawing is an activity that should, if possible, be encouraged in a pre-schooler. It familiarizes him with basic materials, helps fine muscle control, and is an excellent means of self-expression.

5 A young child needs plenty of opportunity to broaden his horizons. Even the simplest outing - perhaps to a nearby park - provides him with a wealth of new and stimulating experiences.

6 A few minutes set aside each day that the parents regularly devote to a child can be extremely valuable, allowing him to talk through any anxieties and assuring him of his parents' love.

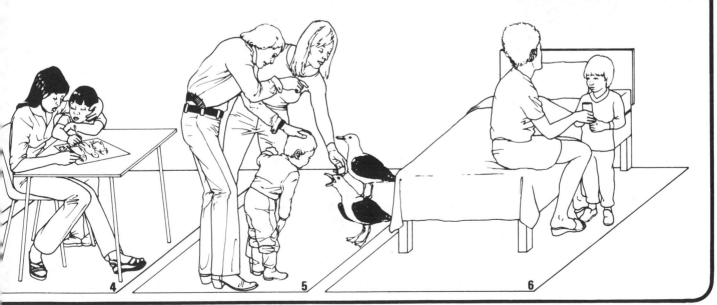

©DIAGRAM

ONE CHILD OR MORE

THE FIRST CHILD

Even if he is later joined by brothers and sisters, every first child begins life as an only child. This means that for some time at least he enjoys a great deal of his parents undivided attention. His activities and reactions take on a concentrated significance, and he may be the sole object of their anxieties, hopes, and expectations. If the first child remains an only child, wise parents will take care not to develop an overanxious or overprotective attitude toward him. It is important that an only child should not spend all his time with adults. Time spent in the company of other children is his best way of learning about human relationships at his own level.

ADDING TO THE FAMILY

Parents can do a great deal to minimize a child's possible jealousy and displeasure when a new baby joins the family. Even a young child should be told about the new baby in advance - but not so soon that he becomes bored with waiting. Feeling his mother's abdomen may help make the situation more real.
Any important event or necessary change in the child's life - like starting nursery school or moving from a crib to a bed - should not be timed to coincide with the baby's birth as this will increase his sense of insecurity.
Before introducing the new baby the mother should make a point of first devoting a little time to the older child - perhaps giving him a reassuring cuddle and a small gift, and listening to his news.

BROTHERS AND SISTERS

It is unrealistic of parents always to expect their children to live side by side in perfect harmony. While still deriving considerable pleasure from being part of a family, each child is an individual - and will sometimes behave like one at his siblings' expense. It is, however, extremely rare for jealousy or resentment between brothers and sisters to grow into the kind of bitterness that damages family relationships.

The age gap between children is a significant factor in their relationship. A gap of just a year or two may mean that they will be good play companions in middle childhood, while a longer gap sometimes means a relationship relatively free of jealousy.
Twins often have an extremely close relationship with each other. They are, however, still individuals, and parents should avoid treating them as a unit.

Aspects of sibling behavior:
a Playing together
b Playing individually
c Interrupting the other's play
d Open confrontation

a   b   c   d

LIFE AND DEATH

SEXUAL CURIOSITY

A typical child's interest in sexual matters fluctuates according to his age.

Around three, most children begin to show an interest in the genital differences between the sexes. By four, a child may respond to stress by grasping the genitals and urgently needing to urinate. A year later, at five, he is typically more modest.

Around the age of six, he starts to show a marked awareness of the physical differences between the sexes, and may indulge in mild exhibitionism or sex play. Sexual curiosity is less obvious in most seven-year-olds, but around the age of eight an interest in "peeping," sharing smutty jokes, and discussing sex information with friends may develop.

By nine, many children seek out pictures on sexual topics in books, and indulge in some childish sex swearing.

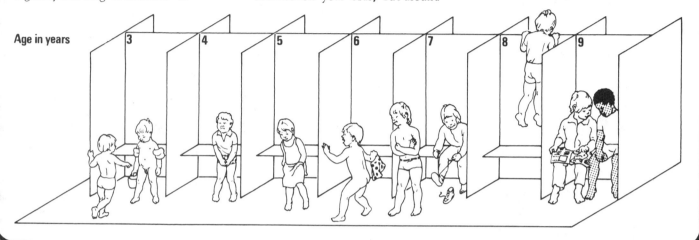

Age in years 3 4 5 6 7 8 9

THE FACTS OF LIFE

A child's questions about sex are just one aspect of his need to learn about and come to terms with the world around him.

If possible, parents should answer these questions simply and truthfully as they arise, just as they would answer a question about any other topic.

The temptation to give too much information should be avoided; if a child wants to know more he will generally ask. It is also unwise to postpone answering a question out of embarrassment, as this may make a child think that his natural interest is really an unhealthy curiosity in a taboo subject.

The earliest questions generally begin at the age of about three or four and usually concern babies - where they come from, how they get out, and, later, how they got there. Replies should be simple and direct, and at a level the child can understand.

In later childhood, menstruation and wet dreams should be explained to both boys and girls in the context of growing up and becoming able to produce babies.

For a parent who is uncertain of how to cope with a child's questions, a great many books are available to help him. In addition, sex education is now included in the curriculum of most schools. Sex education, however, is not simply a matter of answering a child's questions. The relationship between the parents, and the family approach to topics such as nudity in the home are also influential in the formation of a child's attitudes to sex.

TALKING ABOUT DEATH

A child of any age should be encouraged to ask questions and talk openly about death, even if his attitudes seem immature or illogical. It is generally wise to tell the truth about a death as simply as possible. An explanation that grandfather has "gone to sleep" or "gone away" for ever may make the child fearful of sleeping or of journeys, in case he too may die.

A young child often seems to accept death very calmly, and may be upset more by the grief of those around him.

An older child may ask more thoughtful questions about his own or his parents' deaths - will dying hurt, when will it happen, and why? In these cases, the best reassurance an adult can give is to show that he himself is not overanxious or preoccupied with thoughts of death.

©DIAGRAM

SAFETY OUTSIDE THE HOME

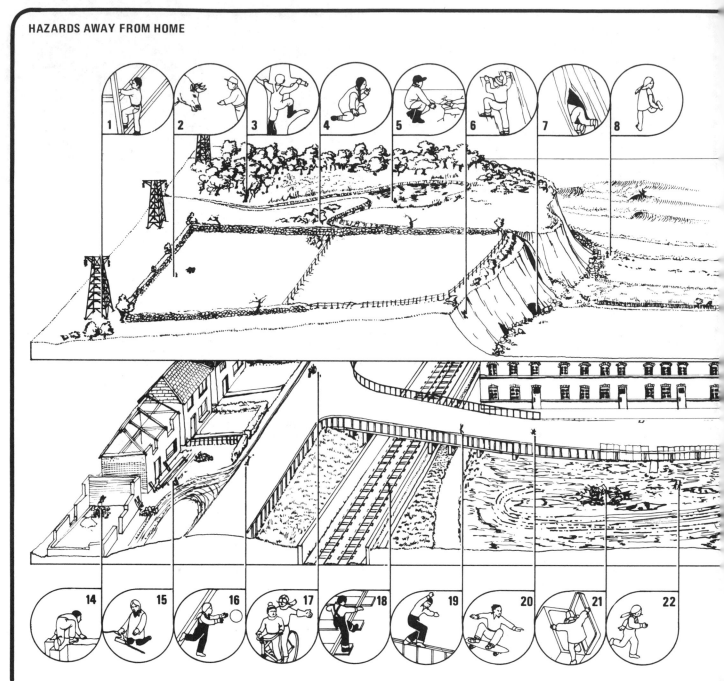

KEEPING SAFE

As soon as children are old enough to go out alone, they are faced with an enormous number of potential hazards. Parents now have the problem of encouraging their children's independence and self-reliance, while at the same time making them aware of how to cope with the many possible dangers they may meet.

WATER SAFETY

Water can be a major hazard. Ideally all children should be taught to swim as soon as possible, and should then be strongly discouraged from swimming where it is dangerous. Life jackets are a wise precaution for young children and nonswimmers playing near water, and are a must for water activities like sailing.

SAFETY WHEN CLIMBING

Most children love climbing and it is sensible that they should be taught to climb as safely as possible - always keeping at least three limbs "mountaineer-style" in contact with whatever is being climbed. At the same time they should be actively discouraged from climbing rotten trees or dangerous cliff faces, and from

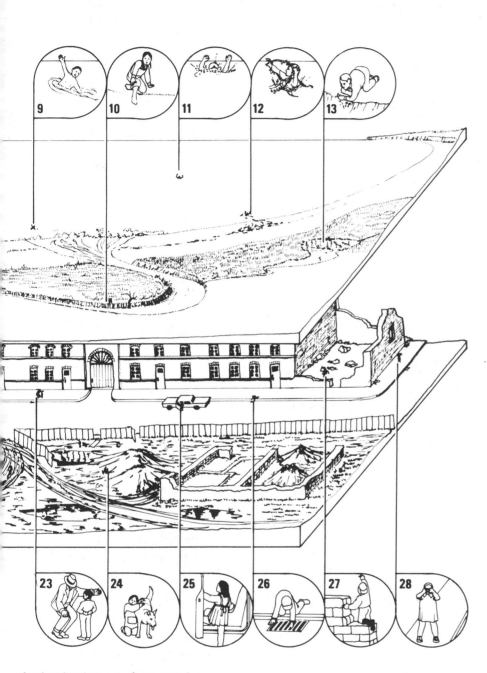

COMMON OUTDOOR HAZARDS

1 Climbing electric pylons
2 Approaching dangerous animals
3 Climbing unsafe trees
4 Eating unidentified berries or fruits
5 Sliding or playing on thin or untested ice
6 Climbing cliffs, rock faces, or quarries
7 Exploring caves or tunnels
8 Stepping barefoot on sharp stones or broken glass
9 Dangerous currents
10 Paddling in a fast-flowing or dirty river
11 Swimming too far from land
12 Becoming entangled in seaweed
13 Leaning out too far over the edge of a pond, lake, or river
14 Climbing high walls
15 Playing with dangerous tools and implements
16 Running out into the road while playing
17 Two riding on one bicycle
18 Walking on railroad tracks
19 Balancing on dangerous railings
20 Playing in the road on a skate board or roller skates
21 Playing with discarded items such as an ice box
22 Running with a lollipop-stick in the mouth
23 Talking to strangers and accepting gifts from them.
24 Stroking unknown animals, especially dogs
25 Accepting rides from strangers
26 Playing in dirty gutters
27 Playing on demolition sites
28 Going too far from home

playing in places such as quarries, building sites, and demolition areas.

ROAD SAFETY

Teaching children the rules of road safety is an obvious and important precautionary measure that parents should take. It is also vital to stress the dangers of accepting car rides or going for walks with strangers.

©DIAGRAM

GOING TO SCHOOL

STARTING SCHOOL

Some children look forward eagerly to starting school and enjoy it from the very first day. Many others, however, find the experience overwhelming, frightening, or simply disappointing, and may take a little longer to settle down.

During their first weeks at school it is quite common for children to show signs of strain. They may be generally irritable and ready to cry, have nightmares, begin nailbiting, or return to thumb sucking or bedwetting.

This is normally a short phase; if it persists beyond the early weeks the parents should see the child's teacher.

In a very few cases the parents' reluctance to let go of a child may make it more difficult for him to adjust to his new life.

PREPARING FOR SCHOOL

Beginning school is one of the most important events in a child's life. Inevitably it entails significant changes in daily routine, and there are several ways in which parents can help a child with this big step.

Preparation begins ideally at birth as a child from a secure and loving family background generally finds it easier to adapt to school life.

The pre-school child should be encouraged to take an interest in a wide variety of topics. Parents can help by talking to him, reading him stories, and by taking time to answer his many questions with care.

In addition, he should be made familiar with the basic "tools" of the educational trade - books, pencils, paints, and paper.

Playing with other children outside his own home helps accustom a child to being separated from his parents, and visiting the school before he becomes a full-time pupil is invaluable if it can be arranged. Otherwise, taking him past the gate or the school yard may help alleviate some anxieties.

Simple practical preparation is also extremely helpful. The child should be able to dress and undress himself and take care of and recognize his own possessions - labeling them with his name or a patch of color may help.

He should be able to listen to and carry out simple instructions, and should be familiar with the basics of road safety.

STRUCTURE, SUBJECTS, AIDS

Since most parents were at school, significant changes have taken place in the internal structure of schools, in the subjects taught, and in the methods and aids used.

STRUCTURE

The traditional single-teacher system is still most common for pupils in their first years at school. This system - in which one teacher is responsible for all the subjects taught to a class - gives young children an opportunity to develop a deep relationship with an adult outside the family.

A subject-teaching system - in which each subject is taught by a different teacher - allows older children to benefit from specialist teaching in different fields.

A more recent development is team teaching, in which several teachers share responsibility for a group of children. Within this system a child with a learning difficulty can receive individual attention.

"Streaming," or "tracking," is commonly used as a means of providing a style and level of teaching appropriate to the needs of different pupils. Some schools divide children into general ability groups, in which they are taught for all subjects. Others favor a system in which children are divided into groups, or "sets," for individual subjects according to their ability in each.

CURRICULUM

The subjects in the daily curriculum vary from school to school. Though determined to some extent by the facilities available, the curriculum at any school is generally designed to cater as fully as possible for the needs of its pupils at different ages. In most schools, the traditional academic subjects such as math, sciences, and languages dominate

the curriculum - with arts, technical subjects, and sporting activities also included.

Increasing concern to emphasize the link between one subject and another, together with a growing enthusiasm for sociological and environmental topics, has in recent years led to the development of a combined-subject approach to lesson planning.

TEACHING AIDS

Mechanical teaching aids play an important part in modern education. Movies and programs on television and radio provide material about specialist topics and are used as a basis for further class teaching.

Tape recorders are particularly valuable in language teaching; some schools have laboratories in which each child has his own machine and can, under the teacher's supervision, progress at his own pace.

PROBLEMS AT SCHOOL

DISLIKE OF SCHOOL

Dislike of school may be shown in various ways. In a mild form, the child may be tearful, or complain of headaches or stomachaches. At its most serious, it may amount to absolute refusal to go to school - school phobia.

Reasons for dislike of school include the following.

a) Parental influences may have convinced the child that school is unpleasant or frightening.

b) When at school, the child may make poor progress in his work and perhaps fail to match up to the expectations of his parents.

c) He may be worried about catching up after an absence, or be afraid of a particular teacher.

d) He may be the object of bullying or teasing, or he may feel that he is unpopular.

In most cases, punishment for school-refusing is useless. The child should be encouraged to go to school every day, and the reasons behind the problem should be investigated.

BAD BEHAVIOR

Persistent bad behavior at school is damaging for a child and may also have a disruptive effect on his fellow pupils.

The following are examples of underlying causes of behavioral difficulties at school.

a) A child may be bored with easy lessons, or struggling to keep up.

b) The problem may result from a desire to attract attention - a common side effect of neglect or even rejection at home.

c) A child may have a problem that needs professional treatment.

PARENTS AND SCHOOL

Parents can help their school-age child most constructively by taking a real interest in his activities at school. This can mean simply listening to him describe the day's events, or may involve a deeper commitment - in a parent-teacher association, or by helping as a volunteer in the school library or on trips.

"Open" days are a useful guide to the way a school operates and offer a valuable opportunity for discussing with the teacher a child's particular problems or difficulties. It is generally wise to cooperate with the teacher if possible; open disagreement between teacher and parent can severely test a child's loyalties.

At home, parents can encourage their child by seeing that he has a quiet place to do his homework.

SPECIAL SERVICES

Many schools now include on the staff some personnel specially trained to deal with pupils' individual problems. At the center of this network is the teacher who is responsible for referring pupils to the appropriate staff member. The health unit may include a nurse who gives day-to-day health care and copes with emergencies, and a doctor who makes regular visits to give checks or advice. A social worker or guidance counselor may be available to help with problems arising from the pupil's family situation such as unemployment or lack of money, and they may refer pupils with severe emotional problems to a psychologist or psychiatrist. Specific learning difficulties can be helped by specialist staff such as speech therapists and remedial reading teachers.

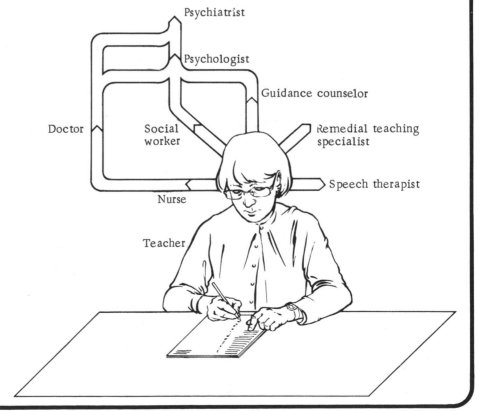

©DIAGRAM

HOBBIES AND INTERESTS

HOBBIES AND INTERESTS

The development of hobbies and special interests usually takes place after the child has begun school. An invaluable form of recreation and self-expression, hobbies can also be an important source of self-education and can in some cases influence career choice.

The range of pursuits that may interest a child is limitless. Some are individual occupations, others group activities; some may seem worthwhile and others pointless; some rely mainly on money and others on the child's own resourcefulness.

For many children, interest in a particular topic is stimulated by their friends' choice of hobby.

In other cases, the parents' own leisure activities may be a decisive influence.

COLLECTING

At some time most children enjoy collecting as a hobby. The objects they collect cover a wide range, and favorites include model cars, dolls, coins, stamps, photographs, sport cards, rocks, seashells, and flowers.

Displaying or mounting these objects may be a major part of the child's enjoyment, and special storage - an album, drawer, or shelf - helps encourage tidiness. Though collecting is essentially an individual pursuit, some schools and neighborhoods organize societies for children with specialty interests such as stamps or coins.

Stamp collecting   Shell collecting

CLUBS AND SOCIETIES

Clubs, societies, and other group activities are an important leisure interest for many children, particularly after the age of about seven.

Most organizations for younger children rely on a common interest, and among the activities especially popular with seven- to ten-year-olds are Brownies and Cub Scouts. In addition, many schools arrange extra-curricular clubs and societies for rather more specialized leisure interests.

For older children and teenagers group activities such as youth clubs or camps are especially important as they offer an opportunity to meet and make friends outside the context of the family home, and can in this way play a valuable part in the child's social and emotional development.

MUSICAL ACTIVITIES

A young child's enthusiasm for music can be encouraged from an early age with a recorder, or melodica, or a special junior xylophone or glockenspiel. Later, piano or singing lessons, or instruction in an orchestral instrument or guitar may be appreciated and can lead to participation in enjoyable group activities such as ensembles, orchestras, and choirs.

Listening to music should be an integral part of a child's enjoyment and parents can encourage his interest by providing the opportunity to listen to music of every kind on disk, tape, or radio, and, ideally, at a live performance.

ARTS AND CRAFTS

Hobbies connected with arts and crafts have wide appeal to children of many different ages Particularly popular with younger children are activities such as drawing, painting, simple cookery, and model making. As manipulative skills develop and his creativity begins to be reflected in these activities, the child's interest may extend to other hobbies including carpentry, sewing, embroidery, and modeling.

Many art and craft interests can easily be encouraged in the home and need not necessarily be a burden on the family budget. "Starter" craft kits containing all the necessary materials for a particular activity are now widely available, if rather expensive, but many hobbies make use of comparatively cheap materials such as paint and glue, natural objects such as flowers, or household items like paper containers, and scraps of fabric.

Clay modeling

Model-making

Needlework

SCIENTIFIC INTERESTS

From the age of about seven, some children develop a deep interest in a variety of scientific subjects including natural history, astronomy, chemistry, and photography. Interest can be encouraged with special equipment - perhaps a good science set, magnifying glass, microscope, or camera - and reference books appropriate to the child's age.

SPORTING ACTIVITIES

Sporting activities are an important leisure pursuit for many children. Great enjoyment may be derived from taking part in individual and team activities, while being a spectator or a follower of a particular team can also be a lot of fun.

Especially popular are team activities which offer the chance of physical exertion combined with the enjoyable social aspects of belonging to a group.

Most schools and many community organizations arrange teams and matches, but for many children an informal ball game organized spontaneously is just as good.

As well as being an excellent form of amusement a childhood interest in sport may continue into adult life, and by encouraging regular exercise can help promote general fitness.

©DIAGRAM

CHILDHOOD ANXIETIES

CHILDHOOD ANXIETIES

Even if they are never involved in a major crisis, all children experience anxiety at some time or another. Often the cause of a child's anxiety is easy to recognize. Many "first times," such as a first visit to the dentist, first bus trip alone, or first time away from home, fall into this category. Also very common are age-old childhood fears like fear of the dark, fear of animals, and fear of strangers. Some children regularly become anxious before examinations or tests at school, or worry that their classmates do not like them.

Many childhood anxieties seem unnecessary or illogical to an adult, but the problem is no less real to a child who is upset. In such cases a parent should take care never to ridicule the child, but should gently reassure him and attempt to convince him that his fears are groundless.

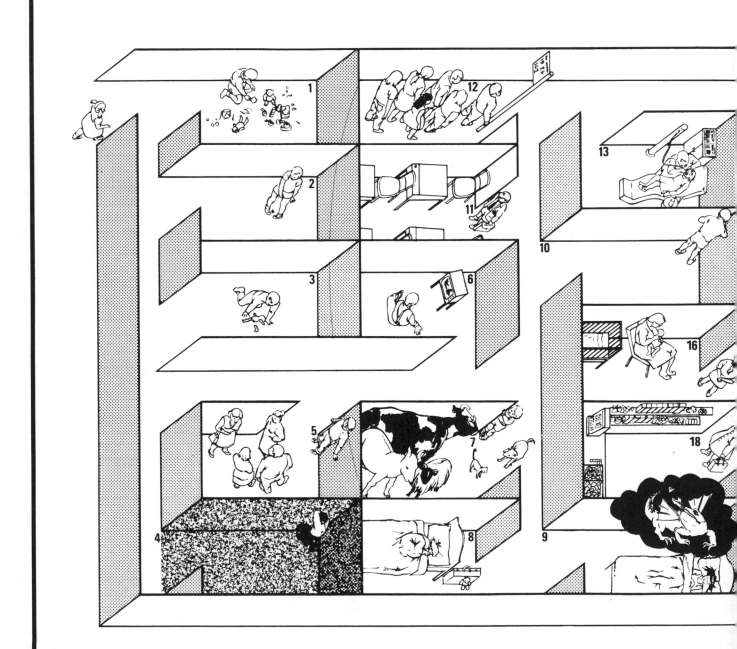

Some children have a naturally anxious and timid temperament, and frequently become upset at apparently trivial occurrences. A parent will need a lot of patience to boost his child's confidence and help him come to terms with his worries.

Other children appear never to be worried by anything, but parents should watch for signs of hidden anxiety. Taking time to talk quietly with a child is the best way of ensuring all is well.

Some events in a child's life are almost certain to cause feelings of anxiety and insecurity. Among the most common are the arrival of a new baby, and the first few weeks at a new school.

In cases such as these an understanding parent will anticipate the child's worries and do all he can to ease the child through each potentially upsetting situation as it arises.

COMMON CAUSES OF ANXIETY
1 Losing a toy
2 Toilet accident
3 Breaking a toy
4 Fear of the dark
5 Fear of strangers
6 Violence on television
7 Fear of animals
8 Spending a night away from home
9 Nightmares
10 Getting lost
11 Starting school
12 Making a bus or train trip alone
13 Visit to the dentist or doctor
14 Wearing inappropriate clothes
15 Arguing with a friend
16 Arrival of a sibling
17 Forgetting equipment needed for school
18 Losing money on way to stores
19 Accidentally breaking a window
20 Changing schools
21 Death of a pet
22 Being late
23 Examinations or tests
24 Being reprimanded without cause
25 Losing a friend

©DIAGRAM

BEHAVIOR DIFFICULTIES 1

UNDERLYING CAUSES

Every child displays behavior difficulties at some time or another - expressing anxiety at different situations, or behaving in a manner that is unacceptable to others.

Very often, "bad" behavior has no deep-rooted psychological cause. In some cases the child is simply too young to appreciate that what he is doing is wrong. Sometimes, moreover, his difficult behavior is part of a general developmental pattern - temper tantrums, for example, play a vital part in the emotional development of every toddler. In older children, high spirits and a love of fun and adventure may lead to trouble and clashes with authority that the child never really intended.

In other cases, however, a child's difficult behavior may derive from emotional disturbance - either a temporary upset or a longer-term more serious problem.

A variety of emotional problems may be found to lie behind a child's behavior difficulties. Among the most common are: insecurity resulting from family disharmony or a broken home; jealousy of another member of the family; lack of affection; childhood anxieties disfigurement or disability, or sometimes overweight; over-rigid discipline at home or at school; fear of failure; or feelings of guilt, whether real or imaginary. Sometimes an emotional problem resulting in behavior difficulties proves too deep-rooted to be identified and treated by the parents alone. In these cases, professional help from a counselor, social worker, psychologist, or psychiatrist should provide the child with the guidance he needs.

WARNING SIGNS

Children have many different ways of expressing an emotional upset. In general, a parent should take seriously any severe or persistent behavior problem that is uncharacteristic of a particular child or uncommon in other children of the same age.

Persistent crying, obvious anxiety or depression, a return to "babyish" behavior, apparent lack of interest, unusual rudeness, bedwetting, and sleepwalking are examples of behavior that should be investigated as possible indicators of deeper-rooted emotional problems.

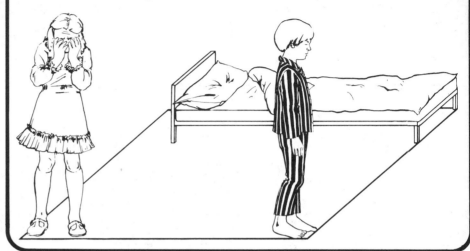

DISCIPLINE

A child's first experience of discipline usually comes from within the family. A parent's aim should be to give the child guidance about expected standards of behavior, encourage him to learn self-control, and help him to acquire a sense of responsibility in maturity.

Lack of parental discipline may cause a child to become insecure and confused as he finds that behavior tolerated at home is unacceptable outside the family. Excessively strict discipline, on the other hand, risks curbing a child's spirit of adventure and may cause him to become fearful of ever trying anything new. In some cases, unnecessarily harsh restrictions on freedom in early childhood lead to open and complete rebellion against authority in later years.

Disciplining a child is not simply a matter of giving out punishments - although sometimes even the most peace-loving parent will feel that some form of punishment is necessary. In fact, one of the most important elements in discipline is parental example: a child learns more from modeling his behavior on that of his parents than he does from raised voices or corporal punishment.

Expectations about a child's behavior must always be in line with his abilities at different ages. It is useless, for example, to expect a one-year-old to feed himself politely.

It is also essential to remember that each child is an individual, and must be treated accordingly. Even within the same family, children have very different temperaments and require varying methods of disciplinary guidance.

PASSING PHASES

Certain types of behavior that play a perfectly normal part in a young child's development can sometimes cause parents to become unnecessarily alarmed.

For example, reluctance to participate in cooperative play (**a**) and refusal to share toys (**b**) are passing phases in a typical developmental pattern. Both will be outgrown without parental interference. Similarly, habits such as nail biting, nose picking, thumb sucking (**c**), genital exploration (**d**), eating strange things (**e**), and temper tantrums (**f**) are common in toddlers and generally require investigation only if they persist well into childhood.

Head banging, cot rocking, breath holding, and air swallowing are rather more alarming aspects of behavior that may be indulged in if jealousy or insecurity are causing a child to feel in particular need of comfort.

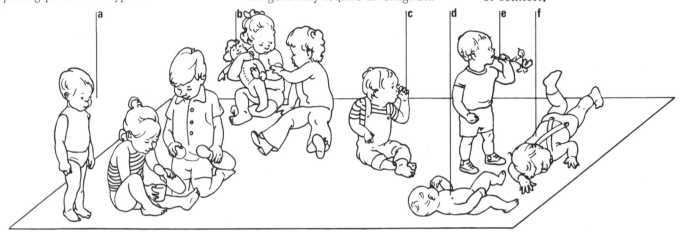

SPEECH PROBLEMS

Late talking is a common cause of anxiety to parents. Usually there is no cause for concern, but it is essential that the child has a specialized hearing test to check for any defect.

Mild stuttering and frequent stumbling over words are very common in young children, occurring because the child's eagerness to speak exceeds his level of verbal fluency. In almost every case these problems disappear without specialized help.

Persistent stuttering in older children is commoner in boys than girls. Emotional tension makes the problem worse, and parents should avoid drawing attention to the child's difficulty. Speech therapy is always helpful for children who stutter, and in many cases results in a complete cure.

BEDWETTING

Bedwetting is a common problem that may cause considerable anxiety to a child and his parents. In most cases the child would, in time, simply grow out of the problem, but various steps may be taken to help a child remain dry.

CAUSES

Staying dry through the night is a skill that some young children find particularly difficult. In these cases the problem usually results from a small or sensitive bladder in combination with a large urine output.

In other children bedwetting is a temporary lapse, following a prolonged period of night dryness. Often occurring between the ages of four and seven, this type of bedwetting is often the result of a short-term emotional upset such as starting school or the birth of a new baby in the family. Only vary rarely is bedwetting a symptom of severe emotional disturbance.

TREATMENT

In every case it is sensible to restrict fluid intake in the late afternoon and evening.

Because urine production is greatest in the first couple of hours after bedtime, waking the child and taking him to the toilet at this time is often all that is needed to keep his bed dry. A special pad and bell alarm is often helpful in severe cases. A drop of urine on the pad, placed between two undersheets, causes the bell to ring - and wakes the child in time to go to the toilet. An alarm of this kind can greatly boost the confidence of an anxious child.

In rare cases, a doctor may feel that drugs or psychiatric help offer the best chance of solving a child's particular bedwetting problem.

©DIAGRAM

BEHAVIOR DIFFICULTIES 2

BEHAVIOR PROBLEMS
A variety of childhood behavior problems - common and less common - are shown here.

1) LYING

Lying takes two basic forms - stories told for effect, and untruths told to hide a misdeed. The telling of fantastic stories is a normal stage in the development of a young child - a sign of a blossoming imagination. In older children, this type of lying usually involves "improving" a real event. If it occurs very frequently it may be a sign that the child needs to be reassured that great deeds are not necessary to win the esteem of others.

All children tell lies to cover up their misdeeds. In young children it is a sign that they now understand that certain things are not permitted.

An older child who often lies may merely be more adventurous than most - or may be seriously afraid of losing his parents' love if his misdeeds are discovered.

2) SHYNESS

Even outgoing children sometimes feel shy - after moving to a new neighborhood, for example, or starting at a new school. Shyness of this type rarely persists, particularly if the child is gently helped to find his feet. Severe and persistent shyness may be more of a problem, although in these cases, too, the problem often disappears with time and sympathetic handling. A severely shy child should be encouraged to make friends of his own age. Sometimes a child finds it easier to cope in his own territory and inviting another child into the home to play may be helpful. It is particularly important that parents do not allow the very shy child to become too dependent on themselves - he needs to feel loved and secure, but must also learn to stand up for himself.

3) IRRATIONAL FEARS

All children suffer at times from irrational fears. Fear of the dark, fear of strangers, fear of animals, fear of being left alone in quite safe situations, are all examples of common childhood fears. Usually fears such as these are short-lived or, even if persistent, do not in fact cause the child very great disturbance provided that a reassuring parent is comparatively close at hand.

In a few children, however, extreme fearfulness is a sign of some deeper emotional disturbance that may need professional help. Such children may panic at normal, everyday things like a furry toy or pet, be exceptionally afraid of any new experience, be over-anxious about getting hurt in relatively safe activities, or be constantly afraid of failure in very simple tasks.

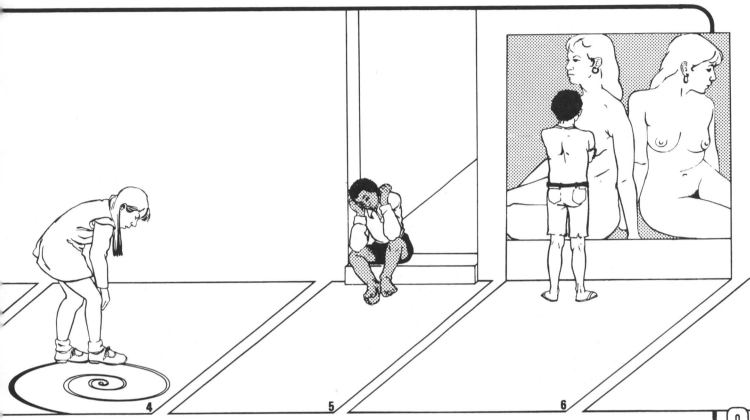

4) COMPULSIVE BEHAVIOR
Children of different ages draw
comfort from the performance of
ritual actions. Complicated
bedtime rituals are
common in young children and have
an importance beyond their value
as a delaying tactic. Between the
ages of seven and ten a great
variety of rituals designed to
ward off "bad luck" are extremely
common - not stepping on sidewalk
cracks, for example, or rapidly
touching different parts of the
body on seeing an ambulance.
Actions such as these are a normal
part of any child's development.
Sometimes, however, a child's
compulsive behavior is a sign of
emotional insecurity. Rocking
back and forward, head banging, an
obsessive interest in neatness or
cleanliness, or compulsions about
food may mean that a child is
subconsciously asking for help.

5) LETHARGY
Although periods of lethargy may
become quite common with the
approach of adolescence, this
type of behavior is less typical of
younger children and may be a
sign that all is not well.
A child who is sickening for an
illness is unlikely to want to
join in the boisterous games of
his friends. Usually the cause of
the problem soon becomes
obvious as other symptoms appear.
If a child is physically quite
healthy, frequent lethargy may
be a sign that he is in need of
help in coming to terms with an
emotional problem. This is
particularly likely if generally
withdrawn behavior is combined
with other signs of strain such as
absentmindedness, apparent
lack of deep feelings, excessive
daydreaming, nightmares, or an
inability to make decisions.

6) INTEREST IN SEX AND VIOLENCE
Some interest in sexual topics is
perfectly natural and normal
among young children.
Childish sex games and the use of
sex-based swear words may be
disturbing to a parent, but are
unlikely to represent more than a
passing - and important - phase
in a child's development.
Violence, too, seems to some
parents to play an unnaturally
large part in the lives of their
children. The acting out of
aggressions in games of cowboys,
soldiers, gangsters, etc appears to
be a fundamental part of growing
up and coming to terms with more
sophisticated social relations.
Only if a child's interest in sex
or violence is particularly
marked - and part of a wider
picture of disturbed behavior - is
it at all likely that there is any
real cause for parental concern.

G

©DIAGRAM

BEHAVIOR DIFFICULTIES 3

MORE BEHAVIOR PROBLEMS

7) TANTRUMS AND TEMPER

Temper tantrums are common in young children, particularly at ages two to four. They result from the child's inevitable lack of self-control, in combination with a developing realization of himself as an individual with wishes of his own. Punishment will certainly have no effect; leaving the child alone to calm down will sometimes do the trick.

Displays of temper usually become less frequent as a child gets older and becomes better able to control himself. But, as with adults, there will always be some things that annoy him: a desire to do something that is forbidden, or frustration at his lack of skill in a particular activity.

In a few cases, excessive and frequent shows of temper may need investigation as indicators of deeper-rooted stress in a child.

8) CHEATING

As with many other childhood behavior problems, cheating can have a variety of causes.

Most worrying is the classic case of a child who, deeply unsure of himself, cheats in an attempt to win the admiration of others. The parents of a child who cheats persistently should ask themselves very seriously whether they are not pushing the child too hard; and must take pains to assure him that their love in no way depends on good results at school.

Occasional cheating may be more simply explained: the child failed to do a piece of work - either deliberately or not - and fell back on cheating to avoid discovery.

Another phenomenon quite common among older children is a simple cheating "epidemic" - in which many children cheat for a time just to follow the crowd.

9) STEALING

Most young children "steal" toys or candies from each other - a perfectly normal form of behavior in the early years before a child can properly appreciate the concept of ownership.

Even after they are old enough to know better, many children still steal occasionally. Often the lapse is only temporary - perhaps the child had spent all his pocket money but desperately wanted candy or ice cream.

Another common form of stealing among older children follows a group dare - going into a store and taking something, just to prove it can be done.

Cases of repeated stealing, however, should be carefully investigated as it is possible that the child is using stealing as a means of drawing attention to more serious emotional problems or needs.

10) SMOKING, ALCOHOL, DRUGS
The widespread availability of
cigarettes, alcohol, and in some
cases drugs, makes it virtually
impossible for a parent to ensure
that his child does not have some
opportunity to try them.
Experiments with these adult
"vices" are common even among
pre-adolescents - perhaps as a
group activity, or perhaps a child
may help himself when his parents
are out. In many cases, these
earliest experiments fortunately
prove unpleasant to the child who
is then quite happy to postpone
his next attempt until he is older
and more mature.
Family attitudes and example vary
a great deal in these matters, and
will obviously affect the child.
It is important in any case that
the child knows his parents' views,
and also that he appreciates when
he is breaking the law.

11) FIGHTING
Many children resort to physical
battle in disagreements with
their friends. To a child with
an immature ability to argue,
actions frequently speak louder
than words. Sometimes the conflict
results more from surplus energy
than from any deep disagreement.
In almost every case there is no
cause for worry - the adversaries
are reconciled within minutes of
the outbreak of hostilities.
Cases of bullying and cruelty
toward younger children are,
however, very much more serious.
A child who consistently fights
with others who are physically
weaker than himself is almost
certainly expressing deep
feelings of hostility that he
cannot understand. Only if he is
helped to come to terms with his
inner conflicts can he learn to
control his aggression.

12) DESTRUCTIVE BEHAVIOR
Feelings of anger often provoke
destructive behavior in young
children; the child throws his toy
on to the floor - and breaks it,
much to his amazement. Even
the deliberate breaking of
another child's toy in a sudden
fit of temper probably signifies
no more than a passing storm.
More worrying is a child who
repeatedly or calculatedly
destroys the belongings of other
children; a child who behaves in
this way almost certainly needs
help with a more serious problem.
Also worrying to parents is a
child who indulges in acts of
vandalism. Activities such as
breaking windows are often
carried out for dares and may not
indicate any emotional problem -
even so, the child risks trouble
with the police and needs to be
kept in check.

©DIAGRAM

CRISES IN THE FAMILY

COPING WITH CRISES

When a crisis of any kind hits a family, a child is almost invariably affected, either directly or indirectly.

Occasionally a crisis situation can be anticipated. If a child or parent has to go into the hospital, for example, there may be an opportunity to prepare the child by explaining the event beforehand.

Events such as a serious accident or the death of a member of the family will cause disturbance in even the most stable home. Times such as these are distressing for everyone and, despite his own worries or grief, a parent should try to treat a child with as much understanding as possible.

Other crises in a child's life result from the development of disharmony within the family.

Constant arguing by the parents, physical cruelty, separation, and divorce, are all examples of domestic strife that can have an indirect but profound effect on a child's state of mind.

Not all cases of parental stress are caused by family disharmony. Unemployment, poor housing, and large families can also result in the kind of parental stress that is communicated to a child - at worst, in the form of battering.

Sometimes a child encounters a crisis situation outside the home - for example in cases of social or racial prejudice, civil disturbance, or personal attack.

In some crisis situations the strain on the family as a whole is so great that the needs of a child may be unintentionally neglected. At times like these, support from friends or in the form of professional advisory and counseling services may prove of the greatest value to everyone concerned.

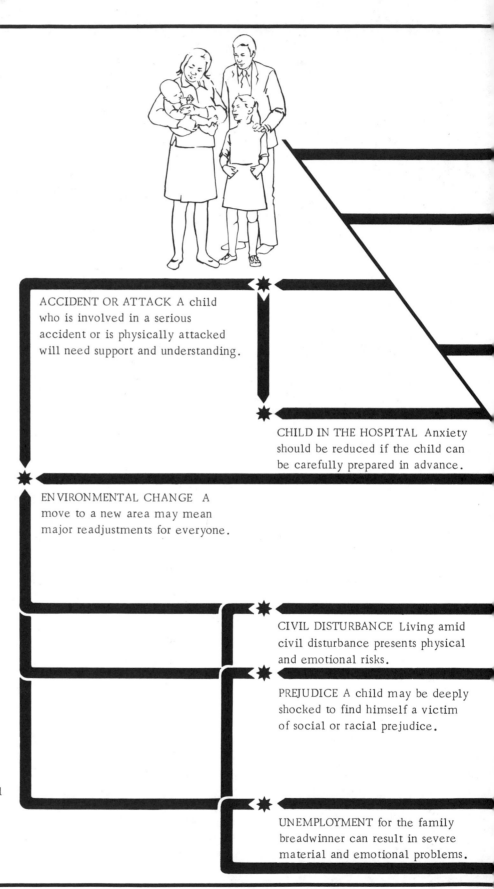

ACCIDENT OR ATTACK A child who is involved in a serious accident or is physically attacked will need support and understanding.

CHILD IN THE HOSPITAL Anxiety should be reduced if the child can be carefully prepared in advance.

ENVIRONMENTAL CHANGE A move to a new area may mean major readjustments for everyone.

CIVIL DISTURBANCE Living amid civil disturbance presents physical and emotional risks.

PREJUDICE A child may be deeply shocked to find himself a victim of social or racial prejudice.

UNEMPLOYMENT for the family breadwinner can result in severe material and emotional problems.

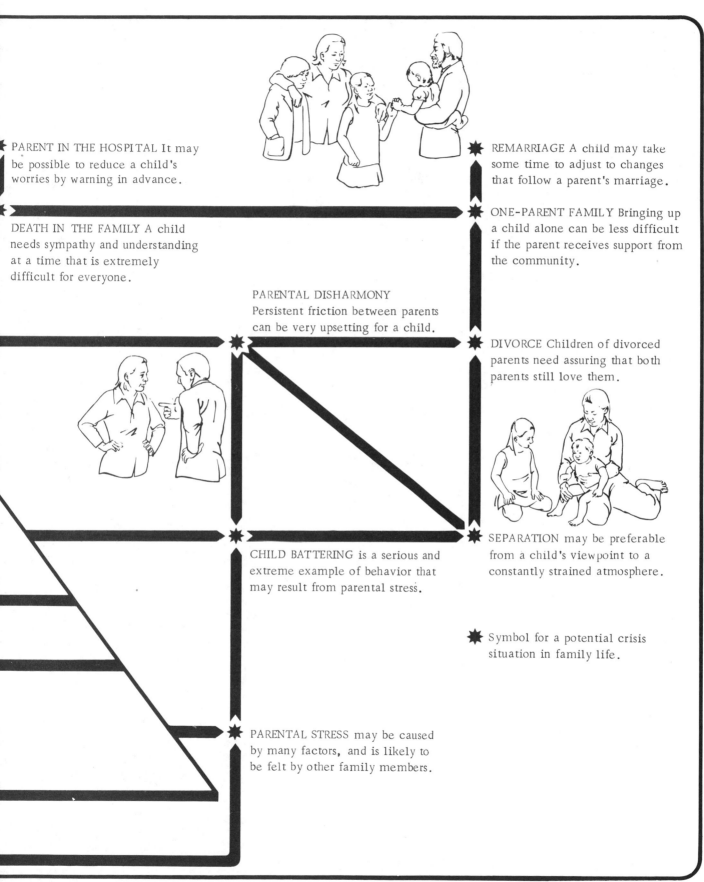

PARENT IN THE HOSPITAL It may be possible to reduce a child's worries by warning in advance.

DEATH IN THE FAMILY A child needs sympathy and understanding at a time that is extremely difficult for everyone.

PARENTAL DISHARMONY Persistent friction between parents can be very upsetting for a child.

CHILD BATTERING is a serious and extreme example of behavior that may result from parental stress.

PARENTAL STRESS may be caused by many factors, and is likely to be felt by other family members.

REMARRIAGE A child may take some time to adjust to changes that follow a parent's marriage.

ONE-PARENT FAMILY Bringing up a child alone can be less difficult if the parent receives support from the community.

DIVORCE Children of divorced parents need assuring that both parents still love them.

SEPARATION may be preferable from a child's viewpoint to a constantly strained atmosphere.

Symbol for a potential crisis situation in family life.

©DIAGRAM

TOWARD EMOTIONAL MATURITY

TOWARD MATURITY

Physical developments during adolescence bring a child's body to maturity. Inside his body the child will also have to mature emotionally, and leave behind a time when he was cared for by others. Soon he will have to care for himself, and then perhaps for his own family. This is exciting, but it can be frightening too. Interests and abilities may suddenly change. A boy who was small and physically weak may develop and shine on the sports field. Academic ability may rush ahead, or suddenly halt.

In their teens, young people no longer feel that their parents must be respected or obeyed without question. Quarreling is common; the teenager is testing his parents to find out if he really accepts their attitudes.

Above all, adolescence is a time when the opposite sex must be met in a new way. Traditionally boys and girls do not think highly of each other in the years before puberty. This now changes. A girl who was happy and confident in her pre-teen years may become shy with boys, feel that she is unacceptable to them, and become miserable. Boys, too, need support. There are new social customs to be learned and new sexual needs to be coped with. In adolescence young people test themselves. Only in this way can they learn who they are and what they will become. They are not sure yet if they are extroverted or shy, lighthearted or serious. Will their abilities and temperament draw them to practical work, a profession, business, or the arts? The late teens see the formation of moral and political views; young people conceive the sort of ideal world in which they would like to live.

Before the changes and turmoils of adolescence the pre-teen child is characteristically happy and self-confident.

Teenagers' exploration of their own opinions and identity may lead to family rows as parental standards are challenged.

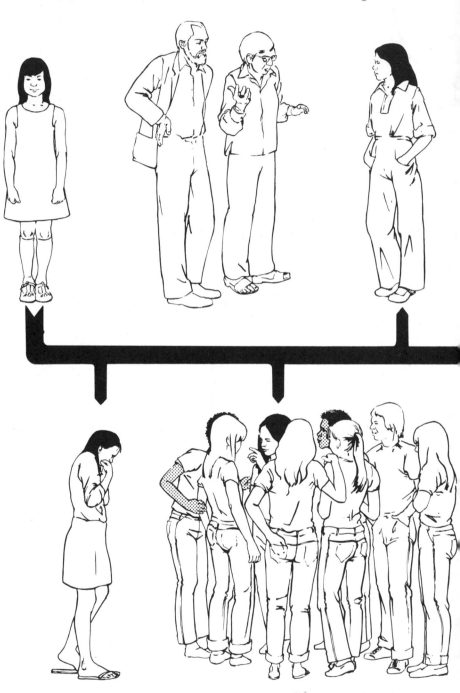

When their friends seem to be developing faster than they are, teenagers can often feel deeply uncertain of themselves.

Young teenagers often have a strong sense of togetherness. They need each other's support, and dress alike to win this.

Teenagers also challenge school authority. In this way they test their strength and show that they will soon be adults.

Growing up can be frightening. Teenagers go through periods of lethargy and depression, wishing that changes were not so rapid.

Mixed in with the rebellion of adolescence is often a strong concern for others and a need to help them in the tasks of life.

Teenagers want to know if forbidden things are really dangerous and shocking. Many try drink, smoking, and drugs.

Through dating, boys and girls meet each other on a more serious level. It is an essential part of getting to know the opposite sex.

Emerging from the trials of adolescence, the young adult seeks to make the world a better place for everyone.

©DIAGRAM

15 Play

Some adults mistakenly believe that children's play is a pleasant but essentially unproductive means of passing time. This is very far from the truth.

Left: Artists of the future in a 19th-century magazine illustration (The Mansell Collection)

TIME FOR PLAY

THE VALUE OF PLAY

Play is vital to a child's mental and physical development. It is one of his principal means of learning - through which he first masters basic skills, begins to express himself, and learns to get along with others.

Four broad types of play can be distinguished - active physical play, manipulative play, creative and imaginative play, and social play. Obviously some activities can be included in more than one category. The relative importance of different types of play varies with a child's age.

The need for play exists even in a very young baby, and persists in some form throughout childhood. Young children play most happily near the center of family activity. For older children who do not want an adult constantly at hand, it is a good idea, if possible, to have a special play area where boisterous games are permitted and toys do not always have to be tidied away. Materials for play need not be costly toys. Many simple household items make ideal playthings and have the advantage of encouraging creative and imaginative play.

The importance of commercially produced toys, however, should not be underestimated. Reputable manufacturers have spent a lot of time researching the developmental needs of children of different ages. A well chosen, carefully designed toy will provide a child with much more than entertainment. When buying a toy it is also important to pay careful attention to aspects of safety.

TYPES OF PLAY

ACTIVE PHYSICAL PLAY

This type of play begins in the womb and continues all through childhood. Involving the larger motor muscles, it teaches control, coordination, and balance, and leads ultimately to all kinds of adventure and sporting activities.

MANIPULATIVE PLAY

This teaches the control of the finer motor muscles, particularly those in the hands. It begins soon after birth as a baby learns to grasp, lift, and squeeze, and leads eventually to mastery of skills like feeding, dressing, drawing, and writing.

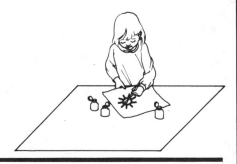

IMAGINATIVE AND CREATIVE PLAY

This type of play is especially significant as it teaches self-awareness, encourages the development of imaginative powers, and allows the safe working out of emotions such as fear and aggression. Imaginative play becomes most important when the child has mastered the basic skills that enable him to express himself.

SOCIAL PLAY

This type of play is an essential part of a child's emotional development. By playing first alongside other children and later with them, a child learns about cooperation, communication, and sharing. Social play thus helps him to build satisfactory relationships with others, and eventually to take up his role in the community.

IMPROVISED PLAYTHINGS

Shown here are a few examples of simple household objects that can be used for stimulating play.

1 Plastic bowls and saucepans for filling and emptying.

2 Unbreakable bottles, mugs, and measuring cups for water play.

3 Saucepan lids and wood spoons for banging.

4 Empty boxes for building and model making.

5 Sealed can of stones for rattling.

6 Flour and water dough for modeling.

7 Used bobbins for building or stringing together.

8 Pencil and paper.

9 Old clothes for dressing up.

10 Scissors, glue, and scraps of fabric for collage work.

11 String for cats' cradles.

12 Thread and buttons for threading.

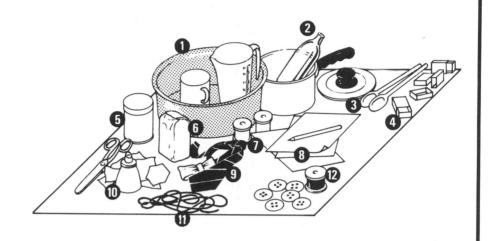

SAFETY OF TOYS

Certain features may make toys unsafe for children who are too young to realize the danger. Toy hazards include the following.

1 Eyes attached with sharp wire.

2 Contents of rattle toxic, or small enough to cause choking.

3 Small wheels to pull off and cause choking or suffocation.

4 Toxic paint.

5 Splinters from rough wood surface.

6 Inflammable doll's clothes.

7 Sharp metal edges.

8 Danger of trapping fingers in clockwork mechanism.

9 Sharp edges on broken plastic toy.

10 Toxic metal parts.

11 Risk of shock from electrically operated toy.

12 Danger of trapping fingers in trigger mechanism.

13 Possibility of an explosion.

©DIAGRAM

THE FIRST TWO YEARS

PLAY: BIRTH TO TWO

For the first two years of life, a child spends a large proportion of his waking hours at play. But this play is never aimless, nor just a means of filling in time - it is an essential and integral part of his development, and the basis on which much of his future learning will be built.

A baby's first play activities include kicking, and investigating parts of his body. He also watches, and later copies, the expressions and actions of others.

From about three months, a baby also needs stimulation in the form of simple playthings. At this stage, it is useful to offer objects of different sizes, shapes, colors, and textures, but they must first be checked for suitability and safety. A mobile suspended above the crib will amuse a young baby and provide excellent visual stimulation. Also valuable are noise-making toys such as bells or rattles, which offer the child a "reward" for his effort.

When a baby can crawl, every item in the home becomes a potential plaything for him to explore and enjoy. As his curiosity should ideally be encouraged and not construed as "naughtiness," it is wise to put away any objects that are valuable or potentially dangerous. A playpen may be useful for short periods, but should only be used if the child is happy to play there.

Even before his first birthday a child starts to benefit from being read to and told simple stories. This is vital for language development. It also gives a toddler a valuable opportunity to enjoy the undivided attention of another member of the family.

TYPES OF PLAY

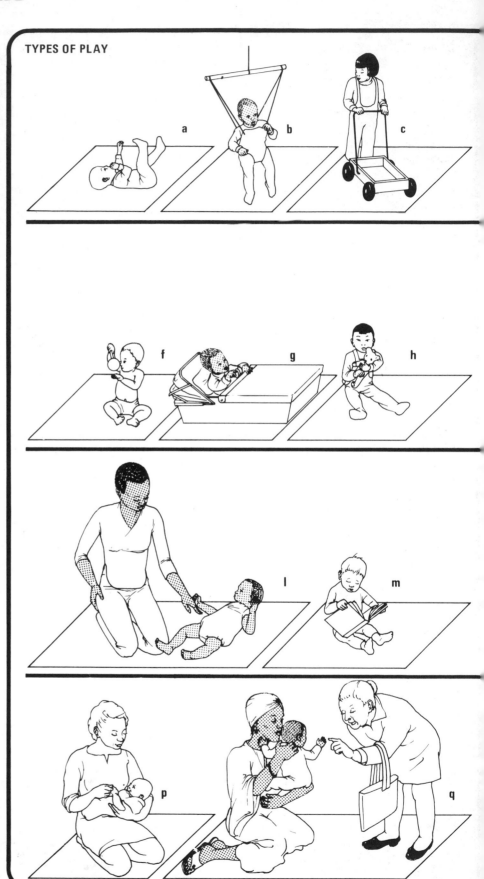

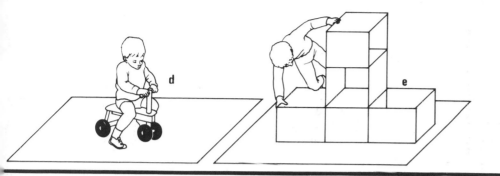

ACTIVE PHYSICAL PLAY

A baby's first form of active physical play involves kicking, reaching, and stretching (**a**). An older baby may enjoy a "bouncer" (**b**). Wheeled toys (**c** and **d**) are popular when the child can walk, and climbing activities (**e**) are useful for strengthening limbs and increasing muscular control.

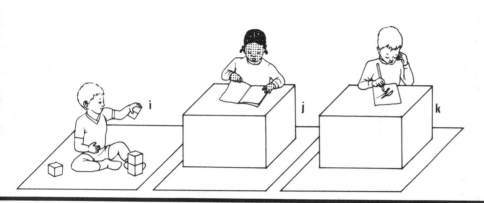

MANIPULATIVE PLAY

For most babies, clutching an adult's finger is the earliest experience of manipulative play. A rattle (**f**), beads (**g**), a squeaky toy (**h**), and blocks (**i**), encourage an older child to grasp, lift, and build, and teach fine muscle control. Turning pages (**j**) and scribbling (**k**) are useful preparation for future imaginative and creative play.

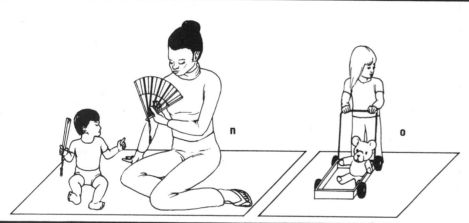

IMAGINATIVE AND CREATIVE PLAY

Among the earliest forms of imaginative play is the imitation of sounds and expressions, often copied from a parent (**l**). Picture books (**m**) and the identification and association of familiar objects (**n**), stimulate the child's imagination. Simple day-to-day events provide a basis for "acting out" situations (**o**).

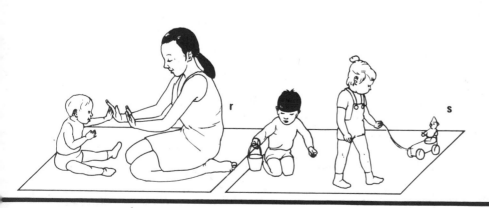

SOCIAL PLAY

A baby's first social play consists of the simple responses such as smiling that result from contact with his mother (**p**); sometimes this sociability extends to strangers (**q**). Older babies enjoy games with person-to-person contact such as pat-a-cake (**r**). By age two, most toddlers start to enjoy playing alongside others (**s**) - though not yet with them.

© DIAGRAM

TWO TO FIVE YEARS

PLAY: TWO TO FIVE

Between the ages of two and five play continues to be a vital and integral part of learning. Experiences gained at this time provide a basis from which learning in school can develop.

Active physical play increases in importance as the child devotes more of his time to outdoor activities and learns to run, jump, skip, hop, and climb.

By the end of the period, manipulative skills are much better developed and the child starts to use them in an imaginative and creative way. For example, a young child may enjoy handling a modeling material and may appreciate its feel and texture, but an older child will use it to create objects from his imagination.

Playing with other children becomes important in these years, and pre-school groups may be useful in this respect.

At first, toddlers will play alongside each other, or play contentedly on the edge of a group of older children. This "parallel" play is gradually superseded by "cooperative" play in which the child actually includes others in his activities.

Toys and books remain extremely important as tools of learning.

Natural materials such as sand, water, and even mud, make ideal playthings for a young child.

Manufactured toys should be fairly simple; elaborate toys that make little call on the imagination have very little value.

Powers of concentration are limited even in a five-year-old, and regular changes of activity are needed. To help a child over this, a few toys - perhaps birthday or Christmas gifts - can be kept on one side for giving as "surprises" when the child is ill or disappointed.

TYPES OF PLAY

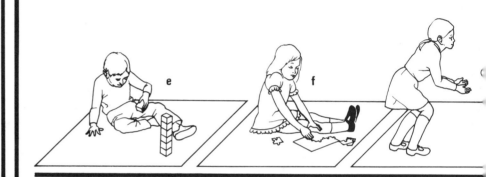

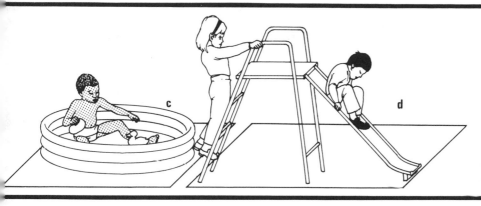

ACTIVE PHYSICAL PLAY

From two to five active play is very important. A tricycle (**a**) or scooter (**b**) are favorite toys, and a wading pool (**c**) may be popular (though supervision is vital). Larger pieces of equipment such as a slide (**d**) are usually found in playgrounds and may be bought or borrowed for garden use.

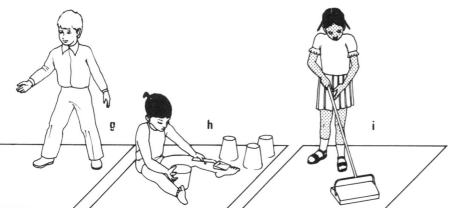

MANIPULATIVE PLAY

At this stage activities that challenge a child's increasing dexterity such as building (**e**), jigsaws (**f**), and throwing and catching (**g**) are very popular. Water and sand play (**h**) are fun and help teach basic scientific principles. Using "household equipment" (**i**) increases a child's manipulative skills and exercises the imagination.

IMAGINATIVE AND CREATIVE PLAY

A child starts to get the most out of this kind of play when he has mastered basic manipulative skills. Modeling (**j**) and drawing (**k**) are two important expressions of creativity, and dressing up (**l**) can add an extra dimension to dramatic play. Some children may now begin to appreciate a musical instrument as a means of self-expression (**m**).

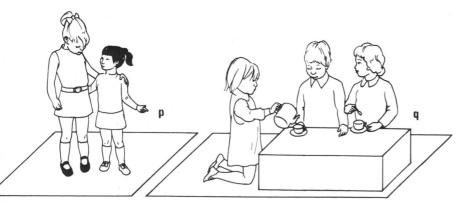

SOCIAL PLAY

The parallel play of infancy is giving way to cooperative play. From the age of about four, simple board and card games (**n** and **o**) are popular, and an older child may be happy to join a younger child in his play (**p**). Elaborate group activities such as "tea parties" (**q**) have a valuable social function as well as being a major form of imaginative play.

© DIAGRAM

FIVE TO TWELVE YEARS

PLAY: FIVE TO TWELVE

After the age of five the role of playthings as instruments of learning gradually becomes less important. Play for school-age children is more significant as a means of recreation, balancing the educational and developmental demands of school. Another change at this time is the development of long-term hobbies and interests from simpler play activities.

Active physical play continues in a variety of forms, ranging from simple outdoor pursuits like tree-climbing to organized sporting activities that appeal to a child's growing competitive spirit.

A vast selection of "playthings" is available for active play in this period, but in general a child's natural inventiveness will allow him to play contentedly for some time without special equipment. Manipulative skills develop considerably in these years, and creative pastimes assume greater importance. Craft sets and construction kits are popular and may help establish a lifelong interest in a subject.

An expensive manufactured set, however, is not always essential: paper, card, wood, empty containers, glue, and paint are just a few of the items that can be used for model making, collage, and similar crafts.

Imaginative and creative play is at its most significant at this period and helps the process of concept learning.

Group activities are especially valuable from the age of about seven. As well as being a lot of fun, being a member of a group or belonging to a club or society may help a child come to terms with some of the complexities of human relationships.

TYPES OF PLAY

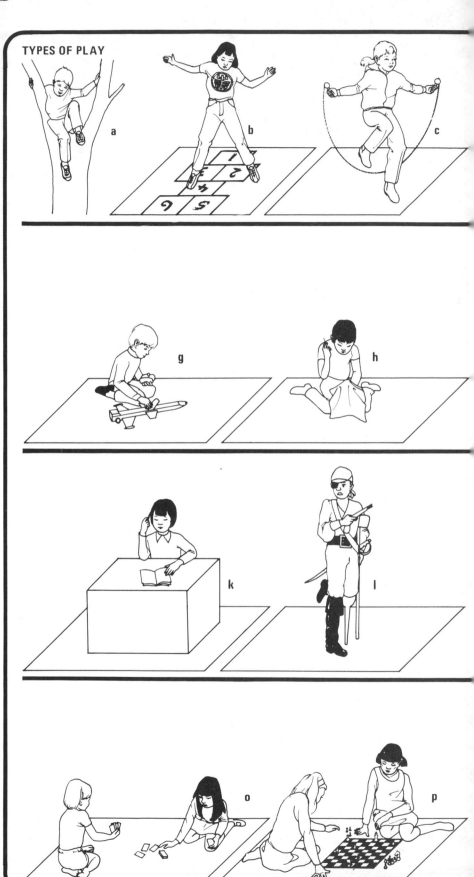

ACTIVE PHYSICAL PLAY

Outdoor pursuits such as tree climbing (**a**) or simple games like hopscotch (**b**) are popular throughout this period. A variety of equipment is available for active play, ranging from skipping ropes (**c**) and hoops (**d**) to scooters (**e**), bicycles, roller skates, and skate boards (**f**).

MANIPULATIVE PLAY

A child's rapidly developing manipulative skills may be applied to an almost endless range of activities in this period. Model making (**g**), embroidery (**h**), and gardening (**i**) are examples of popular pastimes. Particular favorites among "playthings" appropriate for this age group are science and magic sets (**j**).

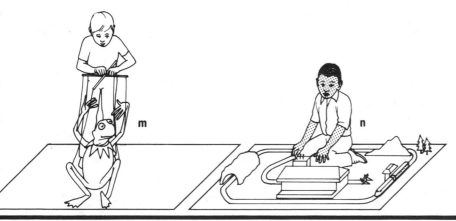

IMAGINATIVE PLAY

Between the ages of five and 12, creative art and craft activities and literary pursuits (**k**) have widespread appeal. In this age group, many children may also enjoy dramatic play through the medium of "acting out" games (**l**) or with puppets (**m**). Elaborate equipment such as a scale version of trains (**n**) can also be a useful basis for imaginative play.

SOCIAL PLAY

After starting school a child becomes increasingly involved with others and starts to enjoy sharing his favorite activities and pastimes. More complex card games (**o**) and board games (**p**) encourage shared play, but most children seem automatically to seek each other out for group activities like a walk, a bike ride, or a ball game (**q**).

©DIAGRAM

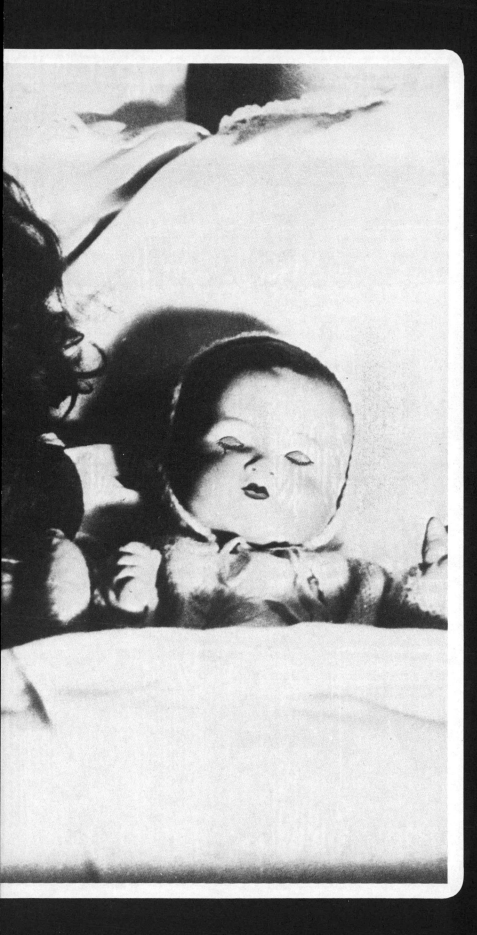

16 The sick child

A variety of illnesses are common in childhood —
most of them mild, others more serious.

Left: Sleep, the great healer
(The Mansell Collection)

THE ONSET OF ILLNESS

CHRONIC OR ACUTE

An illness may be described as being either acute or chronic. An acute illness is one that comes on suddenly, intensifies sharply, and usually lasts a short time. Such an illness is not necessarily serious - the common cold is a good example.

A chronic illness is one that lasts for a considerable time without any rapid developments in the patient's condition. In a chronic infectious illness the unaided body is unable or slow to destroy the agent of infection. Most chronic illnesses, however, can be cured or helped by medical treatment. As is the case with acute illnesses, the term chronic has no bearing on the severity of the disease or its symptoms.

CONGENITAL OR LATER

BEFORE BIRTH

Some conditions are inherited: hemophilia and sickle-cell anemia for instance. Others are acquired during fetal development or during birth, for instance some kinds of brain damage and deafness due to German measles in the pregnant mother. Both groups are termed congenital if present at birth.

IN LIFE

Most diseases arise during life. These include nutritional diseases; diseases due to environmental factors such as cold, heat, damp, or fog; allergies; infestation of skin, hair, or internal organs by parasites; and infection of the body by disease micro-organisms.

Most infections result from hostile micro-organisms entering the body through nose, mouth, ears, eyes, urogenital openings, or broken skin.

SYMPTOMS AND SIGNS

Disturbing symptoms and signs accompany some of the ailments common among children. Many result from poisons released by agents of infection in the body before its defenses can crush them. In many infections a child sweats, shivers, runs a high temperature, and is tired, flushed, apathetic, and lacks appetite. Rashes or spots appear in scarlet fever, measles, and German measles. Diarrhea and vomiting may stem from a variety of causes. Sore throats come with diphtheria, mononucleosis, and poliomyelitis. Aching and shivering come with mumps, chicken pox, German measles, and stress. But many of these symptoms and signs can also have trivial causes and vanish overnight.

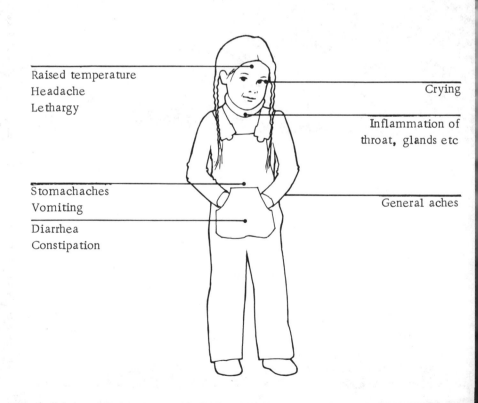

Raised temperature
Headache
Lethargy

Stomachaches
Vomiting

Diarrhea
Constipation

Crying

Inflammation of throat, glands etc

General aches

AGENTS OF INFECTION

Many diseases are due to one of six groups of infective agents: bacteria, viruses, rickettsias, fungi, and protozoan and metazoan parasites.

1 Bacteria are microscopic one-celled organisms, some of which normally live harmlessly on skin, in the nose, mouth, throat, and lungs, and in intestines. Reduced bodily resistance allows some to multiply and cause sore throats or other ailments. Illness can occur if bacteria normally in one part of the body get into another. But most disease bacteria enter the body from outside. Diphtheria, scarlet fever, tuberculosis, and whooping cough are bacterial diseases.

2 Viruses are smaller than bacteria and live as parasites, active and reproducing only in other living cells, which they break down. Viral diseases include the common cold, chicken pox, influenza, measles, mumps, and smallpox.

3 Rickettsias are germs found in certain fleas, lice, mites, and ticks. Rickettsias cause typhus, scrub typhus, and Rocky Mountain spotted fever.

4 Fungi are non-green plants. Tiny fungi cause ringworm and some lung diseases.

5 Protozoan parasites are one-celled animals some of which can get inside the body to cause diseases including amebic dysentery, malaria, and sleeping sickness.

6 Metazoan (many-celled) parasites include tapeworms, roundworms, fleas, and lice.

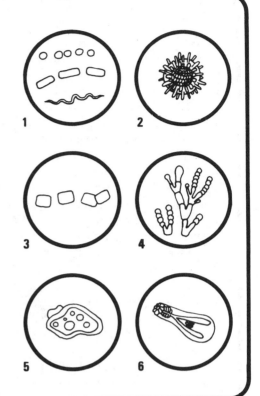

SPREADING DISEASE

1 Disease is often spread by bacteria or viruses airborne in droplets breathed, coughed, or sneezed out by infectious people and breathed in by others. (A sneeze can hurl 20,000 droplets 15ft:4.6m.) People sometimes contract anthrax and tuberculosis by breathing in dusty air bearing old, dried bacterial spores.

2 Some skin conditions spread by skin-to-skin contact.

3 Infected soil or dust entering a cut can cause tetanus or gangrene.

4 Pets may harbor and transmit diseases, for example: dogs, tapeworms; parrots, psittacosis; guinea pigs, encephalitis; turtles, salmonella poisoning; horses, glanders; goats, brucellosis; cats, toxoplasmosis; various mammals, rabies.

5 Food or water contaminated by germs at source, or by poor personal hygiene, can cause diseases such as brucellosis, cholera, typhoid, dysentery, and poliomyelitis.

6 Flies can infect food with bacteria if it is left uncovered.

7 Bites by parasitic insects can also transmit disease.

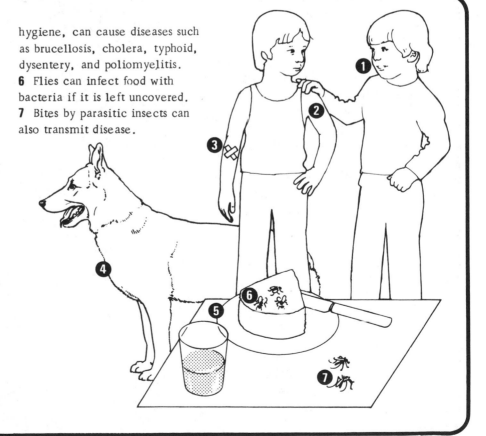

©DIAGRAM

THE FIGHT AGAINST INFECTION

DEFENSE MECHANISMS

The body fights infection in three ways: preventing the entry of foreign organisms; attacking those that get inside the body; and neutralizing those it cannot kill.

The body's main outer barrier is the skin, a sheath that guards underlying tissues. Antiseptic substances in sweat exuded from the skin kill many germs.

Different openings in the skin have special defenses. For instance, lacrimal fluid containing bacteria-combating lysozyme bathes the eyes at each blink.

Tears also wash out foreign bodies from the eyes.

Salivary glands help to combat infectious substances entering the mouth. Adenoids and tonsils make lymphocytes.

The body's openings and internal passages are lined with mucous membranes. Coated with antiseptic substances in a layer of mucus, these act as physical barriers and traps.

Inside the body certain organs produce special defenses.

The stomach secretes acids that attack bacteria in swallowed food. The liver filters harmful substances from blood flowing through it, and creates clotting substances that help wounds heal.

Spleen, bone marrow, and lymph glands all make white blood cells that circulate around the body and attack invading organisms.

Neutrophils produced in bone marrow are white cells that engulf, kill, and digest bacteria.

Macrophages - large, bacteria-engulfing cells - turn the lymph nodes in armpits, neck, and groin into bacterial traps and filters.

Local infection may trigger blood inflow, creating swelling, pain, and pus (white blood cells and bacteria).

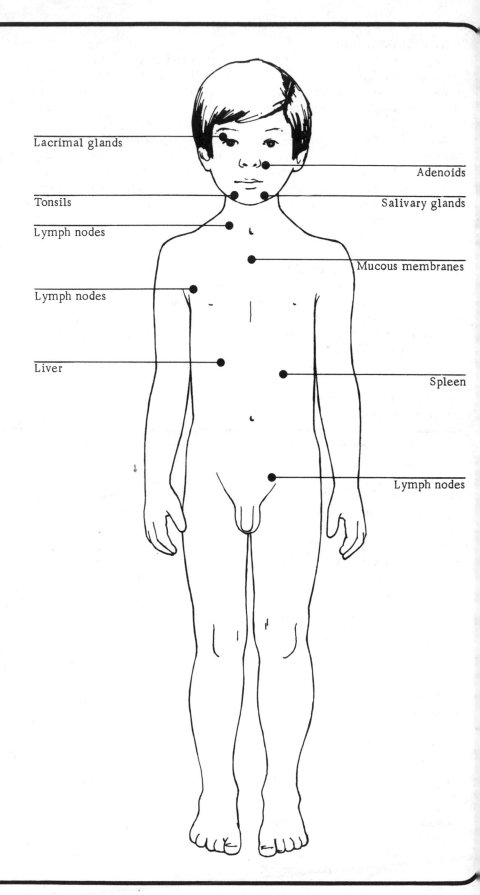

Lacrimal glands

Adenoids

Tonsils

Salivary glands

Lymph nodes

Mucous membranes

Lymph nodes

Liver

Spleen

Lymph nodes

NATURAL ANTIBODIES

Antibodies are large, complex protein molecules produced by plasma cells. They combat antigens (alien proteins that have invaded body fluids). Antibodies attack the protein sheaths of bacteria and viruses by interlocking with and thus inactivating them. Antibodies that neutralize the toxins (poisons) produced by alien bacteria are known as antitoxins.

In early life, production of antibodies is concentrated in the thymus gland. Later, spleen, lymph nodes, and bone marrow become the chief producers. Antibodies are specific in their actions: in most cases, one type of antibody will react only to one type of germ.

TREATING INFECTION

The body's defenses do not easily subdue all infections, but many can be crushed by antibiotics and sulfonamide drugs.

Antibiotics are chemicals made by micro-organisms. They halt the growth of or kill bacteria, fungi, and rickettsias by interfering with nutrient absorption and cell formation. Some act against only a small group of microbes: others (broad spectrum antibiotics) attack a wide range of targets. Much used antibiotics include penicillin, streptomycin, and tetracycline. Diseases susceptible to antibiotics include dysentery, endocarditis (bacterial infection in the heart), scarlet fever, tularemia, tuberculosis, typhoid, and many minor infections. Sulfonamides (or sulfa drugs) are synthetic chemicals all of which contain the elements sulfur, hydrogen, nitrogen, and oxygen. They work by blocking the production of certain chemicals that bacteria need in order to grow.

Some sulfa drugs are used against general infection of the body, others for local action in the intestinal tract and for certain bladder infections.

Sulfa drugs have proved useful against blood poisoning, dysentery, meningitis, pneumonia, and some other diseases. But for most purposes, antibiotics have replaced sulfa drugs. Neither attacks viruses, and some bacteria have evolved resistance to antibiotics.

IMMUNITY

Immunity is the body's original or induced ability to withstand invasion by disease organisms such as certain fungi, bacteria, and viruses.

The chief source of immunity is the lymphatic system, including the spleen and lymph nodes which manufacture antibodies attacking specific alien proteins.

Other sources include interferons - special proteins that curb the spread of viruses inside the body. Unlike antibodies, interferons are not restricted to a single type of target.

The body's production of antibody or white blood cells to combat an alien, antigenic substance is known as an immune reaction. Someone protected by antibodies against a particular disease is said to have immunity to that disease.

NATURAL IMMUNITY

The human body is naturally immune to many plant and animal diseases. But natural immunity to human disease largely varies with inherited differences in individuals' antibody output.

ACTIVE IMMUNITY

This is the immunity the body builds by making antibodies to fight invading organisms. One attack by a disease may produce lifelong immunity against fresh attacks by that disease.

Active immunity to some diseases can be artificially induced by vaccination: introducing weakened agents of disease. Vaccination using live germs is known as inoculation.

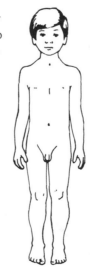

PASSIVE IMMUNITY

An unborn child acquires some antibodies from his mother, and after birth receives antibodies in her milk. The result is a passive natural immunity. This lasts only a few months.

Artificial passive immunity can be given to protect instantly against an established disease. Serum with antibodies against the disease is injected into the patient's bloodstream.

© DIAGRAM

VACCINATION AND IMMUNIZATION

WHY VACCINATE?

Vaccination can protect children against many once-common killing and disabling diseases. Vaccination against smallpox, tuberculosis, and diphtheria alone has saved millions of lives.

But the resulting decline in such diseases has persuaded many parents that vaccination is unnecessary. Others, worried by suggested links between whooping cough vaccine and encephalitis, think vaccination risky. In fact the risk of death or bodily damage from whooping cough far outweighs that from vaccination.

If the proportion of unvaccinated children rises, the risk of grave epidemic disease among them increases too. So, unless doctors advise against it, seek a full vaccination program for your child.

MEDICALLY OBTAINED IMMUNITY

Chief sources of artificial immunity are vaccines, specific kinds conferring protection against specific diseases. They work by exploiting the same natural process in which the body builds natural immunity against disease organisms that get in by chance. Hostile substances in a vaccine provoke the body to make antibodies against them. Some vaccines cause usually trivial and transient symptoms.

Vaccines may be made in various ways to employ:
1) an organism akin to one that causes disease (eg cowpox virus to immunize against smallpox);
2) dead organisms of the disease, or inactivated disease organisms (eg in influenza, Salk polio, and whooping cough vaccines);
3) weakened live disease organisms (BCG, German measles, Sabin polio vaccines);
4) A toxoid - a non-poisonous substance derived from a toxin (poison) produced by the disease organism (eg a toxoid that persuades the body to make antitoxins against diphtheria).

Some vaccines give lifelong protection. Less effective and less durable immunity may be won by injecting a ready-made antibody or antitoxin (eg against tetanus or diphtheria).

Vaccination for diseases other than those named may be needed where they prevail, eg yellow fever, typhoid, and cholera in certain tropical countries. Influenza vaccines are available but not urged for general use.

IMMUNIZATION PROGRAM

Illustrated here is a typical program of vaccinations during childhood. German measles vaccination is particularly recommended for girls, to prevent damage to an unborn child if the disease is contracted during a future pregnancy.

DPT   Diphtheria, whooping cough, and tetanus
DT   Diphtheria and tetanus
BCG   Tuberculosis
  Injection
  Oral vaccine

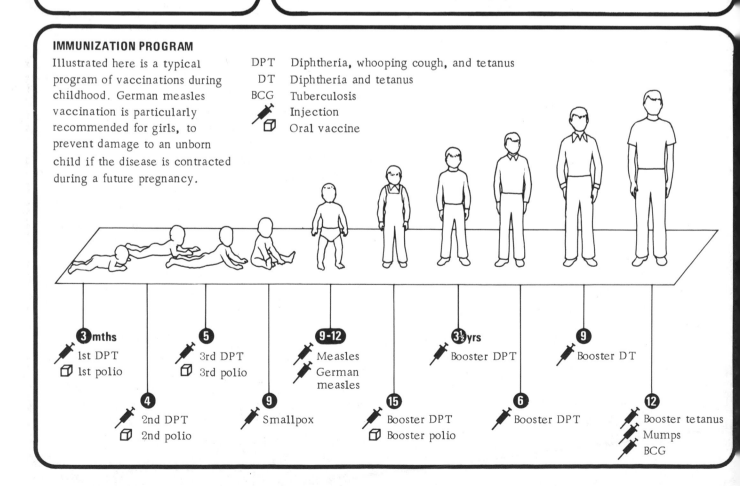

3 mths
1st DPT
1st polio

4
2nd DPT
2nd polio

5
3rd DPT
3rd polio

9-12
Measles
German measles

9
Smallpox

15
Booster DPT
Booster polio

3½ yrs
Booster DPT

6
Booster DPT

9
Booster DT

12
Booster tetanus
Mumps
BCG

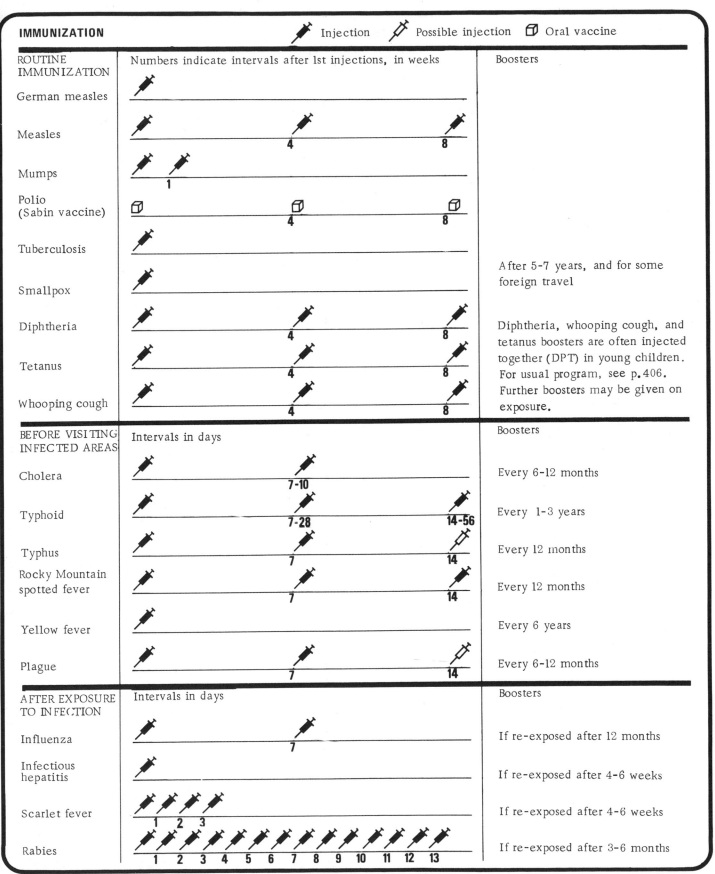

IMMUNIZATION   🖊 Injection   🖋 Possible injection   ▱ Oral vaccine

ROUTINE IMMUNIZATION	Numbers indicate intervals after 1st injections, in weeks	Boosters
German measles		
Measles	4 8	
Mumps	1	
Polio (Sabin vaccine)	4 8	
Tuberculosis		
Smallpox		After 5-7 years, and for some foreign travel
Diphtheria	4 8	Diphtheria, whooping cough, and tetanus boosters are often injected together (DPT) in young children. For usual program, see p.406. Further boosters may be given on exposure.
Tetanus	4 8	
Whooping cough	4 8	

BEFORE VISITING INFECTED AREAS	Intervals in days	Boosters
Cholera	7-10	Every 6-12 months
Typhoid	7-28 14-56	Every 1-3 years
Typhus	7 14	Every 12 months
Rocky Mountain spotted fever	7 14	Every 12 months
Yellow fever		Every 6 years
Plague	7 14	Every 6-12 months

AFTER EXPOSURE TO INFECTION	Intervals in days	Boosters
Influenza	7	If re-exposed after 12 months
Infectious hepatitis		If re-exposed after 4-6 weeks
Scarlet fever	1 2 3	If re-exposed after 4-6 weeks
Rabies	1 2 3 4 5 6 7 8 9 10 11 12 13	If re-exposed after 3-6 months

©DIAGRAM

SKIN CONDITIONS

COMMON EARLY RASHES

Rashes are common in the first few months of life. These rashes are usually insignificant and reflect the reaction of the young baby's skin to external conditions. Treatment may be needed, however, if the rash looks infected.

PRICKLY HEAT appears on the neck and shoulders of babies in hot weather, and may spread to the chest, arms, and face. It forms a rash of pink pimples surrounded by blotches of pink skin. The pimples may blister. It is usually caused by the baby being dressed too warmly.

MILD FACE RASH may occur as tiny white pimples, small red pimples that take longer to go away, or as rough red patches. The cause is unknown. Such rashes will clear up in time without treatment.

DIAPER RASH, see p. 404.

DISEASE RASHES

Many rashes in childhood are insignificant, but others are signs of infectious diseases. To avoid mistakes in identification the doctor should always be consulted if a child with a rash appears in any other way unwell.

1 MEASLES rash develops on the third to the fifth day of the illness, after which the child usually begins to feel better. The rash is of dark red spots that merge into blotches. It usually begins behind the ears and then spreads over the body.

2 GERMAN MEASLES produces a light pink rash that begins on the neck and face and gradually spreads over the body. Often the spots merge to give a flushed appearance. The rash is often the first sign of illness and lasts only a few days.

3 CHICKEN POX rash is often the first sign of illness. Commonest on face, scalp, and chest, and sometimes found in the mouth, spots appear over a three-day period. The rash consists of dark red pimples on which blisters develop a few hours after appearance. These burst easily and scabs form. Unless they are scratched - when permanent scars may result - the scabs fall off to leave pink scars that soon fade. Calamine lotion reduces itching.

4 SCARLET FEVER rash appears on the second day of illness, spreading over the body from damp areas like the groin, armpits, and sometimes the back. The rash is made up of tiny red spots on a flushed skin. The area around the mouth remains white. After a week, the skin over the spots begins to peel.

SKIN INFECTIONS

RINGWORM is a very contagious fungus infection affecting the skin and sometimes the nails. Scaly, crusted lumps form circular patches that clear in the center and spread out from the site of the infection. On the scalp, ringworm often causes the hair to break off short. Ringworm can now be effectively treated with drugs.

ATHLETE'S FOOT is caused by a fungus that thrives in wet, warm conditions. The skin between the toes becomes white, soft, wet, and and itchy. The condition is made worse by perspiration. Good general foot care is essential, and fungicide ointments should be used.

IMPETIGO is a highly contagious bacterial infection. It usually starts on the face with a pimple that develops into a brown crusty scab. Impetigo spreads very rapidly, and antibiotics may be needed.

COLD SORE is a virus infection that tends to recur. A stinging sensation gives rise to clusters of blisters that dry up in about 10 days. Commonest near the mouth, it may occur in the genital area.

WARTS are small harmless growths caused by a virus. On the hands they often disappear without treatment, but on the feet (verrucas) they should be treated, by cutting or chemically, since they are painful and easily spread to others.

BOILS are painful, pus-filled lumps caused by bacterial infection of a hair follicle, a sebaceous or sweat gland, or a break in the skin. They are most common where the skin is rubbed by clothing, and usually burst after several days. Most need only a protective dressing, but a doctor should be consulted if the child is very young or if several boils occur.

ACNE is an infective skin condition common in adolescence.

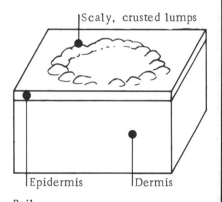

Ringworm

Scaly, crusted lumps

Epidermis   Dermis

Boil

Pus-filled cavity

Core   Dead tissue

1 Measles

2 German measles

3 Chicken pox

4 Scarlet fever (first sites)

SKIN CONDITIONS

ECZEMA is an allergic skin condition in which patches of rough, red, scaly skin cause intense irritation. It is most common in children with a family history of eczema or other allergic conditions such as hay fever or asthma. Causes include foods, and external irritants such as soap or wool. It may be aggravated by nervous tension.

In a young baby, eczema most often begins on the cheeks or forehead. Later, it may occur anywhere on the body, especially in the elbow creases and behind the knees.

The condition often improves or disappears as the child gets older. Cortisone ointment may be prescribed in severe cases.

HIVES, or nettle rash, is another form of allergic skin reaction. Characterized by painful, itchy skin wheals, it may be caused by certain types of food, drug, or insect bite. Hives may also result from emotional tension.

The condition is not normally serious, and soothing lotions will ease the child's discomfort. Medical attention must, however, be sought if swellings that may affect breathing occur in the area of the mouth and throat.

CHILBLAINS are itching, red swellings that occur on the extremities as a result of poor circulation in cold weather. Scratching may lead to infection. Warm clothing is the best form of protection. Treatment may be with drugs, or by improving circulation through exercise.

SWOLLEN RED HANDS may occur in babies in cold weather. The condition disappears once the baby is warm again.

SUN BURN, see p. 525.
FROSTBITE, see p. 525.

Possible locations

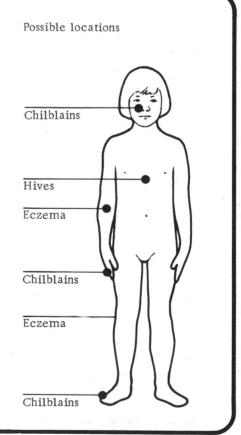

Chilblains

Hives

Eczema

Chilblains

Eczema

Chilblains

©DIAGRAM

INTERNAL DISORDERS

DIGESTIVE PROBLEMS

Digestive problems in children have many causes. Signs include vomiting, stomachache, diarrhea and constipation. (For problems in babies, see pp. 438-9.

In general, the doctor should be consulted if the parent is at all worried by the child's condition, particularly if warning signs occur together, if the child has a fever, or if abdominal pain is at all persistent.

STOMACHACHE Many young children complain of a stomachache when they really mean that they feel unwell, perhaps if they are about to vomit. Other children develop a quite genuine stomachache when they are tense or worried. A common cause of stomachache in children is the onset of another illness such as a cold, throat infection, or influenza. Other causes are appendicitis (K19), and intestinal infections that also cause sickness and diarrhea.

DIARRHEA is seldom significant in itself. If it occurs with vomiting, however, there is a danger of dehydration if the condition persists. The doctor should be called at once to a baby, or to an older child who is unable to take fluids.

CONSTIPATION may sometimes be a sign of illness - if the child is eating or drinking less, or if fluid has been lost through fever or vomiting. One problem, usually a passing phase, is that young children "hold back" their bowel movements until they are hard and difficult to pass. More usually, however, constipation is a condition imagined by parents.

APPENDICITIS

The appendix is a small closed tube leading off the large intestine. In appendicitis it becomes inflamed, usually as a result of a blockage. Sometimes the condition develops very rapidly, becoming critical in 24 hours or so. Often the first symptom is pain around the navel, which then changes to a pain in the right of the abdomen. Tenderness in the area over the appendix - McBurney's point - is a distinctive feature. Nausea is common, and vomiting and fever may occur.

It is important to call a doctor at once if appendicitis is suspected; without prompt medical attention the appendix may burst, causing peritonitis which can be fatal.

Do not give laxatives or treat for indigestion. Cases of acute appendicitis are treated by prompt surgical removal of the appendix.

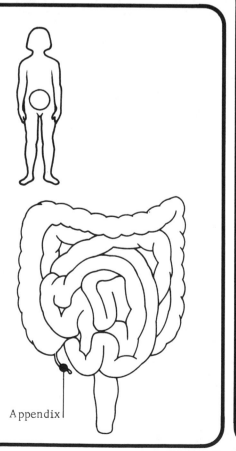

Appendix

RESPIRATORY PROBLEMS

PNEUMONIA is acute inflammation of the lungs. Symptoms include fever, chest pains, and a harsh, dry cough. The infection may be patchy, usually around the bronchial tubes (bronchopneumonia), or may affect an entire lung (lobar pneumonia). When body resistance is low, pneumonia may be caused by bacteria usually present in the mouth and throat. Bronchopneumonia is often a complication of another illness, or may be caused by foreign matter such as food in the lungs. Treatment is with antibiotics or sulfonamide drugs.

BRONCHITIS is acute or chronic inflammation of the bronchi, but other parts of the respiratory tract are often also affected. Mucus gathers and causes wheezy breathing and coughing. Sometimes bronchitis follows a cold, influenza, measles, whooping cough, or chicken pox. Bronchitis may itself lead to pneumonia. Drugs, inhalations, and physiotherapy are used in treatment.

CYSTIC FIBROSIS is a rare disease in which malfunctioning mucus-secreting glands cause severe respiratory problems. Antibiotics are used against infection, and breathing may be helped by physiotherapy and use of a mist tent or respirator.

ASTHMA is a chronic disorder of the bronchial tubes causing attacks of wheezing, coughing, and difficult breathing. It may be due to infection, but is more usually an inherited allergic reaction. Allergic asthma is made worse by emotional stress. Antibiotics may be used against infection, and desensitizing injections against allergies. An attack may be controlled with a special inhaler, and breathing exercises may be useful.

JAUNDICE

Jaundice is a sign of disorder rather than an illness itself. In jaundice the skin and sometimes the whites of the eyes appear yellow due to the presence in the blood of an excess of bile pigment. This pigment is produced in the liver by the normal breakdown of red blood cells. The excess bile pigment in the blood in jaundice may be due to a number of causes, requiring different types of treatment. For this reason it is important to consult a doctor as soon as jaundice is suspected. Other signs of jaundice are dark urine and pale bowel movements. Jaundice is quite common in newborn babies, when the most usual cause is simple inefficiency of the liver. Jaundice of this type usually disappears after a few days, but special treatment is sometimes needed. In very rare cases jaundice in a baby is due to congenital defect: the baby is born with no duct to carry bile away from the liver. A further cause of jaundice in infants is erythroblastosis fetalis, resulting from rhesus incompatability between mother and baby.

Rare in infants but more common in older children is jaundice due to infection. Most common is infectious hepatitis, transmitted by food or drink handled by a carrier of the hepatitis virus. It may occur in epidemics, especially where hygiene is poor. Preventive vaccination may be given. Less common is serum hepatitis, transmitted by infected blood used in transfusion or by contaminated medical instruments.

a Liver
b Gall bladder
c Bile duct
d Duodenum

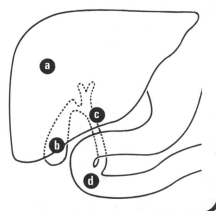

URINARY TRACT PROBLEMS

Infections may occur anywhere in the urinary tract - kidneys, ureters, bladder, or urethra. They are fairly common in young children, when they may be difficult to diagnose since it is not always apparent that the child is finding urination painful or is urinating abnormally often. Other symptoms - such as vomiting and fever - may have many causes. For this reason a urine test is usual whenever a child has an unexplained fever, or if any fever persists.

In some cases infection of the urinary tract results from an infection elsewhere in the body, usually in the throat or ears. This type of urinary tract infection is, however, now less common than formerly because of the use of antibiotics before the initial infection has had a chance to spread. Urinary tract infections may also be caused by the entry of bacteria from below - most commonly in girls because the urethra is shorter than in boys and because the urinary tract opening is nearer the anus.

Infection of the urinary tract is more likely if there is any abnormality in the system. To avoid permanent damage to the kidneys it is important that any abnormality is discovered and treated as soon as possible. Any child who has had a urinary tract infection should be kept under careful observation for some months. A child who suffers from persistent or recurrent urinary tract infections should be X-rayed to find any abnormality.

Difficult urination - straining to urinate or a dribbling urine flow - indicates an abnormally narrow passage or small opening, and requires urgent treatment to prevent permanent kidney damage.

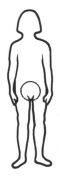

Possible sites of disorder in the urinary system

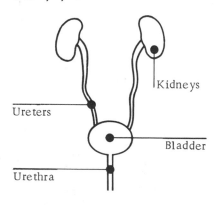

Kidneys

Ureters

Bladder

Urethra

©DIAGRAM

INFESTATIONS

FLEAS, LICE, MITES

Some types of parasite that live by biting the skin of animals and humans to suck their blood cause irritation and may spread disease.

a FLEAS The flea that usually lives on man (Pulex irritans) causes severe irritation. Some animal fleas also bite man - most serious are the bites of the rat flea, which spread typhus and bubonic plague. Eggs are laid in floorboard cracks, beds, and on pets. Fleas are best controlled by strict cleanliness.

b BODY LICE live and lay their eggs in the seams of clothing. Infested clothing should be washed and sterilized in boiling water. Body lice spread typhus, and bites may be sources of other infection.

c HEAD LICE live on the scalp, laying tiny, white, sticky eggs (nits) on the hair. It is easiest to detect head lice by finding the eggs. There may also be itching red spots where the hair meets the back of the neck. Head lice are not known to spread disease, but bites may become sites of infection. The scalp should be shampooed several times with medicated soap, and the hair combed with a fine-toothed comb. Insecticides should not be used.

d MITES Bites by the inch mite cause scabies - groups of scabbed pimples frequently occurring on the backs of the hands, the wrists, penis, and stomach. Scabies is contagious and requires medical treatment. The usual remedy is benzylbenzoate, painted on the skin. Chiggers, or harvest mites, are found in some countries and cause itching, blotching, and blisters. They can be removed with soap and water.

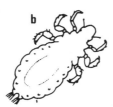
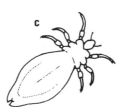
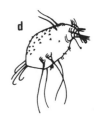

PARASITIC WORMS

Several types of parasitic worm live in the intestines of humans.

a THREADWORMS, or pinworms, are fairly common in children. About $\frac{1}{4}$ in (6mm) long, they resemble tiny white threads. They live in the intestine but come out at night to lay their eggs around the anus, where they cause itching. Mild stomach pains, nausea, and diarrhea may also occur, and worms can be seen in the feces. Eggs are spread on sheets and on the hands after scratching. Treatment is easy, but first consult a doctor.

b COMMON ROUNDWORMS invade intestines, liver, and lungs. They are up to 4in (10cm) long and look like earthworms. There may be no symptoms unless a great many worms are present, in which case an obstruction of the bile duct may occur. Microscopic eggs are passed in the feces, and are spread in contaminated food. A doctor will prescribe effective treatment.

c TAPEWORMS Several types of tapeworm are found in man, usually caught from inadequately cooked beef, fish, or pork. Tapeworms may grow 30ft (9m) long, and attach themselves to the intestinal wall by suckers or hooks on the head. Body segments break off and are passed in the feces. Only the larva of the pork tapeworm develops in man, producing severe complications. Effective drugs can be prescribed.

d HOOKWORMS Commonest in tropical countries, hookworms also occur in the southern states of the USA. About $\frac{1}{2}$ in (1.27cm) long, they attach themselves to the wall of the intestine to suck blood. Symptoms are unusual appetite, constipation alternating with diarrhea, anemia, and malnutrition. Eggs are passed out in the feces. Larvae from contaminated soil enter the body by burrowing into bare feet. Treatment is by drugs, high-protein diet, and iron supplements.

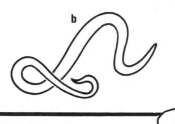
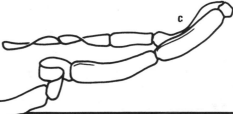
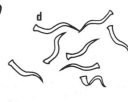

OTHER DISORDERS

GENITAL PROBLEMS

Genital problems that may occur in boys include the following.
HYPOSPADIAS is a congenital deformity of the penis in which the opening of the urethra is on the shaft and not at the tip. It is usually discovered at birth and to prevent infection can be corrected by a simple operation.
HYDROCELE In this condition an excess of fluid protecting the testes causes swelling of the scrotum, often on both sides. In babies, the swelling usually disappears without treatment. If it occurs later in life, it may be treated by drawing off some of the fluid, or by surgery.
UNDESCENDED TESTES The testes are formed in the abdomen and usually descend into the scrotum before birth or soon after. Sometimes, however, the testes fail to descend naturally until later, or surgery may be needed to bring them down.
It is not always easy to tell if the testes have descended because they are attached to muscles that can draw them back into the groin for protection against injury or cold. (The best time to check them is when the child is in a warm bath. Do not handle them, and avoid causing alarm.)
If one or both testes have never been seen in the scrotum when the child is two years old, a doctor should be consulted. Surgery is usual at age six if they remain undescended. If the testes are in the groin they can be brought down easily by surgery: if they are in the abdomen it may be necessary to remove them to prevent trouble in the future.

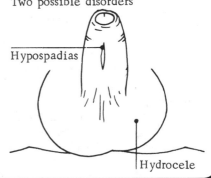

Normal penis and scrotum

Two possible disorders

Hypospadias

Hydrocele

HERNIAS

A hernia occurs when an organ protrudes through a weak point in the muscle around it. Two types of hernia are common in children.
UMBILICAL HERNIA, visible as puffing out of the navel when the child cries, occurs when a small portion of the intestine protrudes through a gap in the abdominal wall left by the umbilical vessels. An umbilical hernia rarely causes trouble, and the gap usually closes after a few weeks or months. Surgery may be recommended if the hernia is still present when the child is two.
INGUINAL HERNIA is potentially more serious and is usually corrected by simple surgery soon after it is discovered, often in the first year. It is most common in boys, occurring when part of the intestine is pushed during crying or straining through the inguinal canal into the groin or scrotum.

In girls, inguinal hernia produces a swelling in the groin. Occasionally an inguinal hernia becomes strangulated - trapped so that its blood supply is cut off, causing pain and vomiting. Immediate surgery is required in such cases because of the risk of gangrene or peritonitis.

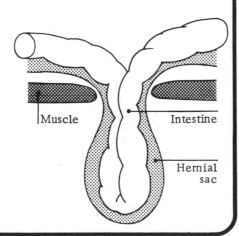

Muscle

Intestine

Hernial sac

INFANTILE CONVULSIONS

Some young children are prone to infantile convulsions that have no connection with epilepsy.
Also called febrile convulsions, they are usually brought on by high fever. (They are best avoided by keeping fever down.)
A convulsion lasts at most for a few minutes, but may be alarming. The child loses consciousness, and starts to twitch and shake. His eyes roll, his teeth are clenched, his breathing is heavy, and he may froth at the mouth or wet himself. A convulsion always ends in sleep. The child must not be left alone until the convulsion ends, because of the risk of inhaling vomit (his head should be turned to one side if he vomits). His limbs should be kept away from danger but no attempt made to restrain him or to hold down his tongue. The doctor should be called after the child has gone to sleep.

©DIAGRAM

GENERAL POINTS

These pages give simple hints about solving some of the basic problems likely to crop up in nursing a sick child. A few general points are included here.

1) Lethargy, loss of appetite, or slight fever indicate the onset of many childhood illnesses.

2) If such a condition develops, put the child to bed; if it worsens, consult a doctor, who may prescribe medicine, and advise on diet and overall care.

3) Some illnesses are infectious and require quarantine.

4) Sick children may need help with simple, everyday actions.

5) They need solicitude but not pampering.

6) Cheerful patients often recover faster than gloomy ones, so as far as possible keep your sick child occupied.

LOWERING TEMPERATURE

Normal body temperature is said to be 98.6°F (37°C). But temperature fluctuates even in healthy bodies. In fact a temperature as high as 99.5°F (37.5°C) or as low as 97.7°F (36.5°C) may be normal for some individuals. Also body temperature varies during the day. Then, too, a child's temperature may be readily raised by excitement, an infant's by incessant crying. Thus a high temperature need not mean illness.

But a temperature above 100°F (37.8°C) indicates a fever. A temperature of 104°F (40°C) indicates a high fever.

To lower a child's temperature give aspirin (unless the child is allergic to it) and plenty of water or fruit juice. For high fever, a tepid sponge applied to the arms and face may help.

TAKING A TEMPERATURE

The temperature can be taken in the mouth, rectum, or armpit (the latter gives a reading approximately 1°F:0.5°C below the others). Do not take the temperature soon after the child has eaten or drunk anything, or had a bath.

First rinse the thermometer in cold water, and wipe it dry. Place it in the appropriate position (for an oral reading the bulb should be under the tongue, and the mouth kept closed). Remove the thermometer after two minutes. Holding the end opposite the bulb, read the silver column of mercury against the scale. Make a note of the temperature. Shake down the mercury column. Rinse the thermometer in cold water, dry it, and stand it in antiseptic.

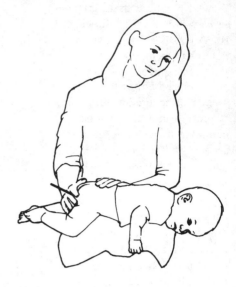

Taking the temperature with a rectal thermometer

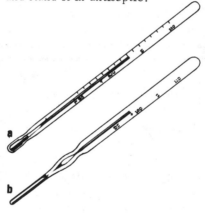

Types of thermometer:
a Oral
b Rectal

Taking the temperature with an oral thermometer in the mouth

Using a sharp wrist action to shake down the mercury column

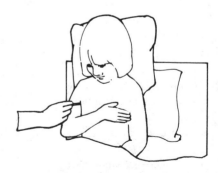

Taking the temperature with a thermometer in the armpit

PULSE RATE

The heart's pumping action makes arteries expand and contract as blood pulses through them. Heart rate can be easily judged by finding the pulse rate of the artery in the wrist just below the thumb. To feel a patient's pulse, press your finger tips lightly but firmly on the wrist and move them gently until you feel the artery beating. Because the ball of the thumb has a pulse avoid using your thumb.
The average pulse rate for an adult is is 65-80 beats a minute. This increases with exertion, during infection, or under emotional stress. Children always have a high pulse rate. A baby's may be 120-140 per minute. A six-year-old child's may be 90. Pulse rhythm may vary as the child breathes in and out.

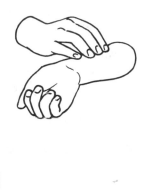

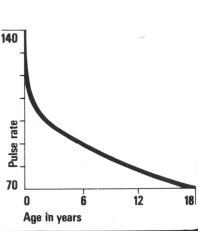

RESPIRATION RATE

Respiration rate is discovered by counting the number of times the chest rises and falls in a minute. It is easily changed by emotion, and is best measured when a person is asleep or unaware that his respiration rate is being checked. Respiration rate, like pulse rate, decreases with age.

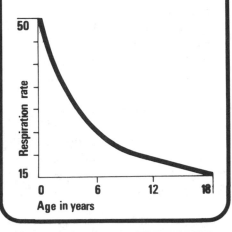

GIVING MEDICINE

Medicines can be dangerous if they are misused, and it is vital always to use them with care. Never give a medicine to anyone for whom it was not prescribed.

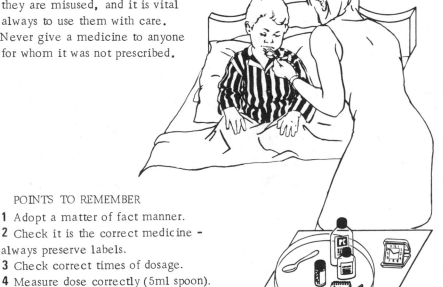

POINTS TO REMEMBER
1 Adopt a matter of fact manner.
2 Check it is the correct medicine - always preserve labels.
3 Check correct times of dosage.
4 Measure dose correctly (5ml spoon).
5 Non-dissolving pills can be crushed and added to jelly or honey.
6 Make a checklist of doses given.
7 Make sure bottle tops are always replaced securely.
8 Finish course if told to do so.

COPING WITH VOMITING

Vomiting can be due to motion sickness, obstruction of the digestive tract, food poisoning, indigestion, worry, tonsillitis, or infections including scarlet fever. Attacks occur most often in early infancy.
A vomiting attack is unpleasant to watch and can be frightening for the child. Deal with it calmly. Hold a bowl for the child and support him while the attack lasts. Then let him rinse his mouth with water. Wash his lips and face with a cloth. Change his bedclothes if they were soiled by the attack. Stay with him for a while in case he vomits again. Don't give him anything to drink for at least a couple of hours. Inform the doctor at once if the child is in pain.

©DIAGRAM

HOME CARE 2

INCREASING COMFORT

A sick child may easily become lonely, miserable, and bored. Try to make his room bright and attractive. If he is well enough to sit up, build a soft wall of pillows to support his back. Remove crumbs and wrinkles from his sheets. These need changing often, and if bedwetting is a problem, put a rubber sheet below them. A bedside table placed in easy reach should hold favorite books, magazines and games, a radio, and drinking water. If he is strong enough, a short period sitting on a chair or on the bed will help relieve monotony. Be solicitous but practical: resist continual attention seeking, but spend set times with him. In a long illness avoid showing too much concern.

GIVING A BED BATH

The patient should be undressed and covered with a blanket. First wash face and neck, then arms and armpits. Roll the blanket to the waist and wash chest and belly. Cover the washed parts after drying thoroughly. Uncover one leg at a time and wash from feet to groin. Turn the patient on his side to wash his back. Let the patient wash whatever parts he is able to.

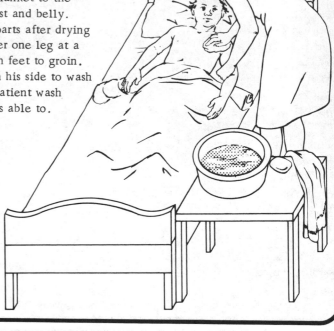

INVALID'S PROGRESS

1 Rest and sleep are among the most effective medicines for the really sick body. At this stage a child wants little more than a comfortable bed, a noise-free bedroom, and understanding.

2 When the child is well enough to sit up and eat meals, provide them on a tray, one course at a time, the food pre-cut if necessary. Make sure the food is tasty and attractive.

3 Early on in his recovery a child prefers physically undemanding amusements: soft toys to hug; books or comics to glance at; a radio to listen to. His attention span may be brief.

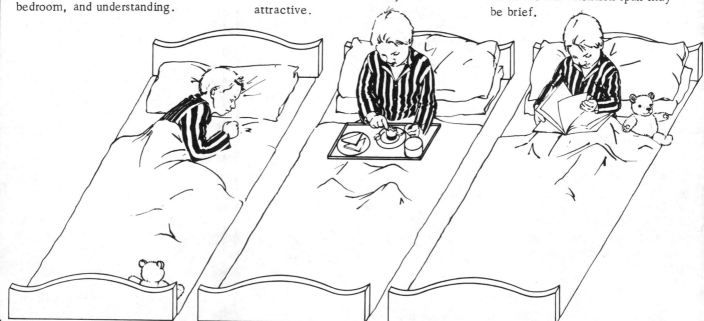

DIET IN ILLNESS

In many illnesses appetite drops off and some foods are not readily digested. Ask your doctor's advice on feeding if possible. The following hints are only general guides and some illnesses need special diets.

At the start of a high fever frequently offer water, soft drinks, and (if wanted) milk with no cream. In a day or two, if he is hungry, even a feverish child may cope with cereal, toast, soft-boiled egg, custard, ice cream, or cookies. Children convalescing after fever can usually tackle meat, fish, and vegetables.

After vomiting, wait two hours before allowing a sip of water. Later give half a glass, and, after several hours, a cookie or a little cereal or skim milk.

In simple colds, extra fluid may be helpful to children with diminished appetites.

In convalescence the child's appetite returns as he recovers. Meanwhile, forcing him to eat everyday foods before his digestive system is disease-free may cause revulsion and can even trigger long-term food fads.

Strict dietary rules apply to some medical conditions. Children with celiac disease, for example, cannot absorb gluten and should not eat bread, cakes, sausages, ice cream, or other foods that contain gluten. Patients convalescing after a liver or kidney disease may need a high-protein diet. Adolescents need extra calcium for building bones.

Foods suitable for inclusion in the diet after one or two days of fever

4 Convalescence can be boring for a child and visits cheer him up. But they should not be too long and the visitors should not be exposed to infection. If in doubt about the risk, ask a doctor.

5 A child on the way to recovery enjoys constructive, imaginative play: model making, sewing, making scrapbooks. It is more fun if you can help. Some schoolwork can be done in bed too.

3 When the child can get up for a time, he enjoys leaving his bedroom to watch television or sit with other family members. Make sure he is not in a draft. Use a blanket for warmth.

©DIAGRAM

CHILD IN THE HOSPITAL

GOING TO THE HOSPITAL

Almost every child has to visit the hospital at some time, either as an outpatient, or for a longer stay for medical treatment or an operation. The unfamiliarity of the hospital and its staff, and the separation from his family, can make even the shortest visit a frightening experience for a child. Parents should therefore use every opportunity to familiarize the child with the idea of the hospital in advance by taking him there to visit a friend or relative.

A simple explanation of the equipment that may be used and the treatment that he is likely to receive can be helpful, and a young child can be encouraged to "act out" some aspects of the treatment on his toys.

HOSPITAL ADMISSION

Admissions to the hospital fall into two broad categories - routine and emergency.

ROUTINE ADMISSIONS

In the case of routine admissions the hospital admission is arranged some time in advance - which has the incidental advantage of giving the parents sufficient time and opportunity to prepare the child. Among children, the commonest reason for routine admission to the hospital is for the correction of a congenital malformation such as cleft palate or a faulty heart valve. Where possible, doctors try to avoid admitting very young children to the hospital, but in some cases surgery has the best chance of success if it is carried out at a particular stage of the child's development.

EMERGENCY ADMISSIONS

An emergency admission to the hospital is potentially more disturbing for a child because of the lack of preparation time. Perhaps the most useful thing that a parent can do to help in these circumstances is to try and appear as calm as possible. Reasons for emergency admission include the diagnosis of an illness such as appendicitis that requires urgent treatment, or a particular development in a disease that requires the child to be kept under constant supervision.

An alarming number of emergency admissions among children, however, are the result of accidents - many of which could be avoided with better safety precautions at home and outdoors.

ADMISSION PROCEDURES

A thorough examination is customary when a child is admitted to the hospital.

The child's weight, height, temperature, pulse, and blood pressure are normally recorded, and a urine test, blood test, and X-ray may be performed.

The parents will be asked routine questions about the child's earlier illnesses and previous visits to the hospital. This helps the doctor build up a general picture of the child's medical history.

If the child is to be admitted, he should be allowed to take with him a few favorite toys, and, if he has one, his "comforter." When the child undresses, it may be less alarming for him if his clothes are left nearby even if he cannot wear them.

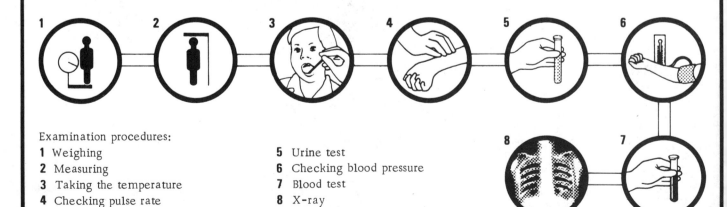

Examination procedures:
1 Weighing
2 Measuring
3 Taking the temperature
4 Checking pulse rate
5 Urine test
6 Checking blood pressure
7 Blood test
8 X-ray

VISITING

A hospital's visiting schedule depends on its organization and staff arrangements.

In some hospitals visiting is normally allowed only during special visiting hours arranged to fit in with the daily routine of each section of the hospital. These restricted hours may make visiting difficult for some parents because of their hours of work or commitments at home, so many hospitals are now introducing unrestricted visiting in children's sections. This enables parents to visit at times that suit their particular circumstances.

A small child left alone in the hospital may feel frightened, overwhelmed, and even abandoned, and his loneliness and confusion may be reflected in a rather indifferent or sulky reaction to his parents' first visit. In general, visits should ideally be kept short and frequent, and parents should if possible use the time to play with their child.

A visit from brothers and sisters will reassure a sick child and may also help them become familiar with the hospital situation. Recognizing the severe emotional stress that a stay in the hospital can impose on a child, some hospitals now provide facilities for parents to remain with their child throughout his stay. If this is the case, a parent can help the child not just by playing with him and reading to him, but also by assisting in aspects of his day-to-day care such as bathing, changing, and administering medicine.

OUTPATIENT VISITS

A short visit to a clinic or the hospital may be necessary for a variety of reasons.

Before admission to the hospital for surgical or medical treatment a visit to a specialist as an outpatient is usual.

Also, many routine tests or examinations, and some minor surgical procedures are frequently performed in the outpatients' department.

The child should be warned in advance of his visit to the hospital, and, as in the case of an admission, he should be given some idea of what to expect.

A favorite book, game, or toy should be taken, as delays between appointments are sometimes unavoidable even in the best-run sections.

HOSPITAL PERSONNEL

The children's section of a hospital is staffed with personnel specially trained in child care.

The treatment of a child in the hospital is supervised by one doctor -either a pediatrician specializing in childhood disorders of every kind, or a specialist in a particular type of disorder.

Routine aspects of medical care and general welfare in a pediatric section are the responsibility of the nursing staff.

Physical and occupational therapists may be involved in the recovery program, and where X-rays are necessary, these will be studied by a radiologist.

A dietitian, a play supervisor, and a teacher are now employed in most pediatric sections, and a medical social worker is usually available to discuss problems with both the child and his family.

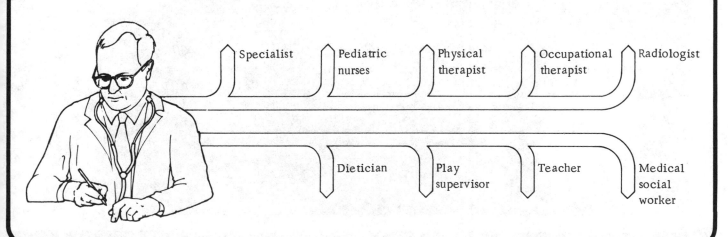

Specialist · Pediatric nurses · Physical therapist · Occupational therapist · Radiologist · Dietician · Play supervisor · Teacher · Medical social worker

© DIAGRAM

499

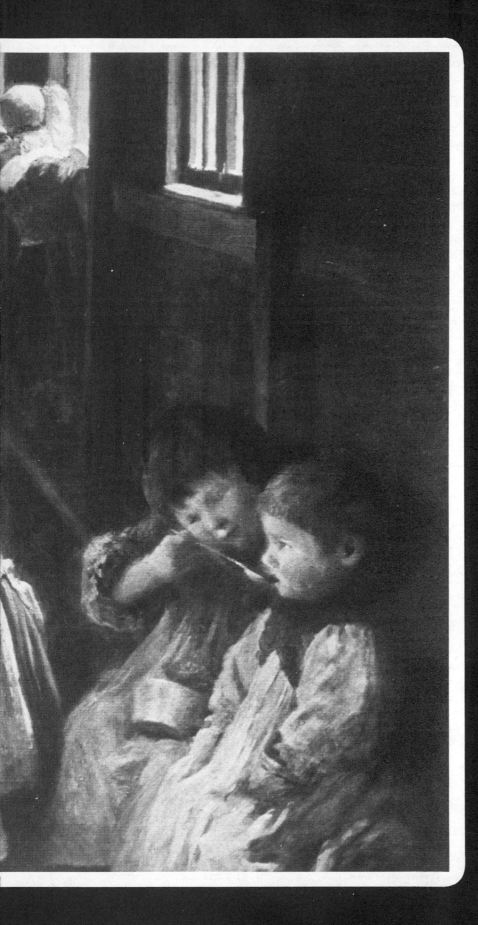

17 Children in need of special care

Some children need special care and attention — from their parents, from the community, and sometimes also from professional personnel.

Left: Suppertime in a French orphanage at the turn of the century (The Mansell Collection)

PHYSICAL HANDICAP 1

RESTRICTED MOBILITY.

Restricted mobility may be a temporary or a permanent condition, or the degree of handicap may vary with the occurrence of remissions in some illnesses. The nature of the disorders that result in restricted mobility is very varied. The biggest single cause of restricted mobility in children is cerebral palsy, although the number of children surviving with handicaps due to spina bifida has increased in recent years. Other causes of restricted mobility in children include muscular dystrophy, limb abnormalities, and some types of arthritis. Thanks to effective immunization programs, however, poliomyelitis is now uncommon. In addition to illnesses, accidents can result in either temporary or permanent problems of mobility.

HELPING THE HANDICAPPED CHILD

A wide range of help has been developed to meet the needs of children with physical handicaps that restrict mobility.
In some cases, such as certain bone malformations and injuries, surgery can ease or even cure the cause of the mobility problem.
In others, notably cerebral palsy, drugs may be used to check some of the symptoms of disorder.
Very often, however, help for the child with restricted mobility must be concentrated on helping him to make the best possible use of whatever movement he has, through medical services such as physical therapy, occupational therapy, special schooling, and the use of various types of aid. Ultimately the amount of controlled movement that a child with a mobility handicap will be able to achieve depends on the nature and severity of his particular problem. But specialized help can produce striking results.
Many children with a mobility handicap benefit from attending a special school, equipped with a full range of specialized facilities and equipment. In other cases it may be possible and preferable for a child to attend the regular neighborhood school.
Coping with the needs of a physically handicapped child in the home can place a severe strain on other members of the family. Help and advice is available from social workers, and from charity organizations associated with particular disabilities. In some areas, excellent community support schemes are now in operation.

SPECIAL AIDS

With the needs of the physically handicapped in mind, various aids have been developed to simplify some routine tasks which might otherwise prove impossible.
An appropriately designed wheelchair can make a significant difference to a handicapped child's life, but many different designs are available and a doctor's advice is essential to avoid an expensive mistake.
For certain types of handicap, feeding equipment such as suction-based plates, dishes with lips, and specially-shaped cutlery encourage valuable independence. Devices that can be operated with minimal movement - perhaps only a toe - have been perfected and are now applied to a wide range of equipment such as page-turners, typewriters, and tape recorders.

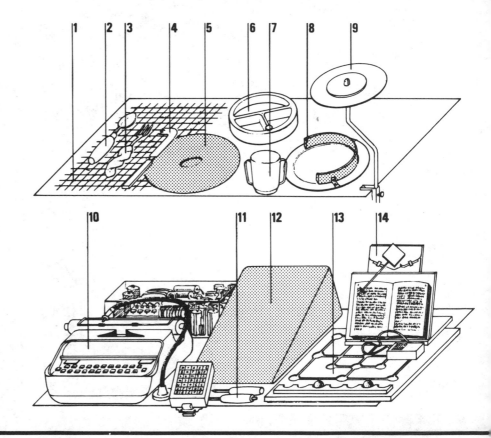

PHYSICAL THERAPY

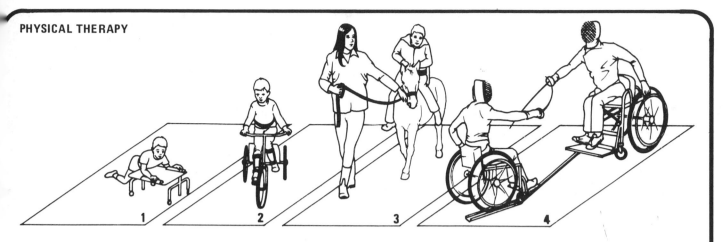

Physical therapy can help a child with restricted mobility in two important ways.

First, it can teach him to make good use of existing movement, and perhaps help him to cope with routine tasks such as feeding and dressing that might otherwise prove impossible.

Second, it can teach him to use certain aids appropriate for his needs which may help compensate for lack of mobility or control of particular parts of the body. For example, the crawler (**1**) has been developed to enable a physically handicapped small child to improve muscle control while getting around and exploring his environment. Many older children enjoy and benefit from riding a tricycle (**2**) or, with the aid of a special saddle and supports, a pony (**3**). Swimming, too, is a valuable and enjoyable form of exercise for the physically handicapped, while a variety of other sporting activities, such as fencing (**4**) and archery, can be enjoyed by children in wheelchairs.

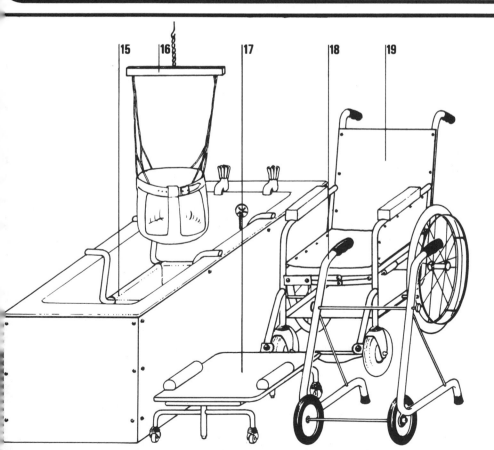

Selection of aids suitable for children with different disabilities.

1 Non-slip place mat
2 Large-handled spoon
3 Fork with malleable handle
4 Knife with curved blade
5 Non-slip mat for plate
6 Suction-based deep divided dish
7 Double-handled cup
8 Detachable lip for plate
9 Adjustable rotating feeding plate
10 Electronic typewriter controlled by "suck and blow" input
11 Thick-grip pencils
12 Foam rubber play wedge
13 Game with deeply recessed board
14 Page-turner
15 Bath seat
16 Hoist
17 Crawler trolley
18 Walking frame
19 Wheelchair

©DIAGRAM

PHYSICAL HANDICAP 2

IMPAIRED HEARING

Impaired hearing in an infant can result from a genetic defect, from damage caused by an infection such as German measles early in pregnancy, from the effect of some drugs, or from brain damage.

Since a hearing defect interferes with the development of language it is vital that the problem is identified as soon as possible. There are a number of warning signs. A child with impaired hearing continues to "babble" long after others are using recognizable words. Also, he may be later than average in walking, less adventurous and imaginative in his play, and often very noisy.

If the child is able to hear anything at all he will be fitted with a hearing aid. At home, parents can do a great deal to supplement the work of experts. The child's hearing aid must be in good working order, and should be worn for as much of the day as possible. Constant aural stimulation is vital, and to make it effective the child should be encouraged to pay attention to the sound source. Good "watching" conditions should be created, and there should be as few visual distractions as possible. When talking to the child about an object a parent should if possible hold the object near his own face so that the child can see the object and his parent's facial movements at the same time. The child will also benefit from feeling the vibrations of his parent's chest or throat during speech, and from the sensation of breath against his hair while the parent sings gently in his ear.

SPECIAL TEACHING
Special pre-school groups, special schools, or special classes in regular schools will help the child with impaired hearing to communicate with confidence.

VISUAL HANDICAP

Among the the causes of visual handicap in infancy are an infection such as German measles early in pregnancy, brain damage at the time of delivery, or an infection in the post-birth period. A child who is blind (with no sight at all) or partially sighted (with some residual vision, however little) will need a great deal of special care and attention to help him cope in a world where there is such dependence on sight.

HELP AT HOME
A visually handicapped child will benefit from added stimulation from the earliest age. Crawling, walking, and talking must be strongly encouraged by parents and skilled workers because of the lack of visual incentive. All the objects in daily use such as feeding implements and clothes should be identified by the parents by name as the child handles them, and investigation by feeling should also be encouraged.

Independence can be developed by ensuring that feeding equipment and clothes are simple enough for the child to manage for himself. Conversation based on reality is invaluable to a visually handicapped child, preventing him from feeling isolated and helping him to learn. If possible, he should be told of an impending loud noise or sudden action to prevent him from being startled.

SCHOOLS
Excellent special schools cater for the needs of visually handicapped children. They aim to encourage the effective use of any residual vision, to teach the use of various aids, to help with any additional handicap, and to ease the child's social adjustment by fostering his sense of personal worth.

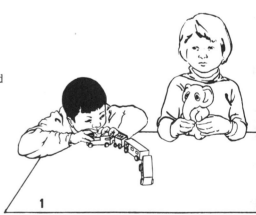

AIDS
A visually handicapped child particularly enjoys and benefits from "noisy" toys and those with hard edges or an intriguing shape (**1**). A special drawing frame (**2**) has been developed to enable the child to feel a shape or word that

Speech reading (lip reading) and special signs are often taught as valuable aids to comprehension and communication.

Among the equipment used in the teaching of children with impaired hearing are simple objects like mirrors (1) and balloons (2) that help the child to see or feel his developing powers of speech, as well as more complex aids like tape recorders and headphones (3).

he has created: a series of raised dots appears as the child gently presses on the board with a ballpoint pen. Learning to type (3) allows the visually handicapped child to communicate by letter with sighted friends, and may also help him find employment later.

Braille books (4) allow a blind child to read for himself; large-type books are available for partially sighted children. A tape recorder (5) is invaluable in education and for entertainment. Special "talking books" are also much appreciated.

OTHER PROBLEMS

In addition to children needing special care because of a more obvious physical handicap affecting mobility, hearing, or sight, there exists another broad group of children who must also be singled out for particular attention. Included in this group are children with chronic physical conditions that require medical attention and special care over a prolonged period of time.

A variety of conditions require regular visits to a hospital for check-ups or treatment, and in some cases of congenital disorder a series of operations may be needed over a period of years. The regular administration of drugs plays an important part in controlling conditions such as epilepsy, and daily injections may be needed to correct hormonal deficiencies as in the case of diabetes.

Certain conditions require that special attention be paid to diet - for example restricting the carbohydrate intake of diabetics, and excluding gluten from the diet of sufferers of celiac disease. Other children require what may be called a cushioned environment. A child with hemophilia, for example, must be protected from situations in which he may become cut or bruised, and over-exertion can be dangerous for children with sickle-cell anemia. Children with heart disorders however, are generally allowed to do as much as they feel is possible.

Caring for a child with a chronic physical problem such as those outlined here involves the parent in added responsibilities - not least in helping the child to lead as normal a life as possible in the circumstances.

© DIAGRAM

MENTAL HANDICAP

MENTAL HANDICAP

Some degree of mental retardation occurs in approximately three per cent of the population. Causes of this type of handicap include chromosomal abnormalities, as in Down's syndrome (mongolism), and brain damage, due to a variety of causes before, during, or after birth. Over 200 causes of retardation have in fact been discovered by researchers, but in the majority of cases it remains impossible to discover the precise cause of mental handicap.

The term mentally retarded is used to describe persons with a wide range of ability. A person who is only mildly retarded will very probably be able to lead a normal, independent life as an adult. Someone who is profoundly retarded, however, may be unable to perform even quite simple everyday tasks. Between these two extremes are persons in need of varying degrees of special care either in the home or sometimes in special centers or institutions.

Whatever the nature and degree of a child's handicap it is preferable for it to be diagnosed as soon as possible. Some types of mental handicap are evident at birth because there are distinguishing physical characteristics, as in the case of mongolism. Other types of handicap become apparent only when the child is older, perhaps when he starts school and finds it impossible to cope without the help of special teaching.

It is even more difficult to test the intelligence level of a child who is mentally retarded than it is to test that of a child who has no handicap. Careful observation over a considerable time should, however, lead to the development of a program of care and training that will help each child achieve his maximum potential.

CARE AND EDUCATION

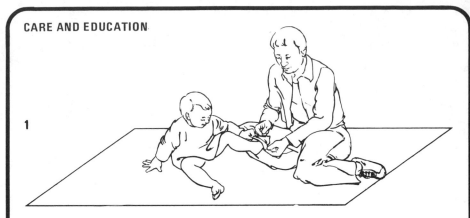

A mentally handicapped child may learn to dress first with help (**1**), and then alone (**2**).

It is essential that a mentally handicapped child should be treated in accordance with his own particular needs - emotional, educational, and social.

A child who is only mildly retarded should be able to cope at a normal school, possibly spending some time in special classes where the emphasis is on developing social, educational, and vocational skills needed to live independently as an adult.

Until comparatively recently it was common practice for more severely retarded children to enter an institution at an early age, and to remain there for the rest of their lives. The last two decades, however, have seen a shift away from institutionalized care - with its risk of emotional deprivation - toward care for the mentally handicapped within the family and the community as a whole. Within this general picture, it is obvious that recommendations for individual cases must depend both on family circumstances and the degree of a particular child's handicap.

Caring for a severely mentally handicapped child in the home obviously places considerable strain on other members of the family. Effective professional and community support are essential if the child's best interests are to be taken into account.

Many moderately retarded children benefit from attending a special day school or training center, where they will if possible be prepared for future employment in a sheltered work situation.

In severe cases expectations will be lower, but systematic training programs can often teach basic skills such as self-feeding, dressing, language development, and social responsiveness.

EMOTIONAL DISTURBANCE

EMOTIONAL DISTURBANCE

An emotionally disturbed child is not likely to complain openly about his worries. Instead he may become depressed - losing interest in activities, seeing few friends, and being unusually quiet. Other children become overanxious - crying a lot, sleeping badly, and clinging to their parents. Unruly behavior, too, may be a sign of disturbance, as the child tries to draw attention to his problems. Children with phobias transfer their real fears to situations or objects with symbolic value. Other children suffer from hysterical illnesses - converting an emotional problem into a physical one. Often parental understanding is sufficient to help a child through a difficult period, but in prolonged or severe cases the child may be in need of professional help.

AUTISM

Autism is an uncommon mental condition in which the child is completely absorbed in himself. He appears indifferent to other people, rejects affection, and refuses to communicate. Often he will sit for hours, lost in a world of fantasy, perhaps playing with his fingers or a scrap of paper. At other times he is hyperactive, behaving in an unruly and difficult manner. Many autistic children can go for days with very little food or sleep.

The cause of autism is unknown, but research suggests that the problem may be biochemical. Special education and therapy have brought dramatic improvements in some cases; the first requirement is usually to break down the child's barrier to communication, so helping prepare the way for the development of more normal patterns of behavior.

TREATMENT

Therapy for a young child usually includes play (**1**) and painting (**2**).

Specialists in the treatment of emotionally disturbed children include psychiatrists (medical doctors with additional training in psychiatry), psychologists (nonmedical professionals with graduate and postgraduate training in psychology), social workers, and nurses involved in mental health programs. Occasionally an emotionally disturbed child will be admitted into a hospital or residential unit for treatment, but in most cases it is preferable for the child to receive treatment while continuing to live at home. Special centers providing day care and treatment of various kinds have been set up in some areas to cater for the needs of emotionally disturbed children. Other children continue to attend their regular school, but visit a child psychiatrist or psychologist for treatment sessions at intervals depending on their own particular needs.

Many disturbed children benefit from psychotherapy. In the case of a young child, the emphasis is on play and painting; the therapist interprets these activities as the child's personal language, and through them helps the child to face and work through his problems. With an older child whose use of language is more developed, the therapist encourages the child to talk freely; in this way the therapist coaxes the child's problems to the surface so that the child can be helped to understand and come to terms with his difficulties. Sometimes a child's emotional problems are part of a general picture of difficulties within his family, and family counseling and other help from social workers may be needed.

© DIAGRAM

FAMILY PROBLEMS

FAMILY PROBLEMS

Children can be deeply affected by family problems. Sometimes the family is under pressure from social problems such as poverty, poor housing, and prejudice.
In other cases the problems have their origins within the family itself - in individual personalities, or in difficult relationships between family members. Very often, pressures from outside the family cause pressures within it. One-parent families have their own particular problems, as do families who foster or adopt a child.
In general it is useless and potentially harmful to pretend that all is well when this is obviously not the case. Families under pressure need support from friends, and perhaps also from professional agencies.

NEGLECT

Neglect may be physical or emotional, immediately obvious or more difficult to recognize. In any case it is likely to have a profound effect on the child - many neglected children grow up to neglect their own families.
Very often, neglect is part of a more general picture of family difficulties. In some cases the parents were emotionally unprepared for raising a family, or are confused by the economic demands of running a home. These same stresses may lead to physical violence against their children as well as chronic neglect.
Parents who are in need of help themselves are often unable to care properly for their children, even though they want to. Support from social workers is essential in all cases of child neglect.

The diagram shows trends in US dependency and neglect cases (based on US Department of Health, Education, and Welfare statistics).

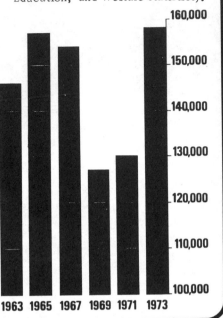

1963 1965 1967 1969 1971 1973

DIVORCE

In a great many countries the number of children whose parents divorce is increasing every year. Special care is needed to ensure that these children suffer as little as possible from the break-up of their parents' marriage.
Many of the parents who now choose divorce would once have opted to "stay together for the sake of the children." In practice, the children rarely benefited. An atmosphere of tension in the home can easily be detected by a child, and there is a strong temptation for parents to draw the children into their own quarrels.
Divorce does not have to mean the loss of a parent; most couples make arrangements for sharing time with their children. If a parent remarries, the child may, after a period of readjustment, come to enjoy and benefit from his new "extended" family.

The diagram (based on US divorce statistics) shows the increasing number of children whose parents are obtaining a divorce.

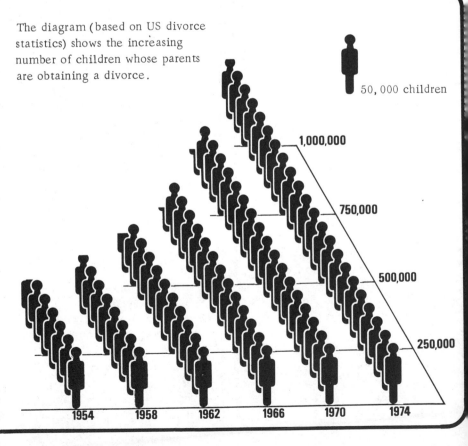

50,000 children

1954 1958 1962 1966 1970 1974

CHILD ABUSE

Every year courts handle many cases in which children have been physically injured by their parents. The problem is not a new one, but its survival into modern, more enlightened times is cause for serious concern. Child abuse, or "battering" as it is now commonly called, occurs in every community. It is found in families at all economic and educational levels, occurring in homes that are clearly in chaos and in others that are apparently well-regulated. Very often, cases of child abuse are discovered when a child is brought to the doctor after an "accident" - and increased medical awareness has helped in this respect. Child battering rarely results from willful cruelty - but is rather a sign that the parent needs urgent community help.

Based on a study by the Children's Division of the American Humane Society, the diagram below shows the proportions of persons brought into court for causing physical injury to a child.

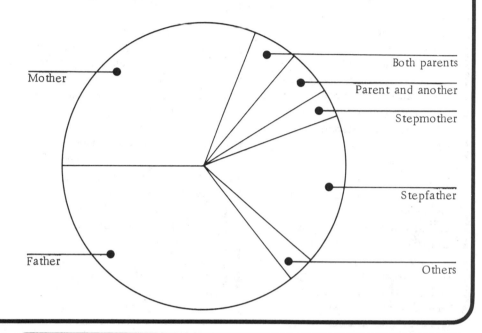

ONE-PARENT FAMILIES

A parent who for whatever reason must bring up a child alone has many additional problems to face. Even problems common to any parent can be an extra strain when there is no partner to give support. Many one-parent families have money problems - the physical and emotional needs of the children may make it difficult to take a well-paid job, or it may prove impossible to do any work outside the home. More nurseries and playgroups for young children are urgent requirements, and provision is needed for older children after school or during vacations. Lone parents also face emotional problems - feelings of isolation, or behavior problems in their children caused by insecurity. Support from friends, self-help organizations, and the community can prove of tremendous value to lone parents and their children.

FOSTERING, ADOPTION, RESIDENTIAL CARE

Sometimes family problems are such that it is impossible or undesirable for children to remain, either temporarily or permanently, in the homes of the natural parents.
FOSTERING If the period of separation is to be temporary, attempts will be made by social workers to place the child in a foster home. Foster parents are carefully screened, and every effort made to unsettle the child as little as possible. Problems arise if the child is moved too often, or if he becomes too attached to his foster parents at the expense of his natural parents to whom he must return.

ADOPTION When the separation from the parents is to be permanent, the child is usually taken first to a reception home. Every effort will then be made to find an adoptive home for him. Adoption of a child is a permanent arrangement, and great care is taken to do everything possible to ensure that the adoption will be a success: the physical and mental health of the child is assessed, and the prospective parents will be searchingly questioned. After the adoption, support will be given by social workers. An adopted child should always be told early on of his adoption - a chance discovery of the fact is much worse than being told the truth at some suitable natural moment.
RESIDENTIAL CARE Only if it is impossible to place him in a suitable family is a child likely to spend much time in a residential children's home. Within these homes, every effort is made to provide as natural a family environment as possible - with small groups of children in the care of residential houseparents.

©DIAGRAM

SOCIAL PROBLEMS

SOCIAL DISADVANTAGE

Some children are in need of special care, from within the family and without, to prevent them from suffering from the effects of social disadvantage – effects that may last for the rest of their lives.

In many cases family poverty is the key problem. Inadequate diet may lead to poor health, which in turn may cause absences from school and poor academic performance. Poor housing, too, has physical and social effects. Parents who are forced to work long hours have less time and energy to devote to their children, while unemployment can strain all family relationships. Many disadvantaged children grow up to lead successful and happy lives, but numerous difficulties must first be overcome.

INTEGRATION

Mutual understanding and respect are needed to hasten the integration of different national, racial, and religious groups within the community as a whole. Differences in social customs can prove confusing to a child, who may encounter different sets of rules at home and elsewhere. Dietary regulations are an obvious example of this type of problem; harder to cope with, perhaps, are different views on polite or impolite behavior. Encouraging children to respect the cultural traditions of others is one of the simplest, most positive ways for parents to help fight destructive prejudice.

Language is one of the major barriers to integration, but in general children tend to cope better than their parents with this

particular problem. Most young children manage to pick up a new language comparatively easily, although special language teaching may be needed at first.

HOUSING PROBLEMS

Inadequate housing can affect the health and happiness of children as well as their parents.

Children who are brought up in slum dwellings obviously face greater risk to health from cold, damp, and vermin. Overcrowding frequently makes matters worse. Ambitious rehousing schemes – often involving moves to large apartment blocks in different parts of town - have not, however, been an unqualified success.

Separation from old friends and difficulty making new ones can cause problems for children and parents, and the lack of community feeling in many new housing developments is a major social loss. Very often the greatest single problem for children is inadequate provision of play space – both indoors and out - and to avoid serious social problems planners must take this into account.

POVERTY

The inhabitants of the United States are among the richest in the world. But here, as in other countries, a great many people live in poverty.

The diagram below shows the proportion of US children aged under 18 who in 1973 were members of families in poverty. The "poverty index" on which it is based was developed by the US Social Security Administration. This index takes into account family income, family size, farm or nonfarm residence, and changes in the Consumer Price Index.

In 1973, 13.1% of children in metropolitan areas were in poverty, compared with 16.6% of children in nonmetropolitan areas.

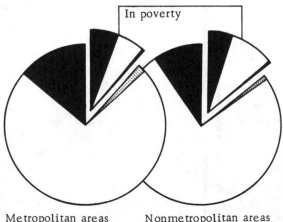

In poverty

Metropolitan areas   Nonmetropolitan areas

The diagram shows proportions of US children in poverty in 1973 (based on US Bureau of the Census statistics).

☐ White children
■ Black children
▨ Other children

JUVENILE CRIME

Recent years have seen alarming increases in the incidence of juvenile crime in many countries. Sometimes delinquent behavior is no more than an isolated incident in a child's normal pattern of development. In other cases, the problem is more deeply rooted - with incidents of varying severity occurring over a period of time. Very often, delinquent behavior is a response to unsatisfactory family or social circumstances, and this must be taken carefully into account by the authorities when deciding the best course of action in each particular case.

Many juvenile offenses stem from boredom, and one positive way in which a community can help tackle the problem of juvenile crime is to see that adequate entertainment facilities are provided.

DELINQUENCY TRENDS

The number of delinquency cases handled in juvenile courts in the USA has shown a marked increase over the past 20 years. Especially striking is the increase in the number of girls now appearing in the courts. Similar increases are to be found in other countries.

The diagram shows cases handled by juvenile courts (from statistics published by the US Department of Health, Education, and Welfare).

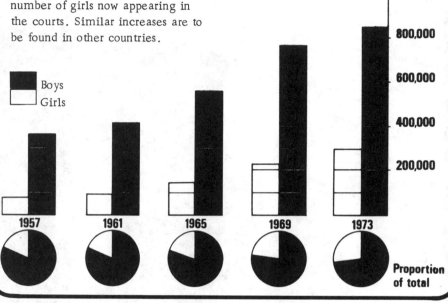

TYPES OF CRIME

The majority of crimes committed by juveniles are crimes against property. In the USA in 1975, property crimes - including burglary, larceny-theft, and motor vehicle theft - accounted for 42% of arrests among persons under 15 years of age. In the same year, violent crime - including murder, forcible rape, robbery, and aggravated assault - accounted for only 3% of the total number of arrests within this same age group. Runaways accounted for 11% of arrests in this age group in 1975, and violations of local curfew and loitering laws - another specifically juvenile category - for a further 4%.

The diagram shows principal causes of arrest among those aged under 15 (based on FBI statistics for 1975).

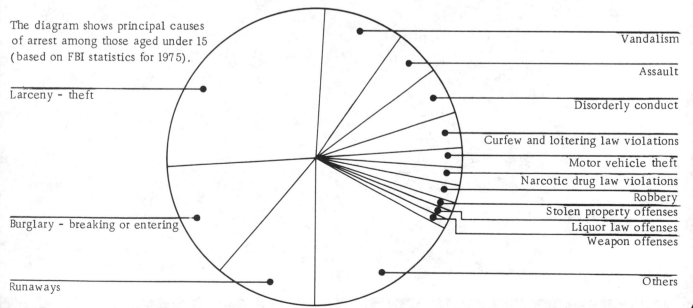

© DIAGRAM

18 First aid

With a first aid kit and a knowledge of basic first aid techniques, a parent can cope with numerous emergencies.

Left: Red Cross aid for Persian Gulf children (Central Press Photos Ltd)

HOME MEDICAL KITS

FIRST AID KIT

Every home should have this kit: close at hand, in a portable box or can, and unlocked - but out of reach of children. It should be complete in itself - not dependent on kitchen scissors, for example.

Its medicines and lotions should be clearly labeled, its dressings kept well wrapped. A basic first aid pamphlet (or copies of this chapter) and a notepad and pencil should be kept inside, and

emergency phone numbers pasted to the lid. It should be sealed with adhesive tape, to keep it clean and dry and help keep out children. Similar kits should be kept in cars, boats, and campers.

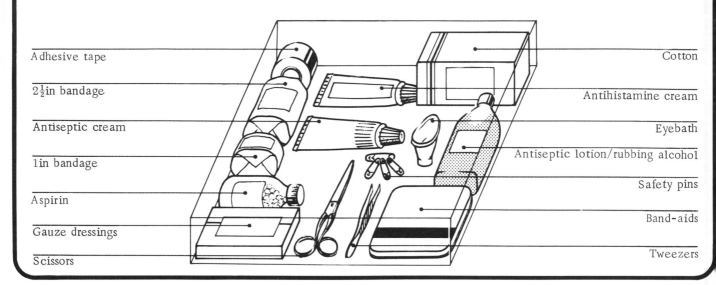

Adhesive tape

2½in bandage

Antiseptic cream

1in bandage

Aspirin

Gauze dressings

Scissors

Cotton

Antihistamine cream

Eyebath

Antiseptic lotion/rubbing alcohol

Safety pins

Band-aids

Tweezers

EMERGENCY ITEMS

Here are shown some common home articles that may be useful in an emergency. It is a good idea to have all of them near at hand. Towels, handkerchiefs, and tissues are useful for cleaning or

covering wounds. Vinegar may be applied to wasp stings, and bicarbonate of soda to other stings and burns. Salt and water is a useful antiseptic. Olive oil may

be used for insects in ears. A needle can be quickly sterilized by holding over a lighted match before using to remove stones from wounds.

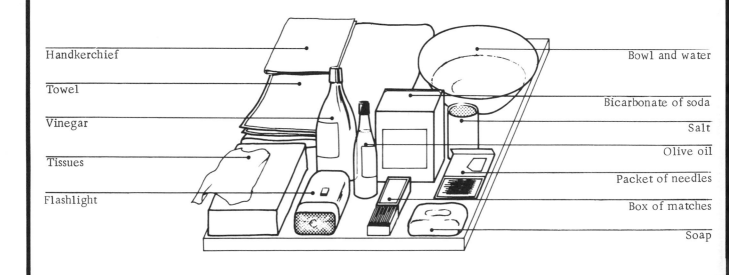

Handkerchief

Towel

Vinegar

Tissues

Flashlight

Bowl and water

Bicarbonate of soda

Salt

Olive oil

Packet of needles

Box of matches

Soap

HOME TREATMENT KIT

This includes all the contents of the first aid kit, plus additional items (mainly medicines and lotions). It can be divided into compartments: one for wound cleaning and dressing, one for bandages and instruments, one for medicines and creams. (Do not include any prescribed medicines, though: these should be locked away separately.) As with a first aid kit, all items should be clearly labeled and first aid instructions and notepad and pencil added. But you should still have a separate – more portable – first aid kit, as shown opposite.

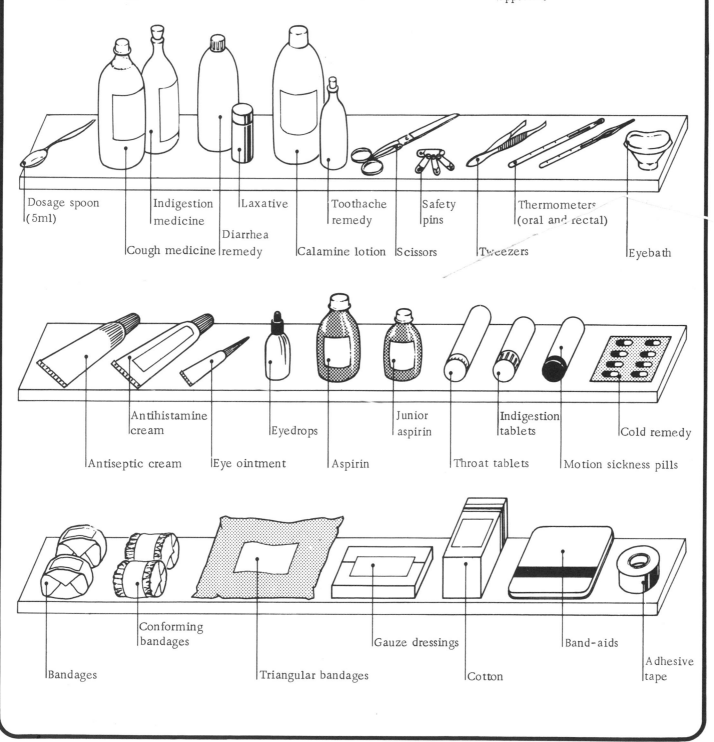

Dosage spoon (5ml)

Indigestion medicine

Laxative

Cough medicine

Diarrhea remedy

Toothache remedy

Calamine lotion

Safety pins

Scissors

Thermometers (oral and rectal)

Tweezers

Eyebath

Antihistamine cream

Antiseptic cream

Eyedrops

Eye ointment

Aspirin

Junior aspirin

Throat tablets

Indigestion tablets

Motion sickness pills

Cold remedy

Conforming bandages

Bandages

Triangular bandages

Gauze dressings

Cotton

Band-aids

Adhesive tape

©DIAGRAM

WOUNDS

STOPPING BLEEDING

Press a pad of clean cloth against the wound. If bleeding continues, add thicker cloths on top, and use more pressure. Keep the injured area still, and calm the child. If a limb is badly cut (but not broken), it helps to raise it above the level of the rest of the body.

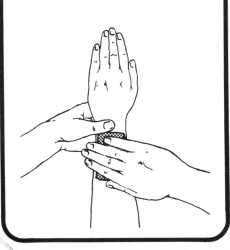

WASHING A WOUND

Wash your own hands first. Clean the skin around the wound (wiping away from the wound), then the wound itself. Use soap, running water, antiseptic, and several fresh swabs (preferably of sterile gauze). Get out all loose dirt, etc; but anything embedded should be left for expert attention.

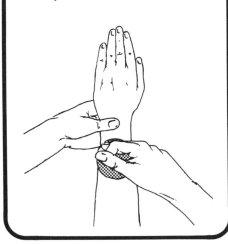

BANDAGING A FINGER

Use sterile gauze as dressing, then cover with a long roll of narrow bandage. Run the bandage from base to tip of the finger, and back down the other side (**1**). Then wrap it around the finger (**2**), split the end (**3**), and tie (**4**). A finger stock (**5**) helps keep the bandage clean and secure.

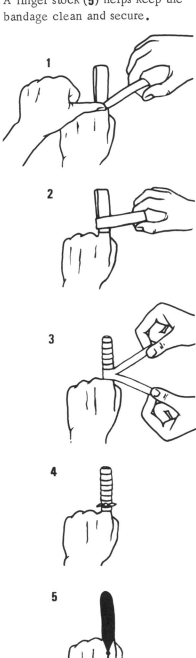

DRESSING A WOUND

Apply mild antiseptic, using sterile gauze. When dry, cover wound and surrounding skin with a piece of sterile gauze, handling it only by the corners. Add surgical cotton on top, and keep this dressing in place with bandage or adhesive tape. Alternatively, use a prepacked sterile dressing and bandage.

When bandaging, start at the narrowest point (**1**). Overlap the first few turns, then work up. Bandage firmly but not too tightly. Finish with a safety pin (**2**) or adhesive tape, or split and knot the bandage end. In an emergency, a clean handkerchief or other cloth can be used.

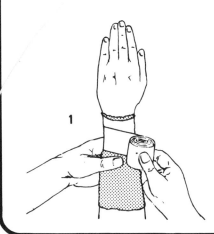

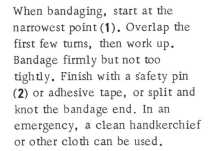

ELBOW OR KNEE BANDAGE
Place the bandage against the elbow (or knee), point upward (**1**). Then wrap one of the bottom ends around the joint (**2**), and repeat with the other bottom end (**3**). Tie these two ends together, not too tightly (**4**), and finish by tucking the bandage point down over the knot (**5**).

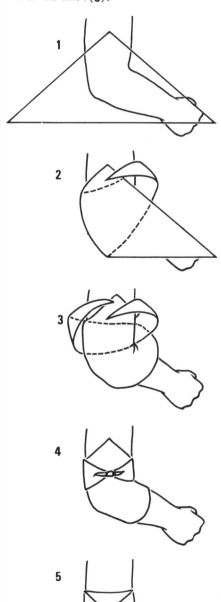

INTERNAL BLEEDING
This can result from broken bones or ruptured internal organs. Blood may be coughed or vomited up, or be visible in the urine or feces, or trickle from nose or ear. Often, however, it is trapped in body tissues (which may swell) or in body cavities like the abdomen or chest (which may become painful). In any case of internal bleeding the victim shows rapidly developing symptoms of "shock" (pale face, cold, clammy skin, restlessness, rapid pulse, etc). Urgent medical attention is needed. Meantime take the measures against shock, and at intervals make notes for the doctor of the patient's pulse. Also note the color of any blood from the mouth (bright red, frothy blood is probably from the lungs, dark red or black from the stomach).

WHEN TO CALL A DOCTOR
In the case of wounds, get urgent medical help for:
a) any internal bleeding;
b) any external bleeding that will not stop;
c) any external bleeding that only stops after considerable blood loss;
d) a bleeding nose or ear after a blow on the head.
Keep the patient still and quiet, and reassure him, till help comes. Also get medical attention without undue delay for:
a) any wound that has something embedded in it;
b) any puncture wound - one that is deeper than it is long (eg one made by a nail or knife point);
c) any wound from an animal bite;
d) any other wound that you think may be a tetanus risk.

BRUISES AND BLACK EYES
A cold wet cloth on the damaged area should reduce pain and swelling. The cloth can be kept in place with waterproof material and a bandage. If after some days pain persists, see a doctor (for example, a black eye may conceal a fractured brow).

BLEEDING NOSE
Make the child sit quietly as shown, nostrils held pinched together. Tell him to dribble any saliva in the mouth - swallowing movements can disturb blood clotting in the nose. After 10 minutes, if there is still bleeding, try plugging the nostrils with sterile gauze. Slight nosebleeds are common in childhood, but see a doctor if they become persistent.

©DIAGRAM

517

RESUSCITATION

ARTIFICIAL RESPIRATION

Mouth-to-mouth respiration can be used in almost any case where breathing has stopped.

1) Lay victim on his back. Turn his head to one side, and clear any debris from his mouth (**a**).

2) Turn his head up again, and put a folded coat under his shoulders. With one hand under the neck, and the other on the crown of the head, tilt the head back as far as possible. Then pull the chin up till the head is fully tilted back (**b**). This position ensures that the tongue does not obstruct the windpipe.

3) Put your mouth firmly over the victim's mouth, pinch his nostrils shut, and blow hard enough to make his chest rise (**c**). (With a small child, put your mouth over his nose and mouth, and use shallow breaths.)

4) Remove your mouth and listen for exhaled breath (**d**). Then repeat the blowing in: once every 3 seconds for children (once every 5 to 6 seconds for adults).

5) If no air is exhaled, check the victim's head and chin position, and check that his tongue is not blocking his throat. Try again.

6) If still no exhalation, put the victim's head down for a moment over your lap, and slap him sharply between the shoulder blades, to dislodge any blockage. Wipe the mouth clear.

a

b
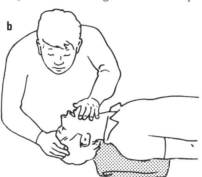

c
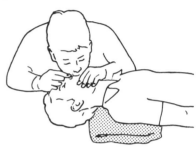

DROWNING

Start artificial respiration at the earliest possible safe moment (eg in shallow water, in a boat, or at the water's edge). Do not try to drain water from the lungs: any that comes up will probably be from the stomach. Just clear the mouth of water, seaweed, etc, and give artificial respiration till breathing starts.

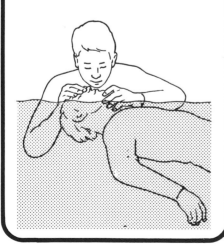

FAINTING

If someone feels faint, make him lie down or sit as shown, and breathe deeply till he feels better. If he faints, lay him down on his back, head low, legs raised. Loosen tight clothing (especially at the neck), and let him come round in his own time. If the fainting lasts more than a minute or two, keep him warm and get medical attention.

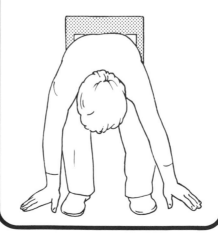

FITS (CONVULSIONS)

In a fit, the victim's body seems jerked by uncontrollable spasms. His head may be thrown back, his lips turn blue, his eyes roll up, his mouth froth. Do not try to restrain him, or throw cold water over him, or pick him up to rush for help. But do guide his movements, remove furniture, and lay him on the ground so that he cannot hurt himself. (If it can be done without force, put a rolled handkerchief between his teeth to stop him biting his tongue.) Also keep his airway clear by loosening clothing at the neck and turning his head to one side so saliva drains out. If you can, guide him into the recovery position (see p.519). A fit usually lasts only a few minutes. Afterward, put him to bed and get medical advice. Give no food or drink. A single fit in a child is commonly a sign of fever and sponging with tepid water may help.

7) Don't give up till the victim starts to breathe. Many have revived after hours of artificial respiration.

8) When he is breathing strongly, keep him warm and get help. Don't let him get up. Put him in the recovery position (N18) if you think he may vomit.

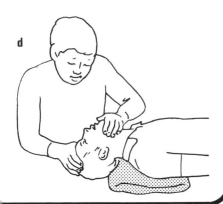

d

RECOVERY POSITION

Death following unconsciousness is often from suffocation or pneumonia, due to the victim inhaling his own saliva or vomit. To prevent this, if someone is unconscious but breathing, place him in the recovery position shown: stomach down; head turned to one side; and arm and leg on that side pulled up till the thigh is at right angles to the body, and the hand level with the jaw. Pull the chin forward and up, so that the tongue cannot block the throat.

Loosen the collar, and see the mouth is clear of debris, blood, or mucus; remove false teeth. Do not put a pillow under the head. In fact, if possible, raise the legs and body slightly above head level, so fluids drain away from the lungs. Look for hidden bleeding beneath clothes or body; deal with any external wounds. Then cover with one blanket and watch closely till help comes. Give nothing to drink even if consciousness returns.

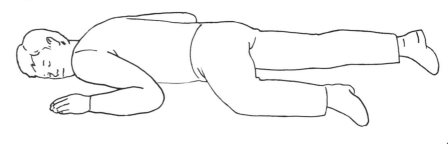

ELECTRIC SHOCK

ELECTRIC SHOCK
Act fast - every second counts.
1) Break the victim's contact with the current in the quickest SAFE way (see below).
2) Check his breathing, and use artificial respiration if necessary (continue for hours if need be: recovery is still possible).
3) If breathing but unconscious, put in recovery position.
4) Give first aid to any burns.
5) Get help urgently.
BREAKING ELECTRIC CONTACT
a) Pull out plug, turn off current at fuse box, or pull away appliance by cord; or
b) pull at a DRY, LOOSE part of the victim's clothing; or
c) push or pull at the body with any dry non-metallic object.
But DO NOT touch the victim's body; and be sure you are standing on a dry surface and touching only dry materials.

SHOCK

Shock is of two kinds:
a) nervous shock, due to emotional trauma or severe pain; and
b) surgical shock, due to loss of body fluid (eg from bleeding, burns, or repeated vomiting or diarrhea). The second is much more dangerous. The victim is pale, with a cold or clammy skin, or rapid pulse, and rapid shallow breathing. He is often restless and apprehensive, with nausea and thirst. He may faint.
You should act quickly.
1) Lay the casualty down. Use the recovery position if vomiting seems likely. Otherwise keep his head slightly raised and turned to

one side, as shown.
2) Deal with the physical cause of shock (eg try to stop any bleeding).
3) Get medical help.
You can also loosen the patient's clothing at neck, chest, and waist, and cover him with a sheet or thin blanket. Moisten his lips if he is thirsty, and if possible note his pulse and breathing rates.
Do not:
a) warm the patient;
b) move him unless forced to;
c) give him anything to drink until he has been seen by a doctor.

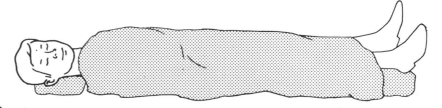

©DIAGRAM

FOREIGN BODIES

CHOKING

Put the child over your knee, or over a chair back; or, if he is small enough, pick him up bodily. Then give three or four firm slaps between the shoulder blades. If this does not work, get medical help at once, and give artificial respiration (N13) if necessary.

Also – but only as a desperate measure – try reaching in and pulling the object out with your fingers.

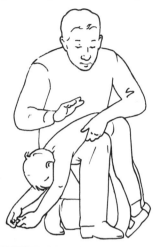

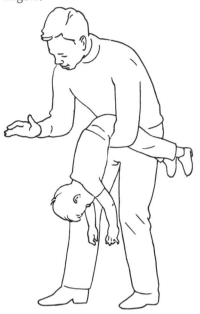

FOREIGN BODY SWALLOWED

If the object is smooth, small, and rounded, it should cause no trouble. Just give normal diet, and examine the child's bowel movements for a few days, to make sure that the object has passed through. See a doctor, though, if the child is under two, or if he seems to become unwell. But if the object is sharp or pointed; or if there is any chance that it might have been inhaled into the lungs, not swallowed; then give nothing to eat or drink, and get medical help.

REMOVING RINGS

Rings can get stuck because they are too small, or because the finger has swollen due to injury or infection. Smearing the finger liberally with soap (a) may allow the ring to be pulled off. If not, try binding thread tightly around the forefinger (b) for a short distance above the ring. Pull the ring up onto the bound part, unwind the thread behind it, and bind again above where the ring now is. Continue until you have worked the ring off the finger. But do not leave the binding on for any length of time, or else the blood supply to the finger will be affected.

GRAZES

Remove any loose dirt, etc, with moist sterile swabs, or with tweezers sterilized in boiling water for five minutes. Then treat as a wound. But do not try to get out anything that is stuck or embedded: just wash the surrounding area, dry, put on a sterile dressing, and get medical attention.

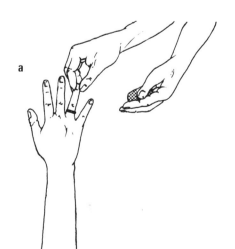

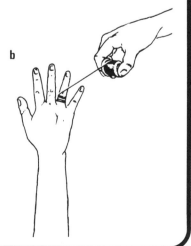

FOREIGN BODY IN THE EYE

Stop the child from rubbing the eye. If the object is sharp or hot, or if the eye is bleeding, do nothing, but get medical attention at once. But if it is probably just a speck of dust, bring the upper eyelid down over the lower, as shown (**a**), while the child turns his eye upward, or wash the eye, using an eyebath or eyedropper.

If these fail, look for the speck, and try to remove it with a moist cotton swab, or the moistened corner of a handkerchief (**b**). But if the foreign body does not move with the gentlest touch, stop, and get medical attention. Bandage the eye meanwhile to ease pain.

If any chemical gets in a child's eye, immediately flush it out with large amounts of water, making sure that the eye is wide open and the lids pulled back (**c**). Keep the child's head turned, so any chemical washed out does not run into the other eye. Do this for 15 minutes, then get medical help.

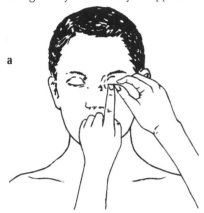

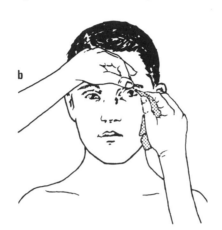

SLIVERS (SPLINTERS)

If the sliver protrudes, wash your hands, wash the area well, and pluck out the sliver with sterilized tweezers, needle, or knife point. Press the skin so that a spot of blood comes from the wound, then wash well or apply a mild antiseptic. Cover with sterile dressing if necessary. Deeply embedded slivers or inflamed wounds need medical attention.

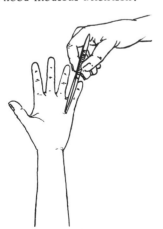

FOREIGN BODY IN THE NOSE OR EAR

FOREIGN BODY IN THE NOSE
If the object is small and smooth, a sneeze will usually dislodge it: use pepper to set off sneezing (**a**). If this does not work, get medical advice. Do not try violent nose-blowing, or probe into the nostril, or pour water or oil into the ears.

FOREIGN BODY IN THE EAR
If the child has an insect in his ear, put in a few drops of lukewarm olive or mineral oil. This will stop the frightening buzzing, and may even wash the insect out. However with any other object in the ear, do nothing, except to tilt the head to one side, as shown (**b**), to see if the object falls out. If not, get medical help; any probing may damage the ear.

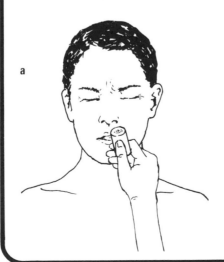

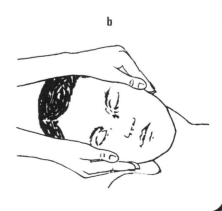

©DIAGRAM

POISONING

DANGEROUS ITEMS

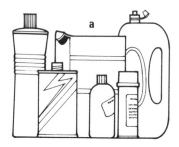    

a Some household chemicals, eg insecticide, rat poison, and weedkiller; kerosene, gasoline, benzine, turpentine, and any cleaning fluid; liquid furniture and auto polish; lye, and alkalis used for cleaning drains, bowls, etc; oil of wintergreen, ammonia, bleach, washing soda, and detergents; mothballs, and lead-based paint (which should never be used on indoor surfaces).
b Some grooming articles, eg perfume, cosmetics, hair tonics, nail polish and remover.
c Many prescription and some non-prescription drugs.
d Alcohol.
e Food containing bacteria; and misused food (eg salt given to a baby in mistake for sugar).
f Some common house plants (and water they stand in), eg soleander, diefenbachia, poinsettia.

POSSIBLE SIGNS

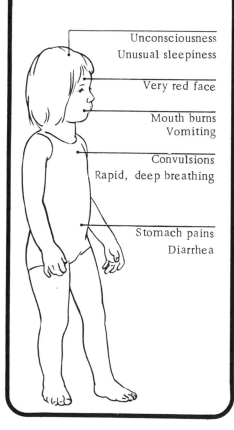

Unconsciousness
Unusual sleepiness

Very red face

Mouth burns
Vomiting

Convulsions
Rapid, deep breathing

Stomach pains
Diarrhea

WHAT TO DO

1) If breathing is failing, give artificial respiration (see p. 524). (NB: use mouth to nose respiration if there is any chance of poison still in the mouth.)
2) If breathing but unconscious, put in recovery position (see p. 519).
3) If conscious but likely to vomit, put in recovery position (see p. 519).
4) Question the victim, if possible, and look for evidence of the poison (empty bottles, scattered pills, an odor, pills in mouth).
5) Call emergency help.
6) Decide whether to make the patient vomit; see N30.
7) Keep the patient warm till help comes. Do not leave.
8) Keep a sample of the poison and any vomit.
Still get attention even if: the patient vomits, and then seems all right; the poison seems to have had no effect; you are not sure if any poison was taken.

INDUCED VOMITING

Make the patient vomit ONLY if:
a) he is fully conscious; and
b) he is not convulsing; and
c) you know for certain that the poison is not an acid, alkali, or liquid petroleum product.
To induce vomiting: either tickle the back of the throat with your fingers; or give two tablespoons of salt or mustard in a glass of warm water. Before he vomits make the patient lie on his front, with his head lower than his body and over a bowl. Induce repeated vomiting if possible.
If it is safe to induce vomiting, it is also safe to give bland fluid (eg water or milk) afterward to dilute any remaining poison.

BITES AND STINGS

ANIMAL BITES

Calm the victim. Use running water on the wound to flush out saliva. Then wash wound well for five minutes with sterile swabs, using soap and water (not strong antiseptics such as iodine). Rinse, dry, and dress the wound, and get immediate medical attention. Animal bites are seldom serious, but anti-tetanus and/or anti-rabies injections may be needed. (If possible, catch the animal so it can be watched for rabies symptoms.)

PLANTS AND JELLYFISH

NETTLE STINGS Relieve with calamine lotion or antihistamine cream.
POISON OAK OR POISON IVY Wash at once with soap and water. Then wash with rubbing alcohol to relieve itching.
JELLYFISH STING Treat with calamine lotion or antihistamine cream. But if the victim gets short of breath, or faints, get emergency medical attention.

MOSQUITOES, GNATS, ANTS

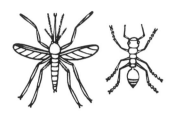

Wash with soap and water. Apply calamine lotion, antihistamine cream, or a paste of bicarbonate of soda and water. Cover any swelling with a cold wet cloth. Do not let the child scratch the bite - this can damage the skin and increases the risk of infection. Mosquitoes and gnats can carry disease so if any complications develop within a few days get medical advice, especially if in a tropical or subtropical country.

FLEAS, LICE, TICKS

Isolated lice and ticks can be loosened by covering them with oil, grease, turpentine, or nail polish. They can then be removed with tweezers (with ticks make sure that you remove the head as well as the body). Crush the creature, and flush away or burn. Clean flea, louse, or tick wounds with soap and water or a mild antiseptic, and apply calamine lotion or antihistamine cream. Get advice about the possibility of transmitted diseases such as tick fever.

BEES, WASPS, HORNETS

If the sting is still in the skin, scrape it out with a sterilized needle. (Do not pull it out with tweezers or fingernails: you may squeeze more poison into the wound.) If the sting has just occurred, apply antihistamine cream. If not, run cold water over the area, dry, and then apply surgical spirit or a solution of bicarbonate of soda. For a sting in the mouth, give a mouthwash of bicarbonate of soda solution. Get medical help if:
a) the victim shows signs of general distress, eg skin rash, pallor, weakness, nausea, or tightness in chest, nose, or throat;
b) there is a dangerous swelling (eg from a sting in the mouth); or
c) the victim has been stung many times.

SNAKE BITES

If non-poisonous, treat as an animal bite. If poisonous or unknown, tie a tourniquet above the bite, cut two $\frac{1}{4}$in crossed incisions over the fang marks, and suck out the bite well (swallowing the poison is dangerous only if you have an abrasion in the mouth or digestive tract). Get urgent medical help. Keep the victim lying down and quiet, with the bitten area in ice or cold water. Loosen the tourniquet for 2-3 minutes every 15 minutes; suck the wound every 5 minutes. If possible, kill the snake for identification.

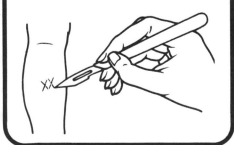

©DIAGRAM

BURNS

TYPES OF BURN

There are three main types:
a) dry burns, caused by fire, over-hot material (eg metal or rubber), electricity, or friction;
b) scalds, caused by over-hot liquid or fat; and
c) chemical burns, caused by acids, alkalis, and some other chemicals.

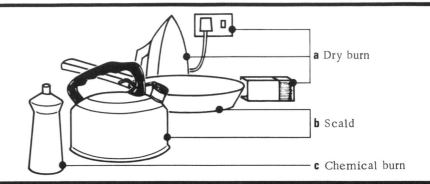

a Dry burn

b Scald

c Chemical burn

SERIOUSNESS OF A BURN

This depends on its area and depth. The area will be obvious. Degrees of depth include:
a) skin reddened, but not blistered;
b) skin blistered;
c) layers of skin destroyed.
The first two are "superficial," the last "deep."
Any deep burn, however small, needs medical attention. But a large superficial burn can be more dangerous, for shock due to loss of body fluid depends on the area of a burn, not its depth. Pain is no guide to a burn's seriousness: a deep burn can destroy nerve ends, so no pain is felt.

CHEMICAL BURNS

Chemical burns are caused by acids (eg hydrochloric acid), alkalis (eg caustic soda), and some other chemicals. With chemical burns, always wash the burn with large amounts of water for up to 10 minutes. Also remove any affected clothing, with gloves if necessary. Then treat as for other burns.

TREATMENT OF BURNS

SMALL SUPERFICIAL BURNS (ie smaller than the size of the victim's palm). Run cold tap water over the burn for a few minutes (a). Wash own hands well; also wash the burn gently if dirty. Dry the burn. If there is no blistering of the skin, a mild, soothing ointment may be applied. If there is blistering, apply nothing, and do not pierce the blisters, simply cover the burn with a sterile non-fluffy dressing and bandage (b).
ALL OTHER BURNS As above, but keep under water longer, apply nothing except a dressing, give liquid to drink, then get medical attention. Do not breathe on the burn or touch it, and do not pull away clothing stuck to it. If large areas are involved, also give treatment for shock (p. 519), and get help urgently. With very large burns, immerse till help arrives (c).

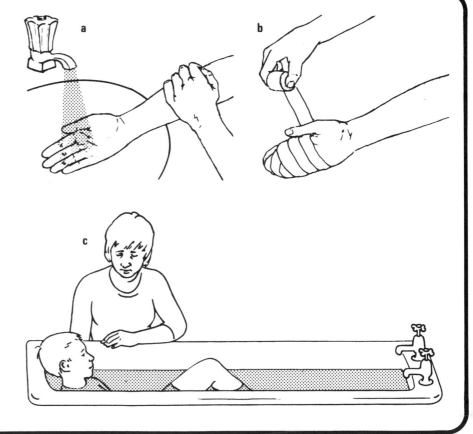

a

b

c

HEAT AND COLD

SUN AND HEAT

SUNBURN

Treat as a burn. If there is no blistering and the burned area is small, apply ointment or lotion. If more severe, apply only a sterile dressing, and get medical advice.

HEAT EXHAUSTION

This develops gradually in very hot and humid conditions, when the body sweats profusely. Loss of body fluid and salt produces shock. Symptoms include muscle cramps, exhaustion, restlessness, a pale face, and cold, clammy skin. Often there is dizziness, headache, nausea, loss of appetite, rapid breathing, and a rapid pulse. Make the victim lie down in a cool darkened area, fan air over him, and apply wet cloths to the head and body. Get him to drink a glass of water containing $\frac{1}{2}$ teaspoon of salt, and repeat this

Heat exhaustion: applying wet cloths

three times at half-hour intervals. If the child is young, or does not recover quickly, get medical help. If fainting or unconsciousness occur, treat immediately as heatstroke.

HEATSTROKE

This is similar to heat exhaustion but more sudden and severe. The victim is red-faced, with hot, dry skin and a high temperature (eg 104°F: 40°C). Breathing is noisy, the pulse strong but fast. Stupor or unconsciousness are common. The urgent need is to get the body temperature down. Strip the victim, and immerse in cold water - or keep pouring cold water over him. Once temperature is below 102°F (38.8°C), wrap in cold wet sheets in the recovery position (p.519). Fan air over him. Get medical help.

COLD

FROSTBITE

Warm the victim gradually at room temperature, and give warm food and drink (not alcohol). Thaw out the frostbitten parts slowly: eg cover frostbitten ears or nose with a gloved hand, place frostbitten fingers in the armpits under the clothing. Do not apply heat directly to the frostbitten part, or rub it, or immerse it in hot water, or apply snow. Finally, begin to move the frostbitten part very gently. Get medical attention.

EXPOSURE

Remove wet clothing, wrap the victim in dry blankets, and get him to warm conditions. If possible, place him in a tub of warm water (not too hot). Dry, place in a warm bed, give warm drinks, and get medical help.

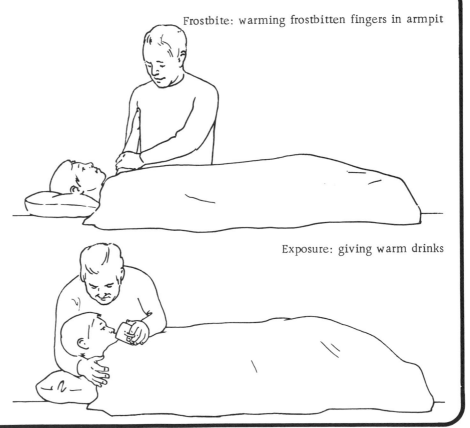

Frostbite: warming frostbitten fingers in armpit

Exposure: giving warm drinks

© DIAGRAM

JOINT AND MUSCLE PROBLEMS

SPRAINS

Sprained ankles are common. The ligaments of the joint are stretched or torn, causing swelling and pain, which increase if the foot is used. A cold compress may reduce the swelling, and firm bandaging relieve pain, but it is best to get medical attention as well in case of fracture. To bandage, surround the joint (**a**) with a thick layer of cotton (**b**) and then bandage firmly as shown (**c**). On top, apply a second layer of cotton (**d**) and bandage again (**e**). Rest foot until swelling goes down.

a

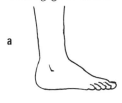

b

c

d

e

PULLED MUSCLE

This is overstretching or tearing of muscle fiber, due to a sudden movement or to handling heavy weights. There is a sudden sharp pain, then pain whenever the damaged part is moved. Make the victim comfortable, with the injured part supported, and get medical attention.

CRAMP

This is sudden painful contraction of a muscle, brought on by chilling (as in swimming), or badly coordinated movement, constriction, or loss of salt and body fluid (eg through sweating). If it occurs, treatment involves forcibly contracting the opposite set of muscles, so that those causing trouble relax by reflex. The drawings show correct procedures for cramp in foot or calf (**a**) and hand (**b**).

a

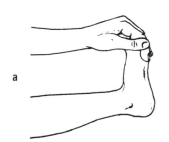

b

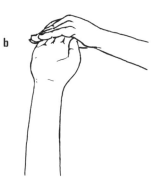

DISLOCATION

The joint is immovable and very painful, and looks deformed. Do not use force, or try to put the bone back in place. Support the limb in a comfortable position (a dislocated leg can be bound to the good leg, in a lying position). Get medical attention. Watch the limb for impaired circulation.

a Normal joints

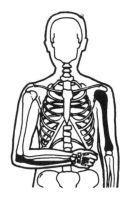

b Dislocated joints

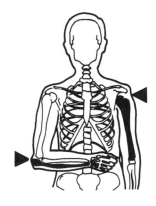

c Signs of dislocation

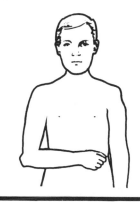

FRACTURES

FRACTURES

Childhood fractures sometimes go unnoticed and untreated. This can be because a child's nervous system may not register the pain very acutely. Also children's fractures are often "greenstick" ones, with no complete break in the bone. So watch for other signs of a fracture: tenderness, swelling, and bruising. A broken limb is also often misshapen and uncontrollable. If you do suspect a fracture:

a) do not let the child use or move the affected part, and do not move or straighten it yourself;
b) stop any bleeding, and lightly cover any protruding bone with sterile dressings;
c) keep the child warm, and treat for shock if necessary; and
d) get medical help.

Do not try to push a protruding bone back in, or clean its wound. And do not move the child, except to avoid further immediate danger - in which case use a stretcher. Moving someone with a broken neck or back is especially dangerous. The only exceptions to the rule about moving are for a broken wrist, arm, or collarbone, when you can move the child to transport or to a warm place. But first support the arm well with a sling.

TYPES OF FRACTURE

The main types are:
a) closed - the skin surface is not broken;
b) open - the bone is exposed to the air (ie it protrudes, or there is a deep wound over it);
c) "greenstick" - the bone is bent or not completely broken;
d) splintered - part of the bone is shattered;
e) complicated - some other part of the body (eg a blood vessel, or a nerve) has also been damaged in the fracture.

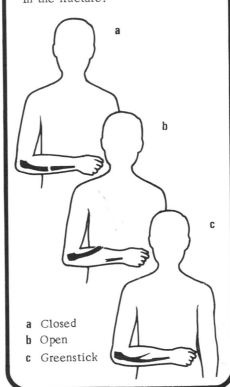

a Closed
b Open
c Greenstick

ARM SLING

Using a triangular bandage:
a put the bandage between the chest and forearm, with the point out beyond the elbow and the top round behind the neck;
b bring the bottom up in front, tie to the top, and fasten in the point.
IMPROVISED SLINGS can be made with belts (c), scarves, neckties, or pinned-up sleeves.

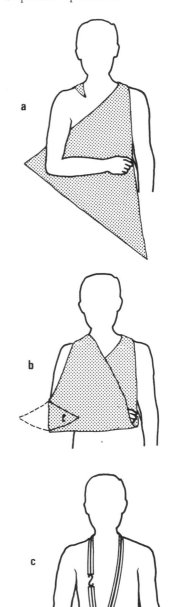

SPLINTS

A splint is used to immobilize a broken limb, if medical help is not quickly available. Wood, metal, or stiffly rolled newspaper can be used, and the splint is bound to the limb above and below the break - firmly, but not too tightly. Alternatively, a broken leg can be bound to the good leg to steady it, and a broken arm to the chest.

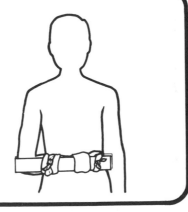

© DIAGRAM

527

REFERENCE MAN AND WOMAN 1

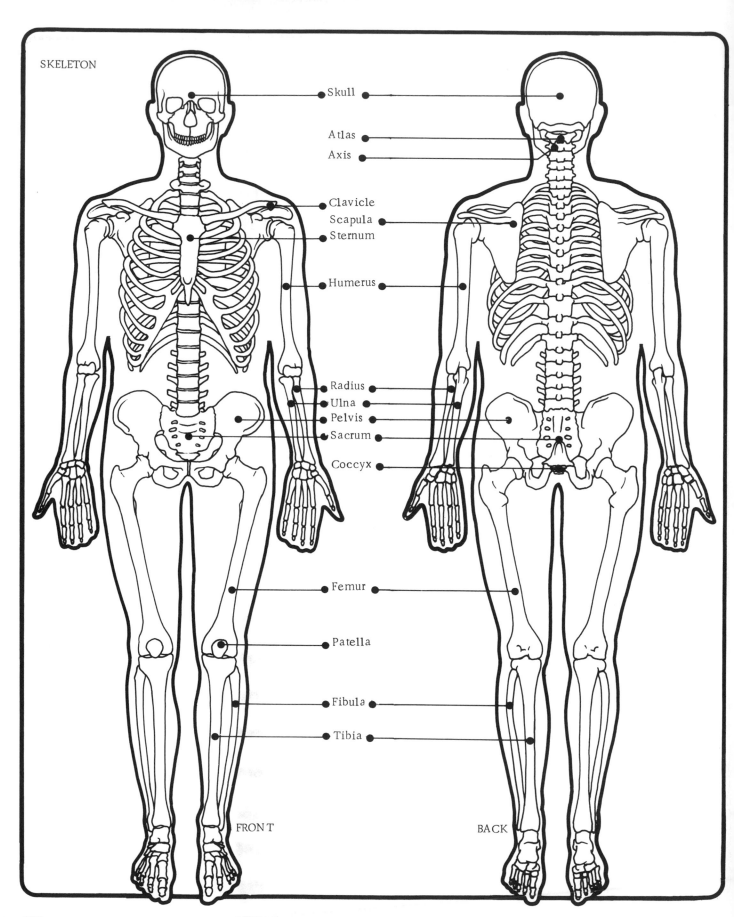

SKELETON

Skull

Atlas

Axis

Clavicle
Scapula
Sternum

Humerus

Radius
Ulna
Pelvis
Sacrum

Coccyx

Femur

Patella

Fibula

Tibia

FRONT

BACK

MUSCLES

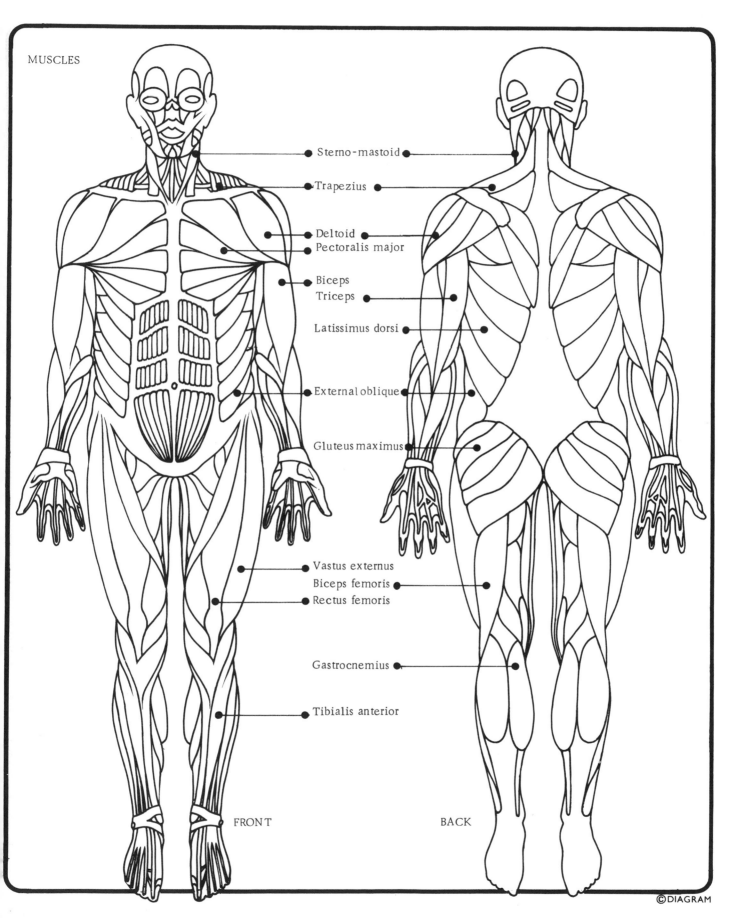

Sterno-mastoid

Trapezius

Deltoid
Pectoralis major

Biceps
Triceps

Latissimus dorsi

External oblique

Gluteus maximus

Vastus externus
Biceps femoris
Rectus femoris

Gastrocnemius

Tibialis anterior

FRONT

BACK

©DIAGRAM

REFERENCE MAN AND WOMAN 2

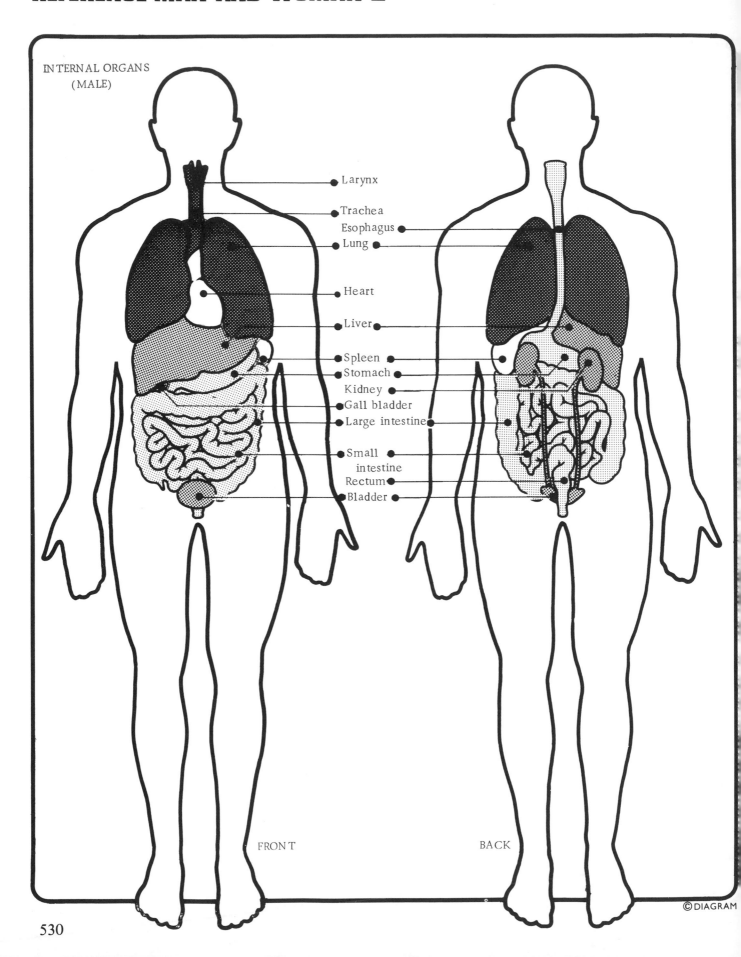

INTERNAL ORGANS
(MALE)

Larynx

Trachea

Esophagus

Lung

Heart

Liver

Spleen

Stomach

Kidney

Gall bladder

Large intestine

Small intestine

Rectum

Bladder

FRONT

BACK

©DIAGRAM

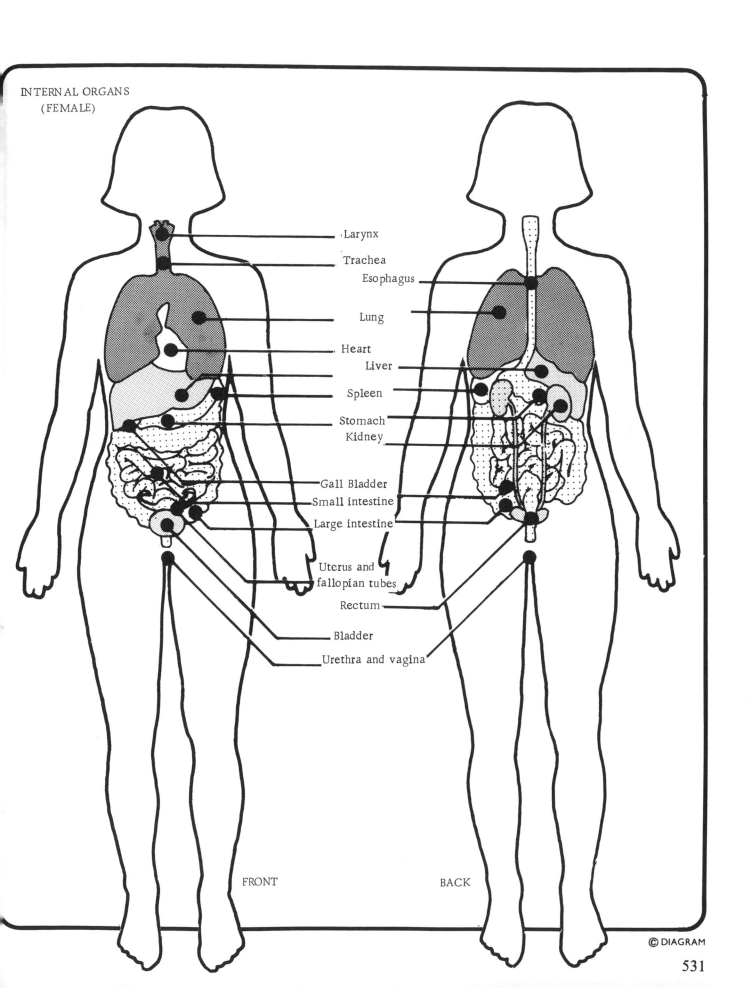

INTERNAL ORGANS
(FEMALE)

Larynx

Trachea

Esophagus

Lung

Heart

Liver

Spleen

Stomach

Kidney

Gall Bladder

Small intestine

Large intestine

Uterus and
fallopian tubes

Rectum

Bladder

Urethra and vagina

FRONT

BACK

© DIAGRAM

REFERENCE MAN AND WOMAN 3

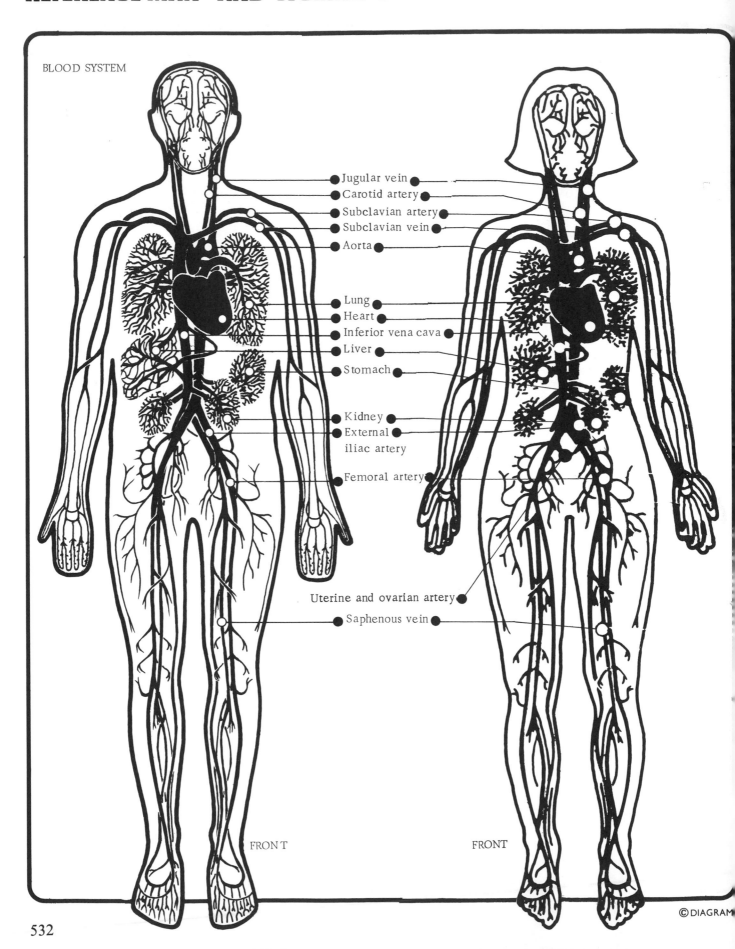

BLOOD SYSTEM

Jugular vein
Carotid artery
Subclavian artery
Subclavian vein
Aorta

Lung
Heart
Inferior vena cava
Liver
Stomach

Kidney
External
iliac artery

Femoral artery

Uterine and ovarian artery
Saphenous vein

FRONT

FRONT

532

©DIAGRAM

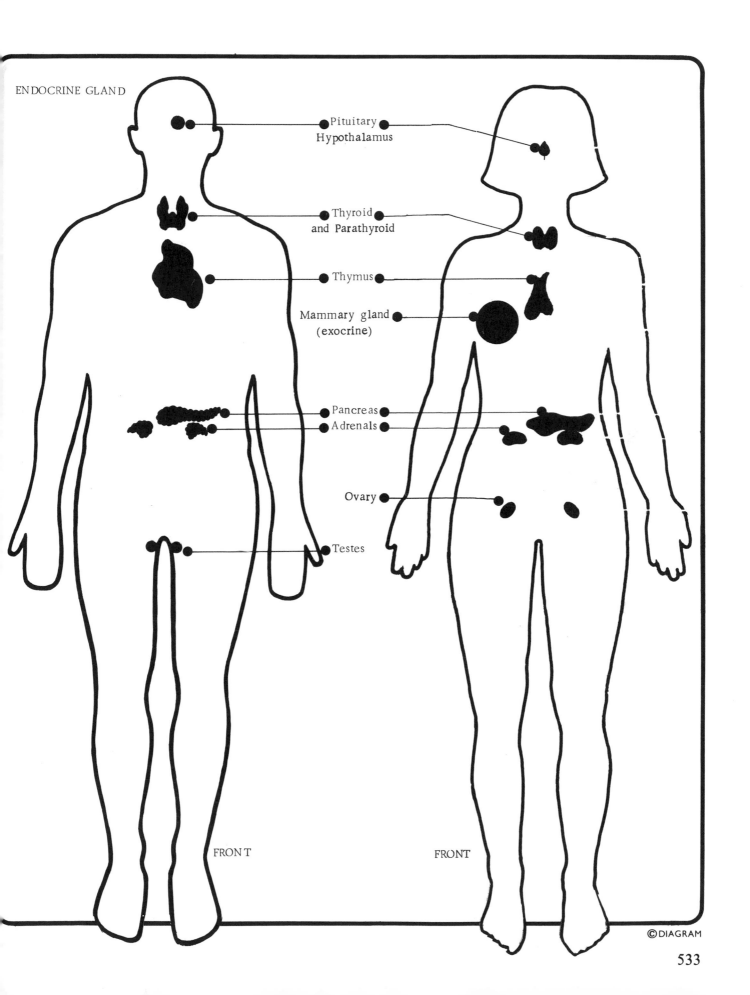

ENDOCRINE GLAND

Pituitary
Hypothalamus

Thyroid
and Parathyroid

Thymus

Mammary gland
(exocrine)

Pancreas
Adrenals

Ovary

Testes

FRONT

FRONT

©DIAGRAM

533

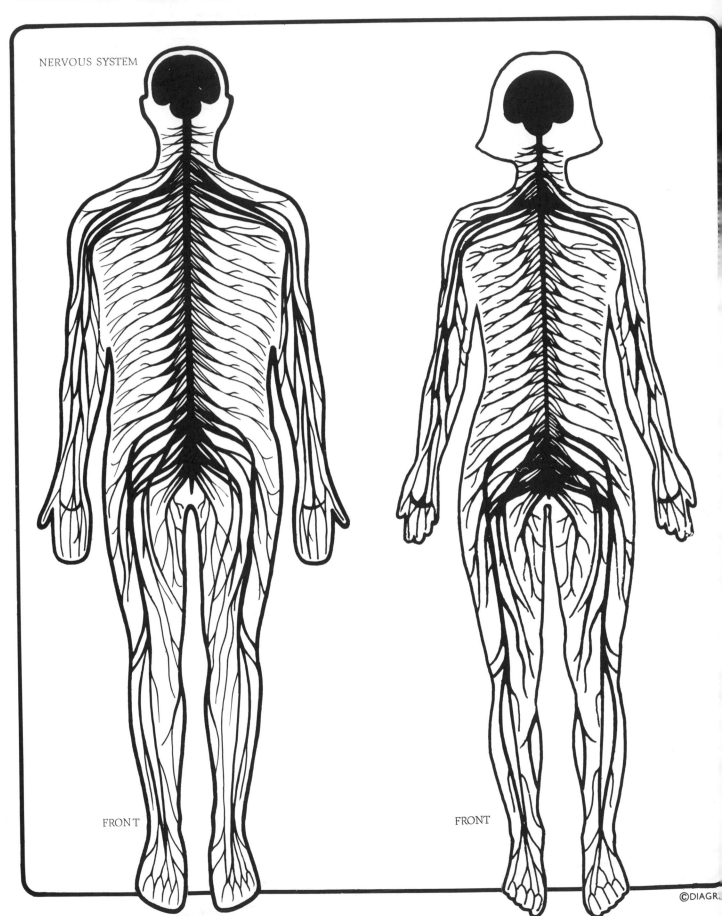

NERVOUS SYSTEM

FRONT

FRONT

©DIAGR

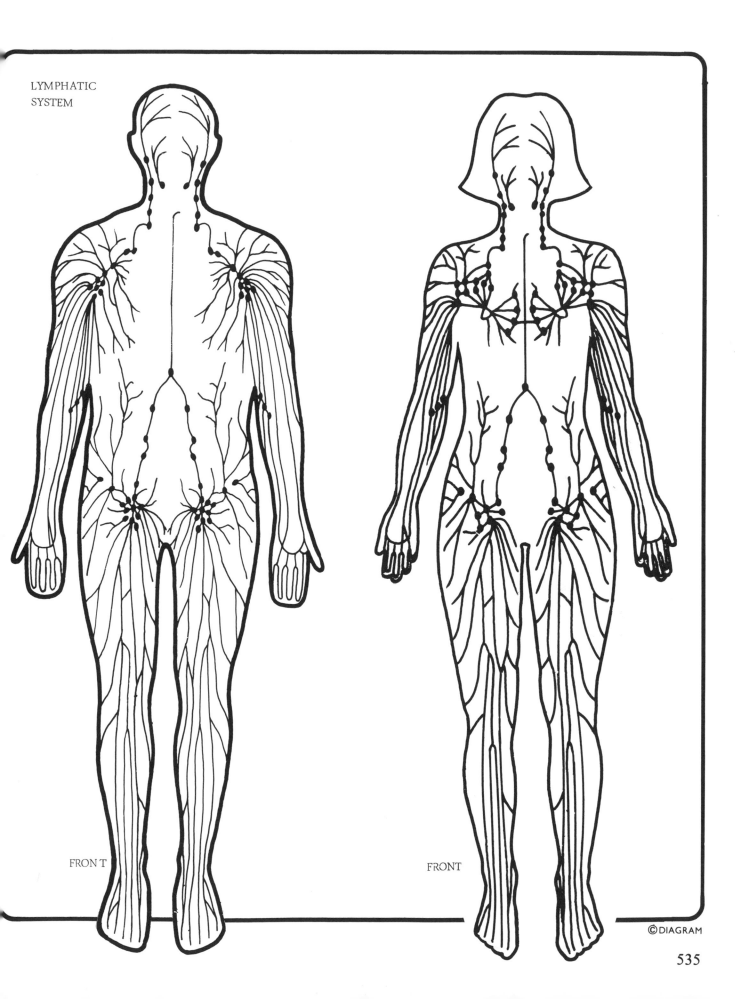

LYMPHATIC
SYSTEM

FRONT

FRONT

©DIAGRAM

535

INDEX

INDEX

INDEX